Handbook of Experimental Pharmacology

Volume 132

Springer
Berlin
Heidelberg
New York
Barcelona
Budapest
Hong Kong
London
Milan
Paris
Singapore
Tokyo

Antithrombotics

Contributors

W. Bell, P. Carmeliet, L. Chi, D. Collen, R.J.G. Cuthbert, C. Esmon, K.P. Gallagher, D. Ginsburg, T. Hara, L.A. Harker, T.A. Hennebry, M. Hollenberg, R.D. Hull, J.A. Jakubowski, R.E. Jordan, F. Kee, D. Keeling, S.D. Kimball, S. Kunitada, M.A. Lauer, L. Leblond, A.M. Lincoff, B.Lucchesi, T.E.Mertz, T. Nagahara, J. Pearson, G.F. Pineo, S. Rebello, J.R. Rubin, A.C.G. Uprichard, S. Watson, H.F. Weisman, P.D. Winocour

Editors

Andrew C.G. Uprichard and Kim P. Gallagher

Springer

Andrew C.G. Uprichard, M.D.
Vice President, Drug Development

Kim P. Gallagher, Ph.D.
Vascular and Cardiac Diseases

Parke-Davis Pharmaceutical Research
2800 Plymouth Road
Ann Arbor, MI 48105
USA

With 66 Figures and 27 Tables

ISBN 3-540-64691-4 Springer-Verlag Berlin Heidelberg New York

Library of Congress Cataloging-in-Publication Data
Antithrombotics / contributors, W. Bell . . . [et al.]; editors, Andrew C.G. Uprichard and Kim P. Gallagher.
p. cm. — (Handbook of experimental pharmacology; v. 132)
Includes bibliographical references and index.
ISBN 3-540-64691-4 (hardcover: alk. paper)
1. Fibrinolytic agents. 2. Thrombosis. 3. Blood—Coagulation. I. Bell, William Robert, 1937– . II. Uprichard, Andrew C.G., 1957– . III. Gallagher, Kim P., 1950– . IV. Series.
QP905.H3 vol. 132
[RM340]
615′.1 s — dc21
[615′.718]
98-29079
CIP

Cover design: *design & production* GmbH, Heidelberg

Typesetting: Best-set Typesetter Ltd., Hong Kong

Production Editor: Angélique Gcouta

SPIN: 10566139 27/3020 – 5 4 3 2 1 0 – Printed on acid-free paper

Preface

When Pedro Cuatrecasas first asked us if we would be interested in editing an edition of the *Handbook of Experimental Pharmacology* on antithrombotics, neither of us really had any idea what we were letting ourselves in for. Looking back almost 3 years later, it would be wrong to remember only the trials and tribulations of the undertaking, for it has been an honor and privilege to be associated with the distinguished contributors and their learned works and we have both benefited greatly from the experience.

In putting this edition together, we have made every attempt to portray a balanced picture of academic and industrial research, from the United States, Europe and Asia, which will be of benefit to physicians, scientists, and students alike. We have also tried to maintain a balance between the traditional preclinical and clinical sciences with newer, mainly molecular, technologies, and current areas of research interest. We are particularly grateful to our authors, some of whom amazed us with their responsiveness, while others required a certain amount of prodding! Recognition should be given to those who stood in at the last minute when long-promised manuscripts failed to materialize: a gesture for which we are particularly grateful. We should note our appreciation also for the forbearance of all the scientists for putting up with our picayune attention to detail for grammar, syntax, citations and references: our hope is that the final product will be viewed as of such a high standard as to have justified all the effort.

It would be remiss not to acknowledge the significant contribution of many others who played a central role in the preparation of this book. Teresa Quinno has acted as a Parke-Davis point person throughout, although she would be the first to acknowledge the help of Regina Tounsel, Loretta Lomske and Nicole Olsen. Our thanks also go out to Doris Walker and her staff at Springer-Verlag for their professional assistance. Maggie Uprichard and Mary Gallagher deserve credit for putting up with long hours and sometimes fragile temperaments.

We both have benefited scientifically from the privilege of helping to compile this edition of the *Handbook*. Our grasp of the specialty has been strengthened from our many interactions with opinion leaders in the area and it is our hope that this work will benefit readers from varied walks in life. Perhaps some day we would do it again.

ANDREW C.G. UPRICHARD
KIM P. GALLAGHER

List of Contributors

BELL, W., Hematology, Nuclear Medicine and Radiology, 1002 Blalock, The Johns Hopkins University School of Medicine, 600 N. Wolfe Street, Baltimore, MD 21287-4928, USA

CARMELIET, P., Center for Molecular and Vascular Biology, University of Leuven Campus Gasthuisberg, Herestraat 49, University of Louvain, Louvain, B-3000, Belgium

CHI, L., Parke-Davis Pharmaceutical Research Division, 2800 Plymouth Road, Ann Arbor, MI 48105, USA

COLLEN, D., Center for Molecular and Vascular Biology, University of Leuven, Campus Gasthuisberg, Herestraat 49, Louvain, B-3000, Belgium

CUTHBERT, R.J.G., Department of Haematology, Altnagelvin Hospital, Londonderry, Northern Ireland, BT47 6SB

ESMON, C., Howard Hughes Medical Institute, Oklahoma Medical Research Foundation, 825 N.E. 13th Street, Oklahoma City, OK 73104, USA

GALLAGHER, K.P., Vascular and Cardiac Diseases, Parke-Davis Pharmaceutical Research Division, Warner-Lambert Company, 2800 Plymouth Road, Ann Arbor, MI 48105, USA

GINSBURG, D., Department of Internal Medicine and Human Genetics, University of Michigan, 4520 MSRB I, 1150 West Medical Center Drive, Ann Arbor, MI 48109-0650, USA

HARA, T., New Products Research Laboratories II, Daiichi Pharmaceutical Co., LTD, Tokyo R&D Center, 16-13 Kitakasai 1-Chome, Edogawa-Ku, Tokyo 134, Japan

HARKER, L.A., Division of Hematology and Oncology, Department of Medicine, Emory University School of Medicine, 1639 Pierce Drive, Woodruff Memorial Bldg 1003, Atlanta, GA 30322, USA

HENNEBRY, T.A., Department of Medicine/Tower 110, The Johns Hopkins University Hospital, 600 N. Wolfe Street, Baltimore, MD 21287-4928, USA

HOLLENBERG, M., Faculty of Medicine, Departments of Pharmacology & Therapeutics and Medicine, The University of Calgary, 3330 Hospital Drive NW, Calgary, Alberta, Canada T2N 4N1

HULL, R.D., Thrombosis Research Unit, 601 South Tower, Foothills Hospital, 1403-29th Street NW, Calgary, AB T2N 2T9, Canada

JAKUBOWSKI, J.A., Clinical Research and Biomedical Operations, Centocor Inc., 200 Great Valley Pkwy, Malvern, PA 19355, USA

JORDAN, R.E., Preclinical Research, Centocor Inc., 200 Great Valley Pkwy, Malvern, PA 19355, USA

KEE, F., Department of Epidemiology and Public Health, Queens University of Belfast, Mulhouse, Royal Victoria Hospital, Grosvenor Road, Belfast BT12 6 BJ, Northern Ireland

KEELING, D., Oxford Haemophilia Centre, The Churchill, Headington, Oxford 0X3 71J, UK

KIMBALL, S.D., Bristol-Myers Squibb, Pharmaceutical Research Institute, PO Box 4000, Princeton, NJ 08543-4000, USA

KUNITADA, S., New Products Research Laboratories II, Daiichi Pharmaceutical Co., LTD, Tokyo R&D Center, 16-13 Kitakasai 1-Chome, Edogawa-Ku, Tokyo 134, Japan

LAUER, M.A., Department of Cardiology, Desk F-25, The Cleveland Clinic Foundation, 9500 Euclid Avenue, Cleveland, OH 44195, USA

LEBLOND, L., Cardiovascular Pharmacology, BioChem Therapeutics, Inc., 275 Armand Frappier Blvd, Laval, Quebec, Canada H7V 4A7

LINCOFF, A.M., Department of Cardiology, Desk F-25, The Cleveland Clinic Foundation, 9500 Euclid Avenue, Cleveland, OH 44195, USA

LUCCHESI, B., Department of Pharmacology, Medical School, University of Michigan, 1301 C Medical Science Research Bldg III, Ann Arbor, MI 48109-0632, USA

MERTZ, T.E., Vascular and Cardiac Diseases, Parke-Davis Pharmaceutical Research Division, Warner-Lambert Company, 2800 Plymouth Road, Ann Arbor, MI 48105, USA

NAGAHARA, T., New Products Research Laboratories II, Daiichi Pharmaceutical Co., LTD, Tokyo R&D Center, 16-13 Kitakasai 1-Chome, Edogawa-Ku, Tokyo 134, Japan

PEARSON, J., University of Michigan, 4520 MSRB I, 1150 West Medical Center Drive, Ann Arbor, MI 48109-0650, USA

PINEO, G.F., Thrombosis Research Unit, 601 South Tower, Foothills Hospital, 1403-29th Street NW, Calgary, AB T2N 2T9, Canada

REBELLO, S., Department of Pharmacology, Medical School, University of Michigan, 1301 C Medical Science Research Bldg III, Ann Arbor, MI 48109-0632, USA

RUBIN, J.R., Biomolecular and Structural Drug Design, Parke-Davis Pharmaceutical Research Division, Warner-Lambert Company, 2800 Plymouth Road, Ann Arbor, MI 48105, USA

UPRICHARD, A.C.G., Drug Development, Parke-Davis Pharmaceutical Research Division, Warner-Lambert Company, 2800 Plymouth Road, Ann Arbor, MI 48105, USA

WATSON, S., Department of Pharmacology, Oxford University, Mansfleld Road, Oxford 0X1 3QT, UK

WEISMAN, H.F., Clinical Research and Biomedical Operations, Centocor Inc., 200 Great Valley Pkwy, Malvern, PA 19355, USA

WINOCOUR, P.D., BioChem Therapeutics, Inc., 275 Armand Frappier Blvd, Laval, Quebec, Canada H7V 4A7

Contents

CHAPTER 2

New Developments in the Molecular Biology of Coagulation and Fibrinolysis

CHAPTER 3

Epidemiology of Arterial and Venous Thrombosis

CHAPTER 8

Platelet Membrane Receptors and Signalling Pathways: New Therapeutic Targets

CHAPTER 9

Heparin and Other Indirect Antithrombin Agents

CHAPTER 12

Anticoagulant Therapy with Warfarin for Thrombotic Disorders

CHAPTER 13

Oral Thrombin Inhibitors: Challenges and Progress

CHAPTER 1

The Coagulation Pathway and Antithrombotic Strategies

L. LEBLOND and P.D. WINOCOUR

A. Introduction

In this chapter, we present a summary of the blood coagulation process and the status of current antithrombotic strategies. Due to space limitations, pertinent reviews are cited whenever possible from which the reader can obtain additional primary references.

Blood clotting is the mechanism by which higher organisms arrest blood loss following vascular injury. Rapid cessation of blood flow is achieved first by the formation of a hemostatic plug through the adhesion and aggregation of circulating platelets followed closely by the generation of an insoluble fibrin network. The activation of platelets releases numerous proteins and small molecules that accelerate and increase platelet plug formation. Platelets also amplify the coagulation reactions by providing a scaffold on their membrane surface and contribute importantly to actions that speed wound healing and repair processes. Another series of proteases make up the fibrinolytic system, which removes the fibrin clot in conjunction with the repair process. The complex and dynamic processes of hemostasis and fibrinolysis are carefully balanced under normal conditions. Abnormalities in any of the hemostatic or fibrinolytic components upset the balance, leading to excessive bleeding or thrombosis.

B. The Coagulation Pathway

The process of blood coagulation involves an ordered series of biochemical reactions that transforms circulating blood into a gel through conversion of soluble fibrinogen to insoluble fibrin. In a series of stepwise reactions, certain circulating plasma proteins (coagulation factors) are converted from their inactive into active forms through the limited proteolysis of one or two peptide bonds. Ultimately, thrombin generated through this process cleaves fibrinogen to form fibrin. Calcium ions and phospholipids play critical roles in many of the coagulation reactions. The phospholipids are mainly provided on the surface of the activated platelets where many of the calcium dependent reactions occur. The activated platelets also provide the membrane surfaces upon which coagulation enzymes can be anchored, assembled and expressed.

I. The Cascade/Waterfall Model

In the 1960s, the process of blood coagulation was described as a cascade or waterfall model (MACFARLANE 1964; DAVIE and RATNOFF 1964). The coagulation reactions were divided into two pathways, the intrinsic and extrinsic (Fig. 1). Both pathways fuse in a common pathway to generate the enzyme thrombin, which converts fibrinogen to fibrin (for review see BERRETTINI et al. 1987; DAVIE et al. 1991; DAVIE 1995).

The intrinsic pathway begins with contact activation, involving factors XII (fXII) and XI (fXI), prekallikrein, and high-molecular-weight kininogen (HMWK). These four contact factors do not require the presence of calcium for their reactions and they are all readily absorbed onto negatively charged surfaces. The initial stage involves the activation of fXII to fXIIa by a process involving a surface-mediated conformational change and by the action of kallikrein formed from prekallikrein. Factor XIIa is responsible, in turn, for the activation of fXI in the presence of the cofactor HMWK. Contact activation proceeds with activation of factor IX (fIX) by fXIa in the presence of calcium. The process continues with formation of the fVIIIa/fIXa complex which activates factor X (fX).

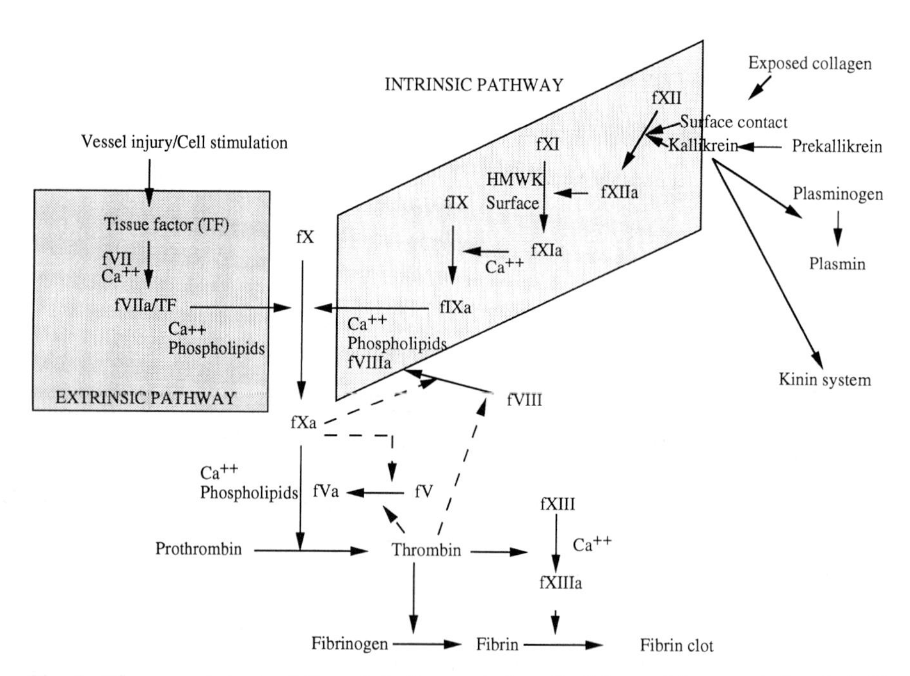

Fig. 1. The cascade/waterfall model of coagulation. In this model, coagulation is initiated by either the intrinsic or the extrinsic pathways, both leading to the formation of fibrin via a common pathway involving fXa, phospholipids and calcium. The dotted lines indicate fXa and thrombin's positive feedback effects of fV and fVIII activation

The extrinsic pathway begins with trauma to tissue, causing exposure of tissue factor (TF) and activation of factor VII (fVII) in the presence of calcium. The process required for fVII activation *in vivo* is uncertain. The fVIIa/TF complex then converts fX to fXa. Thus, the intrinsic and extrinsic pathways converge at activation of fX. Another complex (prothrombinase complex) is formed by fVa and fXa which, in the presence of phospholipids and calcium, convert prothrombin to thrombin. Fibrinogen is cleaved by thrombin to form fibrin. Then factor XIIIa (fXIIIa), produced by thrombin induced activation of fXIII, stabilizes fibrin to form a very strong fibrin clot. The formation of fVa and fVIIIa from their inactive cofactors is accomplished by both fXa and thrombin in a positive feedback loop.

II. The Revised Model

The intrinsic and extrinsic pathways of the cascade model describe the overall process of hemostatic reactions *in vitro* reasonably well, but they do not accurately describe the process of hemostasis *in vivo* (for review see Broze 1995a; Luchtman-Jones and Broze 1995; Mann and Kalafatis 1995; Rapaport and Rao 1995). A deficiency in one of the contact factors required for the initiation of intrinsic coagulation (fXII, HMWK, or prekallikrein), for example, is not associated with abnormal bleeding in humans. Therefore, the intrinsic pathway does not appear to be essential for normal blood coagulation. Even the role of fXI is uncertain, since not all patients with a fXI deficiency show bleeding problems (Seligsohn 1993). Deficiencies in prothrombin, fV, fVII, fVIII, fIX or fX on the other hand are associated with severe bleeding, indicating that these factors are essential for normal hemostasis. These clinical data, and the laboratory observations (Osterud and Rapaport 1977) that fVIIa with TF can activate both fIX and fX, made it necessary to revise the importance ascribed to the intrinsic pathway in the initiation of coagulation. The demonstration that thrombin can activate fXI (Gailani and Broze 1991), and the characterization of an endogenous inhibitor of tissue factor-mediated coagulation, tissue factor pathway inhibitor (TFPI) (for review see Broze 1995b), also suggested that the coagulation cascade needed rethinking.

In the revised model of blood coagulation (Fig. 2), TF rather than the "contact" factors is responsible for the initiation of coagulation. According to the revised model, coagulation is divided into two stages rather than two pathways: an initiation stage and an augmentation stage. The "initiation stage" begins with exposure of TF to circulating fVII or fVIIa. The fVIIa/TF complex catalyzes the formation of fXa and fIXa. While some of the fXa will continue to prime the activation process to produce thrombin by binding with its cofactor Va, other fXa will bind to TFPI. The fXa–TFPI complex acts like an internal brake on the fVIIa/TF complex limiting production of additional fXa and fIXa. The small amount of thrombin generated in the initiation phase, however, "amplifies" the process by activating factors in the old intrinsic

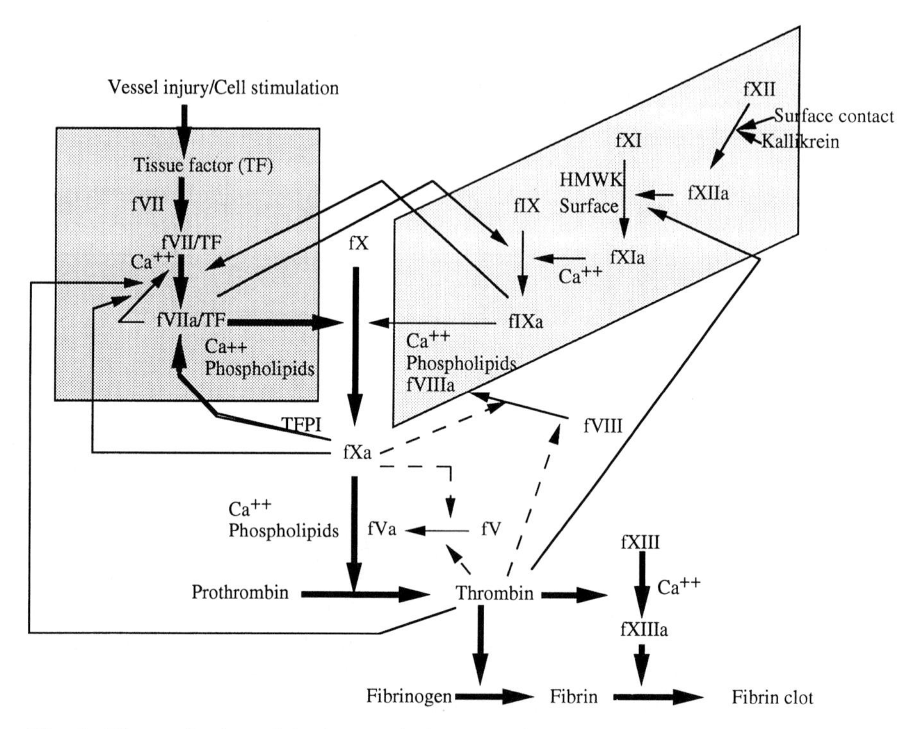

Fig. 2. The revised model of coagulation. In this model, coagulation is initiated by vascular injury/cell stimulation leading to exposure of tissue factor (TF) to the fVII circulating in the blood. TF/fVIIa complex activates fX and fIX. Bold arrows show the "initiation stage" of coagulation. Thrombin so generated "amplifies" additional generation of thrombin through its activation of fXI, fVIII and fV which produces more fXa. TFPI mediates a fXa-dependent inhibition of fVIIa/TF activity. To avoid further complexity in this figure, the surface contact activation system before fXII has not been shown. The dotted lines indicate fXa and thrombin's positive feedback effects on fV and fVIII activation

pathway (fXI, fVIII, fV) (BROZE and GAILANI 1993). This positive feedback loop increases fXa generation and sustains coagulation.

Unresolved, though, are questions concerning the interactions under physiological conditions between fVIIa/TF and the fIXa/fVIIIa/phospholipid-catalyzed activation of fX (RAPAPORT and RAO 1995).

III. Structure-Activity Relationships of Coagulation Proteases

During the last 20 years, almost all the proteins known to participate in the coagulation cascade have been isolated and characterized by molecular biology techniques (FURIE and FURIE 1988; PATTHY 1993). Site-specific mutagenesis has made it possible to identify critical and important regions in the proteins. Crystallographic structural determinations and MRI spectroscopy

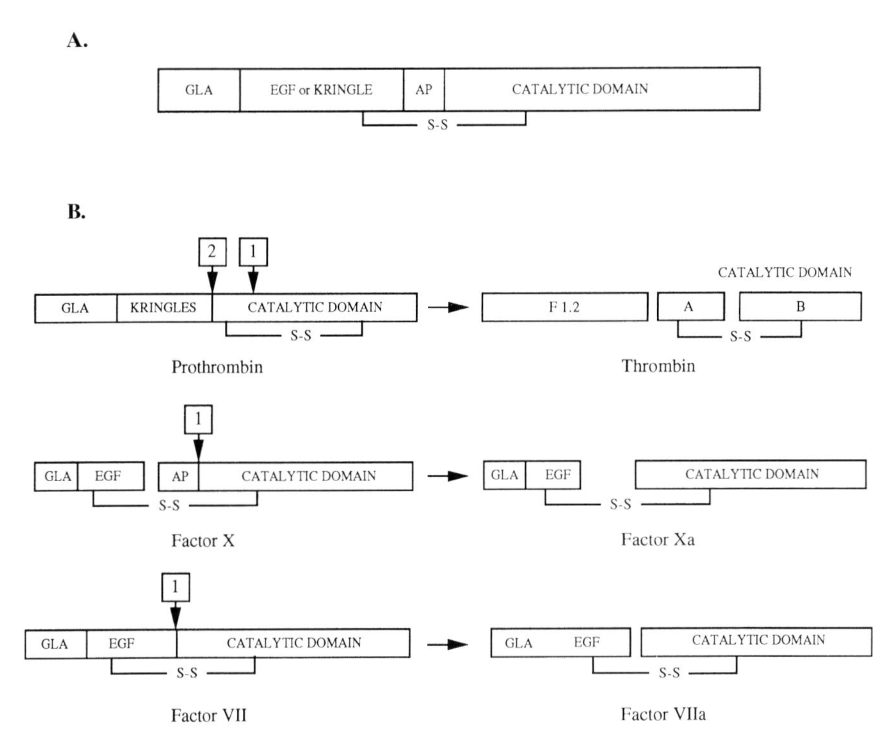

Fig. 3. A. The structural domain of the vitamin K-dependent coagulation factors where: GLA = region containing γ-carboxyglutamic acid residues, EGF = region containing sequences homologous to the human epidermal growth factor (fVII and fX) or Kringle = the kringle domains in prothrombin, AP = region of the activation peptide and CATALYTIC DOMAIN = region containing the serine protease catalytic triad. **B**. Schematic representation of prothrombin, fX and fVII activation. Sites of proteolytic cleavage associated with zymogen activation are indicated by arrows

have greatly helped in understanding the properties and interactions of these proteins (Stubbs and Bode 1994, 1995). The blood coagulation proteases are a family of proteins with diverse functional properties but common structural elements (Fig. 3). Analysis of their amino acid sequences demonstrated that their catalytically active regions are homologous with those of other trypsin-like serine proteases, i.e., they all contain the catalytic triad composed of serine, histidine, and aspartic acid. Nonproteinase regions, such as kringle loops (e.g., in prothrombin) and epidermal growth factor (EGF) domains (e.g., in fIX, fX, fVII, protein C), are involved in interactions with cofactors, substrates and inhibitors that regulate enzyme activity and activation. Coagulation enzymes also possess γ-carboxyglutamic acid domains (Gla) that are synthesized in a vitamin K-dependent reaction in hepatocytes. The Gla domains, located at the amino-terminal portion of the molecules, confer the ability to bind to membranes containing phosphatidylserine or other negatively charged phospholipids in a calcium dependent manner (KALAFATIS et al. 1994), with calcium forming a bridge between the Gla residues and the negatively charged membrane phospholipids. In this review, we will concentrate on the structure-activity relationships of thrombin, fXa and the fVIIa/TF complex.

1. Thrombin

Thrombin is a glycosylated, trypsin-like serine protease with lysine- or arginine-directed specificity which acts as the key terminal protease in the coagulation cascade. In blood, thrombin is produced from its inactive zymogen, prothrombin, by the action of the prothrombinase complex (prothrombin, fXa, fVa, Ca^{2+} and phospholipids) during the final stage of the blood clotting cascade (KALAFATIS et al. 1994). Prothrombin is converted to thrombin by the cleavage of two peptide bonds and the removal of the Gla and the kringle domains (Fig. 3). Thrombin so generated is released from the membrane bound prothrombinase complex into solution and thereafter functions at all levels of hemostasis (FENTON 1986). Thrombin converts the soluble plasma protein, fibrinogen, into the insoluble protein fibrin, which forms the matrix of blood clots. Thrombin also activates fXIII, which catalyzes covalent cross-linking of fibrin, stabilizing the clot. It promotes and amplifies clot formation by activating other coagulation factors including fV and fVIII. Thrombin stimulates platelet aggregation through proteolytic cleavage of the thrombin receptor. In addition to prothrombotic functions, thrombin mediates an antithrombotic effect by activating protein C after binding to the cell surface protein, thrombomodulin (TM). In conjunction with protein S, activated protein C inactivates fVa and fVIIIa, putting a brake on the amplification reactions and thereby modulating production of additional thrombin.

Thrombin's activity is controlled by endogenous protein inhibitors, such as α_2-macroglobulin, the serpins antithrombin III (ATIII) and heparin cofactor II (HCII), and the protease nexin I. Inactivation of thrombin by serpins requires the catalytic participation of heparin or other endogenous sulfated polysaccharides. Through interaction with its specific receptor, thrombin promotes the proliferation of endothelial cells, the liberation of tissue plasminogen activator (t-PA), and the contraction and dilation of blood vessels. Thus, thrombin plays a role in tissue repair, fibrinolysis and control of blood flow, which contribute to the overall hemostatic process.

Thrombin can interact with a number of proteins with a high degree of specificity (GUILLIN et al. 1995). The recent X-ray crystallographic structural determinations of human α-thrombin and its complexes with hirudin (STONE and MARAGANORE 1992), fibrinopeptide A, prothrombin fragment F2 (TULINSKY 1996) and thrombin-receptor peptides (MATHEWS et al. 1994), as well as functional studies with mutant thrombins (TSIANG et al. 1995), have provided important information on how thrombin functions. The specificity of thrombin for its many substrates and ligands may be ascribed to multiple interactions with both the active site cleft and distinct exosite regions: the fibrinogen anion binding exosite (exosite 1), the heparin binding site (exosite 2) and the Na^+ binding loop (GUINTO et al. 1995). Exosites display distinct binding modes with different molecules which modulate thrombin's different functions. The anion-binding exosite 1 is involved in the recognition of fibrin/fibrinogen, HCII, TM, thrombin receptor and glycoprotein Ib (GpIb). In addition, functional studies (NISHIOKA et al. 1993) with recombinant mutant thrombin have indicated that the binding sites for fibrinogen, protein C and TM may overlap, but are not identical. Exosite 2 interacts with heparin, the chondroitin sulfate moiety of TM and prothrombin activation fragment 2. Both binding exosites are involved in the recognition of fV and fVIII (ESMON and LOLLAR 1996). The ability of thrombin to work either as a procoagulant or an anticoagulant enzyme is driven by the binding of Na^+ to a single site (WELLS and DI CERA 1992). Na^+ ion binding to thrombin induces a conformational change in the molecule that changes the substrate specificity of thrombin. A slow-to-fast transition is important because fibrinogen binds to the fast form with greater affinity and is cleaved with higher specificity, while the slow form activates the anticoagulant protein C more specifically (AYALA and DI CERA 1994). Thus, the slow-to-fast transition helps to determine whether thrombin exerts antithrombotic or prothrombotic effects.

2. Factor Xa

Responsible for the cleavage of prothrombin, fXa is a two-chain molecule, consisting of a light chain containing the Gla domain followed by two EGF-like repeats linked through a disulfide bond to the heavy chain containing the serine protease domain (Fig. 3). Activation of fX common to both intrinsic and extrinsic pathways occurs on a phospholipid membrane surface in the presence of calcium. In the extrinsic pathway, fVIIa associates with membrane-bound TF. This complex in turn associates with fX, which results in removal of an activation peptide from the amino-terminal end of the heavy chain and release of fXa. In the intrinsic pathway, fIXa performs the same cleavage in the presence of fVIIIa (for review see FURIE and FURIE 1988 and KALAFATIS et al. 1994); fVIIIa acts possibly by aligning fIXa and fX on a membrane phospholipid in the optimal positions for their interaction.

Recent crystallographic structural determinations of des-Gla fXa (lacking Gla domain) (PADMANABHAN et al. 1993b) and of des-Gla fXa with the synthetic inhibitor DX-9065 (BRANDSTETTER et al. 1996) have provided information on the organization of the catalytic and the second EGF domains of the fXa molecule. The secondary structure and overall architecture of the fXa protease domain is very similar to that of thrombin. The absence in fXa of the Tyr60A–Thr60I insertion loop present in thrombin permits relatively easy access to the active site of fXa by macromolecular substrates. In this respect, fXa resembles trypsin and chymotrypsin much more closely than thrombin. The surface of fXa also shows a distinctive pattern of charged residues which may act as binding exosites. The fXa counterpart of thrombin's fibrinogen binding exosite, for example, is rich in acidic residues and may represent the fVa binding site for formation of prothrombinase complex. A cluster of positively charged residues is present in the area comparable to the heparin binding site of thrombin, although it probably serves a different function in fXa, such as substrate binding. Predominant interaction sites of fXa for fVa appear to reside in the catalytic and second EGF domains (STUBBS and BODE 1995). In addition to prothrombin, fXa also cleaves other protein substrates involved in blood coagulation, such as fVII, fVIII, and fIX (DAVIE et al. 1991). Factor Xa is also the ligand for the EPR-1 receptor (effector cell protease receptor) identified on leukocytes, monocytes (ALTIERI 1994), pancreatic cancer cells (KAKKAR et al. 1995) and endothelial cells (NICHOLSON et al. 1996). Although the affinity is lower than the fVa/fXa interaction, binding of fXa to EPR-1 promotes prothrombin activation. Binding of fXa to EPR-1 also generates intracellular signals involved in activation and proliferation of lymphocytes (ALTIERI and STAMNES 1994) and possibly other cell types.

3. Factor VII/Tissue Factor Complex

TF is a membrane-anchored single polypeptide of 263 amino acids that consists of an extracellular domain, a single transmembrane segment and a small cytoplasmic domain (SPICER et al. 1987), which may be involved in cell signaling (ROTTINGEN et al. 1995). TF is localized to sites immediately surrounding the vasculature and is not expressed by endothelial cells under normal conditions (DRAKE et al. 1989). FVII is a single-chain vitamin K-dependent protein that acts as a serine protease similar to other coagulation factors that circulate in the plasma (Fig. 3) (DAVIE et al. 1991). The coagulation cascade is activated when fVII in the blood binds to TF on the surface of subendothelial cells exposed as a result of injury, or expressed on stimulated endothelial cells or monocytes. Binding of fVII to TF markedly heightens the susceptibility of fVII to cleavage of the single peptide bond that converts it to the active form, fVIIa. Activation to fVIIa when bound to TF can be achieved either by autoactivation (YAMAMOTO et al. 1992) or by trace concentrations of fVIIa, fXa and fIXa in the plasma (RAO et al. 1996; BUTENAS and MANN 1996). Thus, the extent to which the TF/fVIIa complex is formed is determined mainly by the amount of TF available. Binding of fVIIa to TF also causes a marked increase in the catalytic activity of fVIIa for its natural macromolecular substrates, fXa and fIXa. The fVIIa/TF then catalyzes the next step in the coagulation cascade, formation of fXa or/and fIXa.

X-ray structures of the fVIIa/TF complex have shown that fVIIa wraps around the extracellular domain of TF and adopts an extended conformation with the Gla domain proximal to, and the catalytic domain distal to, the cell membrane (BANNER et al. 1996; MULLER et al. 1996 and for review see KIRCHHOFER and NEMERSON 1996). TF does not change its conformation upon binding to fVIIa, suggesting that TF forms a rigid scaffold for immobilizing the flexible fVIIa and positions the active center at the correct distance from the cell membrane. The contact area between the two molecules is extensive and the contact residues found in the crystal structure of the complex correlate well with data from mutational analysis. The main binding site on the TF molecule is located at the interface between domains 1 and 2 (BANNER et al. 1996), while the EGF-1 domain and the catalytic domain of the fVIIa are important for binding to TF (CHANG et al. 1995). The catalytic domain of fVIIa shares the structures common with all the serine proteases and the residues involved in substrate recognition and catalysis are highly conserved both in sequence and spatial relationship. The surface loops, however, differ from those previously determined in the structure of fXa or thrombin.

Factor VIIa is highly resistant to plasma protease inhibitors, compared to other coagulation serine proteases (KONDO and KISIEL 1987). Once activated, most serine proteases have plasma half-lives measured in seconds to minutes, while fVIIa has an in vivo half-life of about 2 h (SELIGSOHN et al. 1978). This led MILLER et al. (1985) to propose that plasma may contain trace levels of fVIIa at all times, which may be important in priming the clotting cascade. Thus, small amounts of fVIIa may complex with TF to convert neighboring fVII/TF to fVIIa/TF by autoactivation, and/or generate an initial burst of fXa that would back activate TF-bound fVII. If such trace amounts of fVIIa do, in fact, represent the very first active protease in the clotting cascade, then elevated plasma fVIIa levels might reasonably be expected to contribute to hypercoagulable states (OFOSU et al. 1996). Significant increases in fVII activity are associated with age, but no correlation has been established as yet between the amount of TF and fVII activity (ALBRECHT et al. 1996).

Strictly speaking, the central role of TF in blood coagulation *in vivo* is largely supported by indirect observations, since TF deficiency has never been identified in man. Studies of TF disruption in mice suggest, however, that TF deficiency may be lethal due to a coagulation disorder or to a loss of vascular integrity (TOOMY et al. 1996; CARMELIET et al. 1996). Severe congenital deficiencies in fVII can produce bleeding diatheses similar to classic hemophilias (for review see TUDDENHAM et al. 1995). The amount of TF required to initiate hemostatic coagulation and the levels which produce thrombosis remain to be established.

C. Physiological Regulators

There are a number of physiological mechanisms by which the coagulation pathways are regulated (Fig. 4). These include: (1) direct inhibition of the coagulation enzymes in fluid phase, mainly by serine protease inhibitors (serpins, a_2-macroglobulin), (2) changes in thrombin's substrate specificity following active-site-independent interaction with the endothelial cell surface receptor, TM, that switches thrombin from a procoagulant into an anticoagulant enzyme, and (3) inhibition of fVIIa/TF pathway by TFPI. The importance

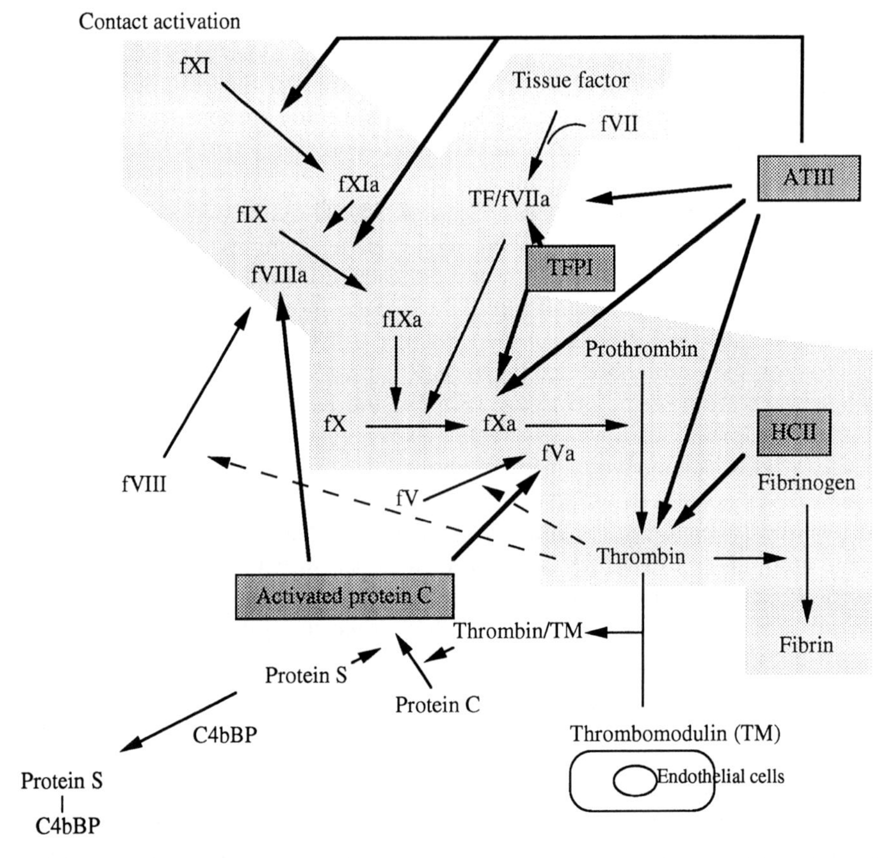

Fig. 4. Coagulation is regulated by physiological regulators such as antithrombin III (ATIII), heparin cofactor II (HCII), tissue factor pathway inhibitor (TFPI), and the protein C/S system. The protein C/S system is activated by the thrombomodulin (TM)/ thrombin complex, resulting in the inactivation of factors VIIIa and Va. Sites of action of the physiological regulators are indicated by heavy arrows

of these complementary control mechanisms is already demonstrated by the fact that patients with deficiency in ATIII (for review see OLDS et al. 1994) or protein C (for review see ESMON 1987; GRIFFIN et al. 1993) or with activated protein C resistance (SWENSSON and DAHLBACK 1994; BERTINA et al. 1995; DAHLBACK 1995) are at high risk for developing thrombosis. However, thromboembolic disease resulting from a physiological deficiency of TFPI activity has yet to be identified (ABILDGAARD 1995).

I. Antithrombin-III (ATIII) and Heparin Cofactor-II (HCII)

Direct control of thrombin and other coagulation enzymes, including fIXa, fXa, fXIa, and fXIIa, is achieved mainly by circulating proteinase inhibitors including ATIII, HCII, and α_2-macroglobulin, of which ATIII appears to be the most efficient. ATIII is a single chain glycoprotein of molecular weight 58 kDa whose primary structure has been elucidated by protein and cDNA sequencing (for review see OLSON and BJORK 1994; OLDS et al. 1994). It is synthesized in the liver and the human plasma concentration is approximately 2.3 μM. ATIII, like other members of the serpin superfamily, blocks the active site of serine proteases by forming a covalent complex with heparin and the enzyme. The complex formation is accompanied by changes in the conformation of both enzyme and inhibitor. The ATIII-thrombin complex is rapidly cleared by the liver or is present in the plasma in association with the adhesion protein vitronectin (PREISSNER et al. 1996). The complex formation between ATIII or HCII and various coagulation factors is greatly stimulated by the presence of heparin or heparan sulfate proteoglycans intercalated in the surface membranes of endothelial cells (for review see OLSON and BJORK 1994). Most of the serine proteases of the coagulation system are inhibited by ATIII but inhibition of thrombin and fXa are probably the most important in regulating coagulation. In the case of fVIIa, the inactivation is only detectable when the protease is complexed with TF (RAO et al. 1993; LAWSON et al. 1993). HCII inhibits only thrombin, but its physiological significance remains unclear.

II. Tissue Factor Pathway Inhibitor

Tissue factor pathway inhibitor (TFPI), also known as extrinsic pathway inhibitor (EPI) or lipoprotein-associated coagulation inhibitor (LACI), is a 42-kDa single chain glycoprotein that consists of a negatively charged N-terminus, three tandem Kunitz-type inhibitory domains, and a positively charged C-terminus (WUN et al. 1988). The specificity of the inhibitor domains has been studied by site-directed mutagenesis (GIRARD et al. 1989). The first Kunitz-type protease inhibitor (KPI) domain interacts with the active site of fVIIa, while the second KPI domain interacts with the active site of fXa, although other parts of the molecule are also involved in the interaction with fXa (GIRARD et al. 1989; PETERSEN et al. 1996). The third KPI domain and the carboxy-terminal portion of the molecule are apparently without inhibitory effect *per se*, but they have been reported to facilitate the binding of TFPI to the cell surface, lipoproteins, glycosaminoglycans and heparin (WESSELSCHMIDT et al. 1993). This portion is essential for the optimal anticoagulant effect of TFPI.

TFPI inhibits fVIIa/TF activity in an fXa dependent manner. It involves the formation of a quaternary complex containing fXa-TFPI-fVIIa/TF in which the second Kunitz domain of TFPI binds fXa and the first Kunitz domain binds fVIIa (BROZE et al. 1988; GIRARD et al. 1989). This inhibitory complex could result from the initial binding of fXa to TFPI, with subsequent binding of the fXa-TFPI complex to fVIIa/TF or, alternatively, TFPI could bind to a preformed fXa-fVIIa/TF complex (BROZE 1995b). At physiological concentrations, TFPI mediates the feedback inhibition of the fVIIa/TF complex, but it does not inhibit the activity of fVIIa/TF until the complex has produced some fXa and fIXa.

Under normal conditions, TFPI circulates in plasma primarily in a complex with lipoproteins, and its normal plasma concentration is approximately 2 n*M* (NOVOTNY et al. 1989). The endothelium is considered to be the main source of TFPI (BAJAJ et al. 1990), but platelets also contain a small fraction of the total TFPI and release the inhibitor upon activation (NOVOTNY et al. 1988). TFPI belongs to the family of heparin-binding proteins. Heparin administration results in a two- to eightfold increase in the plasma concentration of TFPI (SANDSET 1988). The source of this additional TFPI is thought to be the endothelium, where TFPI may be bound to heparan sulfate or to other glycosaminoglycans at the endothelial surface and from which it may be displaced by heparin.

TFPI appears to be the primary regulator of fVIIa/TF catalytic activity during hemostasis. Its role seems to be the inhibition of the fVII/TF complex, which is probably essential for maintaining a normal hemostatic balance (LINDAHL 1995). After initiation of coagulation and progression to the amplification stage, the initial stimulus for clotting must be removed or constrained. Thus, the initiation stage is probably short-lived, due primarily to the action of TFPI.

III. Protein C/S-Thrombomodulin Complex

The protein C anticoagulant pathway appears to be triggered when thrombin binds to the endothelial cell receptor, thrombomodulin (TM). This complex activates protein C to generate the anticoagulant enzyme, activated protein C (APC). In complex with protein S, APC inhibits coagulation by inactivating two regulatory proteins, fVa and fVIIIa (Fig. 4) (ESMON et al. 1986, ESMON 1992 and 1995). This pathway plays a critical role in the negative regulation of blood coagulation, as indicated by the fact that total deficiencies in protein C or protein S are associated with severe and life-threatening thrombotic complications (ESMON and FUKUDOME 1995). In addition, deletion of the TM gene in mice results in embryonic lethality (HEALY et al. 1995).

TM is an integral membrane glycoprotein of 90 kDa molecular mass located on the surface of the vascular endothelium (SUZUKI et al. 1987). TM binds thrombin in a 1:1 complex and converts the enzyme from a procoagulant to an anticoagulant protease. The TM-dependent alteration in substrate specificity from fibrinogen to protein C appears to result from a combination of an allosteric change in the active site conformation of thrombin (DANG et al 1995) and an overlap of the TM and fibrinogen binding sites on thrombin's anion binding exosite 1 (SUZUKI et al 1990; NISHIOKA et al. 1993). Three distinct regions of TM interact with thrombin (the last two of the six EGF-like repeats and the chondroitin sulfate moiety) but EGFs 4–6 are necessary to accelerate protein C activation (for review see ESMON 1995). The

TM-thrombin complex converts protein C to its activated form, which proteolytically destroys cofactors fVa and fVIIIa and thereby inhibits generation of thrombin. TM also inhibits thrombin-induced activation of platelets (ESMON et al. 1983) and blocks thrombin binding to its receptor(s) on the platelet surface (ESMON 1995). The physiologic role of the protein C-TM system as a natural anticoagulant mechanism of the blood vessel wall has been substantiated by studies in animals and humans (ESMON 1987; CONARD et al. 1993). The anticoagulant role of TM is especially effective in the microvasculature, where the endothelial cell surface area is much greater and appears to correlate with a high concentration of TM.

Protein C and protein S are vitamin K-dependent glycoproteins. Once formed, APC is one of the slowest of the serine proteases to be inactivated and cleared from the circulation. Its half-life is around 15 min and its elimination is carried out by complex formation with either protein C inhibitor or α_1-antitrypsin (ESPANA et al. 1991; HEEB and GRIFFIN 1988). Protein S circulates both in a free form and bound to C4b binding protein (C4bBP), a regulatory protein of the complement system. Only the free form is capable of enhancing APC anticoagulant activity. APC and protein C can both interact with an endothelial protein C receptor (EPCR) (FUKUDOME and ESMON 1994). Interaction between EPCR and protein C appears to alleviate the requirement for protein C interaction with negatively charged phospholipid membranes, but the EPCR/APC complex has an altered enzyme specificity (REGAN et al. 1996). Binding of APC to EPCR blocks the anticoagulant activity of APC. Like TM, EPCR is downregulated by tumor necrosis factor a on endothelium *in vitro*, suggesting that these two receptors might function through a common pathway (STEARNS-KUROSAWA et al. 1996).

D. Platelet and Cellular Contributions

In recent years, the importance of cells and cell surface receptors in coagulation has been delineated. Monocytes, for example, are able to express TF after stimulation by lipopolysaccharides, cytokines, modified low density lipoprotein and growth factors (GECZY 1994). Leukocyte, monocyte and platelet surface receptors specifically bind components of the coagulation pathways on their surface (KALAFATIS et al. 1994). Stimulated or damaged circulating blood cells can provide negatively charged phospholipid surfaces necessary to assemble coagulation complexes. Thrombin (VU et al. 1991), fXa (ALTIERI 1994), protein S (STITT et al. 1995), protein C (FUKADOME and ESMON 1994) and fVII/fVIIa by associating with TF (ROTTINGEN et al. 1995) bind to receptors and can activate intracellular signaling pathways. The precise nature and importance of these signaling pathways are not clear as yet but they may be associated with cellular responses required for tissue repair and growth (STIERNBERG et al. 1993). These interactions may also coordinate coagulation with cellular responses in other systems, such as those involved in inflammation (GECZY

1994), in the immune response (DUCHOSAL et al. 1996) and in the growth and remodeling of the nervous system (for review see SUIDAN et al. 1996).

I. Cell Surface Dependence

The concept of cell surface dependent reactions playing an essential role in the initiation, propagation and regulation of the hemostatic process *in vivo* has gained general acceptance. The cell membrane provides the support that enables enzymes, cofactors, and substrates to combine, facilitating maximum activity of the vitamin K-dependent complexes (MANN and LAWSON 1992). The membrane surfaces are provided either by vascular cells that have been damaged or by accumulation of activated platelets at the site of injury. Two classes of coagulation factors bind to membranes containing acidic phospholipids: the vitamin K-dependent zymogens (prothrombin, fX, fIX, fVII, protein C and protein S) and the coagulation cofactors V and VIII (KALAFATIS et al. 1994). Furthermore, the active forms of these enzymes (fVIIa, fXa and fIXa) bind with affinities similar to their zymogens. Thrombin alone does not bind to membranes, but its large activation peptide, prothrombin fragment 1.2, binds to membranes containing phosphatidyl serine. The sequestering of the various enzyme complexes on the same membrane surface and channeling of intermediate products between them may have potential advantages, such as localization of the response and improvement in the catalytic efficiency of the enzyme. The membrane surface also provides a protected environment that can make coagulation reactions difficult to inhibit by natural inhibitors in the blood.

II. Platelet Participation

Platelets are regarded as the preeminent blood particle involved in physiological hemostasis and pathological thrombosis. Platelets adhere to the altered vascular tissue at the site of injury, then aggregate and form a plug that stops blood loss. Platelet adhesion is mediated by the plasma protein, von Willebrand factor (vWf), that forms a bridge between specific glycoprotein receptors (GpIb/GpIX) located on activated platelets and on the subendothelium (SAKARIASSEN et al. 1979). Conformational changes within the vWf molecule induced by shear stress forces in flowing blood appear to be responsible for regulating vWf affinity for GpIb (for review see KROLL et al. 1996). The importance of vWf in hemostasis is indicated by severe bleeding disorders that result when there is a total absence of vWf (for review see CAEN and ROSA 1995). Platelet adhesion is followed by platelet activation that induces a change in the conformation of the GpIIb/GpIIIa receptors on the platelet surface so that they can bind fibrinogen and other adhesive glycoproteins with high affinity (SIMS et al. 1991). Binding of fibrinogen forms a bridge between activated platelets which helps to consolidate the platelet thrombus. Platelet activation results in a redistribution of negatively charged phospholipids to the platelet surface. This procoagulant surface containing phosphatidylserine and phosphatidylinositol binds vitamin K-dependent coagulation factors, such as fX, via Ca^{2+} bridges to the γ-carboxyglutamic acid residues located within these factors. Upon activation, platelets release fV from their α-granules. FV and fVa bind directly to the platelet surface and provide high-affinity binding sites for fXa. In addition to keeping the reactions localized, the complex formation on the phospholipid surface significantly increases the reaction rates of the prothrombinase complex. Additionally, prothrombin, the substrate of the prothrombinase complex, and fX can interact directly with the activated platelet membrane (SCANDURA et al. 1996). Activated platelets secrete ADP and form thromboxane A_2 (TXA_2), two substances that facilitate recruitment of additional platelets into the growing thrombus. Platelets also release a number of growth factors, such as platelet factor 4 (PF4) and platelet-derived growth factor (PDGF), that influence smooth muscle migration and proliferation and may facilitate the repair process.

III. Vascular Contributions

There remains a great deal to be learned about blood coagulation under conditions that more closely simulate conditions *in vivo*. The coagulation processes differ somewhat in arteries and veins, in large and small vessels, and in vessels with rapid or limited blood flow. Cellular and macromolecular components of circulating blood interact with the endothelial lining. Recent developments make it clear that many antithrombotic and prothrombotic responses of the endothelium depend on the rheology of blood flow (for review see GRABOWSKI and LAM 1995).

The vascular endothelium plays a critical role in controlling blood coagulation. Its luminal surface is the site of procoagulant and anticoagulant activities carried out by cell surface molecules that act as receptors for a number of coagulation and anticoagulation factors (Fig. 5) (GERLACH et al. 1990). Normal vascular endothelium *in situ* does not favor coagulation and thrombosis. Prostacyclin (PGI_2) and nitric oxide (NO), potent inhibitors of platelet and monocyte activation and vasodilators, are released by normal endothelium (for review see VANE 1994). The endothelial surface expresses ecto-adenosine diphosphatase, which degrades adenosine diphosphate (ADP) minimizing platelet aggregation, TM, which serves as a binding site for thrombin to activate protein C, and heparin-like molecules, which serve as cofactors for ATIII (for review see WU and THIAGARAJAN 1996). Normal endothelial cells actively participate in fibrinolysis through their ability to synthesize several components of the system, such as the tissue plasminogen activator (t-PA), the urokinase-type plasminogen activator (u-PA), and plasminogen activator inhibitor 1 (PAI-1). (for review see LIJNEN and COLLEN 1997).

In the early phase of injury, endothelial cells respond with increased secretion of vasoprotective molecules, but when the injury is severe or prolonged, the vasoprotective and antithrombotic properties may be lost. Under

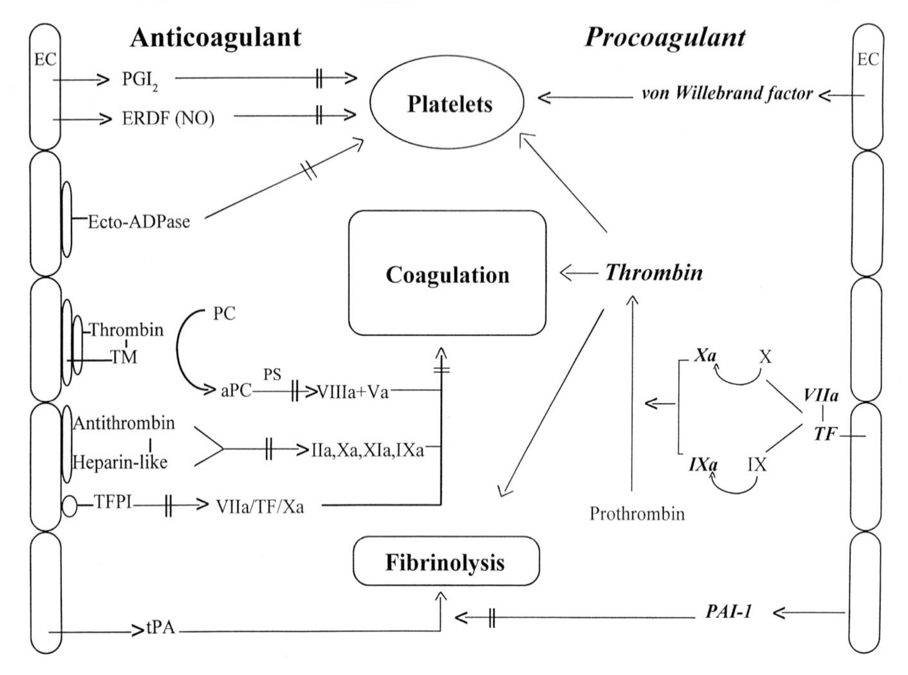

Fig. 5. Vasoprotective and procoagulant properties of endothelial cells
From: WU and THIAGARAJAN (1996) with the permission, from the Annual Review of Medecine

some pathological conditions, such as atherosclerosis, diabetes mellitus and other vascular diseases, endothelial cells may lose their thromboresistance by downregulation of active anticoagulant properties or by the expression of dormant procoagulant pathways. Endothelial cells produce endothelins that are potent vasoconstrictors as well as a number of growth factors, such as vascular endothelial growth factor (VEGF), which may contribute to endothelial generation and vascular repair.

E. Fibrinolysis

Under normal circumstances, the fibrin clot serves a temporary function. It needs to be removed and replaced by repair processes to restore blood flow to organs and tissues. The fibrinolytic system fulfills the blood clot removal role and operates in close balance with the coagulation system. Contact coagulation factors initiate or enhance clot formation, but they also initiate processes leading to clot degradation (Fig. 1). The key enzyme in the fibrinolytic system is the serine protease, plasmin, which degrades fibrin to produce progressively smaller degradation products containing the D-domain of fibrinogen (D-dimer). Plasmin circulates as an inactive proenzyme, plasminogen, that is converted to its active form, plasmin, through the action of plasminogen activators. Plasminogen activators are intrinsic or extrinsic to plasma. Intrinsic activation (contact dependent pathway) of fibrinolysis is initiated upon activation of fXII to XIIa when it interacts with a negatively charged foreign surface in the presence of high molecular weight kininogen and prekallikrein (Fig. 1), initiating coagulation as well as plasmin formation. Kallikrein generated during contact phase activation converts scu-PA (single chain plasminogen activator or pro-urokinase) to tcu-PA (two chain plasminogen activator or high molecular weight plasminogen activator) that considerably enhances the fibrinolytic potential of scu-PA. Extrinsic activators include tissue plasminogen activator (t-PA), a 70-kDa molecule synthesized mainly by the endothelium and which is the main factor responsible for physiologic lysis of clots in the cardiovascular system. Another extrinsic activator is urinary plasminogen activator (u-PA) found originally in human urine, but also present in plasma and secreted by many different cell types.

Fibrinolysis is regulated by the endogenous production and secretion of activators and specific inhibitors. Free circulating plasmin is rapidly inactivated by the specific inhibitors, α_2-antiplasmin and plasminogen activator inhibitor-1 (PAI-1). Any excess of free plasmin needs to be neutralized rapidly since plasmin not only degrades fibrin, but also circulating fibrinogen and other proteins involved in coagulation, such as fV and fVIII, which could cause a serious bleeding tendency. α_2-Antiplasmin, synthesized by the liver, is a member of the serpin family. PAI-1 is a serine proteinase inhibitor, released by the endothelial cells, that reacts rapidly with t-PA and u-PA, forming a stable stoichiometric (1:1) complex. Endothelial cells synthesize both PAI-1 and t-PA.

F. Antithrombotic Strategies

Activation of the clotting system is necessary to prevent blood loss after injury but uncontrolled intravascular activation of coagulation can cause pathological thrombosis. This can lead to the serious clinical consequences of myocardial infarction (MI), stroke, pulmonary embolism, deep vein thrombosis (DVT) and disseminated intravascular coagulation (DIC). These thromboembolic diseases are the leading causes of morbidity and mortality in most industrialized societies.

Although all blood thrombi include fibrin and platelets, their relative ratios vary depending on their location. Thrombi that form in veins have an abundance of fibrin with entrapped red blood cells, but the relative platelet content may be low. Arterial thrombi, which form under conditions of higher pressure and shear forces often associated with endothelial injury, are relatively fibrin-poor but platelet-rich. Therefore, the appropriate target of an antithrombotic agent (inhibition of fibrin formation and/or platelet activity) will depend on the clinical condition (arterial or venous thrombosis) for which it is intended. In addition, since these agents will inhibit both abnormal and normal hemostasis they should provide clinical benefit without an increased risk of bleeding, have no other important side-effects, and have a wide therapeutic range (SIXMA and DE GROOT 1992). They should also have an appropriate half-life, i.e., short duration if used for acute conditions in which parenteral administration may be the ideal route but a long duration (by oral administration) for conditions in which chronic administration is desired. Ideal drugs are not presently available, but the pharmaceutical industries are devoting considerable efforts to develop new antithrombotics that will overcome the limitations of currently used agents (review by FAREED 1996; FAREED et al. 1995; review by KEREVEUR and SAMAMA 1995; review by TURPIE et al. 1995; review by WEITZ et al. 1995). Usually antithrombotics are classified according to their mechanism of action. Antiplatelet agents are designed to work primarily by decreasing platelet aggregation, whereas anticoagulants are designed to work primarily by decreasing thrombin formation or inhibiting thrombin's action after it is formed. However, antiplatelet agents may also function as anticoagulants *in vivo* since activated platelets can facilitate thrombin generation by providing a catalytic surface on which coagulation reactions occur and by releasing an activated form of fV. Likewise, anticoagulants can also act as antiplatelet agents, since thrombin is a potent agonist for platelet activation.

I. Coagulation Factor Inhibitors

Anticoagulant therapy targets either the action of thrombin or its generation (Fig. 6). Current anticoagulant therapy is limited to heparin, low-molecular-weight heparin (LMWH) and warfarin. These agents are described as indirect inhibitors since heparin and LMWH are dependent on endogenous ATIII and warfarin acts indirectly through inhibition of the production of vitamin K-dependent proteins.

Warfarin is the only effective oral anticoagulant currently on the market. Warfarin and related coumarins (dicoumarol-type) reduce thrombin generation by blocking the vitamin-K dependent γ-carboxylation of glutamic acid in the liver to form γ-carboxyglutamic acid (Gla) residues, which are essential for the function of a number of enzymes in the coagulation cascade (for review see VERMEER and HAMULYÁK 1991). Warfarin reduces not only the coagulant proteins, such as prothrombin, fVII, fIX and fX, but the proteins C and S that

Fig. 6. Direct and indirect anticoagulants. The direct anticoagulants block coagulation factors by binding to a coagulation enzyme and preventing the interaction of this enzyme with its substrate. Direct anticoagulants are divided into either direct thrombin inhibitors or thrombin generation inhibitors. Indirect anticoagulants do not interact directly with coagulation enzyme, instead, these agents either enhance the catalysis of the naturally occurring physiological regulators, antithrombin III and heparin cofactor II, or work by blocking the synthesis of vitamin K-dependent coagulation enzymes. For more details about drugs see reference cited in the text

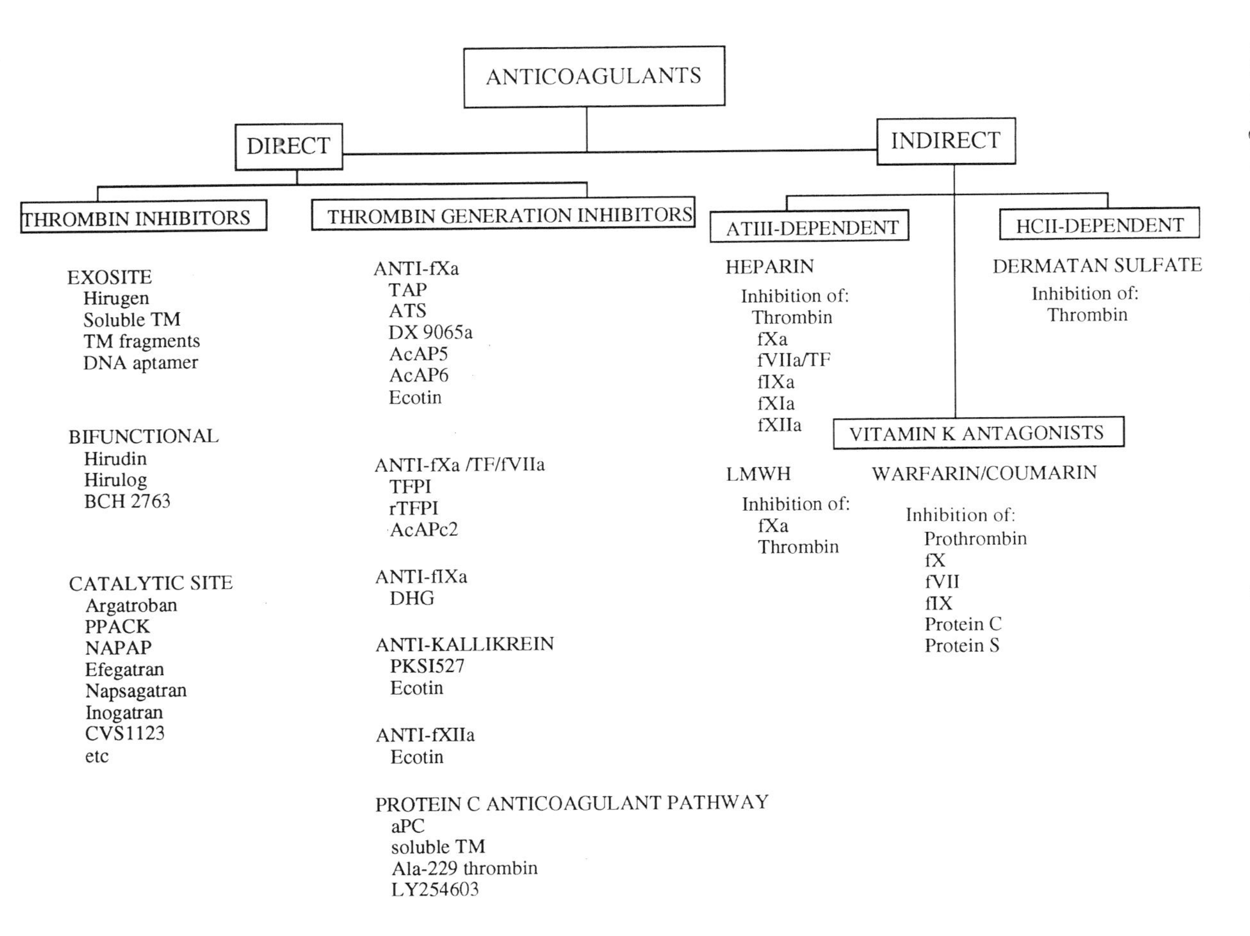
ANTICOAGULANTS
DIRECT
INDIRECT
THROMBIN INHIBITORS
EXOSITE
Hirugen
Soluble TM
TM fragments
DNA aptamer
BIFUNCTIONAL
Hirudin
Hirulog
BCH 2763
CATALYTIC SITE
Argatroban
PPACK
NAPAP
Efegatran
Napsagatran
Inogatran
CVS1123
etc
THROMBIN GENERATION INHIBITORS
ANTI-fXa
TAP
ATS
DX 9065a
AcAP5
AcAP6
Ecotin
ANTI-fXa /TF/fVIIa
TFPI
rTFPI
AcAPc2
ANTI-fIXa
DHG
ANTI-KALLIKREIN
PKSI527
Ecotin
ANTI-fXIIa
Ecotin
PROTEIN C ANTICOAGULANT PATHWAY
aPC
soluble TM
Ala-229 thrombin
LY254603
ATIII-DEPENDENT
HEPARIN
Inhibition of:
Thrombin
fXa
fVIIa/TF
fIXa
fXIa
fXIIa
LMWH
Inhibition of:
fXa
Thrombin
VITAMIN K ANTAGONISTS
WARFARIN/COUMARIN
Inhibition of:
Prothrombin
fX
fVII
fIX
Protein C
Protein S
HCII-DEPENDENT
DERMATAN SULFATE
Inhibition of:
Thrombin

work as endogenous anticoagulants. Since the mechanism of action is the inhibition of *de novo* synthesis of all vitamin K-dependent coagulation factors, warfarin action does not occur immediately after its administration and requires prolonged dosing to establish a stable level of anticoagulation. As a corollary, the anticoagulant effect of warfarin is also not readily reversible. Bleeding is the major clinical problem associated with the use of warfarin and there is marked intrapatient and interpatient variability, which requires periodic monitoring of prothrombin time (PT). The major advantages of warfarin are its oral effectiveness and long duration of action.

Heparin is an undefined mixture of highly sulfated polysaccharide chains ranging in molecular weight from 3000 to 30,000, with a mean of 15,000. Heparin inactivates thrombin, as well as the coagulation factors IXa, Xa, XIa and XIIa, by accelerating their reaction with the physiological inhibitor ATIII (for review see FAREED et al. 1995). Heparin leads to rapid inhibition of thrombin and other active coagulation enzymes. However, heparin has several disadvantages. It needs to be administrated parenterally and it binds avidly to plasma proteins, thereby requiring frequent monitoring due to its unpredictable pharmacokinetics. It is ineffective in patients with ATIII deficiencies. In a small but significant number of patients it can lead to heparin-induced thrombocytopenia. Heparin's anticoagulant function is neutralized by platelet factor 4 (PF4) released from activated platelets. Thrombin bound to fibrin in thrombi or to the exposed subendothelial matrix remains active and is resistant to heparin-ATIII inhibition (WEITZ et al. 1990; BAR-SHAVIT et al. 1989).

Low molecular weight heparin (LMWH) preparations are fragments of standard heparin produced by controlled chemical or enzymatic depolymerization (for review see FAREED et al. 1995). They are heterogeneous in size, ranging in molecular weight from 1000 to 10,000, with a mean of 4000–5000. LMWHs differ from unfractionated heparin in that they have better bioavailability and pharmacokinetics, are relatively resistant to inhibition by PF4, and have an increased ratio of fXa/thrombin inhibition (GREEN et al. 1994). Like heparin, LMWHs are ATIII dependent anticoagulants and act on several coagulation factors. There is less risk of bleeding, however, with LMWHs than with standard heparin. LMWHs are obtained by various manufacturing processes that may result in a certain variability in their pharmacological profile.

Newer anticoagulants have been designed which more specifically inhibit activated coagulation factors and may potentially have fewer undesirable interactions and side effects. They selectively inhibit thrombin or other enzymes higher in the coagulation cascade (Fig. 6).

1. Direct Thrombin Inhibitors

Various structural domains, including the catalytic site and the two exosite binding sites, regulate thrombin's interactions with biological substrates, providing pharmacologists with targets for the development of specific thrombin inhibitors. Direct thrombin inhibitors block either the catalytic site of thrombin or its anion-binding site (recognition site for substrates such as fibrinogen, thrombin receptor, TM, and HCII or both). The most potent and selective inhibitor of thrombin is hirudin, a peptide isolated from the salivary gland of the leech *Hirudo medicinalis* (for review see MARKWARDT 1994), that is now available through recombinant DNA technology. Hirudin complexes with thrombin by apolar interactions with regions adjacent to thrombin's catalytic site and by interactions with the anion-binding exosite (TULINSKY 1996).

The bivalent nature of the hirudin-thrombin interaction has led to the design of hirulog and a hirulog-like class of thrombin inhibitors (FENTON 1992). These molecules literally form a bridge between the active site and the fibrinogen exosite and show effective inhibition in the nano- to picomolar range (MARAGANORE et al. 1990; TSUDA et al. 1994; FINKLE et al. 1998). Hirudin and hirulog have been shown to be very effective antithrombotics in a variety of thrombosis models in animals (see review by BEIJERING et al. 1996). BCH 2763, which mimics the bifunctional binding mode of hirulog and hirudin but is a lower molecular weight compound, has shown good antithrombotic efficacy in animal models of arterial and venous thrombosis (FINKLE et al. 1998; BUCHANAN et al. 1996). The dual mode of interaction of these inhibitors confers on them absolute specificity for the thrombin enzyme. Their major disadvantages are a relatively short half-life and lack of oral bioavailability. Hirudin and hirulog are in advanced stages of clinical development, although in large clinical trials such as GUSTO IIb (1996) and TIMI 9b (ANTMAN 1996) their benefits over heparin were not significant. As treatment for deep vein thrombosis in patients undergoing hip or knee surgery, however, significant benefit was demonstrated (GINSBERG et al. 1994).

Development of specific, active site-directed thrombin inhibitors is a major target for a number of pharmaceutical companies. The attractiveness of this type of thrombin inhibitor resides in the potential for oral bioavailability with a reasonable plasma half-life. During recent years, a number of inhibitors have been developed that interact in various ways with amino acids of the catalytic pocket (Ser195, His57 and Asp102). From the biochemical point of view, these inhibitors have been classified according to their kinetics of inhibition and the type of binding or to the structural origin of the molecule (more details in HAUPTMAN and MARKWARDT 1992). A list of these inhibitors, as well as their mechanisms of inhibition, is given by TAPPARELLI et al. (1993) and KIMBALL (1995). Argatroban, PPACK, NAPAP, efegatran, napsagatran, inogatran and CVS1123 are the best known examples of these inhibitors. Direct thrombin inhibitors have proven to be more effective than aspirin, heparin or GpIIbIIIa receptor antagonists in several different animal models of thrombosis (LEFKOVITS and TOPOL 1994). A number of small molecules, direct thrombin inhibitors are at different stages of development (CALLAS and FAREED 1995; KIMBALL 1995). Oral bioavailability and adequate half-life, however, remain as significant challenges. Argatroban is the only direct thrombin inhibitor that has been approved and that is for parenteral use in humans (Japan).

The fibrinogen anion binding site (exosite 1) on thrombin has also been explored as a potential site for the development of therapeutic agents. Hirugen, a synthetic dodecapeptide comprising residues 53–64 of the carboxy-terminal region of hirudin, soluble TM obtained by recombinant DNA technology, and a single-stranded DNA aptamer composed of 15 nucleotides all bind to exosite 1. These agents prevent thrombin's interaction with macromolecular substrates, such as fibrinogen and the thrombin receptor on platelets, but leave the amidolytic activity towards small substrates unaltered (MAFFRAND 1992; PABORSKY et al. 1993). Soluble TM complexes with thrombin, causes a conformational change in the enzyme that abolishes its procoagulant activity and converts it into a potent activator of protein C. Both hirugen and TM appear to be effective antithrombotics in a variety of experimental thrombosis models (KELLY et al. 1992; ONO et al. 1994). To develop a TM based inhibitor, cyclic peptides derived from the fifth EGF-like domain of TM have been synthesized and shown to have some thrombin-inhibitory activity (LOUGHEED et al. 1995). Recently, a new class of thrombin inhibitors based on single-stranded deoxynucleotides has been described. The single-stranded, 15-mer nucleotide DNA "aptamer" (GS-522) is a potent anticoagulant *in vitro* and has antithrombotic properties in experimental thrombosis models (LI et al. 1994). The DNA "aptamer" may also interact with the putative heparin-binding site of thrombin (PADMANABHAN et al. 1993a) and is currently undergoing further preclinical evaluation for use as an anticoagulant in cardiopulmonary bypass.

Thrombin also has an exosite for high-affinity binding of heparin. In addition to heparin binding, this exosite participates in fV and fVIII activation. The design of inhibitors for this exosite either alone or in conjunction with the catalytic site is conceivable, but to present knowledge has not as yet been attempted.

Direct thrombin inhibitors provide several potential advantages over heparin including few or no direct effects on other coagulation factors, and no reliance on cofactors. They are not neutralized by platelet factors, such as platelet factor 4, and thus remain effective in the presence of activated platelets. An important advantage that the direct thrombin inhibitors have over heparin is that they are effective against clot-bound thrombin (WEITZ and HIRSH 1993). This is important because the catalytic activity of thrombin persists after it is bound to fibrin in a forming thrombus, thereby prompting additional platelet activation and fibrin formation. A potential disadvantage for hirudin or hirudin-like compounds, as well as direct catalytic site inhibitors, is that they are likely to neutralize both the procoagulant and anticoagulant functions of thrombin.

2. Thrombin Generation Inhibitors

Thrombin incorporated into a fibrin clot not only remains able to catalyze conversion of fibrinogen to fibrin, but also continues to amplify its own formation through fV and fVIII activation. This thrombin generation through autoamplification could be interrupted by thrombin generation inhibitors, rather than by inhibitors directly acting on the activity of preformed thrombin. Recent studies with fXa inhibitors or inhibitors working higher up in the coagulation cascade indicate that inhibition of thrombin formation could be as effective in preventing thrombosis as inhibition of thrombin activity directly. Furthermore, it is reasonable to speculate that thrombin generation inhibitors may be effective in clinical conditions in which prevention of thrombus formation is the objective rather than treatment of a preformed thrombus.

a) Factor Xa Inhibitors

A greater understanding of the mechanisms of action of anticoagulant and antithrombotic drugs has focused attention on fXa as a target for intervention. The activated serine protease, fXa, plays a key role in blood coagulation, since its activation occurs at the point of convergence of the extrinsic and intrinsic activation pathways and leads to the final stage of hemostasis, formation of thrombin from the prothrombinase complex. From a theoretical point of view, fXa inhibitors could be highly efficient, since they would interfere with the coagulation cascade at a kinetically important point.

Today heparin and LMWHs are the only drugs on the market with anti-fXa activity. This activity is dependent on ATIII, just as it is for their antithrombin activity, so they are indirect inhibitors of fXa. Recently, naturally occurring direct inhibitors of fXa were isolated from leeches and ticks. Tick anticoagulant peptide (TAP) and antistasin (ATS) are highly selective, potent inhibitors of fXa, working equally well against the free or the bound form of the enzyme, in contrast to heparin and LMWHs (Vlasuk 1993; Tuszynski et al. 1987). Both peptides have been prepared by recombinant technology. DX 9065a is the first nonpeptide, low molecular weight inhibitor of fXa reported to show oral absorption and efficacy in experimental thrombosis models (Hara et al. 1994, 1995). Factor Xa inhibitors may have less deleterious effects on hemostasis than direct thrombin inhibitors. If fXa inhibitors allow limited generation of thrombin to continue through the extrinsic pathway, complete elimination of hemostasis might be avoided, reducing the risk of serious bleeding.

b) Inhibitors of Other Coagulation Factors

The coagulation pathway involves a rapid amplification cascade of proteolytic reactions. Inhibiting these reactions very early in the cascade can be considered an efficient way to reduce the amount of antithrombotic agents needed to exert significant antithrombotic effects and to better target specific thromboembolic diseases. Activation of the extrinsic coagulation pathway appears to play an important role in the initiation of *in vivo* coagulation. This pathway involves the conversion of fVII into fVIIa after its association with TF. TFPI, inhibitors of fVIIa/TF complex formation, and fVIIa inhibitors are able to block the extrinsic coagulation pathway. TFPI, an endogenous coagulation inhibitor, in the presence of calcium and fXa, inhibits the fVIIa/TF complex activity in a retroactive way, since some fXa must be available before the fVIIa/TF complex can be inhibited. Recombinant three domain, two domain glycosylated (TFPI 1–161) and nonglycosylated (1–161) forms of TFPI have antithrombotic effects in experimental animal thrombosis models (KAISER and FAREED 1996; HOLST et al. 1996; ELSAYED et al. 1996). One major advantage of this approach may be the possibility of local therapy. The fVII-VIIa/TF complex is probably formed only at the site of injury, where it remains membrane associated. Monoclonal antibodies against TF or the fVII-VIIa complex can prevent intravascular thrombus formation (PAWASHE et al. 1994; BIEMOND et al. 1995). Smaller molecular weight inhibitors, designed to interact with the catalytic site of fVIIa or against formation of the fVIIa/TF complex, are at the early stage of development (PABORSKY et al. 1995).

Inhibitors of the contact activation pathway (intrinsic pathway) may also play an important role in the regulation of thrombotic disorders. Two inhibitors of this pathway have been reported. PKSI-527 is a highly selective plasma kallikrein inhibitor (KATSUURA et al. 1996) which shows antithrombotic activity in a DIC model in rats. Ecotin, a relatively nonselective serine protease inhibitor found in the periplasm of *Escherichia coli*, is characterized as a potent, reversible inhibitor of fXIIa and plasma kallikrein. However, ecotin is also a very potent fXa inhibitor, as well as being a human leukocyte elastase inhibitor (ULMER et al. 1995).

Other inhibitors are reported to interact with coagulation factors involved in the early steps of the coagulation cascade. Potent small protein anticoagulants isolated from the hookworm, *Ancylostoma caninum*, show either anti-fXa activity (AcAP5 and AcAP6) or activity against the TF/fVIIa complex (AcAPc2) (STASSENS et al. 1996). Inhibition of the fVIIa/TF complex activity by the AcAPc2 anticoagulant actually occurs secondary to inhibition of fXa, similar to TFPI, but AcAPc2 anticoagulant utilizes an exosite that is distinct from that used by fXa. A new glycosaminoglycan, DHG (depolymerized holothurian glycosaminoglycan), is described as exhibiting ATIII- and HCII-independent inhibition of fX activation and HCII-dependent inhibition of thrombin (NAGASE et al. 1995).

Thrombin generation can be reduced through activation of the protein C anticoagulant pathway, which inhibits activated fV and fVIII, thereby suppressing thrombin generation by fXa. Natural and recombinant forms of APC have been used with success as anticoagulant and administration of APC in effective antithrombotic doses is not associated with any prolongation of the bleeding time (GRUBER et al. 1990). Soluble TM, by complexing with thrombin, abolishes its procoagulant activity and converts it into a selective activator of protein C (ONO et al. 1994) that shows antithrombotic activity. In addition, thrombin can be engineered to function *in vivo* as a potent anticoagulant by a single point mutation. Mutation of glutamic acid to alanine at position 229 of the protein's sequence, creating Ala-229 thrombin, substantially shifts thrombin's specificity in favor of the anticoagulant substrate, protein C. *In vivo*, this modified thrombin functions as an endogenous protein C activator (GIBBS et al. 1995). Recently, LY254603 was described as an agent that is able to modulate the substrate cleavage specificity of human thrombin similar to the thrombin point mutation (BERG et al. 1996).

II. Antiplatelet Agents

Platelets also play a particularly key role in arterial thrombus formation, making them an attractive target for clinical conditions involving arterial thrombus formation. Platelet adhesion to the injured wall results in their activation and release of proaggregatory agents, such as adenosine diphosphate (ADP), thromboxane A_2 (TXA_2), and serotonin. These substances are agonists for further platelet aggregation and release, and subsequent platelet recruitment to the injury site leading to thrombus growth. Three strategies to block platelet actions are currently under investigation: inhibition of platelet adherence to the injured vessel wall, inhibition of platelet activation, and inhibition of platelet aggregation (Fig. 7). A summary of antiplatelet agents in development was recently reported by FAREED et al. (1995).

1. Platelet Adhesion and Activation Inhibitors

The first antiplatelet approach is to inhibit platelet adherence to the vessel wall by interfering with the interaction between GpIb and vWf (RUGGERI 1992). Three types of intervention are under development: inhibition of the GpIb-IX binding site for vWf, inhibition of vWf binding sites for GpIB, and inhibition of vWf multimerization. Inhibitors of platelet adhesion remain in early stages of development. Until now, the selective antithrombotic advantages obtained by interrupting GpIB-vWf interaction over inhibitors of other pathways for platelet aggregation remain unclear (HARKER et al. 1994; WEITZ et al. 1995). Many different ligands and receptors mediate platelet adhesion and could offer alternative targets for antithrombotic strategies. A potential problem, however, in this approach is the risk of bleeding, since the formation of a monolayer of adherent platelets in response to vessel injury provides an important hemostatic mechanism to prevent blood loss.

Platelets react rapidly to various endogenous agonists, such as collagen, ADP, arachidonate, TXA_2 and thrombin. Binding between the agonist and its platelet receptor induces signal transduction across the membrane leading to platelet activation. The second potential antiplatelet approach is the prevention of platelet activation by using agents that inhibit specific steps in the signal transduction pathways or the interaction between the agonist and its receptor.

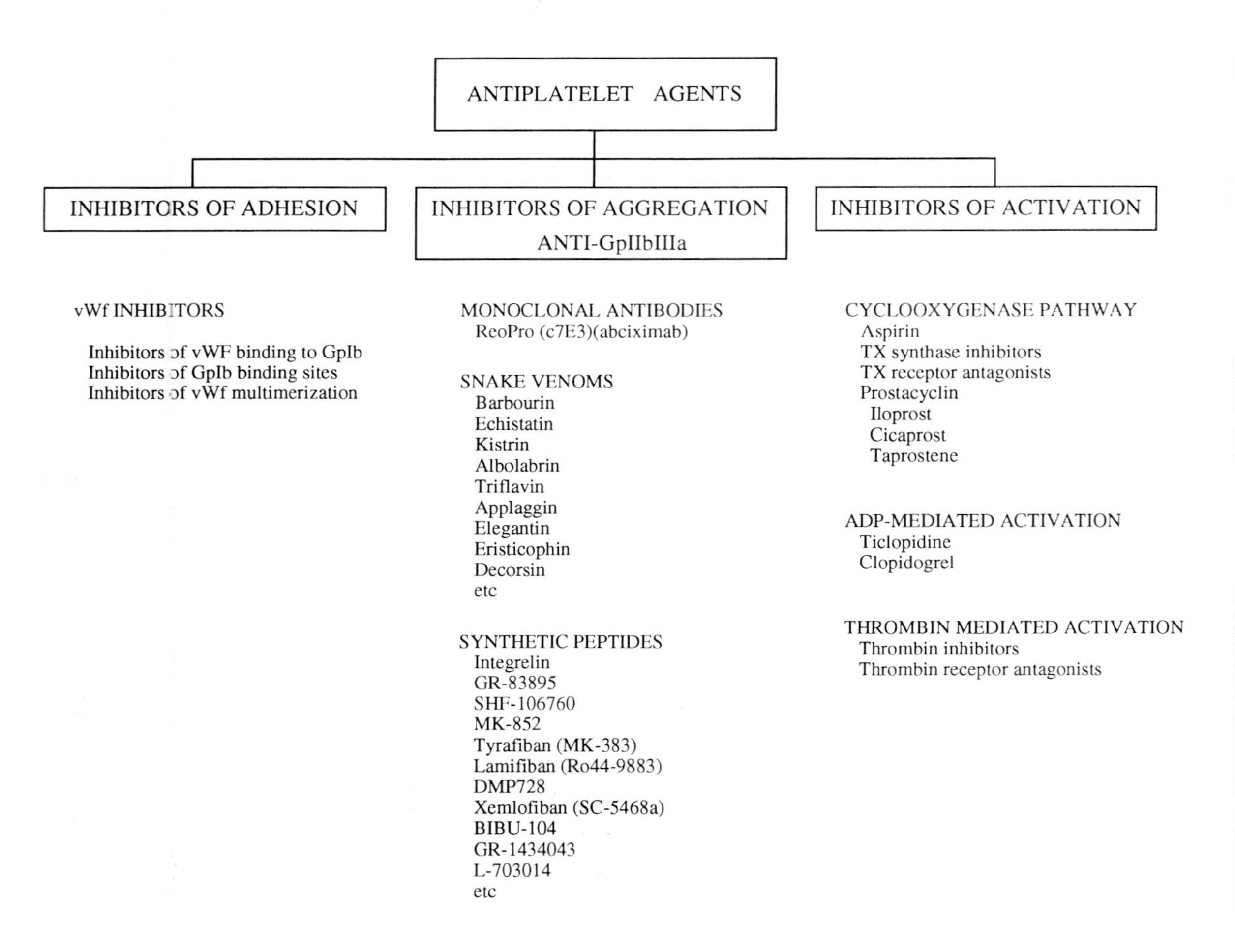
ANTIPLATELET AGENTS
INHIBITORS OF ADHESION
INHIBITORS OF AGGREGATION
ANTI-GpIIbIIIa
INHIBITORS OF ACTIVATION
vWf INHIBITORS
Inhibitors of vWF binding to GpIb
Inhibitors of GpIb binding sites
Inhibitors of vWf multimerization
MONOCLONAL ANTIBODIES
ReoPro (c7E3)(abciximab)
SNAKE VENOMS
Barbourin
Echistatin
Kistrin
Albolabrin
Triflavin
Applaggin
Elegantin
Eristicophin
Decorsin
etc
SYNTHETIC PEPTIDES
Integrelin
GR-83895
SHF-106760
MK-852
Tyrafiban (MK-383)
Lamifiban (Ro44-9883)
DMP728
Xemlofiban (SC-5468a)
BIBU-104
GR-1434043
L-703014
etc
CYCLOOXYGENASE PATHWAY
Aspirin
TX synthase inhibitors
TX receptor antagonists
Prostacyclin
Iloprost
Cicaprost
Taprostene
ADP-MEDIATED ACTIVATION
Ticlopidine
Clopidogrel
THROMBIN MEDIATED ACTIVATION
Thrombin inhibitors
Thrombin receptor antagonists

Aspirin and related drugs (TX synthase inhibitors/TX receptor antagonists, prostacyclin), thienopyridines (ticlopidine and clopidogrel), direct thrombin inhibitors and inhibitors of the thrombin receptor are examples of agents that target platelet activation. Aspirin is the most widely clinically used antithrombotic agent today for long term oral treatment (ROTH and CALVERLEY 1994). Aspirin inhibits platelet aggregation by inhibiting generation of TXA_2 by activated platelets via the irreversible acetylation of the cyclo-oxygenase enzyme. The irreversible nature of inhibition means that TXA_2 production is lost for the life span of the platelet. The effectiveness of aspirin is achieved at a relatively low incidence of severe adverse events. However, platelets can be activated by agonists, such as thrombin and ADP, through other non-cyclo-oxygenase dependent pathways, thus limiting the effectiveness of aspirin to shut down completely the thrombotic process. In addition, aspirin does not inhibit platelet adhesion or granule secretion. TXA_2 formation and action can also be blocked directly by TX synthase inhibitors and TX receptor antagonists. Selective inhibitors of TX synthesis and/or TX receptors have several theoretical advantages over aspirin. These include the selective nature of inhibition, as well as the generation of prostacyclin and prostaglandin D_2 (PGD_2) that are platelet-inhibitory and tissue-protective prostaglandins. Clinical experience with these agents is limited and in general the results have not been encouraging. Prostacyclin is the most potent endogenous inhibitor of platelet aggregation. Prostacyclin and related synthetic compounds, such as iloprost, cicaprost and taprostene, elevate the platelet cyclic adenosine monophosphate (cAMP) content via receptor-dependent stimulation of adenylate cyclase. This results in low cytoplasmic calcium levels, which maintain the platelet in its resting state and inhibit platelet responses to agonists. The major problems with prostacyclin are the receptor-mediated nature of the response and the fact that its effect is not limited to platelets. This may result in nonspecific phenomena (for review see SCHRÖR 1995).

ADP is another important endogenous platelet agonist released at the site of injury by red blood cells, as well as by other platelets. Its mechanism of action is not completely understood. Thienopyridines (ticlopidine and clopidogrel) selectively and specifically interfere with ADP-mediated platelet activation. Thienopyridines do not affect arachidonic acid metabolism in the platelet and do not interfere with the GpIIb/IIIa receptor. Thienopyridines need 3–5 days of oral administration to have full antiplatelet action and antiplatelet effects persist up to 10 days after withdrawal. Thienopyridine therapy is subject to the same limitations as aspirin in the effectiveness of approach: platelets can still be activated through several other mechanisms (for review see SCHRÖR 1995).

Fig. 7. Antiplatelet agents. The main targets for these agents are platelet adhesion, aggregation or activation. For more information on drugs listed in this figure, see FAREED et al. 1995; SCHROR 1995; COOK et al. 1994

Thrombin is the most important platelet agonist generated at the sites of vascular injury. Thrombin-mediated platelet activation can be interrupted by direct thrombin inhibitors (see description earlier) or by interfering with activation of the thrombin receptor. Recent cloning and sequencing of the thrombin receptor gene have opened a new avenue for drug design that will selectively inhibit the effect of thrombin on cells. Thrombin cleaves its receptor, a seven transmembrane G-coupled protein, creating a new tethered ligand, which can be mimicked by synthetic analogues (VU et al. 1991). In targeting the receptor rather than thrombin, inhibitors should inhibit the cellular effects of thrombin without increasing the risk of bleeding seen by inhibiting fibrin formation (see review by BRASS 1995 and by SEILER 1996). Studies in experimental animal models suggest that inhibiting thrombin receptor activation may provide an efficacious approach to limit the vascular response to injury (SEILER 1996). Potent, specific thrombin receptor inhibitors have been characterized recently (BERNATOWICZ et al. 1996) which should permit evaluation of their therapeutic potential in the near future.

2. Fibrinogen Receptor Antagonists

The GpIIb/IIIa receptor plays a crucial role in platelet aggregation and platelet thrombus formation. This receptor is confined to cells of the megakaryocyte/platelet lineage with the exception of some tumor cells (TRIKHA et al. 1996). Platelet activation results in a conformational change in the calcium-dependent heterodimer GpIIb/IIIa. The activated complex is a high affinity receptor for adhesive proteins responsible for the formation of platelet aggregates, including fibrinogen, vWf, vitronectin and fibronectin. All of these ligands bind through their arginine-glycine-aspartic acid (RGD) recognition site (D'SOUZA et al. 1991). Several classes of GpIIb/IIIa inhibitors are available including monoclonal antibodies [7E3, murine/human chimeric monoclonal antibody fragments c7E3 Fab (ReoPro)], small orally available synthetic, peptidomimetic agents, as well as molecules extracted from snake venoms (SCHRÖR 1995; COLLER et al. 1995). Despite the structural diversity of these compounds, they all retain the basic charge relations of the RGD sequence with positive and negative charges separated by approximately 10–20 Å (COOK et al. 1994). Antagonists of GpIIb/IIIa have major advantages over other antiplatelet agents. They are the most powerful antiplatelet agents, since the target is the end product of platelet activation, namely the expression and activation of GpIIb/IIIa receptors on which platelet aggregation depends. These agents act only on platelets and are, therefore, more specific than aspirin or thrombin inhibitors. Blockade of GpIIb/IIIa receptors results in profound inhibition of thrombotic occlusions in animal models and appears to be clinically effective in a variety of thrombotic disease indications (COLLER et al. 1995) although not without some bleeding risk. Orally active compounds are under development and may become antithrombotic drugs in the near future.

III. Thrombolytic Agents

In clinical conditions, such as MI, the therapeutic strategy is thrombus dissolution and prevention of reocclusion. This can be achieved by the pharmacological dissolution of the blood clot via intravenous infusion of plasminogen activators, which activate the blood fibrinolytic system. Several thrombolytic agents are under development: streptokinase, urokinase, recombinant tissue-type plasminogen activator (rt-PA), anisoylated plasminogen activator complex (APSAC), the recombinant single-chain urokinase-type plasminogen activator (rscu-PA or prourokinase), and staphylokinase (VERSTRAETE 1995). Streptokinase is a nonenzymatic protein produced by β-hemolytic streptococci. Urokinase is derived from urine or from human fetal kidney cell lines. Recombinant t-PA is produced through recombinant DNA technology. APSAC, the anisoylated derivative of the Lys-plasminogen streptokinase activator complex, is a pro-enzyme. The catalytic center of the activator complex is temporarily masked by the anisoyl group without affecting the molecule's ability to bind to fibrin. Sustained release of the anisoyl group by deacylation occurs in a controlled manner. Recombinant scu-PA, too, is produced through recombinant DNA technology. Limited hydrolysis in blood converts the single chain molecule into the two chain urokinase-type plasminogen activator (tcu-PA or urokinase) that is associated with a large increase in plasminogen-activating activity. Staphylokinase is a bacterial profibrinolytic agent that forms a complex with plasminogen which, following conversion to plasmin, activates other plasminogen molecules to plasmin (review by COLLEN et al. 1989). Recombinant t-PA and staphylokinase are fibrin-specific thrombolytic agents while streptokinase and urokinase are not (COLLEN and LIJNEN 1995). Non-fibrin-specific thrombolytic agents activate circulating and fibrin-bound plasminogen without distinction. As a result, free circulating plasmin is neutralized very rapidly by α_2-antiplasmin. In this respect, lysis induced by rt-PA and staphylokinase is expected to be more rapid and effective with less risk of bleeding. However, comparative studies between the two groups of thrombolytic agents (streptokinase and rt-PA, staphylokinase and streptokinase) show no difference in the efficacy depending on the categories of thrombotic diseases under treatment (MARDER 1995; COLLEN et al. 1989; LIJNEN et al. 1991; VAN DE WERF 1991).

Thrombolytic agents can also activate platelets, either directly or indirectly via thrombin generation (COLLER 1990), suggesting that the net clinical effect of these agents involves competition between their fibrinolytic effects and their ability to enhance platelet deposition. Therapy may be enhanced by using an antiplatelet agent or a thrombin inhibitor in conjunction with the thrombolytic agent (LOSCALZO 1996). Drug targeting may be another way to achieve more efficacious and safe fibrinolytic agents. Chimeric molecules in which the thrombolytic enzyme is conjugated to antibodies or Fab fragment of antibodies directed against either fibrin, a platelet integrin, or damaged endothelium generally show enhanced clot lysis and appear attractive for improvement of the thrombolytic therapy (review by SAMAMA and ACAR 1995).

Factor XIIIa is the transglutaminase that catalyzes interchain cross-linking reactions and serves to stabilize newly formed fibrin clots. The observation that a factor XIIIa inhibitor accelerated thrombolysis in both the rabbit and dog suggests that inhibitors of this enzyme may have potential in thrombolytic therapy (LEIDY et al. 1990; SHEBUSKI et al. 1990). Factor XIIIa inhibitors can be useful in facilitating lysis and maintaining the lytic state when used with other antithrombotic agents.

IV. Other Strategies

To prevent or reduce the thrombotic process, approaches using agents that are not anticoagulants, antiplatelets or fibrinolytics are available although less well developed. Impairment of fibrin formation by depleting the fibrinogen concentration with snake venom enzymes, such as ancrod, batroxiban or crotalase, is one such approach (review by HARKER et al. 1994). The fibrin formed after the action of the snake venoms on fibrinogen is not cross-linked by factor XIIIa, leading to a fibrin gel that is rapidly lysed by endogenous fibrinolysis. These agents may reduce blood viscosity and inhibit platelet aggregation. Other agents work on endothelial cell modulation. Defibrotide, a polydeoxyribonucleotide derivative, produces an antithrombotic action without producing any systemic anticoagulant effects. Its mechanism of action, however, is not fully understood as yet, and may involve an increase in intracellular cAMP and release of TFPI from endothelial cells (review by FAREED et al. 1995). Manipulating the level of 13-HODE (13-hydroxy-octodecadienoic acid) present in the endothelial cells may reduce the vessel wall thrombogenicity (BUCHANAN and BRISTER 1993). The level of 13-HODE, a product of the linoleic acid and the lipoxygenase pathways, correlates inversely with the thrombogenicity of the endothelial surface. We can also include in this group other agents that work at the endothelial level, such endothelium-derived factors (NO, prostacyclin) that exert a vascular protective effect.

G. Conclusion

The discovery and development of drugs have progressed very rapidly during recent years due to the combined efforts of both academia and pharmaceutical companies. Several factors have contributed to this rapid development. First, molecular biology techniques have made it possible to obtain large amounts of many endogenous coagulant and anticoagulant proteins. Having these proteins available in large quantities has made crystallographic structural determinations possible. Determination of the three-dimensional structure of prothrombin/thrombin, fXa, fIXa, fVIIa/TF, fXIII, ATIII and TFPI has given important clues to the understanding of how these key participants interact with each other and with other components involved in the maintenance of hemostasis. Recombinant DNA technology and site-directed mutagenesis have helped to understand better the structure-function relationships of the components of coagulation and the nature of mutations leading to hereditary diseases, such as hemophilia, and thrombotic disorders. This increased insight into understanding hemostasis at the molecular level has paved the way to the development of more powerful, mechanistic-based antithrombotic agents.

In the coming years, several newer and more potent drugs will be generated by the pharmaceutical industry. These drugs will not only inhibit thrombin or other serine proteases with greater specificity but also will be able to target cellular receptors on specific blood or vessel wall cells.

References

Abildgaard U (1995) Relative roles of tissue factor pathway inhibitor and antithrombin in the control of thrombogenesis. Blood Coagul Fibrinolysis 6:S45–S49

Albrecht S, Kotzsch M, Siegert G, Luther T, Grossmann H, Grosser M et al (1996) Detection of circulating tissue factor and factor VII in a normal population. Thromb Haemost 75:772–777

Altieri DC, Stamnes SJ (1994) Protease-dependent T cell activation: ligation of effector cell protease receptor-1. Cell Immunol 155:372–383

Altieri DC (1994) Molecular cloning of effector cell protease receptor-1, a novel cell surface receptor for the protease factor Xa. J Biol Chem 269:3139–3142

Antman EM (1996) Hirudin in acute myocardial infarction. Thrombolysis and Thrombin Inhibition in Myocardial Infarction (TIMI) 9B trial. Circulation 94:911–921

Ayala Y, Di Cera E (1994) Molecular recognition by thrombin. Role of the slow→fast transition, site-specific ion binding energetics and thermodynamic mapping of structural components. J Mol Biol 235:733–746

Bajaj MS, Kuppuswamy MN, Saito H, Spitzer SG, Bajaj SP (1990) Cultured normal human hepatocytes do not synthesize lipoprotein-associated coagulation inhibitor: evidence that endothelium is the principal site of its synthesis. Proc Natl Acad Sci USA 87:8869–8873

Banner DW, D'Arcy A, Chène C, Winkler FK, Guha A, Konigsberg WH et al (1996) The crystal structure of the complex of blood coagulation factor VIIa with soluble tissue factor. Nature 380:41–46

Bar-Shavit R, Eldor A, Vlodavsky I (1989) Binding of thrombin to subendothelial extracellular matrix. Protection and expression of functional properties. J Clin Invest 84:1096–1104

Beijering RJR, ten Cate H, ten Cate JW (1996) Clinical applications of new antithrombotic agents. Ann Hematol 72:177–183

Berg DT, Wiley MR, Grinnell BW (1996) Enhanced protein C activation and inhibition of fibrinogen cleavage by a thrombin modulator. Science 273:1389–1391

Bernatowicz MS, Klimas CE, Hart KS, Peluso M, Allegretto NJ, Seiler SM (1996) Development of potent thrombin receptor antagonist peptides. J Med Chem 39:4879–4887

Berrettini M, Lämmle B, Griffin JH (1987) Initiation of coagulation and relationships between intrinsic and extrinsic coagulation pathways. In: Verstraete M, Vermylen J, Lijnen R, Arnout J (ed) International Society on Thrombosis and Haemostasis and Leuven University Press, Leuven, pp 473–495

Bertina RM, Reitsma PH, Rosendall FR, Vandenbroucke JP (1995) Resistance to activated protein C and factor V Leiden as risk factors for venous thrombosis. Thromb Haemost 74:449–453
Biemond BJ, Levi M, ten Cate H, Soule HR, Morris LD, Foster DL et al (1995) Complete inhibition of endotoxin-induced coagulation activation in chimpanzees with a monoclonal Fab fragment against factor VII/VIIa. Thromb Haemost 73:223–230
Bolton-Maggs PH, Young-Wan-Yin B, McCraw AH, Slack J, Kernoff PB (1988) Inheritance and bleeding in factor XI deficiency. Br J Haematol 69:521–528
Brandstetter H, Kuhne A, Bode W, Huber R, von der Saal W, Wirthensohn K et al (1996) X-ray structure of active site-inhibited clotting factor Xa. Implications for drug design and substrate recognition. J Biol Chem 271:29988–29992
Brass LF (1995) Issues in the development of thrombin receptor antagonists. Thromb Haemost 74:499–505
Broze GJ Jr (1995a) Tissue factor pathway inhibitor. Thromb Haemost 74:90–93
Broze GJ Jr (1995b) Tissue factor pathway inhibitor and the revised theory of coagulation. Ann Rev Med 46:103–112
Broze GJ Jr, Gailani D (1993) The role of factor XI in coagulation. Thromb Haemost 70:72–74
Broze GJ Jr, Warren LA, Novotny WF, Higuchi DA, Girard JJ, Miletich JP (1988) The lipoprotein-associated coagulation inhibitor that inhibits the factor VII-tissue factor complex also inhibits factor Xa: insight into its possible mechanism of action. Blood 71:335–343
Buchanan MR, Brister SJ (1993) Altering vessel wall fatty acid metabolism: a new strategy for antithrombotic treatment. Sem in Thromb Hemost 19:149–157
Buchanan MR, Brister SJ, Finkle C, DiMaio J, Winocour PD (1996) Inhibition of thrombin generation and platelet activation during cardiopulmonary bypass (CPB) in pigs: relative effects of a direct thrombin inhibitor, BCH 2763, and heparin. Haemostasis 26 3:Abs 131
Butenas S, Mann KG (1996) Kinetics of human factor VII activation. Biochemistry 35:1904–1910
Caen JP, Rosa JP (1995) Platelet-vessel wall interaction: from the bedside to molecules. Thromb Haemost 74:18–24
Callas D, Fareed J (1995) Comparative pharmacology of site directed antithrombin agents. Implication in drug development. Thromb Haemost 74:473–481
Carmeliet P, Mackman N, Moons L, Luther T, Gressens P, Van Vlaenderen I et al (1996) Role of tissue factor in embryonic blood vessel development. Nature 383:73–75
Chang JY, Stafford DW, Straight DL (1995) The roles of factor VII's structural domains in tissue factor binding. Biochemistry 34:12227–12232
Collen D, Lijnen HR (1995) Molecular basis of fibrinolysis, as relevant for thrombolytic therapy. Thromb Haemost 74:167–171
Collen D, Lijnen HR, Todd PA, Goa KL (1989) Tissue-type plasminogen activator. A review of its pharmacology and therapeutic use as a thrombolytic agent. Drugs 38:346–388
Coller BS (1990) Platelets and thrombolytic therapy. N Eng J Med 322:33–42
Coller BS, Anderson K, Weisman HF (1995) New antiplatelet agents: platelet GpIIb/IIIa antagonists. Thromb Haemost 74:302–308
Conard J, Bauer KA, Gruber A, Griffin JH, Schwarz HP, Horellou MH et al (1993) Normalization of markers of coagulation activation with a purified protein C concentrate in adults with homozygous protein C deficiency. Blood 82:1159–1164
Cook NS, Kottirsch G, Zerwes HG (1994) Platelet glycoprotein IIb/IIIa antagonists. Drugs Future 19:135–159
Dahlback B (1995) Inherited thrombophilia: resistance to activated protein C as a pathogenic factor of venous thromboembolism. Blood 85:607–614
Dang OD, Vindigni A, Di Cera E (1995) An allosteric switch controls the procoagulant and anticoagulant activities of thrombin. Proc Natl Acad Sci USA 92:5977–5981

Davie EW (1995) Biochemical and molecular aspects of the coagulation cascade. Thromb Haemost 74:1–6
Davie EW, Ratnoff OD (1964) Waterfall sequence for intrinsic blood clotting. Science 145:1310–1312
Davie EW, Fujikawa K, Kisiel W (1991) The coagulation cascade: initiation, maintenance, and regulation. Biochemistry 30:10363–10370
Drake TA, Morrisey JH, Edgington TS (1989) Selective cellular expression of tissue factor in human tissues. Implications for disorders of hemostasis and thrombosis. Am J Pathol 134:1087–1097
Duchosal MA, Rothermel AL, McConahey PJ, Dixon FJ, Altieri DC (1996) In vivo immunosuppression by targeting a novel protease receptor. Nature 380:352–356
D'Souza SE, Ginsberg MH, Plow EL (1991) Arginyl-glycyl-aspartic acid (RGD): a cell adhesion motif. Trends Biochem Sci 16:246–250
Elsayed YA, Nakagawa K, Kamikubo YI, Enjyoji KI, Kato H, Sueishi K (1996) Effects of recombinant human tissue factor pathway inhibitor on thrombus formation and its in vivo distribution in a rat DIC model. Am J Clin Pathol 106:574–583
Esmon CT (1987) The regulation of natural anticoagulant pathways. Science 235: 1348–1352
Esmon CT (1992) The Protein C anticoagulant pathway. Arterioscler Thromb 12: 135–145
Esmon CT (1995) Thrombomodulin as a model of molecular mechanisms that modulate protease specificity and function at the vessel surface. FASEB J 9:946–955
Esmon CT, Fukudome K (1995) Cellular regulation of the protein C pathway. Semin Cell Biol 6:259–268
Esmon CT, Lollar P (1996) Involvement of thrombin anion-binding exosites 1 and 2 in the activation of factor V and factor VIII. J Biol Chem 271:13882–13887
Esmon NL, Carroll RC, Esmon CT (1983) Thrombomodulin blocks the ability of thrombin to activate platelets. J Biol Chem 258:12238–12242
Esmon CT, Esmon NL, Kurosawa S, Johnson AE (1986) Interaction of thrombin with thrombomodulin. Ann NY Acad Sci 485:215–220
Espana F, Gruber A, Heeb MJ, Hanson SR, Harker LA, Griffin JH (1991) In vivo and in vitro complexes of activated protein C with two inhibitors in baboons. Blood 77:1754–1760
Fareed J (1996) Current trends in antithrombotic drugs and device development. Sem in Thromb Hemost 22 Suppl 1:3–8
Fareed J, Callas DD, Hoppensteadt D, Jeske W, Walenga JM (1995) Recent developments in antithrombotic agents. Exp Opin Invest Drugs 4:389–412
Fenton JW 2d (1986) Thrombin. Ann N Y Acad Sci 485:5–15
Fenton JW 2d (1992) Leeches to hirulogs and other thrombin-directed antithrombotics. Hematol Oncol Clin North Am 6:1121–1129
Finkle C, DiMaio J, Tarazi M, Leblond L, Winocour PD (1998) BCH-2763, a novel potent parenteral thrombin inhibitor is an effective antithrombotic agent in rodent models of arterial and venous thrombosis – comparison with heparin, r-hirudin, hirulog, inogatran and argatroban. Throm Haemost 79:431–438
Fukudome K, Esmon CT (1994) Identification, cloning and regulation of a novel endothelial cell protein C/activated protein C receptor. J Biol Chem 269: 26486–26491
Furie B, Furie BC (1988) The molecular basis of blood coagulation. Cell 53:505–518
Gailani D, Broze GJ Jr (1991) Factor XI activation in a revised model of blood coagulation. Science 253:909–912
Geczy CL (1994) Cellular mechanisms for the activation of blood coagulation. Int Rev Cytol 152:49–108
Gerlach H, Esposito C, Stern DM (1990) Modulation of endothelial hemostatic properties: an active role in the host response. Ann Rev Med 41:15–24
Girard TJ, Warren LA, Novotny WF, Likert KM, Brown SG, Miletich JP et al (1989) Functional significance of the Kunitz-type inhibitory domains of lipoprotein-associated coagulation inhibitor. Nature 338:518–520

Gibbs CS, Coutre SE, Tsiang M, Li WX, Jain AK, Dunn KE et al (1995) Conversion of thrombin into an anticoagluant by protein engineering. Nature 378(6555):413–416
Ginsberg JS, Nurmohamed MT, Gent M, Mackinnon B, Sicurella J, Brill-Edwards P et al (1994) Use of hirulog in the prevention of venous thrombosis after major hip or knee surgery. Circulation 90:2385–2389
Global Use of Strategies to Open Occluded Coronary Arteries (GUSTO) IIb Investigators (1996) A comparison of recombinant hirudin with heparin for the treatment of acute coronary syndromes. N Engl J Med 335:775–782
Grabowski EF, Lam FP (1995) Endothelial cell function, including tissue factor expression, under flow conditions. Thromb Haemost 74:123–128
Green D, Hirsh J, Heit J, Prins M, Davidson B, Lensing AW (1994) Low molecular weight heparin: a critical analysis of clinical trials. Pharmacol Rev 46:89–109
Griffin JH, Evatt B, Wideman C, Fernandez JA (1993) Anticoagulant protein C pathway defective in majority of thrombophilic patients. Blood 82:1989–1993
Gruber A, Hanson SR, Kelly AB, Yan BS, Bang N, Griffin JH et al (1990) Inhibition of thrombus formation by activated recombinant protein C in a primate model of arterial thrombosis. Circulation 82:578–585
Guillin MC, Bezeaud A, Bouton MC, Jandrot-Perrus M (1995) Thrombin specificity. Thromb Haemost 74:129–133
Guinto ER, Vindigni A, Ayala YM, Dang QD, DiCera E (1995) Identification of residues linked to the slow→fast transition of thrombin. Proc Natl Acad Sci USA 92:11185–11189
Hara T, Yokoyama A, Ishihara H, Yokoyama Y, Nagahara T, Iwamoto M (1994) DX-9065a, a new synthetic, potent anticoagulant and selective inhibitor for factor Xa. Thromb Haemost 71:314–319
Hara T, Yokoyama A, Tanabe K, Ishihara H, Iwamoto M (1995) DX-9065a, an orally active, specific inhibitor of factor Xa, inhibits thrombosis without affecting bleeding time in rats. Thromb Haemost 74:635–639
Harker LA, Maraganore JM, Hirsh J (1994) Novel antithrombotic agents. In: Colman RW, Hirsh K, Marder VJ, Salzman EW (eds) Hemostasis and thrombosis: basic principles and clinical practice, 3rd edn. JB Lippincott Company, Philadelphia, pp 1638–1660
Hauptmann J, Markwardt F (1992) Pharmacologic aspects of the development of selective synthetic thrombin inhibitors as anticoagulants. Semin Thromb Hemost 18:200–217
Healy AM, Rayburn HB, Rosenberg RD, Weiler H (1995) Absence of the blood-clotting regulator thrombomodulin causes embryonic lethality in mice before development of a functional cardiovascular system. Proc Natl Acad Sci USA 92:850–854
Heeb MJ, Griffin JH (1988) Physiologic inhibition of human activated protein C by α_1-antitrypsin. J Biol Chem 263:11613–11616
Holst J, Lindblad B, Westerlund G, Bregengaard C, Ezban M, Østergaard PB et al (1996) Pharmacokinetics and delayed experimental antithrombotic effect of two domain non-glycosylated tissue factor pathway inhibitors. Thromb Res 81:461–470
Kaiser B, Fareed J (1996) Recombinant full-length tissue factor pathway inhibitor (TFPI) prevents thrombus formation and rethrombosis after lysis in a rabbit model of jugular vein thrombosis. Thromb Haemost 76:615–620
Kakkar AK, Lemoine NR, Stone SR, Altieri D, Williamson RCN (1995) Identification of a thrombin receptor with factor Xa receptor and tissue factor in human pancreatic carcinoma cells. J Clin Pathol Clin Mol Pathol 48:M228–M290
Kalafatis M, Swords NA, Rand MD, Mann KG (1994) Membrane-dependent reactions in blood coagulation: role of the vitamin K-dependent enzyme complexes. Biochim Biophys Acta 1227:113–129
Katsuura Y, Okamoto S, Ohno N, Wanaka K (1996) Effects of a highly selective synthetic inhibitor of plasma kallikrein on disseminated intravascular coagulation in rats. Thromb Res 82:361–368

Kelly AB, Maraganore JM, Bourdon P, Hanson SR, Harker LA (1992) Antithrombotic effects of synthetic peptides targeting various functional domains of thrombin. Proc Natl Acad Sci USA 89:6040–6044
Kereveur A, Samama MM (1995) Les nouveaux antithrombotiques. Presse Med 24:1777–1787
Kimball SD (1995) Challenges in the development of orally bioavailable thrombin active site inhibitors. Blood Coagul Fibrinolysis 6:511–519
Kirchhofer D, Nemerson Y (1996) Initiation of blood coagulation: the tissue factor/factor VIIa complex. Curr Opin Biotechnol 7:386–391
Kondo S, Kisiel W (1987) Regulation of factor VIIa activity in plasma: evidence that antithrombin III is the sole plasma protease inhibitor of human factor VIIa. Thromb Res 46:325–335
Kroll MH, Hellums JD, McIntire LV, Schafer AI, Moake JL (1996) Platelets and shear stress. Blood 88:1525–1541
Lawson JH, Butenas S, Ribarik N, Mann KG (1993) Complex-dependent inhibition of factor VIIa by antithrombin III and heparin. J Biol Chem 268:767–770
Lefkovits J, Topol EJ (1994) Direct thrombin inhibitors in cardiovascular medicine. Circulation 90:1522–1536
Leidy EM, Stern AM, Friedman PA, Bush LR (1990) Enhanced thrombolysis by factor XIIIa inhibitor in a rabbit model of femoral artery thrombosis. Thromb Res 59:15–26
Li WX, Kaplan AV, Grant GW, Toole JJ, Leung LL (1994) A novel nucleotide-based thrombin inhibitor inhibits clot-bound thrombin and reduces arterial platelet thrombus formation. Blood 83:677–682
Lijnen HR, Collen D (1997) Endothelium in hemostasis and thrombosis. Prog Cardio Diseases XXXIX:343–350
Lijnen HR, Stassen JM, Vanlinthout I, Fukao H, Okada K, Matsuo O et al (1991) Comparative fibrinolytic properties of staphylokinase and streptokinase in animal models of venous thrombosis. Thromb Haemost 66:468–473
Lindahl AK (1995) Tissue factor pathway inhibitor in health and diseases. Trends Cardiovasc Med 5:167–171
Loscalzo J (1996) Thrombin inhibitors in fibrinolysis. A Hobson's choice of alternatives. Circulation 94:863–865
Lougheed JC, Bowman CL, Meininger DP, Komives EA (1995) Thrombin inhibition by cyclic peptides from thrombomodulin. Protein Sci 4:773–780
Lutchtman-Jones L, Broze GJ Jr (1995) The current status of coagulation. Ann Med 27:47–52
MacFarlane RG (1964) An enzyme cascade in the blood clotting mechanism, and its function as a biochemical amplifier. Nature 202:498–499
Maffrand JP (1992) Direct thrombin inhibitors. Nouv Rev Fr Hematol 34:405–419
Mann KG, Lawson JM (1992) The role of the membrane in the expression of the vitamin K-dependent enzymes. Arch Pathol Lab Med 116:1330–1336
Mann KG, Kalafatis M (1995) The coagulation explosion. Cerebrovasc Dis 5:93–97
Maraganore JM, Bourdon P, Jablonski J, Ramachandran KL, Fenton JW 2d (1990) Design and characterization of hirulogs: a novel class of bivalent peptide inhibitors of thrombin. Biochemistry 29:7095–7101
Marder VJ (1995) Thrombolytic therapy: overview of results in major vascular occlusions. Thromb Haemost 74:101–105
Markwardt F (1994) The development of hirudin as an antithrombotic drug. Thromb Res 74:1–23
Mathews II, Padmanabhan KP, Ganesh V, Tulinsky A, Ishii M, Chen J et al (1994) Crystallographic structures of thrombin complexed with thrombin receptor peptides: existence of expected and novel binding modes. Biochemistry 33:3266–3279
Miller BC, Hultin MB, Jesty J (1985) Altered factor VII activity in hemophilia. Blood 65:845–849
Muller YA, Ultsch MH, de Vos AM (1996) The crystal structure of the extracellular domain of human tissue factor refined to 1.7 Å resolution. J Mol Biol 256:144–159

Nagase H, Enjyoji K, Minamiguchi K, Kitazato KT, Saito H et al (1995) Depolymerized holothurian glycosaminoglycan with novel anticoagulant actions: antithrombin III-and heparin cofactor II independent inhibition of factor X activation by factor IXa-factor VIIIa complex and heparin cofactor II-dependent inhibition of thrombin. Blood 85:1527–1534

Nicholson AC, Nachman RL, Altieri DC, Summers BD, Ruf W, Edgington TS et al (1996) Effector cell protease receptor-1 is a vascular receptor for coagulation factor Xa. J Biol Chem 271:28407–28413

Nishioka J, Taneda H, Suzuki K (1993) Estimation of the possible recognition sites for thrombomodulin, procoagulant, and anticoagulant proteins around the active center of α-thrombin. J Biochem 114:148–155

Novotny WF, Girard TJ, Miletich JP, Broze GJ Jr (1988) Platelets secrete a coagulation inhibitor functionally and antigenically similar to the lipoprotein-associated coagulation inhibitor. Blood 72:2020–2025

Novotny WF, Girard TJ, Miletich JP, Broze GJ Jr (1989) Purification and characterization of the lipoprotein-associated coagulation inhibitor from human plasma. J Biol Chem 264:18832–18837

Ofosu FA, Craven S, Dewar L, Anvari N, Andrew M, Blajchman MA (1996) Age-related changes in factor VII proteolysis in vivo. Br J Haematol 94:407–412

Olds RJ, Lane DA, Mille B, Chowdhury V, Thein SL (1994) Antithrombin: the principal inhibitor of thrombin. Semin Thromb Hemost 20:353–372

Olson ST, Bjork I (1994) Regulation of thrombin activity by antithrombin and heparin. Semin Thromb Hemost 20:373–409

Ono M, Nawa K, Marumoto Y (1994) Antithrombotic effects of recombinant soluble thrombomodulin in a rat model of vascular shunt thrombosis. Thromb Haemost 72:421–425

Osterud B, Rapaport SI (1977) Activation of factor IX by the reaction product of tissue factor and factor VII: additional pathway for initiating blood coagulation. Proc Natl Acad Sci USA 74:5260–5264

Paborsky LR, McCurdy SN, Griffin LC, Toole JJ, Leung LL (1993) The single-stranded DNA aptamer-binding site of human thrombin. J Biol Chem 268:20808–20811

Paborsky LR, Law VS, Mao CT, Leung LL, Gibbs CS (1995) A peptide derived from a tissue factor loop region functions as a tissue factor–factor VIIa antagonist. Biochemistry 34:15328–15333

Padmanabhan K, Padmanabhan KP, Ferrara JD, Sadler JE, Tulinsky A (1993a) The structure of alpha-thrombin inhibited by a 15-mer single-stranded DNA aptamer. J Biol Chem 268:17651–17654

Padmanabhan K, Padmanabhan KP, Tulinsky A, Park CH, Bode W, Huber R et al (1993b) Structure of human des(1–45) factor Xa at 2.2 Å resolution. J Mol Biol 232:947–966

Patthy L (1993) Modular design of proteases of coagulation, fibrinolysis and complement activation: implications for protein engineering and structure-function studies. Methods Enzymol 222:10–23

Pawashe AB, Golino P, Ambrosio G, Migliaccio F, Ragni M, Pascucci I et al (1994) A monoclonal antibody against rabbit tissue factor inhibits thrombus formation in stenotic injured rabbit carotid arteries. Circ Res 74:56–63

Petersen LC, Bjørn SE, Olsen OH, Nordfang O, Norris F, Norris K (1996) Inhibitory properties of separate recombinant Kunitz-type-protease-inhibitor domains from tissue-factor-pathway inhibitor. Eur J Biochem 235:310–316

Preissner KT, De Boer H, Pannekoek H, De Groot PG (1996) Thrombin regulation by physiological inhibitors: the role of vitronectin. Semin Thromb Hemost 22:165–172

Rao LV, Rapaport SI, Hoang AD (1993) Binding of factor VIIa to tissue factor permits rapid antithrombin III/heparin inhibition of factor VIIa. Blood 81:2600–2607

Rao LV, Williams T, Rapaport SI (1996) Studies of the activation of factor VII bound to tissue factor. Blood 87:3738–3748

Rapaport SI, Rao LVM (1995) The tissue factor pathway: how it has become a "Prima Ballerina". Thromb Haemost 74:7–17
Regan LM, Stearns-Kurosawa DJ, Kurosawa S, Mollica J, Fukudome K, Esmon CT (1996) The endothelial cell protein C receptor. Inhibition of activated protein C anticoagulant function without modulation of reaction with proteinase inhibitors. J Biol Chem 271:17499–17503
Roth GJ, Calverley DC (1994) Aspirin, platelets, and thrombosis: theory and practice. Blood 83:885–898
Rottingen JA, Enden T, Camerer E, Iversen JG, Prydz H (1995) Binding of human factor VIIa to tissue factor induces cytosolic Ca^{++} signals in J82 cells, transfected COS-1 cells, Madin-Darby canine kidney cells and in human endothelial cells induced to synthesize tissue factor. J Biol Chem 270:4650–4660
Ruggeri ZM (1992) Von Willebrand factor as a target for antithrombotic intervention. Circulation 86:III26–III29
Sakariassen KS, Bolhuis PA, Sixma JJ (1979) Human blood platelet adhesion to artery subendothelium is mediated by factor VIII-von Willebrand factor bound to the subendothelium. Nature 279:636–638
Samama MM, Acar J (1995) Thrombolytic therapy: future issues. Thromb Haemost 74:106–110
Sandset PM, Abildgaard U, Larsen ML (1988) Heparin induces release of extrinsic coagulation pathway inhibitor (EPI). Thromb Res 50:803–813
Scandura JM, Ahmad SS, Walsh PN (1996) A binding site expressed on the surface of activated human platelets is shared by factor X and prothrombin. Biochemistry 35:8890–8902
Schrör K (1995) Antiplatelet drugs. A comparative review. Drugs 50:7–28
Seiler SM (1996) Thrombin receptor antagonists. Semin Thromb Hemost 22:223–232
Seligsohn U (1993) Factor XI deficiency. Thromb Haemost 70:68–71
Seligsohn U, Kasper CK, Osterud B, Rapaport SI (1978) Activated factor VII: presence in factor IX concentrates and persistence in the circulation after infusion. Blood 53:828–837
Shebuski RJ, Sitko GR, Claremon DA, Baldwin JJ, Remy DC, Stern AM (1990) Inhibition of factor XIIIa in a canine model of coronary thrombosis: effect on reperfusion and acute reocclusion after recombinant tissue-type plasminogen activator. Blood 75:1455–1459
Sims PJ, Ginsberg MH, Plow EF, Shattil SJ (1991) Effect of platelet activation on the conformation of the plasma membrane glycoprotein IIb-IIIa complex. J Biol Chem 266:7345–7352
Sixma JJ, de Groot PG (1992) The ideal anti-thrombotic drug. Thromb Res 68:507–512
Spicer EK, Horton R, Bloem L, Bach R, Williams KR, Guha A et al (1987) Isolation of cDNA clones coding for human tissue factor: primary structure of the protein and cDNA. Proc Natl Acad Sci USA 84:5148–5152
Stassens P, Bergum PW, Gansemans Y, Jespers L, Laroche Y, Huang S et al (1996) Anticoagulant repertoire of the hookworm Ancylostoma caninum. Proc Natl Acad Sci USA 93:2149–2154
Stearns-Kurosawa DJ, Kurosawa S, Mollica JS, Ferrell GL, Esmon CT (1996) The endothelial cell protein C receptor augments protein C activation by the thrombin-thrombomodulin complex. Proc Natl Acad Sci USA 93:10212–10216
Stiernberg J, Redin WR, Warner WS, Carney DH (1993) The role of thrombin and thrombin receptor activating peptide (TRAP-508) in initiation of tissue repair. Thromb Haemost 70:158–162
Stitt TN, Conn G, Gore M, Lai C, Bruno J, Radziejewski C et al (1995) The anticoagulation factor Protein S and its relative, Gas6, are ligands for the Tyro 3/Axl family of receptor tyrosine kinases. Cell 80:661–670
Stone SR, Maraganore JM (1992) Hirudin interactions with thrombin. In: Berliner LJ (ed) Thrombin: structure and functions. Plenum Press New York, pp 219–256
Stubbs MT, Bode W (1994) Coagulation factors and their inhibitors. Curr Opin Struct Biol 4:823–832

Stubbs MT, Bode W (1995) Structure and specificity in coagulation and its inhibition. Trends Cardiovasc Med 5:157–166
Suidan HS, Niclou SP, Monard D (1996) The thrombin receptor in the nervous system. Semin Thromb Hemost 22:125–133
Suzuki K, Kusumoto H, Deyashiki Y, Nishioka J, Maruyama I, Zushi M et al (1987) Structure and expression of human thrombomodulin, a thrombin receptor on endothelium acting as a cofactor for protein C activation. EMBO J 6:1891–1897
Suzuki K, Nishioka J, Hayashi T (1990) Localization of thrombomodulin-binding site within human thrombin. J Biol Chem 265:13263–13267
Swensson PJ, Dahlback B (1994) Resistance to activated protein C as a basis for venous thrombasis. N Engl J Med 330:517–522
Tapparelli C, Metternich R, Ehrhardt C, Cook NS (1993) Synthetic low-molecular weight thrombin inhibitors: molecular design and pharmacological profile. Trends Pharmacol Sci 14:366–376
Toomey JR, Kratzer KE, Lasky NM, Stanton JJ, Broze GJ Jr (1996) Targeted disruption of the murine tissue factor gene results in embryonic lethality. Blood 88:1583–1587
Trikha M, Timar J, Lundy SK, Szekeres K, Tang K, Grignon D et al (1996) Human prostate carcinoma cells express functional αIIbβ3 integrin. Cancer Res 56:5071–5078
Tsiang M, Jain AK, Dunn KE, Rojas ME, Leung LL, Gibbs CS (1995) Functional mapping of the surface residues of human thrombin. J Biol Chem 270:16854–16863
Tsuda Y, Cygler M, Gibbs BF, Pedyczak A, Fethiere J, Yue SY et al (1994) Design of potent bivalent thrombin inhibitors based on hirudin sequence: incorporation of nonsubstrate-type active site inhibitors. Biochemistry 33:14443–14451
Tuddenham EG, Pemberton S, Cooper DN (1995) Inherited factor VII deficiency: genetics and molecular pathology. Thromb Haemost 74:313–321
Tulinsky A (1996) Molecular interactions of thrombin. Semin Thromb Hemost 22:117–124
Turpie AG, Weitz JI, Hirsh J (1995) Advances in antithrombotic therapy: novel agents. Thromb Haemost 74:565–571
Tuszynski G, Gasic TB, Gasic GJ (1987) Isolation and characterization of antistasin. An inhibitor of metastasis and coagulation. J Biol Chem 262:9718–9723
Ulmer JS, Lindquist RN, Dennis MS, Lazarus RA (1995) Ecotin is a potent inhibitor of the contact system proteases factor XIIa and plasma kallikrein. FEBS Lett 365:159–163
Van De Werf F (1991) L'infarctus du myocarde et la thrombolyse. La Recherche 22:425–433
Vane JR (1994) The Croonian Lecture, 1993. The endothelium: maestro of the blood circulation. Philos Trans R Soc Lond B Biol Sci 343:225–246
Vlasuk GP (1993) Structural and functional characterization of tick anticoagulant peptide (TAP): a potent and selective inhibitor of blood coagulation factor Xa. Thromb Haemost 70:212–216
Vermeer C, Hamulyák K (1991) Pathophysiology of vitamin K-deficiency and oral anticoagulants. Thromb Haemost 66:153–159
Verstraete M (1995) The fibrinolytic system: from Petri dishes to genetic engineering. Thromb Haemost 74:25–35
Vu TK, Hung DT, Wheaton VI, Coughlin SR (1991) Molecular cloning of a functional thrombin receptor reveals a novel proteolytic mechanism of receptor activation. Cell 64:1057–1068
Weitz JI, Hirsh J (1993) New anticoagulant strategies. J Lab Clin Med 122:364–373
Weitz JI, Hudoba M, Massel D, Maraganore J, Hirsh J (1990) Clot-bound thrombin is protected from inhibition by heparin-antithrombin III but is susceptible to inactivation by antithrombin III-independent inhibitors. J Clin Invest 86:385–391
Weitz JI, Califf RM, Ginsberg JS, Hirsh J, Theroux P (1995) New antithrombotics. Chest 108:471S-478 S

Wells CM, Di Cera E (1992) Thrombin is a Na^+-activated enzyme. Biochemistry 31:11721–11730
Wesselschmidt R, Likert K, Huang Z, MacPhail L, Broze GJ Jr (1993) Structural requirements for tissue factor pathway inhibitor interactions with factor Xa and heparin. Blood Coagul Fibrinolysis 4:661–669
Wu KK, Thiagarajan P (1996) Role of endothelium in thrombosis and hemostasis. Ann Rev Med 47:315–331
Wun TC, Kretzmer KK, Girard TJ, Miletich JP, Broze GJ Jr (1988) Cloning and characterization of a cDNA coding for the lipoprotein-associated coagulation inhibitor shows that it consists of three tandem Kunitz-type inhibitory domains. J Biol Chem 263:6001–6004
Yamamoto M, Nakagaki T, Kisiel W (1992) Tissue factor-dependent autoactivation of human blood coagulation factor VII. J Biol Chem 267:19089–19094

CHAPTER 2

New Developments in the Molecular Biology of Coagulation and Fibrinolysis

P. Carmeliet and D. Collen

A. Introduction

I. The Coagulation System

Preservation of vascular integrity following traumatic or infectious challenges is essential for survival of multicellular organisms. A major defense mechanism involves the formation of hemostatic plugs by activation of platelets and polymerization of fibrin. Initiation of the plasma coagulation system upon exposure of blood to nonvascular cells is triggered by tissue factor (TF), which is expressed by a variety of cells surrounding the vasculature as a hemostatic envelope and which functions as a cellular receptor and cofactor for activation of the zymogen factor VII to the serine proteinase factor VIIa (Edginton et al. 1991). This complex activates factor X directly or indirectly via activation of factor IX, resulting in the generation of thrombin and the conversion of fibrinogen to fibrin (Furie and Furie 1988; Davie 1995). Thrombin and factor Xa produce a positive feedback stimulation of coagulation by activating factor VIII and factor V, which serve as membrane-bound receptors/cofactors for the proteolytic enzymes factor IXa and factor Xa, respectively (Furie and Furie 1988; Davie 1995).

In contrast, thrombin when bound to its cellular receptor thrombomodulin also functions as an anticoagulant by activating the protein C anticoagulant system (Esmon 1992; Dahlbäck 1995). Activated protein C in the presence of its cofactor protein S inactivates factor Va and factor VIIIa, thereby reducing thrombin generation (Esmon 1992; Dahlbäck 1995). Anticoagulation is further provided by antithrombin III, which binds to and inactivates thrombin, factor IXa and factor Xa in a reaction that is greatly enhanced by heparin (Bick and Pegram 1994). Anticoagulation is further secured by tissue factor pathway inhibitor, which directly inhibits factor Xa and, in a factor Xa-dependent manner, produces feedback inhibition of the factor VIIa/tissue factor catalytic complex (Broze 1992). A revised hypothesis of coagulation has been suggested in which factor VIIa/TF is responsible for the initiation of coagulation but, owing to tissue factor pathway inhibitor-mediated feedback inhibition, amplification of the procoagulant response through the actions of factor VIII, factor IX and factor XI is required for sustained coagulation (Broze 1992). Deficiencies of anticoagulant factors or aberrant expression of procoagulant factors have been implicated in hemostasis during inflammation, sepsis, atherosclerosis and cancer (EDGINTON et al. 1991; BICK and PEGRAM 1994), whereas deficiencies of procoagulant factors have been related to increased bleeding tendencies (HOYER 1996; BOLTON-MAGGS 1995).

Evidence has been provided that the coagulation system may also be involved in other functions beyond coagulation including cellular migration and proliferation, immune response, angiogenesis, embryonic development, and cancer and brain function (EDGINGTON et al. 1991; ALTIERI 1978; COUGHLIN 1994). Its precise role and relevance in these processes in vivo remains, however, largely unknown.

II. The Plasminogen System

The plasminogen system is composed of an inactive proenzyme plasminogen (Plg) that can be converted to plasmin by either of two plasminogen activators (PA), tissue-type PA (t-PA) or urokinase-type PA (u-PA) (ASTRUP 1978; COLLEN and LIJNEN 1991; VASSALLI et al. 1991). This system is controlled at the level of plasminogen activators by plasminogen activator inhibitors (PAIs), of which PAI-1 is believed to be physiologically the most important (SCHNEIDERMAN and LOSKUTOFF 1991; WIMAN 1995; LAWRENCE and GINSBURG 1995), and at the level of plasmin by α_2-antiplasmin (COLLEN and LIJNEN 1991). Vitronectin stabilizes PAI-1 in its active conformation and may also localize PAI-1 to specific sites in the extracellular matrix (LAWRENCE and GINSBURG 1995). Other inhibitors include PAI-2, PAI-3, proteinase nexin-1 and α_2-macroglobulin (COLLEN and LIJNEN 1991; BACHMANN 1995). Due to its fibrin-specificity, t-PA is primarily involved in clot dissolution, although it has also been claimed to be involved in ovulation, bone remodeling and brain function (VASSALLI et al. 1991; MARTIN et al. 1993). Cellular receptors for t-PA and Plg have been identified which might localize plasmin proteolysis to the cell surface (HAJJAR 1995; PLOW et al. 1995). u-PA also binds a cellular receptor, the urokinase receptor (u-PAR), and has been implicated in pericellular proteolysis during cell migration and tissue remodeling in a variety of normal and pathological processes including ovulation, trophoblast invasion, angiogenesis, keratinocyte migration, inflammation, wound healing and cancer (BLASI et al. 1994; VASSALLI 1994; DANO et al. 1994). u-PAR binds to vitronectin, whereas PAI-1 controls recognition of vitronectin by u-PAR or the $\alpha_v\beta_3$-integrin receptor, suggesting a role in coordinating cell adhesion (WEI et al. 1996; STEFANSSON and LAWRENCE 1996). As discussed below, it is presently unclear whether or in which conditions binding of u-PA to u-PAR is required in vivo. Plasmin can degrade fibrin and extracellular matrix proteins but can also activate or liberate growth factors from the extracellular matrix (including latent transforming growth factor, basic fibroblast growth factor and vascular endothelial growth factor), and activate other matrix-degrading proteinases (such as the metalloproteinases) (SAKSELA and RIFKIN 1988). Cell-specific clearance of plasminogen activators or of complexes with their inhibitors by low-density lipoprotein receptor-related protein (LRP) or gp330 may modulate pericellular plasmin proteolysis (ANDREASEN et al. 1994).

III. Targeted Manipulation and Adenovirus-Mediated Transfer of Genes in Mice

Novel gene technologies that were developed over the last decade have allowed the manipulation of the genetic balance of candidate molecules in mice in a controllable manner. Targeting of genes via homologous recombination in embryonic stem cells allows the study of the consequences of deficiencies, mutations, and conditional or tissue-specific expression of gene products in transgenic mice (NAGY 1998). Using a novel embryonic stem cell technology (aggregation of embryonic stem cells with tetraploid embryos), it has become possible to generate completely embryonic stem cell-derived embryos in a single step (NAGY and ROSSANT 1996). In addition, the technology allows the bypass of conventional germline transmission, the separation of extra- from intraembryonic phenotypes and the study of homozygous deficient phenotypes of genes that cause embryonic lethality when heterozygous deficient (NAGY and ROSSANT 1996; CARMELIET et al. 1996a).

Viral gene transfer can also be used to manipulate the expression of genes, e.g., via implantation of retrovirally transduced cells (MULLIGAN 1993) or via adenoviral-mediated gene transfer in vivo (SCHNEIDER and FRENCH 1993). In fact, intravenous administration of a recombinant adenovirus results in expression of target genes to plasma levels above 10μg/ml (CARMELIET et al. 1997b). Such studies allow the generation of but also the rescue of disease models, and the evaluation of possible gene-transfer therapies. Table 1 summarizes the phenotypes associated with genetic alterations in the fibrinolytic or coagulation system that result from targeted inactivation in mice compared to those that occur spontaneously in humans.

B. Embryonic Development and Reproduction

I. Coagulation System

1. Tissue Factor and Factor VII

Only limited information is available on the expression of coagulation factors during embryonic development. Tissue factor expression has been identified in the visceral endoderm of the yolk sac, in embryonic epithelia, the heart, and the nervous system and in vascular smooth muscle cells (LUTHER et al. 1996). Intriguingly, expression of prothrombin and factor VII was undetectable in the early embryo, raising the question of whether these molecules might act independently of their currently known ligands (LUTHER et al. 1996).

Following initial differentiation of stem cells into endothelial cells and their assembly into endothelial cell-lined channels (vasculogenesis), the

Table 1. Phenotypes resulting from targeted gene deletions in mice and spontaneous mutations in humans

Deficiency	Mouse	Man
I. Coagulation System		
Tissue Factor (TF)	embryonic lethality due to defective blood vessel formation	unknown
Factor VII (fVII)	– normal embryonic development – perinatal lethality due to bleeding	bleeding
Thrombomodulin (TM)	embryonic lethality possibly due to defective fetomaternal interaction	unknown
Factor V (fV)	postnatal lethality due to severe spontaneous bleeding	bleeding
Fibrinogen (Fbg)	– bleeding associated with trauma – abnormal wound healing – abortion due to maternal bleeding	bleeding
Factor XI (FXI)	normal[1]	minor bleeding
Factor VIII (fVIII)	bleeding	bleeding
II. Fibrinolytic System		
Tissue-type plasminogen activator (t-PA)	– increased thrombotic susceptibility – mild glomerulonephritis – reduced neurotoxicity – abnormal long term potentiation – impaired neuronal migration – aneurysm formation in atherosclerotic vessels	unknown
Urokinase-type plasminogen activator (u-PA)	– increased thrombotic susceptibility – impaired neointima formation – mild glomerulonephritis – impaired macrophage function – reduced decidual vascularization – reduced trophoblast invasion – reduced platelet activation and trapping – reduced tumor invasion – protection against atherosclerotic aneurysm formation	unknown
t-PA : u-PA	– severe spontaneous thrombosis – impaired neointima formation – reduced ovulation and fertility – cachexia and shorter survival – severe glomerulonephritis – abnormal tissue remodeling	unknown

Table 1. (*continued*)

Deficiency	Mouse	Man
Urokinase Receptor (u-PAR)	normal	unknown
t-PA:u-PAR	normal	unknown
Plasminogen (Plg)	– severe spontaneous thrombosis – reduced ovulation and fertility – cachexia and shorter survival – severe glomerulonephritis – reduced neurotoxicity – impaired skin healing – reduced macrophage and keratinocyte migration – impaired transplant atherosclerosis	thrombosis
Plasminogen activator inhibitor-1 (PAI-1)	– reduced thrombotic incidence – no bleeding – accelerated neointima formation – reduced lung inflammation and fibrosis – reduced atherosclerosis	bleeding
Plasminogen activator inhibitor-2 (PAI-2)	normal[1]	unknown
Vitronectin (VN)	normal[1]	unknown
α_2-Macrogobulin	– reduced lung inflammation – severe pancreatitis – increased resistance to endotoxin shock	unknown
Proteinase Nexin-1 (PN-1)	reduced fertility	unknown
LDL receptor-related protein (LRP)	embryonic lethality due to bleeding	unknown

Only those genetic deficiencies in man that correspond to published genetic deficiencies in mice are summarized. The table only displays the abnormal phenotypes which are described in more detail in the text; the possible lack of a phenotype is discussed in the text[1]: initial phenotypic analysis did not reveal any major abnormalities thus far.

embryonic vasculature further develops via sprouting of new channels from preexisting vessels (angiogenesis) (Fig. 1). Once the endothelial cells are assembled into vascular channels, they become surrounded by smooth muscle cells/pericytes that may affect maturation of the blood vessels, not only by providing the fragile primitive blood vessels the structural support required to accommodate the increased blood pressure but also by controlling endothelial cell proliferation, vascular permeability and vascular tone (NELHS and DRENCKHAN 1993) (Fig. 1).

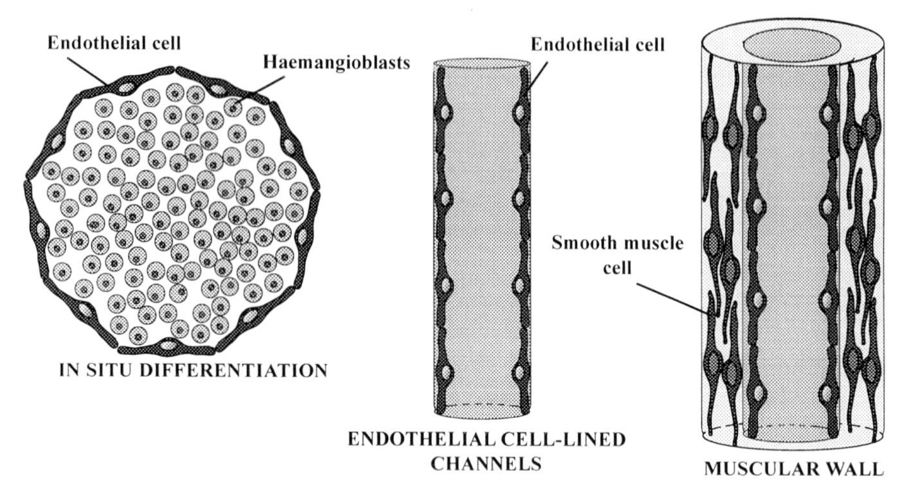

Fig. 1. Formation of blood vessels: early endothelial stem cells (hemangioblasts) differentiate *in situ* into endothelial cells (left panel) and become organized into a primitive vascular plexus consisting of endothelial cell-lined channels (middle panel). Subsequently, these primitive blood v essels acquire a muscular wall to accomodate the increased blood pressure during further embryonic development (right panel)

Targeted inactivation of the *tissue factor* gene resulted in increased fragility of the endothelial cell-lined channels in the yolk sac, which are essential for transferring maternally derived nutrients from the yolk sac to the rapidly growing embryo (CARMELIET et al. 1996b) (Fig. 2a). At a time when the blood pressure increased during embryogenesis (day 9 of gestation), the immature tissue factor deficient blood vessels ruptured, formed microaneurysms and "blood lakes," and failed to sustain proper circulation between the yolk sac and embryo. Secondarily, the embryo became wasted and died due to

Fig. 2. a, Schematic representation of the yolk sac connected to the embryo via the vitello-embryonic blood vessels, mediating transfer of maternally derived nutrients to the embryo proper. **b–d**, The yolk sac of wild type tissue factor ($TF^{+/+}$) embryos at 9.5 days of age, cultured *in vitro* from 8.5 days of age, contains large blood-filled vitello-embryonic vessels (arrows) interconnected with a capillary-like vascular plexus in the yolk sac (b). In contrast, the yolk sac in homozygous TF deficient ($TF^{-/-}$) embryos reveals significant vascular defects, ranging from micro-aneurysmatic "vascular blood lakes" in a early stage (c), to complete desintegration of the whole vascular plexus and vitelline vessels and leakage of blood in the extracoelomic cavity (arrow) ("vascular bleeding") in a preterminal stage (d). Note the enlarged pericardial cavity (arrowheads) in the mutant embryo, indicative of defective vitello-embryonic circulation. **e–g**, Ultrastructural analysis of the yolk sac of wild type ($TF^{+/+}$) (*E,F*) and TF deficient ($TF^{-/-}$) (g) embryos, revealing the presence of normal visceral endoderm cells (e), endothelial cell (En) and an interposed pericyte (P) (*E*). Note the abundant extracellular matrix between the visceral endoderm cells and the endothelium (arrows in F). In contrast, in the mutant yolk sac, pericytes are absent and there is a paucity of extracellular matrix. "C" indicates the lumen of the vascular channels

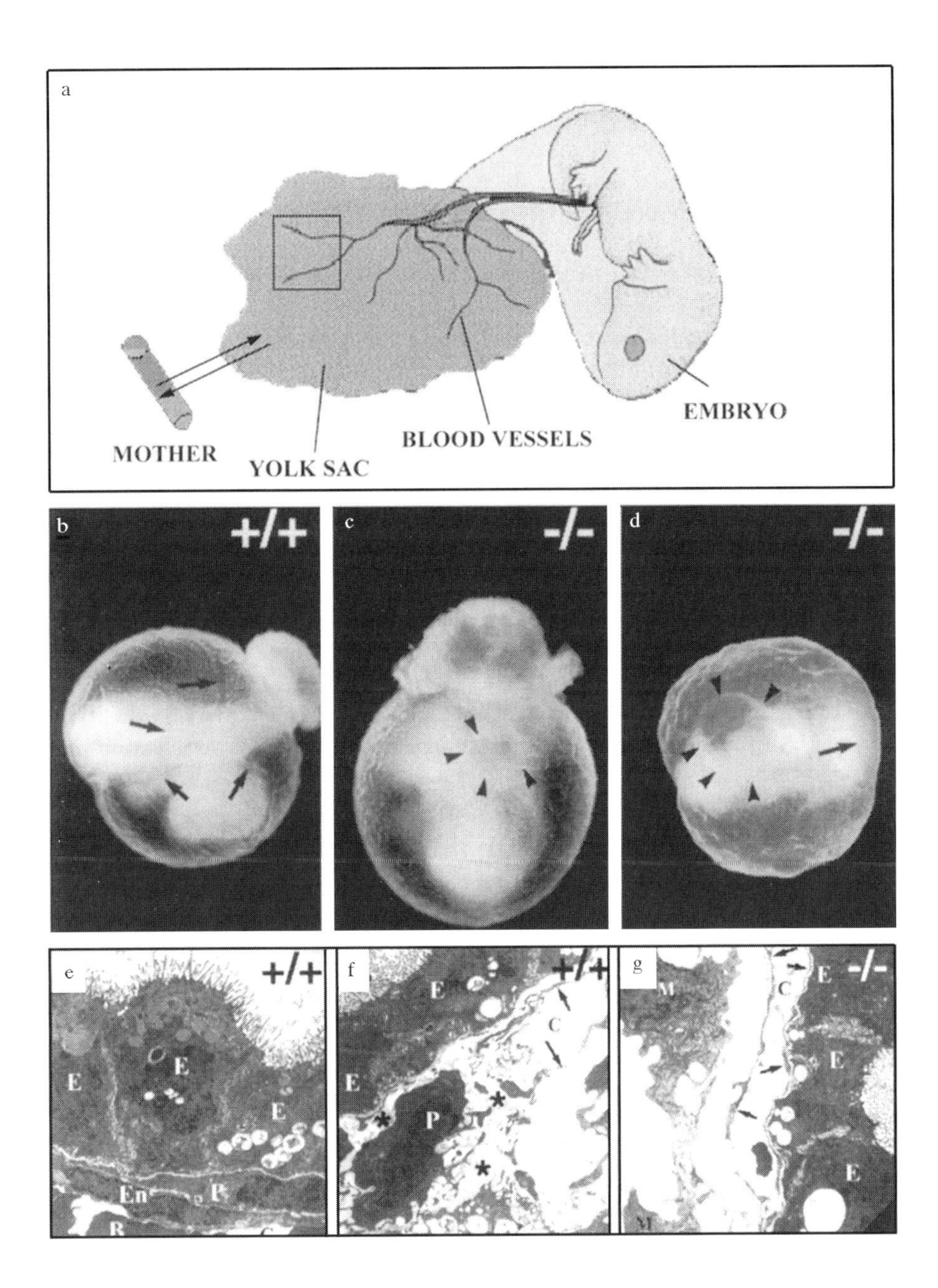
a
MOTHER
YOLK SAC
BLOOD VESSELS
EMBRYO
b
+/+
c
-/-
d
-/-
e
+/+
E
E
E
En
P
R
f
+/+
E
E
C
P
g
-/-
M
C
E
E
E
M

generalized necrosis. Only in advanced stages of deterioration did the immature blood vessels become leaky, resulting in bleeding into the extracoelomic cavity. Since hematopoiesis at 8.5 days of gestation in tissue factor deficient embryos appeared normal, and the proliferation and differentiation potential of tissue factor deficient hematopoietic stem cells in vitro was normal, pallor of the mutant embryos most likely resulted from bleeding and not from defective hematopoiesis (CARMELIET et al. 1996b). Similar observations were made when tissue factor deficient embryos were cultured in vitro (Fig. 2b–d), suggesting that the observed vascular defects in the yolk sac were not merely due to a possible defect in fetomaternal exchange. A role for tissue factor in development is also suggested by the observations of Toomey et al. (1996) and of Bugge et al. (1996).

In order to evaluate which cell type(s) was (were) responsible for the vascular defects, morphological and functional analysis of the different cell types in the yolk sac, e.g., endothelial cells, visceral endoderm cells and mesenchymal (smooth muscle/pericyte-like) cells, was performed (CARMELIET et al. 1996b). Visceral endoderm cell function appeared normal, based on their nutritive properties (uptake of horseradish peroxidase), expression of specific markers (hepatocyte nuclear factor-4, E-cadherin, α-fetoprotein, vascular endothelial growth factor), ultrastructure and normal proliferation and apoptosis rates. Microaneurysms also did not appear to result from increased plasmin proteolysis by visceral endoderm cells, as evidenced by in situ zymographic analysis of yolk sac sections (CARMELIET et al. 1996b). Endothelial cells were also not likely to be the primary cause of the vascular defects since expression of endothelial cell specific markers (Flt, Flk, PECAM, TIE1, TEK and VE-cadherin) as well as their ultrastructural and growth properties were normal.

In contrast, defective development and/or maturation of mesenchymal cells (smooth muscle cell/pericyte-like cells) appeared to be the likely reason for the vascular defects. These cells surround the endothelium in yolk sac vessels, form a primitive "muscular" wall and provide structural support by their close physical association and increasing production of extracellular matrix proteins (Fig. 2e,f). Microscopic and ultrastructural analysis revealed that deficiency of tissue factor resulted in a 75% reduction of the number of

Fig. 3. ***a***, vascular defects induced by tissue factor deficiency: smooth muscle cell/pericyte-like mesenchymal cells accumulate and differentiate around endothelial cell-lined capillaries in the yolk sac from a wild type ($TF^{+/+}$) embryo beyond 9 days of gestation to sustain the increased blood pressure load. In the tissue factor deficient ($TF^{-/-}$) yolk sac, these mesenchymally derived smooth muscle cell/pericyte-like cels fail to accumulate and differentiate around the endothelial cells. As a result, the visceral endoderm and mesothelial cell layers detach (*arrowheads in left capillaries*), and the fragile capillaries rupture and form microaneurysms and "blood lakes" that ultimately leak blood into the (extravascular) extracoelomic cavity (*arrows in right capillaries*).

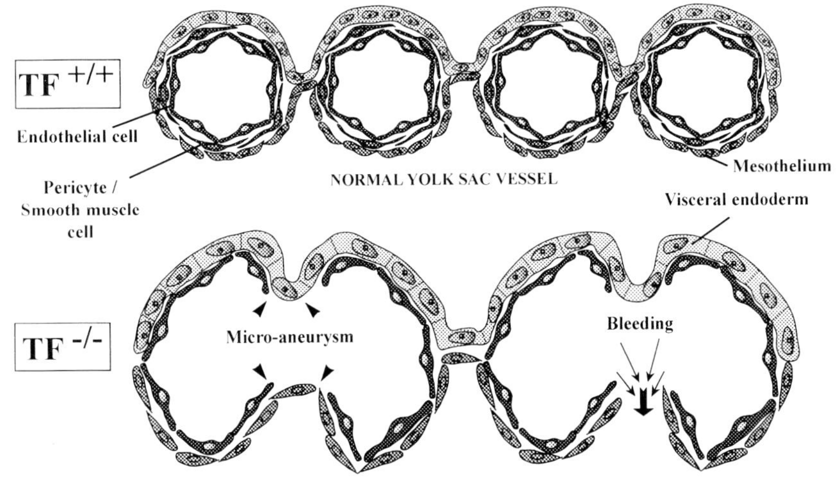

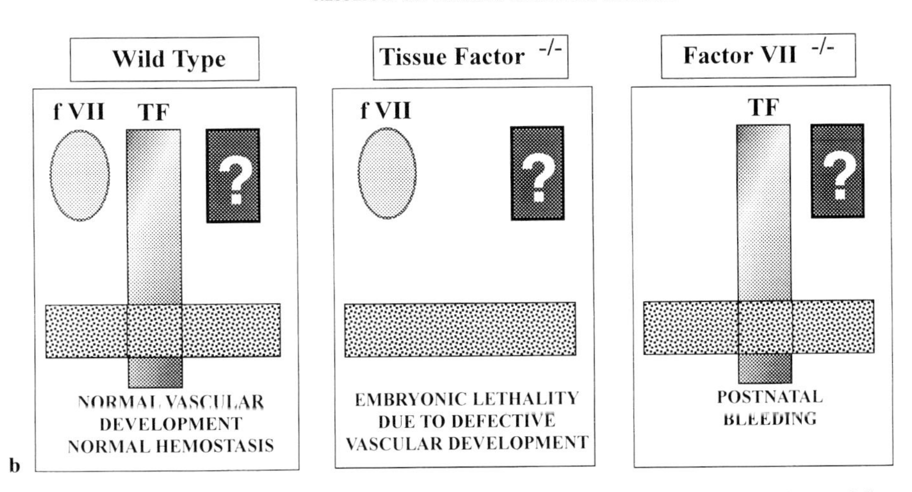

As a result of the vascular defects, blood circulation from the yolk sac to the rapidly growing embryo is compromized, inducing wasting of the embryo due to deprivation of essential nutrients. Notably, leakage of blood from defective vessels ("vascular" bleeding) only occurs at a final stage of necrosis and deterioration. ***b***, Targeting studies of the coagulation system: Tissue factor is a membrane receptor, the extracellular domain of which is sufficient and required for initiation of the coagulation cascade and fibrin formation. The intracellular domain has been implicated in intracellular signaling. It remains to be determined whether tissue factor interacts with factor VII or an as yet unidentified factor to mediate its morphogenic properties during embryogenesis and whether this involves intracellular signal transduction. Indeed, factor VII deficient embryos develop normally until birth, after which they die due to massive bleeding, suggesting that, in these embryos, tissue factor either interacts with another ligand or acts independently. Alternatively, placental transfer of factor VII might rescue factor VII deficient embryogenesis, but transfer of human factor VIIa across mouse placenta was undetectable (see text)

mesenchymal cells, and a diminished amount of extracellular matrix (Fig. 2g) (CARMELIET et al. 1996b). Immunocytochemical analysis further revealed a reduced level of smooth muscle α-actin staining in these cells, suggesting impaired differentiation and/or accumulation (CARMELIET et al. 1996b). Because these primitive smooth muscle cells provide structural support for the endothelium, the vessels in the mutant embryos are too fragile and break open at a time during development, when the blood pressure is increased due to more vigorous heart contractions and increased blood cell viscosity (Fig. 3a). Inappropriate vascular fragility as a result of peri-endothelial mesenchymal cell/pericyte defects, variably resulting in bleeding, has also been observed in mice deficient in platelet derived growth factor (PDGF) (LEVÉEN et al. 1994), transforming growth factor-β (SURI et al. 1996) and angiopoietin-1 (DICKSON et al. 1995). Pericytes have also been implicated in adult diabetic retinopathy where pericyte "drop out" results in the formation and rupture of microaneurysms and blindness (NELHS and DRENCKHAN 1993).

At 9.5 days of gestation, vascular defects were not detected within the embryo proper. It is possible that vessels in the yolk sac require a higher structural support at an earlier stage of vascular development than the blood vessels within the embryo. One explanation may be that the intraembryonic vessels are embedded within embryonic tissue, which provides better structural support than the paucity of cells in the yolk sac. This hypothesis is further supported by the observations that yolk sac vessels display smooth muscle α-actin staining at an earlier stage than the dorsal aorta within the embryo, suggesting earlier maturation/differentiation of smooth muscle cells in the yolk sac (CARMELIET et al. 1996b).

Unresolved questions are how tissue factor exerts this morphogenic action, i.e., via intracellular signaling as suggested previously (ROTTINGEN et al. 1995), and/or whether fibrin formation occurs and is essential during early vascular development, as suggested by others (BUGGE et al. 1996a). Thus far, there is no conclusive evidence that coagulation factors pass the fetomaternal barrier and that the early stage embryo has a functional clotting system. There was also no evidence for fibrin deposits in the visceral yolk sac from our

---

Fig. 4a,b. Schematic representation of a molecular model of the development of a smooth muscle cell layer around the endothelial cell-lined channels during the recruitment (a) and maturation (b) phase. Vascular development in the yolk sac is shown as an example. ***a***, Undifferentiated mesenchymal cells (MC) produce angiopoietin-1 that, via interaction with the TIE-2/TEK receptor on early endothelial cells (EC), induces the production of (a) recruitment signal(s), possibly PDGF-BB. In addition, the cellular tissue factor receptor, present on neighbouring cells such as for example on visceral endoderm cells (VE) in the yolk sac, may act as a chemotactic recruitment signal or differentiation factor by itself, or induce the generation of downstream coagulation factors such as thrombin, that can interact with its receptor on mesenchymal cells. ***b***, Once mesenchymal cells are in close contact with the endothelial cells, TGF-β1, produced by the mesenchymal or endothelial cells, may become activated by plasmin,

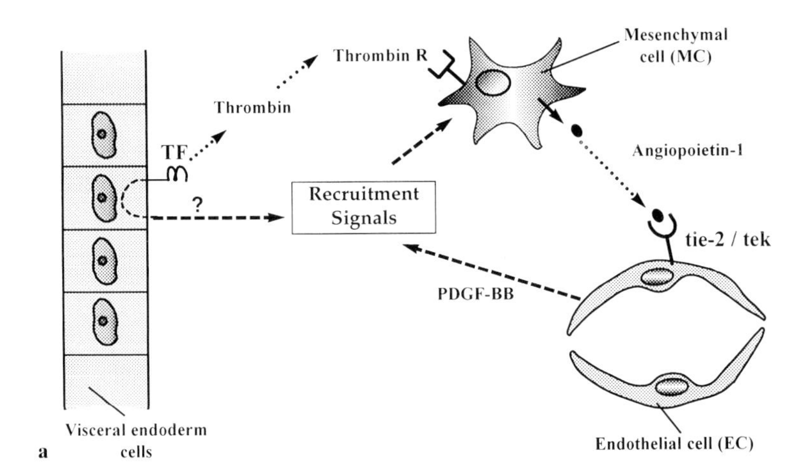

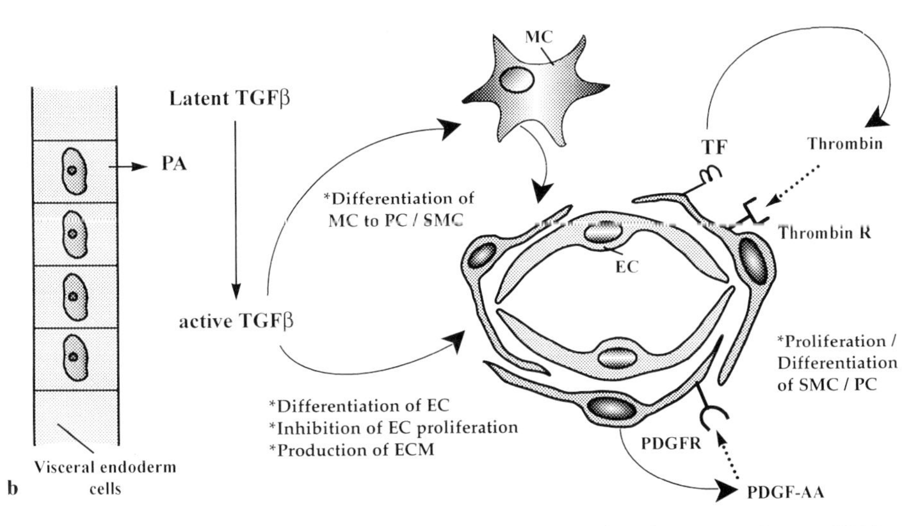

that is converted from plasminogen by the action of plasminogen activator (PA), produced by the visceral endoderm, endothelial or mesenchymal cells. Active TGF-β1 may then induce differentiation of mesenchymal cells to pericytes/smooth muscle cells, control production of extracellular matrix, inhibit endothelial cell growth and regulate their differentiation. PDGF-AA, produced by smooth muscle cells, and tissue factor (possibly via generation of thrombin) may control proliferation and/or differentiation of smooth muscle cells in an autocrine loop ultrastructural analysis of wild-type embryos (CARMELIET et al. 1996b), as would be expected in the case of TF deficient embryos dying secondarily due to a hemostatic defect. Instead, the analysis suggests that TF, produced by visceral endoderm, mesenchymal cells, provides spatial migratory cues for recruitment of mesenchymal cells to envelope the endothelial cells, that it provides a signal for mesenchymal cells to differentiate in pericytes/smooth muscle cells and/or that it affects their growth (Fig. 4). It could do so by generation of downstream coagulation molecules (including factor Xa and thrombin), which both can affect migration/proliferation of vascular cells (ALTIERI 1995; COUGHLIN 1994), by acting as a cellular adhesion molecule, or by affecting novel cellular mechanisms (for example the production of a mesenchymal recruitment signal). Although our analysis revealed no endothelial cell abnormalities, and normal expression of VEGF, a possible (in)direct effect of TF on endothelial cells cannot be excluded in view of the close structural and functional relationship between endothelial cells and pericytes.

It is unknown whether tissue factor (only) interacts with its (currently only known) ligand, factor VII (Fig. 3b). Indeed, analysis of factor VII deficient mice reveals that they develop normally until birth and die in utero 1 day prior to birth or early postnatally due to massive hemorrhaging (ROSEN et al. 1997). Whether this means that tissue factor acts independently of factor VII, or whether embryogenesis of factor VII deficient embryos is rescued by minimal transfer of maternal factor VII, remains to be determined. Our analysis reveals, however, that intravenous injection in pregnant mice of human recombinant factor VIIa induces supraphysiological plasma levels in the mother but only background levels in early stage embryos and that factor VII procoagulant activity is undetectable in plasma from 11.5-day-old factor VII deficient embryos.

2. Thrombomodulin

In mouse embryos, thrombomodulin is present in the parietal endoderm (day 7.5 of gestation) (FORD et al. 1993). Later, the antigen is present in the developing vasculature as well as in nonvascular structures, including the lung buds and the developing central nervous system. Thrombomodulin is found on all endothelial cells of the blood vessels and lymphatics, except on the high endothelium of postcapillary venules, and hepatic sinusoids. Thrombomodulin is also present on megakaryocytes, platelets, syncytiotrophoblasts, synovium, meningial cells, mononuclear phagocytes and mesothelium of the pleura and peritoneum. However, its function in these nonendothelial cells remains unknown.

Transgenic embryos bearing a single copy of the *LacZ* gene, targeted into the endogenous thrombomodulin gene locus via homologous recombination (WEILER-GUETTLER et al. 1996), express this marker gene in the parietal endoderm, in trophoblasts in the mesoderm of the yolk sac and the embryo (7 days of gestation), in the cardiogenic mesoderm, angioblastic head mesenchyme, dorsal aorta and yolk sac vessels (8.5 days), and in all intra- and extra-embryonic vessels (9.5 days). *LacZ* staining was also prominent in nonvascular structures including the leptomeninges, neuroepithelium and lung of 9.5-day-old embryos. In the adult mouse, *LacZ* stained the endothelial cells in the heart, lung and spleen and, less intensely, in the brain, liver, skin, body wall, bones, retina, skeletal muscle, gastrointestinal tract, choroid plexus, kidney and aorta. Reporter gene activity was also documented in nonvascular structures such as brain meninges, interstitium of testes, and the skin. Thus, targeting of transgenes to the thrombomodulin reproduces in most tissues a temporospatial pattern indistinguishable from the endogenous *thrombomodulin* gene.

Gene inactivation of thrombomodulin resulted in overall growth retardation and embryonic death around 9.0–9.5 days postcoitum (HEALY et al. 1995). Notably, removal of thrombomodulin deficient embryos from the maternal decidua rescued the arrested organogenesis, suggesting that defective fetomaternal interactions might compromise the development of the embryo proper (HEALY et al. 1995). Increased fibrin deposition did not seem to affect the barrier function of the Reichert's membrane or the parietal endoderm (HEALY et al. 1995). Whether the embryonic lethality might be related to defective trophoblast function remains to be determined.

3. Thrombin Receptor and Factor V

Thrombin has been implicated in processes beyond hemostasis (COUGHLIN 1994). Indeed, the molecule is mitogenic for fibroblasts and vascular smooth muscle cells, chemotactic for monocytes, induces neurite outgrowth and activates endothelial cells. Many of the cell signaling activities of thrombin appear to be mediated by the thrombin receptor (COUGHLIN 1994). Expression studies have further suggested that the thrombin receptor participates in inflammatory, proliferative or reparative responses such as restenosis, atherosclerosis, neovascularization and tumorigenesis. In addition, in situ analysis indicated that the thrombin receptor is expressed during early embryogenesis in the developing heart and blood vessels, in the brain and in several epithelial tissues (SOIFER et al. 1994).

Targeting of the thrombin receptor resulted in block of embryonic development in approximately 50% of the homozygous deficient embryos around a similar developmental stage as in tissue factor deficient embryos, presumably resulting from abnormal yolk sac vascular development (CONNOLLY et al. 1996). Although the cellular defect was not characterized in detail, vitelline vessels appeared to be defective, resulting in increased fragility of blood vessels with secondary rupture, blood leakage and pallor of the embryo. Enlarged pericardial cavities in the thrombin receptor deficient embryos suggested compromised vitelloembryonic blood circulation. A putative role of thrombin in early blood vessel formation is represented in Fig. 4.

Similar to the thrombin receptor deficient phenotype, deficiency of factor V resulted in embryonic lethality in approximately half of the homozygously deficient embryos, possibly due to vascular defects in the yolk sac (Cui et al. 1996). The leakage of blood from the defective blood vessels in tissue factor, thrombin receptor and factor V deficient embryos ("vascular" bleeding) contrasts with the postnatal bleeding in mice deficient in factor VII, factor VIII (Bi et al. 1995) and fibrinogen (Suh et al. 1995) and in the surviving fraction of factor V deficient mice, which occurs due to defective clot formation following trauma of normally developed blood vessels ("hemostatic" bleeding). Bleeding in the latter mice occurred shortly after birth (factor V, factor VII and fibrinogen) or was associated with injury (factor VIII). Thus, it appears from these targeting studies that several coagulation factors (tissue factor, factor V, thrombin receptor) may participate in morphogenic processes beyond control of fibrin-dependent hemostasis, whereas other coagulation factors (factor VII, factor VIII, fibrinogen) play a predominant role in hemostasis via clot formation.

4. Fibrinogen

Ovulation, embryo implantation and placentation involve tissue remodeling and breaching of intact vessels, requiring proper hemostasis. Somewhat surprisingly, intraovarian bleeding did not occur following ovulation in fibrinogen deficient mice (Suh et al. 1995). However, deficiency of fibrinogen significantly affected embryogenesis (Suh et al. 1995). Indeed, development of fibrinogen deficient embryos in homozygous fibrinogen deficient females was arrested at 9–10 days postcoitum due to severe intrauterine bleeding. There was no evidence of bleeding within developing embryos or their amniotic or yolk sacs as long as the placentas were intact. Rather, the location, volume and absence of nucleated (embryonic) red blood cells within the hemorrhagic areas suggest that hemorrhaging was from a maternal source. It is possible that bleeding was caused by the invasion of embryonic trophoblasts into and disruption of maternal vasculature within the placenta. Abortion was not observed in heterozygous fibrinogen deficient females mated to heterozygous or homozygous fibrinogen deficient males, indicating the importance of the maternal fibrinogen during embryonic development (Suh et al. 1995).

II. Fibrinolytic System

The plasminogen system has been claimed to be involved in ovulation, spermatocyte migration, fertilization, embryo implantation and embryogenesis, and in the associated remodeling of the ovary, prostate and mammary gland (Astrup 1978; Vassalli et al. 1991). Since homozygous deficiencies of several fibrinolytic system components including t-PA and u-PA have not been observed, it was anticipated that inactivation of these genes might cause embryonic lethality. However, transgenic mice overexpressing PAI-1 (ERICKSON et al. 1990; EITZMAN et al. 1996), u-PA (HECKEL et al. 1990) or the amino-terminal fragment of u-PA (SIDENIUS 1993) and mice with single or combined deficiencies of t-PA and/or u-PA (CARMELIET et al. 1994, 1995), t-PA and u-PAR (BUGGE et al. 1996), PAI-1 (CARMELIET et al. 1993a,b), u-PAR (DEWERCHIN et al. 1996; BUGGE et al. 1995a), plasminogen (PLOPLIS et al. 1995; BUGGE et al. 1995b), PAI-2 (DOUGHERTY et al. 1995), vitronectin (ZHENG et al. 1995) or α_2-macroglobulin (UMANS et al. 1995) survived embryonic development and were viable at birth. Thus far, u-PA deficiency has only been found to reduce the rate of trophoblast migration during early embryogenesis and the formation of blood lacunae in late pregnancies possibly due to impaired endothelial cell function (Teesalu, Blasi and Tallerico, personal communication).

Inactivation of the LRP gene resulted in embryonic lethality at mid-gestation, secondary to abdominal bleeding (HERZ et al. 1992). It is presently unclear to what extent lethality in these mice is caused by abnormal plasmin proteolysis since LRP is a multifunctional clearance receptor not only for fibrinolytic system components but also for other unrelated molecules (ANDREASEN et al. 1994). Another interesting but unresolved question is why homozygous but not heterozygous proteinase nexin-1 deficient mice are unable to sire offspring (Botteri and Vander Putten, personal communication).

Mice with single deficiency of t-PA or u-PA (CARMELIET et al. 1994, 1995), u-PAR (DEWERCHIN et al. 1996; BUGGE et al. 1995), PAI-1 (CARMELIET et al. 1993a,b), vitronectin (ZHENG et al. 1995) or α_2-macroglobulin (UMANS et al. 1995) are fertile. Normal fertility was also observed in a transgenic mouse strain expressing an antisense t-PA mRNA with reduction of t-PA activity in the oocytes by more than 50% (RICHARD et al. 1993). Both plasminogen activators appeared, however, to cooperate, since Plg deficient and combined t-PA:u-PA deficient mice were less fertile than wild-type mice or mice with a single deficiency of t-PA or u-PA (CARMELIET et al. 1994; PLOPLIS et al. 1995). In part, this could be due to poor general health and fibrin deposits in the gonads once they became sick and cachectic. However, gonadotropin-induced ovulation was also significantly reduced in healthy 25-day-old female mice lacking both plasminogen activators (LEONARDSSON et al. 1995) or plasminogen (Ny et al., personal communication). The observation that combined t-PA:u-PAR deficient mice are fertile (BUGGE et al. 1996b) suggests that u-PA can still mediate sufficient pericellular proteolysis in the absence of u-PAR to rescue the defective ovulation of combined t-PA:u-PA deficient mice (CARMELIET et al. 1995). Thus, ovulation can occur in the absence of t-PA, u-PA, u-PAR, PAI-1 or α_2-macroglobulin, but is reduced in mice with Plg deficiency or combined t-PA and u-PA deficiency.

III. Integrated View of a Role for the Coagulation and Fibrinolytic System in Vascular Development

It is apparent from the gene targeting studies that the coagulation system participates in vascular development during embryogenesis. The following discussion presents a hypothetical model of the possible role of the coagulation and fibrinolytic factors in vessel formation during embryonic development and their relationship with other vascular growth factors (Fig. 5). Differentiation of endothelial cells (EC in Fig. 5), their organization in vascular tubes and assembly into a network, and the sprouting of new blood vessels are mediated in large part by vascular endothelial growth factor (VEGF), interacting with its cellular receptors Flk-1 and Flt-1 (Carmeliet et al. 1996a; Shalaby et al. 1995; Fong et al. 1995). However, once endothelial cell lined channels are formed, vascular integrity is maintained by organization of a primitive muscular wall, providing the required structural support against increased blood pressure and modulating endothelial cell function. The recently cloned angiopoietin-1, via interaction with its TIE-2/TEK receptor, may activate endothelial cells to produce a recruitment factor for mesenchymal cells (MC in Fig. 5), possibly platelet-derived growth factor (PDGF)-BB (Levéen et al. 1994; Suri et al. 1996). Once the mesenchymal cells contact the endothelium, transforming growth factor-β1 (TGF-β1) may be activated and induce differentiation of the mesenchymal cells into pericytes and smooth muscle cells (SMC in Fig. 5), inhibit endothelial cell proliferation and stimulate extracellular matrix (ECM) deposition (Dickson et al. 1995). Activation of latent TGF-β1 may occur via plasmin, generated from plasminogen by the action of plasminogen activators, which can be produced by endothelial, mesenchymal and visceral endoderm (VE) cells (Saksela and Rifkin 1988). Smooth muscle cells may further control their own growth via autocrine production of PDGF-AA. Tissue factor, and possibly other coagulation factors (including factor V and the thrombin receptor), may participate in early vascular development via recruitment, differentiation and/or proliferation of mesenchymal cells to smooth muscle cells/pericytes, but the precise molecular mechanisms remain to be determined.

C. Health and Survival

I. Coagulation System

Deficiencies of the procoagulant factor V, factor VII, factor VIII and fibrinogen significantly affect survival of mutant mice due to bleeding complications. Within 2 days after birth, approximately 30% of the fibrinogen deficient offspring developed overt intra-abdominal bleeding but surprisingly only 10% of these neonates died (Suh et al. 1995). A second period of increased risk of developing fatal intra-abdominal bleeding occurred between 30 and 60 days, resulting in a 50%–60% survival rate. Survival of fibrinogen deficient mice was

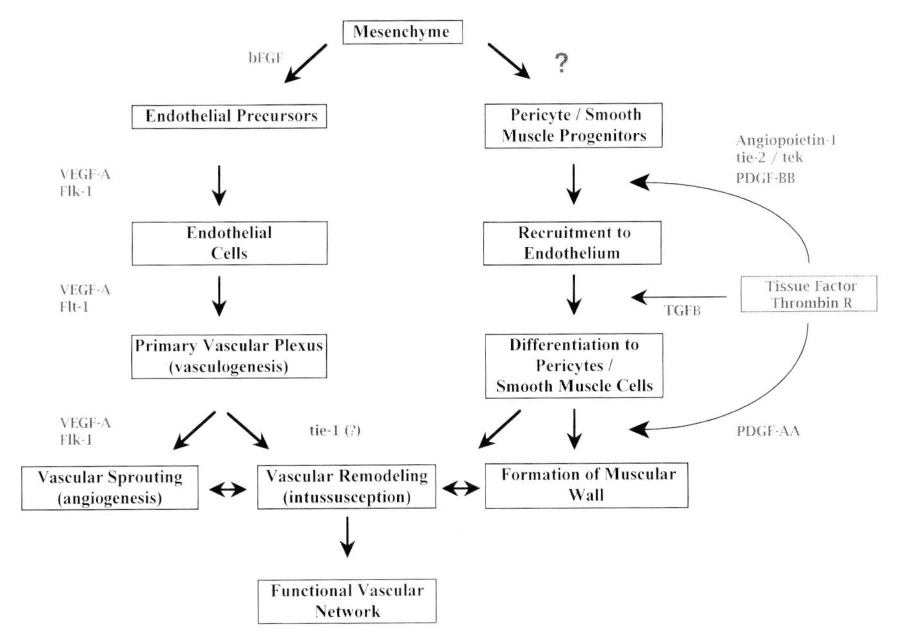

Fig. 5. Integrated view of early blood vessel development during embryogenesis. Mesenchymal cells differentiate to hemagioblasts, in part by the action of basic fibroblast growth factor (bFGF). Further *in situ* differentiation into endothelial cells is mediated by VEGF-A, through interaction with VEGFR-2/Flk-1, whereas formation of the endothelial cell-lined channels and organization into a primitive vascular plexus (vasculogenesis) is dependent on VEGF-A and its receptor Flt-1. VEGF-A and Flk-1 are further involved in the process of angiogenesis, or the sprouting of neovessels from preexisting vessels, in lumen formation and in mediating correct vascular connections. Tissue factor, and possibly the thrombin receptor and factor V, participate in the formation of a primitive smooth muscle cell sheet around the fragile endothelial cell-lined channels, possibly by acting as a recruitment signal or as a differentiation/proliferation factor for mesenchymal cells. Angiopoietin-1 and its receptor TEK play a somewhat similar role in recruiting mesenchymal cells and the remodelling of early blood vessels by intussusception. PDGF-BB might be one of the recruitment signals for the mesenchymal cells, whereas TGF-β1 could participate in differentiation of the mesenchymal cells to pericytes/smooth muscle cells, inhibition of endothelial cell growth and production of extracellular matrix. The role of TIE-1 is less defined, but may relate to maturation and integrity of the microvasculature. P1GF is not essential for embryonic vascular development, but appears to be required for neovascularization during adulthood, for example during wound healing

highly dependent on the genetic background, possibly due to differences in general activity level (Suh et al. 1995). A fraction (approximately 50%) of factor V deficient mice also revealed intra-abdominal bleeding resulting in early postnatal death (Cui et al. 1996). Factor VII deficient mice died due to massive intra-abdominal bleeding within the first 3 days postnatally whereas the remainder of the factor VII deficient neonates died due to intracranial bleeding within 2–3 weeks (Rosen et al. 1997). In contrast, factor VIII deficient mice did not bleed spontaneously but displayed life-threatening bleeding when challenged with trauma during tail cutting (BI et al. 1995).

II. Fibrinolytic System

No effects on health and survival were observed in t-PA-, u-PAR-, PAI-1-, vitronectin- or α_2-macroglobulin deficient mice (CARMELIET et al. 1993a,b, 1994, 1995; BUGGE et al. 1995a,b, 1996; DEWERCHIN et al. 1996; PLOPLIS et al. 1995; DOUGHERTY et al. 1995; ZHENG et al. 1995; UMANS et al. 1995). A small percentage of u-PA deficient mice developed chronic (nonhealing) ulcerations and rectal prolapse but without effect on survival (CARMELIET et al. 1994). Plg deficient (PLOPLIS et al. 1995; BUGGE et al. 1995b) and combined t-PA:u-PA deficient (CARMELIET et al. 1994) but not combined t-PA:u-PAR deficient (BUGGE et al. 1996b) mice developed chronic ulcerations and rectal prolapse, suggesting that sufficient u-PA-mediated plasmin proteolysis can occur in the absence of u-PAR. In addition, these mice suffered significant growth retardation, developed a wasting-syndrome with anemia, dyspnea, lethargia and cachexia and had a significantly shorter life span (Fig. 6). Generalized thrombosis in the gastrointestinal tract, in the lungs and in other organs (including gonads, liver and kidney) might, at least in part, explain their increased morbidity and mortality. Cross-breeding of the Plg deficient mice with fibrinogen deficient mice rescued the increased morbidity and mortality of the Plg deficient mice, suggesting that the role of Plg is primarily mediated by fibrin degradation (BUGGE et al. 1996c).

D. Hemostasis

I. Coagulation System

Deficiency of tissue factor and loss of the thrombin receptor and factor V (in approximately 50% of the embryos) resulted in bleeding due to vascular defects ("vascular bleeding"). This section describes the bleeding resulting from defects in clot formation in the absence of vascular defects ("hemostatic bleeding"). Inactivation of the intrinsic pathway by deletion of factor XI had little effect as mutant animals showed normal survival, fertility and fecundity, but no signs of bleeding (GAILANI et al. 1996). Factor VII deficient mice died due to massive intra-abdominal bleeding within the first 3 days postnatally whereas the remainder of the factor VII deficient neonates died due to intracranial bleeding within 2–3 weeks (ROSEN et al. 1997). Deficiency of factor VII in patients results in severe bleeding when plasma levels of factor VII are below 2% of normal plasma levels (TUDDENHAM et al. 1995). However, lack of early postnatal death in patients with reduced factor VII plasma levels may relate to the fact that factor VII was not completely absent in contrast to the factor VII deficient mice, which completely lack this factor. Factor V deficient mice appeared to suffer a severe bleeding phenotype (resulting in early post-

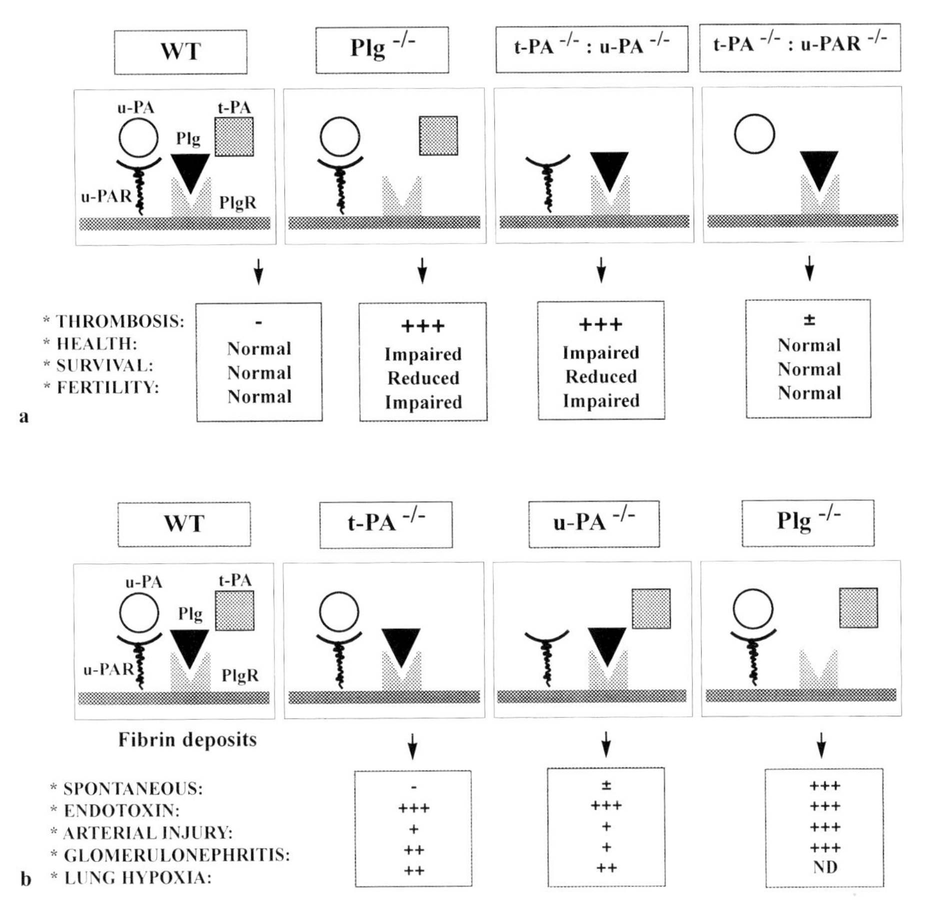

Fig. 6a,b. Role of the plasminogen system in fibrin surveillance. Fibrin deposits in plasminogen system knock out mice before and after experimental challenges. ***a***, Fibrin deposits during unchallenged conditions were occasionally observed in u-PA deficient mice and more extensively in Plg deficient or combined t-PA: u-PA deficient mice, indicating that both plasminogen activators cooperate in prevention of fibrin deposition. Inflammatory and/or traumatic challenges including local injection of proinflammatory endotoxin in the footpad, skin wounding, experimental glomerulonephritis, lung hypoxia and arterial injury induce extravascular fibrin deposits but also intravacular thrombosis in veins or capillaries (endotoxin, lung hypoxia, skin wounding, glomerulonephritis) and arteries (arterial injury). ***b***, Combined deficiency of t-PA and u-PA but not of t-PA and u-PAR results in widespread fibrin deposition, reduced fertility, multiorgan dysfunction with dyspnea, lethargia and cachexia, ultimately leading to premature death, indicating that sufficient pericellular plasmin proteolysis occurs in the absence of binding of u-PA to its cellular receptor

natal death), suggesting critical hemostatic functions of thrombin activation beyond fibrin generation (Cui et al. 1996). The more severe phenotype of factor V deficiency in mice than in humans is consistent with the detection of residual factor V activities in most patients (White 1995). Deficiency of factor

VIII (hemophilia A) in patients predisposes to spontaneous and trauma-induced bleeding into joints and soft tissues (FURIE and FURIE 1988). Mice deficient in factor VIII suffered life-threatening bleeding in association with tail injury but did not appear to bleed spontaneously (BI et al. 1995). The factor VIII deficient mice may constitute useful models for studying the immune response that limits recombinant factor VIII substitution in hemophilic patients as well as for designing possible gene therapy strategies. Surprisingly, the thrombin receptor deficient mice that developed properly did not reveal signs of bleeding, suggesting that other related thrombin receptors may play a significant role in platelet activation and hemostasis (CONNOLLY et al. 1996). Deficiency of fibrinogen resulted in overt intra-abdominal, subcutaneous, joint and/or periumbilical bleeding in the neonatal period (SUH et al. 1995). These are the common sites of spontaneous bleeding events in humans with acquired or congenital coagulation disorders (MONTGOMERY and SCOTT 1993). The bleeding manifestations in adult fibrinogen deficient mice (e.g., hemoperitoneum, epistaxis, hepatic, renal, intraintestinal, intrathoracic and soft tissue hematomas) are generally comparable to those observed in the rare human congenital disorder afibrinogenemia and probably resulted from coincidental mechanical trauma (AL-MONDHIRY and EHMANN 1994). Although the afibrinogenemic murine blood was totally unclottable and platelets failed to aggregate, bleeding was not consistently life-threatening (SUH et al. 1995). Possibly, bleeding was somewhat controlled by the residual thrombin generation and platelet activation. Whether other platelet receptors beyond the GPIIb/IIIa receptor or other ligands than fibrinogen (including vitronectin, fibronectin or von Willebrand's factor) might rescue deficient platelet interactions remains to be determined.

II. Fibrinolytic System

Hemostasis involves platelet deposition and coagulation to stabilize hemostatic plugs. Failure to stabilize the clot, e.g., as a result of hyperfibrinolytic activity, might result in delayed rebleeding. A hemorrhagic tendency has indeed been observed in patients with increased plasma t-PA or reduced plasma α_2-antiplasmin or PAI-1 activity levels (FAY et al. 1992; AOKI 1989). Delayed rebleeding might also explain the hemorrhagic tendency in transgenic mice, expressing high levels of plasma u-PA (HECKEL et al. 1990) and in transgenic mice overexpressing GM-CSF, in which increased production of u-PA by peritoneal macrophages occurs (ELLIOTT et al. 1992). In contrast to patients with low or absent plasma PAI-1 levels, PAI-1 deficient mice did not reveal spontaneous or delayed rebleeding, even after trauma (CARMELIET et al. 1993a). Lower plasma PAI-1 levels and the occurrence of alternative PAIs in murine plasma (unpublished data) might explain the less pronounced hyperfibrinolytic phenotype and the species-specific difference in the control of plasmin proteolysis.

E. Thrombosis and Thrombolysis

I. Coagulation System

Deficiencies of coagulation inhibitors in humans predispose to an increased risk for thrombosis (BICK and PEGRAM 1994). Heterozygous thrombomodulin deficient mice were viable and did not appear to develop spontaneous thrombosis, possibly indicating that the mice need to be challenged either genetically (by cross-breeding them with other thrombosis-prone transgenic mice) or physiologically (by administration of proinflammatory reagents, injury, etc.). However, recently generated transgenic mice with a mutated *thrombomodulin*$_{Q387P}$ gene, which reduces interaction with protein C, survived embryonic development and revealed an increased spontaneous thrombotic incidence (H. WEILER et al., personal communication). The recently generated mutant factor V mice (CUI et al. 1995) engineered to have a similar activated protein C-resistance phenotype as in humans (DAHLBÄCK 1995) might be valuable to examine the role of this anticoagulant protein in vivo.

II. Fibrinolytic System

t-PA is believed to be primarily responsible for removal of fibrin from the vascular tree via clot-restricted plasminogen activation (COLLEN and LIJNEN 1991). The role of u-PA in thrombolysis is less well defined. It lacks affinity for fibrin and probably requires conversion from a single-chain precursor to a catalytically active two-chain derivative (COLLEN and LIJNEN 1991). Under what conditions u-PA might participate in fibrin clot dissolution in vivo remains to be determined. Other plasminogen activation pathways, such as the "intrinsic" pathway (involving blood coagulation factor XII, high molecular weight kininogen, prekallikrein and possibly u-PA) (KLUFT et al. 1987) as well as plasminogen-independent mechanisms (PLOW and EDGINGTON 1975), have been proposed to contribute to clot lysis, but their role in vivo remains to be defined.

Deficient fibrinolytic activity, e.g., resulting from increased plasma PAI-1 levels or reduced plasma t-PA or plasminogen levels, might participate in the development of thrombotic events (AOKI 1989). Elevated plasma PAI-1 levels have indeed been correlated with a higher risk of deep venous thrombosis and of thrombosis during hemolytic uremic syndrome, disseminated intravascular coagulation, sepsis, surgery and trauma (SCHNEIDERMAN and LOSKUTOFF 1991; WIMAN 1995). PAI-1 plasma levels have also been elevated in patients with ischemic heart disease, angina pectoris and recurrent myocardial infarction (HAMSTEN et al. 1987). However, the acute phase reactant behavior of PAI-1 does not allow the deduction of whether increased PAI-1 levels are a cause or consequence of thrombosis. To date, abnormal fibrin clot surveillance resulting from genetic deficiencies of t-PA or u-PA has not been reported in humans but quantitative and qualitative deficiencies of plasminogen have been associated with an increased thrombotic tendency (AOKI 1989; ROBBINS 1988).

III. Fibrin Deposits and Pulmonary Plasma Clot Lysis in Transgenic Mice

Microscopic analysis of tissues from u-PA deficient mice revealed occasional minor fibrin deposits in liver and intestines and excessive fibrin deposition in chronic nonhealing skin ulcerations, whereas in t-PA deficient mice no spontaneous fibrin deposits were observed (CARMELIET et al. 1994). Mice with a single deficiency of plasminogen (Plg) or a combined deficiency of t-PA and u-PA, however, revealed extensive fibrin deposits in several organs (including the liver, lung, gastrointestinal tract, and reproductive organs) associated with ischemic necrosis, possibly resulting from thrombotic occlusions (SUH et al. 1995; PLOPLIS et al. 1995; BUGGE et al. 1995b). Fibrin deposits were observed at the same predilection sites and around the same age in Plg deficient as in combined t-PA:u-PA deficient mice, suggesting that t-PA and u-PA are the only physiologically significant plasminogen activators in vivo. These fibrin deposits were absent in mice with a combined deficiency of Plg and fibrinogen (BUGGE et al. 1996c). Interestingly, mice with a combined deficiency of t-PA and u-PAR did not display such excessive fibrin deposits, suggesting that sufficient plasmin proteolysis can occur in the absence of u-PA binding to u-PAR (Fig. 6) (BUGGE et al. 1996a). Transgenic mice overexpressing human PAI-1 under the control of the metallothionin promoter displayed cell, fibrin and platelet-rich venous occlusions in the tail and hindlegs (ERICKSON et al. 1990), whereas mice overexpressing murine PAI-1 under the control of the cytomegalovirus promoter did not suffer such complications (EITZMAN et al. 1996; Ginsburg et al., personal communication). The reason for this discrepancy is at present unclear. Mice with a single deficiency of t-PA or u-PA were significantly more susceptible to the development of venous thrombosis following local injection of proinflammatory endotoxin in the footpad (CARMELIET et al. 1994). Hypoxia also induced ^{125}I-labeled fibrin deposition in t-PA or u-PA deficient mice, but not in wild-type or PAI-1 deficient mice (Pinsky et al., personal communication). The increased susceptibility of t-PA deficient mice to endotoxin and the severe spontaneous thrombotic phenotype of combined t-PA:u-PA or Plg deficient mice could be explained by their significantly reduced rate of spontaneous lysis of ^{125}I-fibrin labeled plasma clots, injected via the jugular vein and embolized into the pulmonary arteries (CARMELIET et al. 1994; PLOPLIS et al. 1995). On the contrary, PAI-1 deficient mice were virtually protected against development of venous thrombosis following injection of endotoxin, consistent with their ability to lyse ^{125}I-fibrin labeled plasma clots at a significantly higher rate than wild-type mice (CARMELIET et al. 1993b). The increased susceptibility of u-PA deficient mice to thrombosis associated with inflammation or injury might be due to their impaired macrophage function. Indeed, thioglycollate-stimulated macrophages (which are known to express cell-associated u-PA) isolated from u-PA deficient mice lacked plasminogen-dependent breakdown of ^{125}I-labeled fibrin (fibrinolysis) or of ^{3}H-labeled subendothelial matrix (mostly collagenolysis), whereas macrophages from t-PA deficient or PAI-1 deficient mice did not (CARMELIET et al. 1994, 1995).

Lipoprotein(a) contains the lipid and protein components of low-density lipoprotein plus apolipoprotein(a) (LIU and LWAN 1994). Extensive homology of apolipoprotein(a) to plasminogen has prompted the proposal that apolipoprotein(a) forms a link between thrombosis and atherosclerosis, but in vitro studies have not yielded conclusive evidence. Transgenic mice overexpressing apolipoprotein(a) displayed reduced thrombolytic potential but only after administration of pharmacological doses of recombinant t-PA, suggesting a mild hypofibrinolytic condition (PALABRICA et al. 1995). Studies using transgenic mice overexpressing lipoprotein(a) extended these findings and revealed that spontaneous lysis of ^{125}I-fibrin labeled pulmonary plasma clots (thus not lysis induced by exogenous administration of recombinant t-PA) was also reduced (Carmeliet et al., unpublished observations).

IV. Adenovirus-Mediated Transfer of t-PA or PAI-1

More recently, we have used adenoviral-mediated transfer of fibrinolytic system components in these "knock-out" mice in an attempt to revert their phenotypes. Intravenous injection of adenoviruses, expressing a recombinant PAI-1 resistant human t-PA (*rt-PA*) gene, in t-PA deficient mice increased plasma rt-PA levels 100- to 1000-fold above normal and restored their impaired thrombolytic potential in a dose-related way (CARMELIET et al. 1997b). Notably, adenoviral t-PA gene transfer increased thrombolysis to significant levels within 4 h and was sustained for more than a week, suggesting that it might be useful for restoring deficient thrombolysis in subacute conditions. Conversely, adenovirus-mediated transfer of recombinant human PAI-1 in PAI-1 deficient mice resulted in 100- to 1000-fold increased plasma PAI-1 levels above normal and efficiently reduced the increased thrombolytic potential of PAI-1 deficient mice (Carmeliet et al., unpublished observations).

Collectively, these gene targeting and gene transfer studies confirm the importance of the plasminogen system in maintaining vascular patency and indicate that t-PA and u-PA are the only physiologically significant plasminogen activators in vivo which appear to cooperate significantly in fibrin surveillance. Interestingly, u-PA appears to play a more significant role than previously anticipated in the prevention of fibrin deposition during conditions of inflammation or injury, possibly through cell-associated plasmin proteolysis. A surprising finding is, however, that u-PA can still exert its biological role (pericellular proteolysis) in the absence of u-PAR. Apparently, the marginal role of u-PAR is related to the ability of u-PA to become *localized around* but not *bound to* the cell surface via interaction with other macromolecules such as fibrin, plasminogen, vitronectin or proteoglycans (CARMELIET et al. 1995).

F. Neointima Formation

Vascular interventions for the treatment of atherothrombosis such as bypass surgery, percutaneous transluminal balloon angioplasty, atherectomy or the in situ application of vascular stents restore blood flow and improve tissue oxygenation but induce "restenosis" of the vessel within 3–6 months in 30%–50% of treated patients (FORRESTER et al. 1991). This may result from remodeling of the vessel wall and/or accumulation of cells and extracellular matrix in the intimal or adventitial layer. Several mechanisms are believed to participate in the intimal thickening as part of a hyperactive wound healing response including thrombosis, proliferation, apoptosis or migration of smooth muscle cells (LIBBY et al. 1992; CLOWES and REIDY 1991). Proteinases participate in the degradation of the extracellular basement membrane surrounding the smooth muscle cells, allowing them to migrate to distant sites. Two proteinase systems have been implicated, the plasminogen (or fibrinolytic) system and the metalloproteinase system, which in concert can degrade most extracellular matrix proteins.

In contrast to the constitutive expression of t-PA by quiescent endothelial cells (COLLEN and LIJNEN 1991) and of PAI-1 by uninjured vascular smooth muscle cells (SIMPSON et al. 1991), u-PA and t-PA activity in the vessel wall are significantly increased after injury, coincident with the time of smooth muscle cell proliferation and migration (CLOWES et al. 1990; JACKSON and REIDY 1992; JACKSON et al. 1993). This increase in plasmin proteolysis is counterbalanced by increased expression of PAI-1 in injured smooth muscle and endothelial cells and by its release from accumulating platelets (SAWA et al. 1992). Expression of components of the fibrinolytic system is also induced in cultured endothelial cells, smooth muscle cells or macrophages as a result of wounding or treatment with growth factors and cytokines that are released after injury (VASSALLI 1994; CARMELIET et al. 1998). t-PA has been proposed to act as an autocrine mitogen after injury (HERBERT et al. 1994). The precise and causative role of the plasminogen system in matrix remodeling, passivation of the injured luminal vessel surface, and migration or proliferation of vascular cells, however, remain to be determined.

We have used an experimental model based on the use of an electric current to examine the molecular mechanisms of neointima formation in mice deficient in fibrinolytic system components (CARMELIET et al. 1997a). The electric current injury model differs from mechanical injury models in that it induces a more severe injury across the vessel wall resulting in necrosis of all smooth muscle cells. This necessitates wound healing to initiate from the adjacent uninjured borders and to progress into the central necrotic region. Microscopic and morphometric analysis revealed that the rate and degree of neointima formation and the neointimal cell accumulation after injury was similar in wild-type, t-PA deficient and u-PAR deficient arteries (CARMELIET et al. 1997b,c,d, 1998). However, neointima formation in PAI-1 deficient arteries occurred at earlier times postinjury (CARMELIET et al. 1997d).

In contrast, both the degree and the rate of arterial neointima formation in u-PA deficient, Plg deficient and combined t-PA:u-PA deficient arteries were significantly reduced until 6 weeks after injury (CARMELIET et al. 1997b,e). Evaluation of the mechanisms responsible for these genotype-specific differences in neointima formation revealed that proliferation of medial and neointimal smooth muscle cells was only marginally different between the genotypes (CARMELIET et al. 1997b–d). Impaired migration of smooth muscle cells could be a significant cause of reduced neointima formation in mice lacking u-PA-mediated plasmin proteolysis since smooth muscle cells migrated over a shorter distance from the uninjured border into the central injured region in Plg deficient than in wild-type arteries (CARMELIET et al. 1997b). As deficiency of u-PAR did not affect arterial stenoses (CARMELIET et al. 1998), the role of u-PAR remains obscure. It has been shown that scavenging of soluble u-PA by a truncated (nonmembrane-anchored) u-PAR impairs cellular invasion (WILHELM et al. 1994). Furthermore, a membrane-anchored form of u-PA catalyzes plasminogen activation on the cell surface with characteristics comparable to those of u-PAR-bound u-PA (LEE et al. 1992), suggesting that *cell surface localization* rather than *binding to* u-PAR is important. Possibly, the kinetic advantage resulting from u-PAR accelerated plasminogen activation may be irrelevant for certain u-PA dependent phenomena that develop over long time periods, or that are compensated by the increased extracellular accumulation of u-PA in u-PAR deficient mice (probably resulting from defective clearance of u-PA) (BUGGE et al. 1995a). Alternatively, u-PA may be localized to the cell surface via binding to other molecules such as fibrin, plasminogen, extracellular matrix or cell adhesion molecules (PLOW et al. 1995; WEI et al. 1996; STEPHENS et al. 1992). Our results also suggest that pericellular proteolysis can still occur in the absence of u-PAR. This is confirmed by the observation that degradation of ^{125}I-labeled fibrin or subendothelial matrix over 8 h is only transiently affected by u-PAR deficiency (DEWERCHIN et al. 1996; CARMELIET et al. 1998) and that thrombosis, sterility and organ dysfunction in combined t-PA:u-PAR deficient mice is significantly less severe than in combined t-PA:u-PA deficient mice (CARMELIET et al. 1994; BUGGE et al. 1996b). Figure 7 schematically represents a hypothetical model of smooth muscle cell function and neointima formation in the absence of u-PA or u-PAR. It should be noticed, however, that the lack of an appreciable effect on neointima formation in u-PAR knockout mice does not exclude a role for this receptor in other biological processes (such as in cancer), since the relevance of u-PAR may depend on the amount and the cell-specific, temporal and spatial expression of u-PAR. Whether, to what extent and under what conditions u-PAR may be more important in other u-PA dependent phenomena such as cancer or angiogenesis remains to be determined.

The involvement of plasmin proteolysis in neointima formation was supported by intravenous injection in PAI-1 deficient mice of a replication-defective adenovirus that expresses human PAI-1, which resulted in more

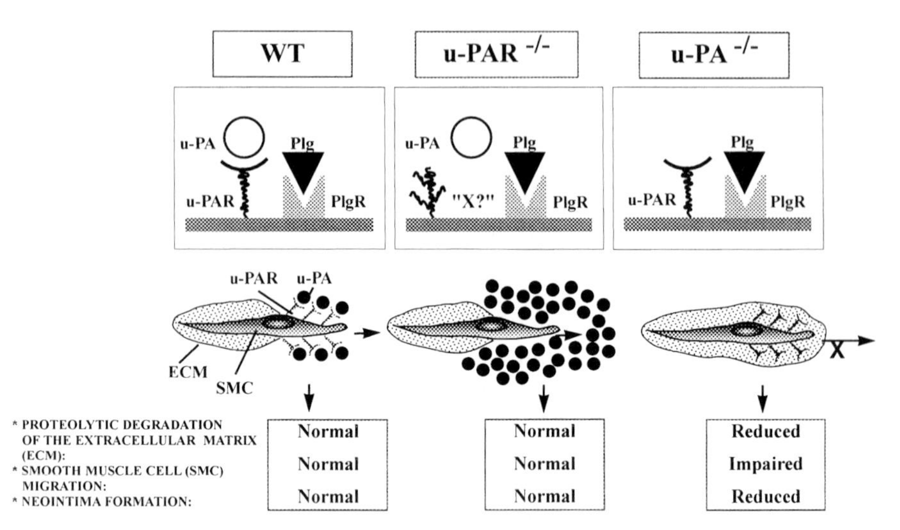

Fig. 7. Hypothetical model of the role of receptor-independent u-PA-mediated plasminogen activation in neointima formation. The smooth muscle cell (SMC) is surrounded by an extracellular matrix (ECM) that needs to be proteolytically degraded to allow cellular migration. In the wild type (WT) smooth muscle cells, u-PA is bound to u-PAR, mediating plasminogen activation and plasmic degradation of the extracellular matrix such that the cell can migrate. In the u-PAR deficient smooth muscle cells, u-PA becomes localized to the cell surface, possibly via interaction with other matrix molecules (denoted as "X?"), allowing sufficient pericellular plasmin proteolysis for the cells to migrate. u-PA might also accumulate to increased levels due to deficient u-PAR-mediated clearance. In contrast, smooth muscle cells that lack u-PA have reduced pericellular plasmin proteolysis and fail to migrate efficiently, resulting in reduced neointima formation. The proposed impairment of smooth muscle cell migration in mice lacking u-PA-mediated plasminogen activation is indirectly suggested by the observations that proliferation of u-PA deficient cells was similar as in wild type and that smooth muscle cells migrated over a shorter distance in the Plg deficient than in wild type arteries

than 1000-fold increased plasma PAI-1 levels and in a similar degree of inhibition of neointima formation as observed in u-PA deficient mice (CARMELIET et al. 1997c). Proteinase inhibitors have been suggested as anti-restenosis drugs. Our studies suggest that strategies aimed at reducing u-PA-mediated plasmin proteolysis may be beneficial for reduction or prevention of restenosis. However, antifibrinolytic strategies should be targeted at inhibiting plasmin proteolysis and not at preventing the interaction of u-PA with its receptor.

G. Atherosclerosis

Epidemiologic, genetic and molecular evidence suggests that impaired fibrinolysis resulting from increased PAI-1 or reduced t-PA expression or from inhibition of plasminogen activation may contribute to the development and/

or progression of atherosclerosis (WIMAN 1995; HAMSTEN and ERIKSSON 1994; JUHAN-VAGUE and COLLEN 1992; SCHNEIDERMAN et al. 1992), presumably by promoting thrombosis or matrix deposition. A possible role for increased plasmin proteolysis in atherosclerosis is, however, suggested by the enhanced expression of t-PA and u-PA in plaques (SCHNEIDERMAN et al. 1995; LUPU et al. 1995; GRAINGER et al. 1994). Plasmin proteolysis might indeed participate in plaque neovascularization, in induction of plaque rupture or in ulceration and formation of aneurysms (SAWA et al. 1992; CARMELIET and COLLEN 1998). A causative role of the plasminogen system in these processes has, however, not been conclusively demonstrated.

Atherosclerosis was studied in mice deficient in apolipoprotein E (apoE) and in t-PA, u-PA or PAI-1, and fed a cholesterol-rich diet for 5–25 weeks. No differences in the size or predilection site of plaques were observed between mice with a single deficiency of apoE or with a combined deficiency of apoE and t-PA or of apoE and u-PA. However, significant genotypic differences were observed in the destruction of the media with resultant erosion, transmedial ulceration, medial smooth muscle cell loss, dilatation of the vessel wall and microaneurysm formation (CARMELIET et al. 1997e). At the ultrastructural level, elastin fibers were eroded, fragmented and completely degraded, whereas collagen bundles and glycoprotein-rich matrix were totally disorganized and scattered. Whereas both apoE deficient and apoE:t-PA deficient mice developed severe media destruction, apoE:u-PA deficient mice were virtually completely protected. Plaque macrophages expressed abundant amounts of u-PA mRNA, antigen and activity at the base of the plaque and in the media, similar to in the atherosclerotic, aneurysmatic arteries in patients (SCHNEIDERMAN et al. 1995; LUPU et al. 1995). Macrophages cross the elastin fibers but only after proteolytic digestion of the elastin, a process that was remarkably enhanced by u-PA. u-PA-mediated plasmin increased degradation by macrophages of ^{3}H-fucose labeled glycoproteins (unpublished observations). Since glycoproteins surround elastin fibers in the aortic wall, their degradation will increase exposure of the highly insoluble elastin to elastases, facilitating thereby in vivo elastolysis. In addition, plasmin promoted degradation of elastin and collagen via activation of matrix metalloproteinases (MMPs) such as stromelysin-1 (MMP-3), gelatinase-B (MMP-9), the macrophage metalloelastase (MMP-12) and collagenase-3 (MMP-13) (CARMELIET et al. 1997e). The expression of several of these metalloproteinases was induced in situ in advanced atherosclerotic plaques by macrophages which also expressed increased amounts of u-PA. Taken together, these results suggest a role of u-PA in the structural integrity of the atherosclerotic vessel wall via triggering activation of metalloproteinases.

In contrast, mice with a combined deficiency of apoE and PAI-1 developed normal fatty streak lesions but, subsequently, revealed a transient delayed progression to fibroproliferative plaques. Whether the increased plasmin proteolytic balance in these mice might prevent matrix accumulation and, consequently, delay plaque progression, or whether more abundant plasmin increased activation of latent TGF-β1 with its pleiotropic role on smooth muscle cell function and matrix accumulation, remains to be determined. Taken together, these targeting studies identify a specific role for u-PA in the destruction of the media that may precede aneurysm formation, and for PAI-1 in plaque progression, possibly by promoting matrix deposition.

Lipoprotein(a) has been proposed to reduce plasminogen activation and to predispose to atherosclerosis by reducing clot lysis (LUI and LAWN 1994). Alternatively, reduced plasmin proteolysis could diminish activation of latent TGF-β1, providing hereby a growth stimulus for smooth muscle cells (GRAINGER et al. 1994). A significant correlation between high levels of apolipoprotein(a) and reduced in situ plasmin activity was observed in atherosclerotic vessels of transgenic mice overexpressing apolipoprotein(a) (GRAINGER et al. 1994). Ongoing experiments in mice deficient in t-PA, u-PA, PAI-1, u-PAR or plasminogen will demonstrate whether plasmin is a significant activator of latent TGF-β1 in vivo.

H. Tissue Remodeling Associated with Wound Healing

Impaired fibrinolysis, resulting from reduced u-PA or increased PAI-1 activity, has been implicated in the deposition of fibrin and of extracellular matrix components in the kidney and the lung during inflammation (BERTOZZI et al. 1990; TOMOOKA et al. 1992). Electron microscopic analysis demonstrated that adult combined t-PA : u-PA deficient mice developed fibrin deposition not only in the intravascular lumen but also in extravascular compartments such as in the alveoli of the lung, the mesangium in the kidney, and the subendothelial space of Disse in the liver (Carmeliet et al., unpublished observations). Furthermore, severe tissue remodeling such as fusion of the podocytes in the glomerulus and endothelial cell necrosis in adjacent capillaries were frequently observed. Notably, extravascular fibrin deposition appeared to precede intravascular thrombosis, possibly suggesting that the triggering event for abnormal fibrin deposition and tissue remodeling is located in the extracellular compartment. These pathological findings are reminiscent of those observed in glomerulonephritis and in the acute respiratory distress syndrome in patients (BERTOZZI et al. 1990; TOMOOKA et al. 1992). The involvement of the plasminogen system in inflammation and wound healing was further extended by the observations that plasminogen deficient and combined t-PA : u-PA deficient mice, and to a lesser extent t-PA deficient or u-PA deficient mice, suffered severe experimental glomerulonephritis, characterized by increased formation of fibrin-rich glomerular crescents after challenge with antiglomerular membrane antibodies (KITCHING et al. 1997). In addition, PAI-1 overexpressing mice suffered a more severe lung injury and deposition of fibrin and collagen-rich matrix after bleomycin challenge (EITZMAN et al. 1996) or hyperoxia (BARAZZONE et al. 1996), whereas PAI-1 deficient (EITZMAN et al. 1996) or α_2-macroglobulin deficient (UMANS et al. 1995) mice were protected against such fibrotic reaction. Deficiency of α_2-macroglobulin also increased the mortality associated with experimentally induced acute pancreatitis, possibly because of uncontrolled proteolysis (UMANS et al. 1995). Plg deficient mice also displayed fibrin-rich gastric ulcerations, in association with infection by pathogenic *Helicobacter* (PLOPLIS et al. 1995; BUGGE et al. 1995). In addition, they suffered delayed and impaired closure of skin wounds (ROMER et al. 1996). Notably, keratinocyte migration appeared to be reduced but, surprisingly, the granulation tissue was normal except for the more abundant presence of fibrin(ogen) and fibronectin at the wound edges (ROMER et al. 1996). In fact, Plg deficient mice, like their wild-type controls, had abundant infiltration of macrophages, neutrophilic granulocytes and fibroblast-like cells, and pronounced neovascularization. However, wound healing in combined Plg:fibrinogen deficient mice was not impaired, indicating that fibrin mediates, to a large degree, the effects of plasminogen deficiency (BUGGE et al. 1996c). Taken together, the plasminogen system appears to play a significant role in the tissue remodeling during wound healing, in part mediated by its role in fibrin surveillance. This notion is supported by the observation that fibrinogen deficient mice had an unusual wound healing response in which the migrating and proliferating cells (primarily fibroblasts) form a thick layer *encapsulating* but not *infiltrating* hematomas (SUH et al. 1995). It is thus possible that fibrin provides a critical initial matrix for the movement of cells into sites of injury.

Different degrees of wound healing responses have been reported, depending on the environmental conditions and infectious challenges. The most significant phenotype occurred in u-PA deficient mice after infection with botryomycosis (SHAPIRO et al. 1997). In contrast to their wild-type littermates, housed in the same environmental conditions, u-PA deficient mice developed a suppurative infection of the skin characterized by the presence of abscesses and granulomas, containing large numbers of polymorphonuclear leukocytes and histiocytes that were surrounded by a capsule of fibrous connective tissue. Such destructive tissue remodeling is indeed more severe than observed in combined t-PA:u-PA deficient or Plg deficient mice (CARMELIET et al. 1994; PLOPLIS et al. 1995), indicating that the phenotypes observed in these knockout mice are determined to a large extent by the infectious or inflammatory challenge.

Despite a well-documented expression pattern of the fibrinolytic system during certain (patho)biological processes, experimental studies have not always confirmed its relevance in vivo. A biological process in which knockout studies have thus far failed to demonstrate a significant role of the plasminogen system is bone remodeling. Plasmin proteolysis might facilitate bone resorption via activation of latent collagenases or TGF-β, or by freeing insulin-like growth factor-1 from its inhibitory binding proteins, or by promoting osteoclast migration (MARTIN et al. 1993). Nevertheless, bone resorption in cultured fetal metatarsals and calvariae from t-PA or u-PA deficient mice was not different from that in wild-type mice (LELOUP et al. 1995). Bone turnover was slightly increased but total body calcium content and radiographic examination of the skeleton in young and old t-PA deficient mice were normal; only a minor degree of cancellous osteopenia (as revealed by dynamic bone histomorphometric analysis of the tibial metaphysis) was observed (BOUILLON et al. 1995). In aggregate, although the present results have not revealed an absolute requirement of the plasminogen system for normal bone remodeling or corneal wound healing, a more subtle role of this system remains possible. It should be noted, however, that lack of an obvious phenotype in a knockout mouse does not exclude a possible role for the target gene during conditions of ectopic or abnormal expression. It is also possible that the knockout mouse may have adapted to the deficiency of the target gene by alternative mechanisms. However, thus far, compensatory upregulation of the residual plasminogen activators has not been observed (CONNOLLY et al. 1996; WILHELM et al. 1994).

I. Conclusions

Studies with transgenic mice over- or underexpressing components of the coagulation or fibrinolytic system not only confirmed the significant role of these proteinase systems in hemostasis and fibrin clot surveillance but also revealed novel insights into the precise role and interaction of the individual molecules. The coagulation and not the fibrinolytic system appeared to play a more essential role in embryonic development than anticipated. Although life without plasminogen is possible, health and survival are severely compromised. Both systems are involved in infection, inflammation and wound healing such as in arterial neointima formation, glomerulonephritis, skin ulcerations, pancreatitis and lung inflammation. Unexpectedly, these studies have revealed a more negligible role of the plasminogen system in bone remodeling and blood vessel development, although this may be due to the type of challenge or analysis used to study these phenotypes. Furthermore, lack of an appreciable effect of a specific gene deletion does not rule out its possible involvement in pathological processes when inappropriately expressed and warrants examination for compensatory mechanisms. These transgenic mice may not only provide suitable models for further elucidation of the relevance of the plasminogen system in other (patho)physiological processes such as atherosclerosis or malignancy, but may also serve as models for the evaluation of new (gene) therapies.

Acknowledgements. The authors are grateful to the members of the Center for Transgene Technology and Gene Therapy and to all external collaborators for their contribution to these studies and to M. Deprez for the artwork.

References

al-Mondhiry H, Ehmann WC (1994) Congenital afibrinogenemia. Am J Hematol 46:343–347

Altieri DC (1995) Xa receptor EPR-1. FASEB J 9:860–865
Andreasen PA, Sottrup-Jensen Ll, Kjoller L, Nykjaer A, Moestrup SK, Petersen CM et al (1994) Receptor-mediated endocytosis of plasminogen activators and activator/inhibitor complexes. FEBS Lett 338:239–245
Aoki N (1989) Hemostasis associated with abnormalities of fibrinolysis. Blood Rev 3:11–17
Astrup T (1978) In: Davidson JF, Rowan RM, Samama MM, Desnoyers PC (eds) Progress in chemical fibrinolysis and thrombolysis, vol 3. Raven, New York, pp 1–57
Bachmann F (1995) The enigma PAI-2. Gene expression, evolutionary and functional aspects. Thromb Haemost 74:172–179
Barazzone C, Belin D, Piguet PF, Vassalli JD, Sappino AP (1996) Plasminogen activator inhibitor-1 in acute hyperoxic mouse lung injury. J Clin Invest 98:2666–2673
Bertozzi P, Astedt B, Zenzius L, Lynch K, LeMaire F, Zapoli W et al (1990) Depressed bronchoalveolar urokinase activity in patients with adult respiratory distress syndrome. N Engl J Med 322:890–897
Bi L, Lawler AM, Antonarakis SE, High KA, Gearhart JD, Kazazaian HH Jr (1995) Targeted disruption of the mouse factor VIII gene produces a model of haemophilia A. Nat Genet 10:119–121
Bick RL, Pegram M (1994) Syndromes of hypercoagulability and thrombosis: a review. Semin Thromb Hemost 20:109–132
Blasi F, Conese M, Moller LB, Pedersen N, Cavallaro U, Cubellis MV et al (1994) The urokinase receptor: structure, regulation and inhibitor-mediated internalization. Fibrinolysis 8 [Suppl 1]:182–188
Bolton-Maggs PH, Hill FG (1995) The rarer inherited coagulation disorders: a review. Blood Rev 9:65–76
Bouillon R, Van Herck E, Verhaeghe J, Carmeliet P (1995) Bone metabolism in transgenic mice, deficient in tissue type plasminogen activator. Xth international congress on calcium regulatory hormones, Feb 1995, Melbourne, Australia, Bone
Broze GJ Jr (1992) Tissue factor pathway inhibitor and the revised hypothesis of blood coagulation. Trends Cardiovasc Med 2:72–77
Bugge TH, Suh TT, Flick MJ, Daugherty CC, Romer J, Solberg H et al (1995a) The receptor for urokinase-type plasminogen activator is not essential for mouse development or fertility. J Biol Chem 270:16886–16894
Bugge TH, Flick MJ, Daugherty CC, Degen JL (1995b) Plasminogen deficiency causes severe thrombosis but is compatible with development and reproduction. Genes Dev 9:794–807
Bugge TH, Xiao Q, Kombrinck K, Flick MJ, Holmback K, Danton (1996a) Fatal embryonic bleeding events in mice lacking tissue factor, the cell-associated initiator of blood coagulation. Proc Natl Acad Sci USA 93:6258–6263
Bugge TH, Flick MJ, Danton MJ, Daugherty CC, Romer J, Dano K et al (1996b) Urokinase-type plasminogen activator is effective in fibrin clearance in the absence of its receptor or tissue-type plasminogen activator. Proc Natl Acad Sci USA 93:5899–5904
Bugge TH, Kombrinck KW, Flick MJ, Daugherty CC, Danton MJS, Degen JD (1996c) Loss of fibrinogen rescues mice from the pleiotropic effects of plasminogen deficiency. Cell 87:709–719
Carmeliet P, Collen D (1998) Physiological consequences of over- or under-expression of fibrinolytic system components in transgenic mice. In: Vadas M, Harlan J (eds) Vascular control of hemostasis. Adv Vasc Biol (in press)
Carmeliet P, Kieckens L, Schoonjans L, Ream B, Van Nuffelen A, Prendergast G et al (1993a) Plasminogen activator inhibitor-1 gene-deficient mice. I. Generation by homologous recombination and characterization. J Clin Invest 92:2746–2755

Carmeliet P, Stassen JM, Schoonjans L, Ream B, van den Oord JJ, De Mol M et al (1993b) Plasminogen activator inhibitor-1 gene-deficient mice. II. Effects on hemostasis, thrombosis and thrombolysis. J Clin Invest 92:2756–2760

Carmeliet P, Schoonjans L, Kieckens L, Ream B, Degen J, Bronson R et al (1994) Physiological consequences of loss of plasminogen activator gene function in mice. Nature 368:419–424

Carmeliet P, Bouché A, De Clercq C, Janssen S, Pollefeyt S, Wijns S et al (1995) Biological effects of disruption of the tissue-type plasminogen activator, urokinase-type plasminogen activator, and plasminogen activator inhibitor-1 genes in mice. Ann NY Acad Sci 748:367–381

Carmeliet P, Ferreira V, Breier G, Pollefeyt S, Kieckens L, Gertsenstein M et al (1996a) Abnormal blood vessel development and lethality in embryos lacking a single VEGF allele. Nature 380:435–439

Carmeliet P, Mackman N, Moons L, Luther T, Gressens P, Van Vlaenderen I et al (1996b) Role of tissue factor in embryonic blood vessel development. Nature 383:73–75

Carmeliet P, Moons L, Stassen JM, DeMoi M, Bouche A, van den Oord et al (1997a) Vascular wound healing and neointima formation induced by perivascular electric injury in mice. Am J Pathol 150:761–777

Carmeliet P, Stassen JM, Collen D, Meidell R, Gerard R (1997b) Adenovirus-mediated gene transfer of rt-PA restores thrombolysis in t-PA deficient mice (submitted)

Carmeliet P, Moons L, Ploplis V, Plow E, Collen D (1997c) Impaired arterial neointima formation in mice with disruption plasminogen gene. J Clin Invest 99:200–208

Carmeliet P, Moons L, Lijnen R, et al. (1997d) Inhibitory role of plasminogen activator inhibitor-1 in arterial wound healing and neointima formation: a gene targeting and gene transfer study in mice. Circulation 96:3180–3191

Carmeliet P, Moons L, Dewerchin M, et al. (1997e) Insights in vessel development and vascular disorders using targeted inactivation and transfer of vascular endothelial growth factor, the tissue factor receptor, and the plasminogen system. Ann NY Acad Sci 811:191–206

Carmeliet P, Moons L, Dewerchin M, Stassen JM, Declercq C, Gerard R et al (1998) Receptor-independent role of urokinase-type plasminogen activator in arterial neointima formation in mice (submitted)

Clowes AW, Reidy MA (1991) Prevention of stenosis after vascular reconstruction: pharmacologic control of intimal hyperplasia – a review. J Vasc Sur 13:885–891

Clowes AW, Clowes MM, Ay YP, Reidy MA, Belin D (1990) Smooth muscle cells express urokinase during mitogenesis and tissue-type plasminogen activator during migration in injured rat carotid artery. Circ Res 67:61–67

Collen D, Lijnen HR (1991) Basic and clinical aspects of fibrinolysis and thrombolysis. Blood 78:3114–3124

Connolly AJ, Ishihara H, Kahn ML, Farese RV Jr, Coughlin SR (1996) Role of the thrombin receptor in development and evidence for a second receptor. Nature 381:516–519

Coughlin SR (1994) Molecular mechanisms of thrombin signaling. Semin Hematol 31:270–277

Cui J, Saunders TL, Ginsburg D (1995) Analysis of factor V function by gene targeting in embryonic stem cells. Blood 86:449a

Cui J, O'Shea KS, Purkayastha A, Saunders TL, Ginsburg D (1996) Fatal haemorrhage and incomplete block to embryogenesis in mice lacking coagulation factor V. Nature 384:66–68

Dahlbäck B (1995) New molecular insights into the genetics of thrombophilia. Resistance to activated protein C caused by Arg506 to Gln mutation in factor V as a pathogenic risk factor for venous thrombosis. Thromb Haemost 74:139–148

Dano K, Behrendt N, Brünner N, Ellis V, Ploug M, Pyke C (1994) The urokinase receptor. Protein structure and role in plasminogen activation and cancer invasion. Fibrinolysis 8 [Suppl 1]:189–203
Davie E (1995) Biochemical and molecular aspects of the coagulation cascade. Thromb Haemost 74:1–6
Dewerchin M, Van Nuffelen A, Wallays G, Bouché A, Moons L, Carmeliet P et al (1996) Generation and characterization of urokinase receptor deficient mice. J Clin Invest 97:870–878
Dickson MC, Martin JS, Cousins FM, Kulkarni AB, Karsson S, Akhurst RJ (1995) Defective haematopoiesis and vasculogenesis in transforming growth factor-ß1 knock out mice. Development 121:1845–1854
Dougherty K, Yang A, Harris J, Saunders T, Camper S, Ginsburg D (1995) Targeted deletion of the murine plasminogen activator inhibitor-2 (PAI-2) gene by homologous recombination. Blood 86:455a
Edgington TS, Mackman N, Brand K, Ruf W (1991) The structural biology of expression and function of tissue factor. Thromb Haemost 66:67–79
Eitzman DT, McCoy RD, Zheng X, Fay WP, Shen T, Ginsburg D (1996) Bleomycin-induced pulmonary fibrosis in transgenic mice that either lack or overexpress the murine plasminogen activator inhibitor-1 gene. J Clin Invest 97:232–237
Elliott MJ, Faulkner-Jones BE, Stanton H, Hamilton JA, Metcalf D (1992) Plasminogen activator in granulocyte-macrophage-CSF transgenic mice. J Immunol 149:3678–3681
Erickson LA, Fici GJ, Lund JE, Boyle TP, Polites HG, Marotti KR (1990) Development of venous occlusions in mice transgenic for the plasminogen activator inhibitor-1 gene. Nature 346:74–76
Esmon CT (1992) The protein C anticoagulant pathway. Arterioscler Thromb 12:135–145
Fay WP, Shapiro AD, Shih JL, Schleef RR, Ginsburg D (1992) Brief report: complete deficiency of plasminogen-activator inhibitor type 1 due to a frameshift mutation. N Engl J Med 327:1729–1733
Fong GH, Rossant J, Breitman ML (1995) Role of the Flt-1 receptor tyrosine kinase in regulating the assembly of vascular endothelium. Nature 376:66–70
Ford VA, Wilkinson JE, Kennel SJ (1993) Thrombomodulin distribution during murine development. Roux Arch Dev Biol 202:364–370
Forrester JS, Fishbein M, Helfant R, Fagin J (1991) A paradigm for restenosis based on cell biology: clues for the development of new preventive therapies. J Am Coll Cardiol 17:758–769
Furie B, Furie BC (1988) The molecular basis of blood coagulation. Cell 53:505–518
Gailani D, Laskey N, Broze GJ (1996) A murine model of factor XI deficiency. Blood 88:469a
Grainger DJ, Kemp PR, Liu AC, Lawn RM, Metcalfe JM (1994) Activation of transforming growth factor-β is inhibited in transgenic apolipoprotein (a) mice. Nature 370:460–462
Hajjar KA (1995) Cellular receptors in the regulation of plasmin generation. Thromb Haemost 74:29–301
Hamsten A, Eriksson P (1994) Fibrinolysis and atherosclerosis: update. Fibrinolysis 8:253–262
Hamsten A, de Faire U, Walldius G, Dahlen G, Szamosi A, Landou C et al (1987) Plasminogen activator inhibitor in plasma: risk factor for recurrent myocardial infarction. Lancet 2:3–9
Healy AM, Rayburn H, Rosenberg RD, Weiler H (1995) Absence of the blood-clotting regulator thrombomodulin causes embryonic lethality in mice before development of a functional cardiovascular system. Proc Natl Acad Sci USA 92:850–854
Heckel JL, Sandgren EP, Degen JL, Palmiter RD, Brinster RL (1990) Neonatal bleeding in transgenic mice expressing urokinase-type plasminogen activator. Cell 62:447–456

Herbert JM, Lamarche I, Prabonnaud V, Dol F, Gauthier T (1994) Tissue-type plasminogen activator is a potent mitogen for human aortic smooth muscle cells. J Biol Chem 269:3076–3080

Herz J, Clouthier DE, Hammer RE (1992) LDL receptor-related protein internalizes and degrades uPA:PAI-1 complexes and is essential for embryo implantation. Cell 71:411–421

Hoyer LW (1996) Hemophilia A. N Engl J Med 330:38–47

Jackson CL, Reidy MA (1992) The role of plasminogen activation in smooth muscle cell migration after arterial injury. Ann NY Acad Sci 667:141–150

Jackson CL, Raines EW, Ross R, Reidy MA (1993) Role of endogenous platelet-derived growth factor in arterial smooth muscle cell migration after balloon catheter injury. Arterioscler Thromb 13:1218–1226

Juhan-Vague I, Collen D (1992) On the role of coagulation and fibrinolysis in atherosclerosis. Ann Epidemiol 2:427–438

Kitching AR, Holdsworth SR, Ploplis VA, Plow EF, Collen D, Carmeliet P et al (1997) Plasminogen and plasminogen activators protect against renal injury in crescentic glomerulonephritis. J Exp Med 185:963–968

Kluft C, Dooijewaard G, Emeis JJ (1987) Role of the contact system in fibrinolysis. Semin Thromb Hemost 13:50–68

Lawrence DA, Ginsburg D (1995) Plasminogen activator inhibitors. In: High KA, Roberts HR (eds) Molecular basis of thrombosis and hemostasis. Dekker, New York, pp 517–543

Lee SW, Kahn ML, Dichek DA (1992) Expression of an anchored urokinase in the apical endothelial cell membrane. Preservation of enzymatic activity and enhancement of cells surface plasminogen activation. J Biol Chem 267:13020–13029

Leloup G, Lemoine P, Carmeliet P, Vaes G (1995) Bone resorption and response to parathyroid hormone (PTH) or 1,25 dihyroxyvitamin D3 (1,25-D) in fetal metatarsals and calvariae from transgenic mice devoid of tissue (tPA)or urokinase (uPA) plasminogen activator or of their inhibitor, PAI-1. Calcif Tissue Int 56:442

Leonardsson G, Peng XR, Liu K, Nordström L, Carmeliet P, Mulligan R et al (1995) Ovulation efficiency is reduced in mice that lack plasminogen activator gene function: functional redundancy among physiological plasminogen activators. Proc Natl Acad Sci USA 92:12446–12500

Levéen P, Pekny M, Gebre-Medhin S, Swolin B, Larsson E, Betsholtz C (1994) Mice deficient for PDGF B show renal, cardiovascular, and hematological abnormalities. Genes Dev 8:1875–1887

Libby P, Schwartz D, Brogi E, Tanaka H, Clinton SK (1992) A cascade model for restenosis. A special case of atherosclerosis progression. Circulation 86(III):47–52

Liu AC, Lawn RM (1994) Lipoprotein(a) and atherogenesis. Trends Cardiovasc Med 4:40–44

Lupu F, Heim DA, Bachmann F, Hurni M, Kakkar VV, Kruithof EKO (1995) Plasminogen activator expression in human atherosclerotic lesions. Arterioscler Thromb Vasc Biol 15:1444–1455

Luther T, Flössel C, Mackman N, Bierhaus A, Kasper M, Albrecht S et al (1996) Tissue factor expression during human and mouse development. Am J Pathol 149:101–113

Martin T, Allan EH, Fukumoto S (1993) The plasminogen activator and inhibitor system in bone remodelling. Growth Regul 3:209–214

Montgomery RR, Scott JP (1993) Hemostasis: diseases of the fluid phase. In: Nathan GD, Oski FA (eds) Hematology of infancy and childhood. Saunders, Philadelphia, pp 1605–1650

Mulligan RC (1993) The basic science of gene therapy. Science 260:926–932

Nagy A (1998) Engineering the mouse genome. Methods Enzymol (in press)

Nagy A, Rossant J (1996) Targeted mutagenesis: analysis of phenotype without germline transmission. J Clin Invest 97:1360–1365

Nelhs V, Drenckhan D (1993) The versatility of microvascular pericytes: from mesenchyme to smooth muscle? Histochemistry 99:1–12
Palabrica TM, Liu AC, Aronowitz MJ, Furie B, Lawn RM, Furie BC (1995) Antifibrinolytic activity of apolipoprotein(a) in vivo: human apolipoprotein(a) transgenic mice are resistant to tissue plasminogen activator-mediated thrombolysis. Nat Med 1:256–259
Ploplis V, Carmeliet P, Vazirzadeh S, Van Vlaenderen I, Moons L, Plow E et al (1995) Effects of disruption of the plasminogen gene on thrombosis, growth, and health in mice. Circulation 92:2585–2593
Plow EF, Edgington TS (1975) An alternative pathway for fibrinolysis. I. The cleavage of fibrinogen by leukocyte proteases at physiologic pH. J Clin Invest 56:30–38
Plow EF, Herren T, Redlitz A, Miles LA, Hoover-Plow JL (1995) The cell biology of the plasminogen system. FASEB J 9:939–945
Richards WG, Carroll PM, Kinloch RA, Wassarman PM, Strickland S (1993) Creating maternal effect mutations in transgenic mice: antisense inhibition of an oocyte gene product. Dev Biol 160:543–553
Robbins KC (1988) Dysplasminogenemias. Enzyme 40:70–78
Romer J, Bugge TH, Pyke C, Lund LR, Flick MJ, Degen JL et al (1996) Impaired wound healing in mice with a disrupted plasminogen gene. Nat Med 2:287–292
Rosen ED, Chan JC, Idusogie E, et al. (1997) Mice lacking factor VII develop normally but suffer fatal perinatal bleeding. Nature 390:290–294
Rottingen JA, Enden T, Camerer E, Iversen JG, Prydz H (1995) Binding of human factor VIIa to tissue factor induces cytosolic Ca_{2+} signals in J82 cells, transfected COS-1 cells, Madin-Darby canine kidney cells and in human endothelial cells induced to synthesize tissue factor. J Biol Chem 270:4650–4660
Saksela O, Rifkin DB (1988) Cell-associated plasminogen activation: regulation and physiological functions. Annu Rev Cell Biol 4:93–126
Sawa H, Fujii S, Sobel BE (1992) Augmented arterial wall expression of type-1 plasminogen activator inhibitor induced by thrombosis. Arterioscler Thromb 12:1507–1515
Schneider MD, French BA (1993) The advent of adenovirus. Gene therapy for cardiovascular disease. Circulation 88:1937–1942
Schneiderman J, Loskutoff DJ (1991) Plasminogen activator inhibitors. Trends Cardiovasc Med 1:99–102
Schneiderman J, Sawdey MS, Keeton MR, Bordin GM, Bernstein EF, Dilley RB et al (1992) Increased type 1 plasminogen activator inhibitor gene expression in athero sclerotic human arteries. Proc Natl Acad Sci USA 89:6998–7002
Schneiderman J, Bordin GM, Engelberg I, Adar R, Seiffert D, Thinnes T (1995) Expression of fibrinolytic genes in atherosclerotic abdominal aortic aneurysm wall. A possible mechanism for aneurysm expansion. J Clin Invest 96:639–645
Shalaby F, Rossant J, Yamaguchi TP, Oertsenstein M, Wu XF, Bretiman ML et al (1995) Failure of blood island formation and vasculogenesis and haematopoiesis in Flk-1-deficient mice. Nature 376:62–66
Shapiro RL, Duquette JG, Nunes I, Roses DF, Harris MN, Wilson EL et al (1997) Urokinase-type plasminogen activator deficient mice are predisposed to staphylococcal botryomycosis, pleuritis and effacement of lymphoid follicles. Am J Pathol 150:359–369
Sidenius N (1993) Expression of the aminoterminal fragment of urokinase-type plasminogen activator in transgenic mice. PhD thesis, University of Copenhagen, Denmark
Simpson AJ, Booth NA, Moore NR, Bennett B (1991) Distribution of plasminogen activator inhibitor (PAI-1) in tissues. J Clin Pathol 44:139–143
Soifer SJ, Peters KG, O'Keefe J, Coughlin SR (1994) Disparate temporal expression of the prothrombin and thrombin receptor genes during mouse development. Am J Pathol 144:60–69

Stefansson S, Lawrence DA (1996) The serpin PAI-1 inhibits cell migration by blocking integrin $\alpha_v\beta_3$ binding to vitronectin. Nature 383:441–443

Stephens RW, Bokman AM, Myöhänen HT, Reisberg T, Tapiovaara H, Pedersen N et al (1992) Heparin binding to the urokinase kringle domain. Biochemistry 31:7572–7579

Suh TT, Holmbäck K, Jensen NJ, Daugherty CC, Small K, Simon DI et al (1995) Resolution of spontaneous bleeding events but failure of pregnancy in fibrinogen-deficient mice. Genes Dev 9:2020–2033

Suri C, Jones PF, Patan S, Bartunkova S, Maisonpierre PC, Davis S et al (1996). Requisite role of angiopoietin-1, a ligand for the TIE2 receptor, during embryonic angiogensis. Cell 87:1171–1180

Tomooka S, Border WA, Marshall BC, Noble NA (1992) Glomerular matrix accumulation is linked to inhibition of the plasmin protease system. Kidney Int 42:1462–1469

Toomey JR, Kratzer KE, Lasky NM, Stanton JJ, Broze GJ Jr (1996) Targeted disruption of the murine tissue factor gene results in embryonic lethality. Blood 88:1583–1587

Tuddenham EG, Pemberton S, Cooper DN (1995) Inherited factor VII deficiency: genetics and molecular pathology. Thromb Haemost 74:313–21

Umans L, Serneels L, Overbergh L, Lorent K, Van Leuven F, Van den Berghe H (1995) Targeted inactivation of the mouse alpha2-macroglobulin gene. J Biol Chem 270:19778–19785

Vassalli JD (1994) The urokinase receptor. Fibrinolysis 8 [Suppl 1]:172–181

Vassalli JD, Sappino AP, Belin D (1991) The plasminogen activator/plasmin system. J Clin Invest 88:1067–1072

Wei Y, Lukashev M, Simon DI, Bodary SC, Rosenberg S, Doyle MV et al (1996) Regulation of integrin function by the urokinase receptor. Science 273:1551–1555

Weiler-Guettler H, Aird WC, Husain M, Rayburn H, Rosenberg RD (1996) Targeting of transgene expression to the vascular endothelium of mice by homologous recombination at the thrombomodulin locus. Circ Res 78:80–187

White GC II (1995) Coagulation factors V and VIII: normal function and clinical disorders. In: Handin RI, Lux SE, Stossel TP (eds) Blood: principles and practice of hematology. Lippincott, Philadelphia, pp 1151–1179

Wilhelm O, Weidle U, Hohl S, Rettenberger P, Schmitt M, Graeff H (1994) Recombinant soluble urokinase receptor as a scavenger for urokinase-type plasminogen activator (UPA). Inhibition of proliferation and invasion of human ovarian cancer cells. FEBS Lett 337:131–134

Wiman B (1995) Plasminogen activator inhibitor 1 (PAI-1) in plasma: its role in thrombotic disease. Thromb Haemost 74:71–76

Yu HR, Schultz RM (1990) Relationship between secreted urokinase plasminogen activator activity and metastatic potential in murine B16 cells transfected with human urokinase sense and antisense genes. Cancer Res 50:7623–7633

Zheng X, Saunders TL, Camper SA, Samuelson LC, Ginsburg D (1995) Vitronectin is not essential for normal mammalian development and fertility. Proc Natl Acad Sci USA 92:12426–12430

CHAPTER 3

Epidemiology of Arterial and Venous Thrombosis

F. KEE

A. Epidemiology: Its Potential and Its Limitations

Epidemiology, at its heart, is a search for patterns in the distribution of disease that can offer pointers to etiology or viable preventive interventions. For most students, their first encounter with the subject involves a recounting of the contribution of John Snow in controlling the cholera epidemic in London in 1854. On the assumption that household sizes were the same, the risk of dying of cholera among those supplied by the Southwark and Vauxhall water company were more than eight times those supplied by Lambeth. His efforts were not without their critics, as was the case when William Farr, the first medical statistician to the General Register Office in London, made predictions about the abeyance of the 1837 smallpox epidemic. Pioneers like these were essentially making inferences about the origins of disease based on their observations of "ecological" associations. Modern epidemiology now draws upon a wider range of study designs to attempt to establish causal relationships. Essentially these designs fall into three broad groups: retrospective designs, as exemplified by case-control studies; prospective designs where a population or a risk factor distribution in that population is followed longitudinally for the occurrence of disease; and experimental designs, as typified by the randomized controlled trial.

Each has particular strengths and limitations. In most case-control studies, for example, the presence or absence of a risk factor is ascertained retrospectively and one is therefore never quite sure whether the exposure or the onset of disease occurred first. Aside from determining the correct temporal relationship, the epidemiologist must also attempt to rule out other explanations for a putative causal association, the three most important being chance, bias and confounding. Most journal editors now recommend that authors predetermine relevant effect sizes and the statistical power required to detect them. The study results can then be appraised for the presence or absence of significant bias or confounding.

One of the most common biases, sometimes known as susceptibility bias, arises when comparisons are made with inappropriate control groups. Clinicians will often choose a particular treatment for patients who are in substantially better (or worse) condition than the patients who are denied that treatment. When the results are compared with other treatments, the groups will differ in their baseline risk of the outcome of interest. Such a proposition, for example, has been suggested as an explanation for the increased risk of venous thromboembolism in users of third generation oral contraceptive pills. Confounding, on the other hand, arises when associations between a putative risk factor and disease actually arise from the correlation between both and a third factor. For example, if a randomly selected group of healthy men, who happened to be carrying a box of matches in their pockets, were followed up for 20 years and their incidence of lung cancer was compared with the general population, then the real effect on risk is more likely to be due to a smoking habit than to carriage of matches.

When the effects of chance, bias and confounding have been examined, the investigator must still apply judgement about the likelihood of a causative association. The following list is an adaptation of factors identified by HILL (1965) to assess the likely role of cause in an association:

Association or causation?
Adapted from Hill (1965) • Strength • Dose-response • Temporality • Consistency • Specificity • Plausibility • Reversibility

These are more a guide than absolute criteria and some are more important than others. What they cannot do, of course, is replace that vital ingredient, common sense. So although the randomized controlled trial can be designed to overcome the mitigation of chance, bias and confounding in the pursuit of cause, their generalizability to a relevant clinical population is still a fundamental predicate for their application to clinical practice. The modern megatrial must not become a technical "fix" and precision in the estimation of effect sizes must not overshadow the need for validity and scientific rigor.

B. The Epidemiological Study of Arterial and Venous Thrombosis

I. Arterial Thrombosis

Because of its higher incidence and case fatality and rather clearer diagnostic signs and symptoms, arterial thrombosis has been subject to much more comprehensive epidemiological inquiry than venous thrombosis over the last 30–40 years. Among the earliest landmark studies were those conducted by William Kannel and colleagues (1987) in Framingham and by Richard Doll and Hill among British doctors (DOLL and HILL 1954). Both of these prospective studies of well-defined and "well" populations were planned just after

World War II. They contributed much to the development of epidemiological method and to our understanding of basic risk factors. However, it was already known that there were wide international variations in coronary heart disease and stroke mortality rates. What was not certain, however, was whether such differences could be attributed to differing underlying incidence, the vagaries of death certification and case ascertainment or to varying case fatality. To address this the World Health Organization sponsored what is still perhaps the largest epidemiological study ever, the MONICA Project (multinational monitoring of trends and determinants in cardiovascular disease). Commencing in the early 1980s, and applying common case ascertainment and diagnostic criteria, this project has now provided detailed incidence and case fatality data for 38 populations from 21 countries (TUNSTALL-PEDOE et al. 1994). Recent results demonstrate that for the age group 35–64 years incidence of myocardial infarction varies 12-fold for men [from 915/100,000 in North Karelia (Finland) to 76/100,000 in Beijing (China)] and nearly 9-fold for women [from 256/100,000 in Glasgow (Scotland) to 30/100,000 in Catalonia (Spain)]. Although rates in much of the developed countries have declined substantially over the last 30 years, the rates in the United States are still twice those in Spain and five times those in Japan. In 1993 alone there were over 2 million hospitalizations in the United States and over 490,000 deaths from coronary heart disease.

Twenty-eight-day case fatality from myocardial infarction varied across the MONICA project from 37% to 81% for men and from 31% to 91% for women. On average, for men, for every coronary death there are at least 1.5 other nonfatal events. For women there was a significant inverse correlation between event and case-fatality rates, suggesting that nonfatal events were being missed in countries where event rates were low. (Event rates are derived from the incidence of fatal *and* nonfatal events; in countries where event rates are low it is more likely that, if there is any under ascertainment, then this will affect nonfatal events disproportionately; hence the denominator of any case fatality rate will be underestimated.)

In the case of arterial thrombosis, there is an intriguing and still unexplained tendency for thrombi to affect certain anatomical regions of the vascular tree with greater frequency than others. Undoubtedly the greatest burden of disease arises from thromboses affecting the coronary and cerebral circulations.

Roughly 30% of all deaths in men and 25% in women are due to coronary heart disease. While rates nationally have fallen over the last 25 years in the United Kingdom (for example, from approximately 450/100,000 in men to 300/100,000), there is still considerable variation in mortality rates both within and between countries. Within the United Kingdom, for example, the male rate varies from approximately 375/100,000 in Northern Ireland to 250/100,000 in southwest Thames (Fig. 1).

While rates in the United States and Australia are even lower than this, many countries in eastern Europe have experienced rising mortality over the

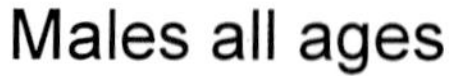

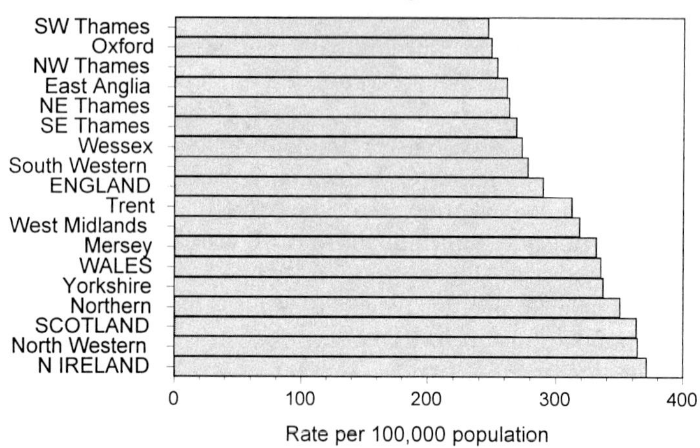

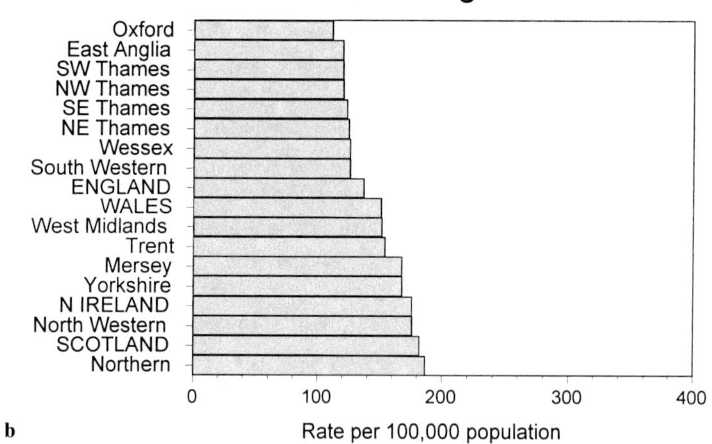

Fig. 1a,b. Standardised Mortality Rates* for Coronary Heart Disease *by sex, RHA and country United Kingdom, 1990.* (OPCS Mortality Statistics: area, Registrar General Northern Ireland and Registrar General Scotland)
*Rates calculated using the European Standard population

same period. However, even among western developed nations, the decline in death rates has not been so marked for that category variously classed as "sudden" or unexpected. The origin of these trends and the contribution of changing risk factors and improved therapeutic interventions are not entirely clear. Nevertheless, the morbidity from coronary artery thrombosis remains considerable. In the 1991 National Health Survey for England, for example,

4% of men and 1% of women reported having had a myocardial infarction while in 1990 it was estimated that coronary heart disease attracted about 4% of all National Health Service Expenditure excluding community health services (DEPARTMENT OF HEALTH, CENTRAL HEALTH MONITORING UNIT 1994).

Pathophysiologically, cerebral thrombosis constitutes the major underlying process in stroke. The incidence of first stroke is approximately 200/100,000/year, though the rate shows marked age variation ranging from 20/100,000 for age groups 55–64 and 2000/100,000 for those aged 85+ years, respectively. About 30% of patients die, most in the first 3 weeks. However, the community prevalence of stroke is approximately 600/100,000 and at least half of these will be physically or cognitively disabled (WADE 1994).

Substantial falls in the mortality from stroke have been occurring throughout most of the developed world, probably from before onset of the decline in coronary heart disease in the 1960s, the fall being somewhat greater among those under 65 years than among all ages combined. As with coronary thromboses, the origins of this decline are unclear. Though there are clearly some common risk factors for stroke and coronary heart disease, some paradoxes in the international pattern of mortality in Europe are evident. Portugal's high levels of stroke mortality contrast with lower levels of coronary heart disease (CHD) mortality (Fig. 2). Conversely, mortality from CHD is relatively high in countries like Ireland and Denmark where stroke mortality is low. It is estimated that in 1990 stroke attracted about 4% of the total National Health Service expenditure.

II. Venous Thrombosis

An epidemiological endeavor on a similar comprehensive scale has never been attempted for venous thrombosis. More often its study in well-defined population samples has been a by-product of a broader community health study such as that in Tecumseh in the early 1970s (COON et al. 1973).

The incidence of deep venous thrombosis in the community is variously reported to be between 60 and 180 symptomatic cases per 100,000/year (CARTER 1994). Clinical signs and symptoms are notoriously unreliable, however, and this is almost certainly an underestimate. Most of these episodes are subclinical and resolve. Others may produce venous valve damage or insufficiency, or propagate proximally. Such proximal thromboses are more likely to embolize to the lungs where they can produce no symptoms, mild or major clinical effects, or in severe cases, death. Typical findings such as leg pain and shortness of breath are absent in more than half of affected individuals including the majority of patients who die of PE. Population mortality rates will therefore be a somewhat unreliable indicator of the mortality burden. Nevertheless, pulmonary embolism remains one of the most important causes of hospital mortality, as borne out by the National Confidential Inquiry into Peri-

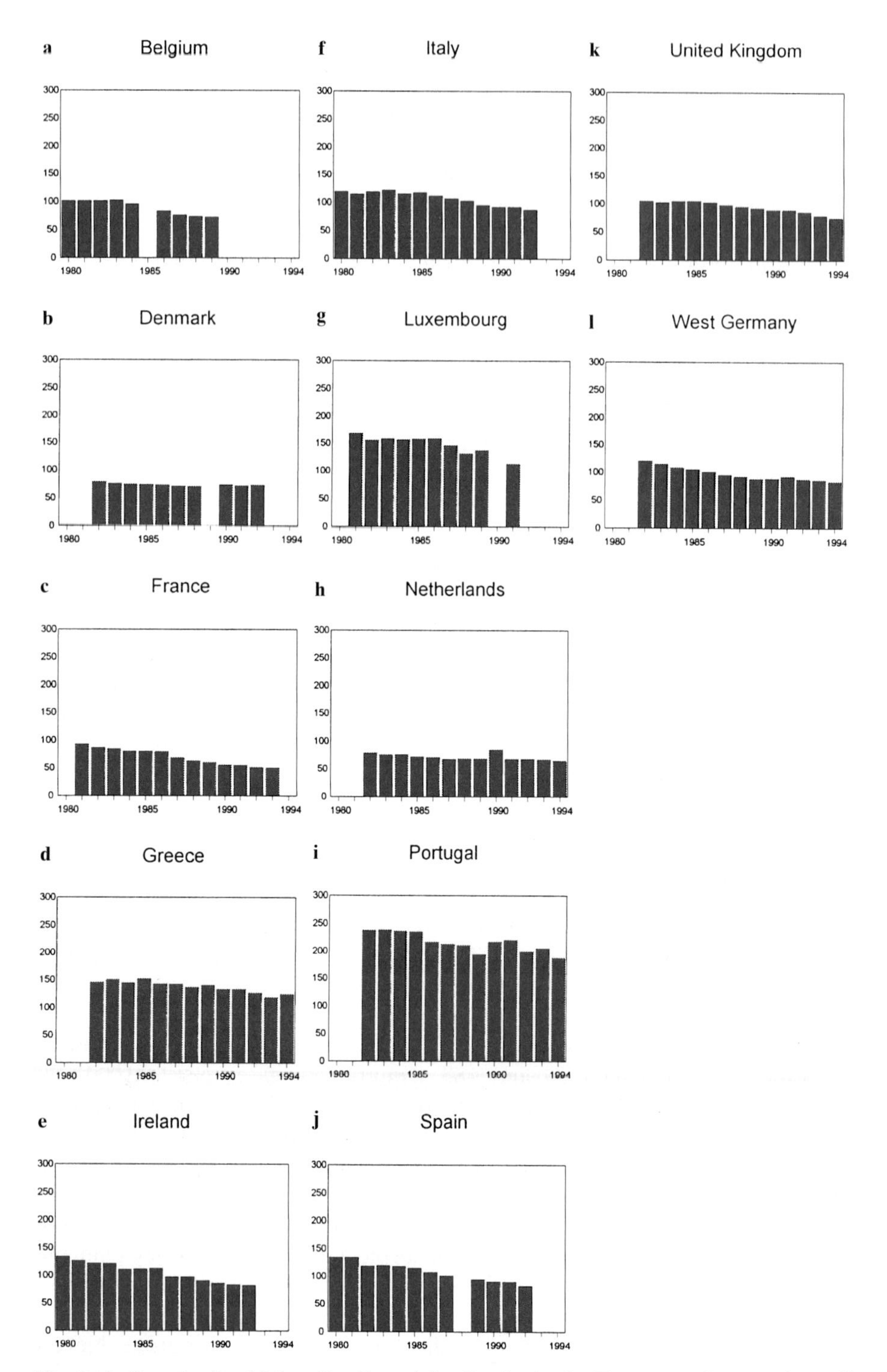

Fig. 2a–l. Standardised Mortality Rates* for Stroke in the European Community. *All persons, all ages, per 100000 population, 1980–1994* (WHO. World Health Statistics Annual 1980–1995)
*Rates calculated using the European Standard population

Operative Deaths and a number of independent necropsy studies. In one recent study from Sheffield (England), 10% of all hospital deaths were from pulmonary embolism (THRIFT CONSENSUS GROUP 1992). In the United Kingdom, the annual costs of the diagnosis and management of venous thrombosis and its sequelae amount to some £600m (HEALTHCARE MANAGEMENT 1993).

While over 100 years ago Virchow could deduce the importance of the interplay between blood flow, blood constitution and vessel wall abnormalities in the genesis of thrombus, rigorously applied epidemiological method in defined population settings has allowed the elucidation of a host of new risk factors for both arterial and venous thrombosis and has begun to cast light on novel pathophysiological processes. Those of major and of recent significance are described in the next section.

C. Risk Factors for Arterial Thrombosis

Traditional methods of epidemiological inquiry, involving surveillance networks such as the MONICA Project or well designed cohort studies, such as in Framingham or the British Doctors Study, have contributed enormously to our understanding of the three main risk factors for atherosclerosis and arterial thrombosis, namely smoking, hypertension and hypercholesterolemia. While individually these factors have large attributable risks for the population (Table 1), it would be naive to assume that we could thus "explain away" the problem. Many new insights are being gained from studies of vascular biology which, step by step, seem to be sketching in ever richer detail to Virchow's triad of disorders of the blood flow, of its constitution or of the vessel wall itself (FUSTER 1994).

Table 1. Attributable risk of coronary death in men and women

	Attributable risk
Men	
Cholesterol	24%
Smoking	40%
Diastolic blood pressure	31%
Body mass index	–
Social class	25%
Women	
Cholesterol	21%
Smoking	37%
Diastolic blood pressure	39%
Body mass index	–
Social class	32%

Source: ISLES et al. (1992)

Extending the frontiers still further will require an even closer alliance between basic scientists and epidemiologists. Perhaps the most obvious "new" paradigm emerges from the integration of the genetic and epidemiological approaches in a case-control study. In many instances, the "new biology" has not undermined but rather underpinned the strengths of the social medicine approach to public health. Ecclesiastes states that there is nothing new under the sun. Every year, it seems, a "new" risk factor for arterial thrombosis gains pre-eminence. Few of these turn out to be new per se but the four examples below illustrate how they are contributing ever finer degrees of detail to a complex web of causation.

I. Fibrinogen

It has now been consistently reported that elevated fibrinogen levels are associated with the incidence of ischemic heart disease (IHD) and probably also of stroke. These results have been recently summarized by MEADE (1995). In the Northwick Park study, for example, a difference of one standard deviation in fibrinogen was associated with a difference in IHD incidence of about 40% in an average follow-up period of 16 years (MEADE et al. 1993). It is now recognized that smoking is a leading cause of high fibrinogen levels and the latter appears to explain a large proportion of the effect of smoking on IHD. Several mechanisms probably underlie the fibrinogen-IHD association. Not only does it increase the blood and plasma viscosity and increase platelet aggregability but it may also contribute to the onset, development and stability of the atheromatous plaque. The fact that these predisposing effects are interrelated is powerfully suggested by the results of several large clinical trials and epidemiological studies. The beneficial effects of statin therapy (for hypercholesterolemia) are manifest early (within 2 years) before plaque regression could occur and thus could not be attributed solely to a decrease in low density cholesterol levels (VAUGHAN et al. 1996). Their early effects are more likely to be hemorrheological. Some of their known effects, for example, are to lower serum fibrinogen and reduce platelet aggregation.

There appear to be many "environmental" influences on levels of fibrinogen in the population, including obesity, the use of the oral contraceptive pill, exercise participation, alcohol intake, season, and of course, smoking. The known individual and environmental determinants probably account for little more than 20% of the population variance in fibrinogen levels but there is an increasing body of evidence that genetic factors are important, as well. For example, regarding a G/A^{-455} polymorphism of the β-fibrinogen gene, the difference in fibrinogen levels between genotype groups is up to 0.6 g/l (or equivalent to one standard deviation in the Northwick Park Study distribution) (GREEN and HUMPHRIES 1994). FOWKES et al (1992) found a higher frequency of the β-fibrinogen BclI 2-allele among patients with peripheral arterial disease compared with healthy controls and estimated that genetic variation accounted for 15% of phenotypic variance. This is almost certainly an underestimate, since HAMSTEN et al. for example, explained 51% of phenotypic variance in genetic terms in a path analysis (HAMSTEN et al. 1987). On the other hand, there was no difference in the genotype distribution between myocardial infarction *cases* and controls in the ECTIM study (Etude Cas-Temoin sur L'Infarctus Myocarde) (SCARABIN et al. 1993). Clearly, therefore, the particular phenotypic expression (for example, intermittent claudication versus myocardial infarction) and the age groups considered may affect the observed associations.

To a certain extent, the *gene* versus *environment* debate has its shelf-life limited by a growing literature on important gene–environment interactions. What is becoming more relevant to this debate, however, is the particular timing and definition of the critical environmental "exposures". Perhaps most controversial, in this regard, is the potential contribution of determinants in utero and in infancy.

II. Fetal-Infant Origins Hypothesis of Ischemic Heart Disease

Since 1987, Professor David Barker and others have elaborated an hypothesis in at least 40 papers and 2 books (BARKER 1992, 1994) that a baby's nourishment before birth and during infancy, as manifest in patterns of fetal, placental and infant growth, "programs" the development of risk factors such as raised blood pressure, fibrinogen concentration or insulin resistance which are key determinants of myocardial infarction risk. Typical of his study designs is a follow-up of 1157 men still living in east Hertfordshire who were born in the county between 1929 and 1930 and who had both birthweight and weight at 1 year recorded by the attending midwife (BARKER et al. 1992). Among 591 men who agreed to participate in the follow-up, mean fibrinogen concentrations measured in middle age fell with increasing weight at 1 year, a trend which was independent of cigarette smoking, alcohol consumption, body mass index and social class.

While none of the Barker studies to date has provided a direct measure of nutritional intake in mothers or babies, the hypothesis has an intriguing appeal. There have been several inconsistent findings, however, notably the different relationships for men and women (PANETH and SUSSER 1995). Low birth weight is associated with insulin resistance more strongly in women than in men while a low weight at 1 year predicts raised fibrinogen and apolipoprotein B levels in men but not in women. Furthermore, it is conceivable that body mass index is an intervening variable, particularly as regards the putative associations between birth weight and insulin resistance. So to adjust for current body mass index is to cancel out the positive effect of birth weight on body mass index (and thence on glucose intolerance), allowing the effect of birth weight in the direction favored by BARKER et al (1992) to remain unopposed.

Other issues which remain largely unaddressed by his studies to date include the lack of control for maternal smoking, the lack of similar findings for twins (VAGERO and LEON 1994) (who almost by definition experience growth retardation in utero), the possibility of selection bias (arising from incomplete follow-up) and the possibility of residual confounding by current social class. BRUNNER et al (1996), for example, in the Whitehall Civil Servants study, found that while height was inversely associated with fibrinogen (a relationship which suggests that in utero and childhood factors may exert long term influence), there remained independent associations both with the subject's father's social class and their own current socioeconomic condition. WILSON et al (1993), on the other hand, showed that Finnish men who were economically disadvantaged both as children and as adults had the highest fibrinogen concentrations. Furthermore, in a recent multicenter case control study from the European Atherosclerosis Research group, healthy students whose fathers had had premature myocardial infarction (before age 55 years) were shorter than age-matched controls (without a parental history), a relationship that was independent of birth weight and body mass index (KEE et al. 1997) (Fig. 3). Given the consistency of the finding across a wide spectrum of social backgrounds in these European populations, an *inherited* contribution to this difference is plausible. Further support for this contention derives from a recent twin study from Sweden, in which the difference in CHD mortality between height-discordant twin pairs was examined (VAGERO and LEON 1994). Although, overall, the shorter twin had 15% higher mortality risk than the taller twin, the linear trend of risk with height was statistically significant for monozygotic but not dizygotic twins (VANDENBROUCKE et al. 1996). Clearly,

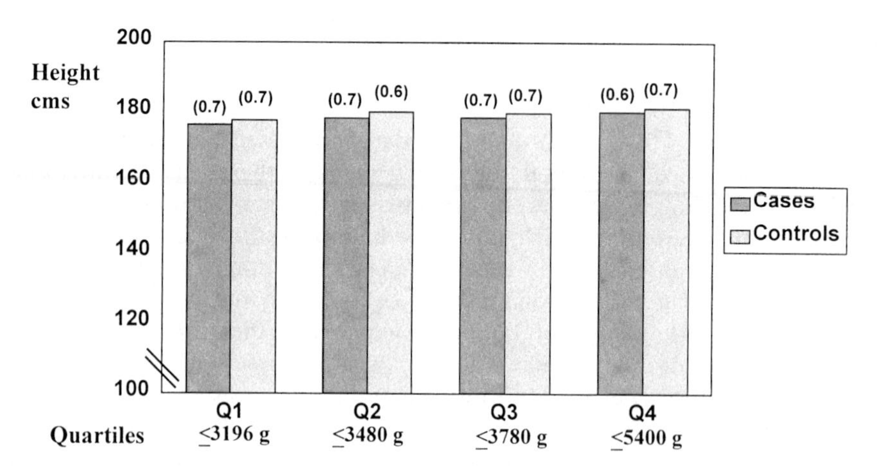

Fig. 3. Mean height (SEM) in cases and controls according to quartiles of birthweight in EARSII (adjusted for age and centre). Association between height and birthweight (cases and controls pooled): p = 0.0001. Heterogeneity of the case/control difference across quartiles of birthweight: NS (KEE et al. 1997)

the contribution of genetic factors to the fetal-origins "programming" hypothesis has yet to be fully clarified.

III. Insulin Resistance

Two Belfast physicians, STOUT and VALLANCE-OWEN, initially suggested a role for insulin in atherogenesis in 1969. REAVEN subsequently postulated that hyperinsulinemia and/or insulin resistance identifies subjects at high risk of cardiovascular disease and seem to cluster in these individuals with dyslipidemia and hypertension, for which he coined the term "syndrome X" (REAVEN 1988). Indeed, in several population studies hyperinsulinemia has been found to be common in groups with high rates of atherosclerotic cardiovascular disease such as Edinburgh men and Asian immigrants to the United Kingdom (LOGAN et al. 1978; McKEIGUE et al. 1991). Such findings have not been universal: Pima Indians, for example, are typically hyperinsulinemic but have a low risk of cardiovascular disease (LIU et al. 1992).

The failure to find a consistent association between hyperinsulinemia and cardiovascular risk suggests that the relationship may not be causal or that its effects are primarily permissive. MODAN et al., for example, found that the excess risk was confined to men with at least one of three conditions: obesity, glucose intolerance or hypertension (MODAN et al. 1991). Certainly a wide spectrum of metabolic disturbances can be demonstrated in insulin resistant subjects. Very low density lipoprotein and triglyceride levels tend to be higher and high density cholesterol lower in such subjects, a pattern which would likely be more atherogenic. However, there is also evidence that insulin resistance is associated with changes that would accentuate thrombosis formation by increasing coagulation and inhibiting fibrinolytic processes. For example, plasminogen activator inhibitor-1 (PAI-1) is associated with recurrent myocardial infarction in young men and is strongly inversely correlated with insulin and triglyceride concentrations and directly correlated with insulin resistance (JUHAN-VAGUE et al. 1991). The risk of myocardial infarction is also likely to be raised by the strong association between insulin resistance and hypertension. That this association is not secondary is suggested by the fact that abnormalities of insulin metabolism can be discerned in normotensive first degree relatives of patients with high blood pressure. Further evidence for the link between deranged insulin metabolism, hypertension and coronary heart disease (CHD) is the recent observation that patients with high blood pressure and ECG evidence of CHD were insulin resistant when compared to a matched group of patients with hypertension who did not demonstrate ischemic ECG changes (SHEU et al. 1992).

While practising physicians need to be aware of the interrelated features of the insulin resistance syndrome, routine determination of fasting insulin levels is probably not justified. Though the syndrome is partially genetically determined, it worsens with age, weight gain and physical inactivity. While there is abundant epidemiological evidence that exercise and weight control reduce mortality risks from coronary heart disease, the principal biological mechanisms underlying the benefit have yet to be determined. Nevertheless, both the maintenance of a normal weight and regular exercise should increase insulin sensitivity so this may be one of several possible mechanisms.

IV. Hyperhomocysteinemia

Recently, a number of studies have demonstrated that homocysteine concentrations are raised in patients with coronary, cerebral and peripheral vascular disease (CLARKE et al. 1991; STAMPFER et al. 1992; LOLIN et al. 1996). In fact, its putative association with cardiovascular risk is not new, having been proposed in 1975 (MCCULLY and WILSON 1975). Homocysteine is a thiol containing amino acid produced by transmethylation reactions which consume *S*-adenosyl methionine (SAM). It is either used to regenerate SAM, involving vitamin B_{12} dependent methionine synthetase (and using 5-methyltetrahydrofolate as the methyl donor) or disposed of by transsulphuration through the action of cystathionine B synthase (CBS). Although the mechanisms for the association between hyperhomocysteinemia and coronary risk are unclear, homocysteine is known to be directly toxic to endothelial cells and can inhibit their release of nitric oxide, thus facilitating platelet aggregation and vasoconstriction (WU et al. 1994). It may also disturb the equipoise between thrombosis and fibrinolysis by inhibiting the processing of thrombomodulin and reducing the activation of protein C.

A number of inborn errors of metabolism may underlie hyperhomocysteinemia. The most common cause of the autosomal recessive condition, homocystinuria (characterized by other skeletal, neurological and ocular abnormalities), is cystathionine B-synthase deficiency, but some cases may clearly be environmentally induced, for example, by deficiencies of B_{12} or folic acid. The heterozygote frequency of CBS deficiency, however, is too low to account for the observed associations between moderate hyperhomocysteinemia and coronary artery disease (DALY et al. 1996). Thus, other genetic abnormalities, such as mutations in the methylene tetrahydrofolate reductase (MTHFR) gene, have been subject to scrutiny. In a recent study of nearly 650 working males from Belfast, homozygosity for the thermolabile MTHFR variant was a major contributing factor to mild hyperhomocysteinemia and in particular to levels above the 95th percentile. Indeed the prevalence of this variant among individuals in the top 5% of the homocysteine distribution was 48%. This is of particular note, because one of the largest studies of the association of homocysteine and myocardial infarction – a prospective study of 14,916 US Physicians – suggested that the 95th percentile was the point at which the risk of MI increased by approximately threefold. Interestingly, the Belfast study also demonstrated that in subjects with serum folate levels below the median, homozygotes had higher homocysteine levels. Furthermore, thermolabile homozygotes as a group had significantly lower folate levels, indicating that there may be at least two mechanisms of gene–environment interaction producing mild hyperhomocysteinemia. In one scenario, homozygosity for the thermolabile variant is associated with higher homocysteine levels among subjects with low folate levels (which might be low for a number of reasons, for example, because of dietary deficiency). On the other hand, homozygotes themselves have a tendency to have lower folate levels and so both mechanisms may interact to produce hyperhomocysteinemia. It is likely that extra dietary folic acid can compensate for the reduced function of the MTHFR and thus reduce homocysteine levels (Boushey et al. 1995). Whether this would lead to a commensurate reduction in cardiovascular risk must be the subject of randomized trials but the design of any such trial would do well to account for genotype as well as phenotype.

Assuming a relative risk of 1.4 for increments of plasma homocysteine of 5 mmol/l and that 10% of the male population have levels of more than 15 mmol/l, then the proportion of potentially preventable coronary heart disease in the population attributable to such high levels is approximately 4.1% (Boushey et al. 1995).

Such calculations might obviously affect public health policy if an acceptable means is found to reduce the exposure of the population to the harmful effects of elevated plasma homocysteine. To a large extent, the calculations have ignored the potentially different responses of those that might carry a genetic predisposition. The suggestion is, nevertheless, that a policy of food fortification with both folic acid and cobalamin (to offset the potentially deleterious unmasking of B_{12} deficiency) could prevent up to 50,000 coronary artery disease deaths in the United States per year.

D. Risk Factors for Venous Thrombosis

On the face of it, it would be logical to attempt to categorize risk factors into abnormalities of (1) blood flow; (2) its constitution or (3) of the vessel wall itself. Such could be either acquired or inherited. In fact, however, seldom does any one of these act alone, the occurrence of thrombosis in individuals and in populations resulting from the concerted action of several risk factors. Although there remain unanswered questions, much has already been written about those risk factors for which medical prophylaxis is most usually attempted. The major findings concerning these are summarized below. Rather more attention is given to the issues about which some controversy is still apparent or which are suggesting novel or fundamental biological mechanisms.

Broadly speaking, the main risk factors can be grouped into postoperative risks and other medical ailments predisposing towards venous stasis. The THRIFT Group (1992) and others (Goldhaber and Morpurgo 1992) have extensively reviewed the evidence and the conditions which enhance the risk of deep venous thrombosis (DVT) and pulmonary embolism (PE) following surgery. Risk increases exponentially over the age of 40 years and the duration and complexity of the operation appear to be important: deep pelvic surgery and hip surgery appear to carry higher risks. The incidence of DVT and PE has been estimated in the control arms of various meta-analyses of prophylaxis but the absolute rate, it should be remembered, will vary according to the method of ascertainment, the case mix and the duration of follow-up. THRIFT consider that low risk patients have a DVT risk under 10%, are usually under 40 years of age and have surgery for less than 30 min. Moderate risk groups have an incidence of DVT of 10%–40% with a risk of fatal PE up to the average risk in hospitalized patients (1%) and such groups would comprise those

- Undergoing major general, urological, gynecological, cardiothoracic, vascular or neurosurgical surgery, especially among those over 40 years or with other risk factors
- With major acute medical illness (e.g. immobilizing myocardial infarction, heart failure, chest infection, cancer or inflammatory bowel disease)
- With major trauma
- With a past history of thrombosis

High risk groups of hospital patients have an incidence of DVT in screening studies of 40%–80% with an above average risk of fatal PE (1%–10%) and would include those

- Undergoing major orthopedic surgery
- Undergoing major pelvic or abdominal surgery for cancer
- Undergoing major surgery or enduring major medical illness with a predisposing condition (vide infra)
- With lower limb paralysis

A number of other conditions (hardly "ailments") have been popularly considered to be associated with deep venous thrombosis. Obesity is frequently cited as a risk factor but its independent association is doubtful. In fact 20 years ago, NICOLAIDES and IRVING (1975) found that the association became nonsignificant when age and past history of thrombosis or major surgery were controlled for and many now consider it an additive risk factor which alone would not justify prophylaxis.

Despite its low absolute incidence, venous thrombosis and pulmonary embolism are still leading causes of death following childbirth, with approximately one episode of clinically recognized PE per 1000 births and one death per 100,000 births (TURNBULL et al. 1989). However, for obvious reasons, studies with rigorous diagnostic ascertainment are very scarce. The main problem is the concern about the use during pregnancy of the various diagnostic tests such as radiofibrinogen scanning or ascending venography. DRILL and CALHOUN (1968) summarized a number of reports of clinically diagnosed DVT and estimated that prepartum the incidence varied from 8 to 15/100,000 women but postpartum the rates were considerably higher, probably over

300/100,000. The absolute reliability of such estimates is suspect but it is reasonable to conclude that the thrombotic risk is much greater in the post-partum period.

While the effects of pregnancy and the puerperium on thrombotic risk are probably due to localized effects on blood flow and the effect of endogenous hormones on the fibrinolytic cascade, there has been much more vigorous debate on the safety or otherwise of the various oral contraceptive pill (OCP) preparations. Oral contraceptives have been linked with a small absolute increase in the incidence of vascular disease in many epidemiological studies although, overall, the risks for most women are deemed to be balanced by the benefits (Stadel 1981). However, in the last 2 years a number of concerns have emerged about the safety of newer "third generation" preparations. At least five major studies have tended to suggest higher thrombosis risks with the newer preparations. In the multinational case-control WHO study, the odds ratio (an approximation to the relative risk for rare outcomes) for venous thromboembolism (VTE) for third generation products (gestodene and desogestrel) was 7.4 compared with 3.6 for second generation preparations (WHO 1995). The Leiden study was conducted between 1988 and 1992 and included 126 cases of verified VTE and 159 age matched controls (Bloemencamp et al. 1995). Compared with nonusers, the odds of VTE for women on second generation OCPs was 3.8 while for those on desogestrel it was 8.7. The GPRD study, on the other hand, was a nonrandomized cohort study of women recruited from 365 English general practices (Jick et al. 1995). Eighty women fulfilled the criteria for VTE, representing a rate per 10,000 women-years of 1.6 in current users of any second generation product OCP, 2.9 in for those using third generation OCPs containing desogestrel and 2.8 for those using a preparation containing gestodene. The Transnational Study was another case-control study conducted during 1993–1995 which enrolled 471 cases of VTE and compared them with 1772 age matched controls (Spitzer et al. 1996). In this study, users of a second generation OCP with <50μm ethinylestrodiol had an odds ratio of VTE of 3.2 while women using a third generation product had an odds ratio of 4.8. Finally, another nonrandom retrospective cohort study, based on the medical records of 540,000 women born between 1941 and 1981 drawn from 143 general practices in the United Kingdom, has recently reported a relative risk of VTE in users of third generation as opposed to second generation OCPs of 1.68 (95% confidence limits 1.04–2.75) (Farmer et al. 1997).

While all five studies had reasonable statistical power, a significant association is not the same as a causal relationship and it is important to be mindful of possible biases that might affect the interpretation of the findings. While most of the studies attempted adjustment for obvious confounders such as age, past history of thrombosis and body mass index, few controlled for familial predisposition. Of perhaps greater significance is the possibility of "prescribing" bias (leading to selection bias) and diagnostic bias (Lidegaard and Milsom 1996). Following the introduction of third generation products, they were marketed as being safer than the older pills. If practitioners thus preferentially prescribed them to higher risk women, then such a group might be expected to have a higher risk of developing VTE independent of the preparation given itself. Similarly, women taking the newer preparations may be more subject to diagnostic investigations when presenting with symptoms of VTE. If this were the case then the observed rate might be inflated. In fact, to greater or lesser extents such biases may have had some bearing on findings of the early studies of the VTE-OCP association (well reviewed by KATERNDAHL et al. 1992). However the strength of the association might be debated, as McPherson has succinctly observed, most of the studies show an effect of broadly similar size and direction despite their differing populations, designs and adjustments. Nevertheless, to place them in context, a woman using a third generation as opposed to a second generation product may have an increased risk of death from venous thromboembolism of 6 per million per year, or roughly equivalent to the added cancer and coronary risk if she smoked 10 cigarettes per year.

Though this is small in absolute terms, every woman deserves to be offered a fully informed choice of contraceptive options. Women with a familial predisposition to thrombophilia certainly must be guarded concerning their OCP choice. In fact, it is becoming increasingly possible to define more accurately the risks for subgroups of women based on their particular genetic constitution. Clarifying various aspects of gene–environment interaction will almost certainly be the major expanding frontier in cardiovascular epidemiology.

An example of this is the recent discovery of how a particular mutation in the gene encoding factor V results in resistance to breakdown by activated protein C and a relative thombophilic state (BERTINA et al. 1994). In a subsequent case control study from Leiden, of 126 women with DVT and 159 controls, the odds ratio for desogestrel-containing contraceptive use was 9.2 (3.9–21.4) among noncarriers and 6.0 (1.9–19.0) among carriers of the mutation (BLOEMENCAMP et al. 1995). However, this latter risk was superimposed on the 8-fold increased risk of venous thrombosis for carriers of the factor V Leiden mutation (implying an almost 50-fold risk for carriers also using desogestrel containing OCPs compared to nonuser noncarriers). In fact the risk of thrombosis in homozygotes is probably several hundred fold. Given that APC resistance is probably the commonest underlying predisposition to venous thrombosis, there has been much speculation about whether screening for this mutation could be justified on public health grounds (DAHLBACK 1996). The benefits and risks would depend on a number of factors including the incidence and fatality of DVT, the prevalence of the mutation (probably between 5% and 15% in the Western world) and the usage rates of various types of OCP. Disregarding the variability in the costs and performance of potential screening tests, approximately 400,000 women would need to be screened before OCP prescription to prevent one fatality. On the other hand, if routine screening during pregnancy was followed by anticoagulation, the number of cases of fatal bleeding might equal or exceed the number of fatal pulmonary emboli that might be prevented. However, taking a personal and family history of deep vein thrombosis when prescribing the OCP or at the first antenatal visit will detect families with a tendency to multiple venous thrombosis and may be worthwhile (VANDENBROUCKE et al. 1996).

E. Future Directions in Epidemiological Research

This chapter, within the space available, has been able merely to highlight how epidemiological inquiry has contributed to an understanding of the etiology, treatment and prevention of arterial and venous thrombosis. Epidemiological inquiry has flourished over the last 40 years and has been able to clarify the contribution of major risk factors to the burden of disease from arterial and venous thrombosis. There are two common ways of expressing this. For example, one might calculate the proportion of disease in the population that would be prevented if a certain risk factor and its consequences were eliminated. Alternatively one might model the likely impact on life expectancy of removing the effects of that risk factor. Tables 1 and 2 illustrate some results pertaining to coronary artery disease and its major risk factors. What is evident from Table 2 is that the elimination of no single risk factor, with perhaps the exception of smoking, will have a huge impact on longevity. The basis of the calculation of the attributable risks shown (Table 1), however, is an assumption of no major interaction effects. This is unlikely to reflect reality. In trying to unravel the web of causation, the spider always seems to be a few steps ahead. Classical epidemiology has attempted to measure the associations between endpoints and easily measurable antecedents such as serum fibrinogen or cholesterol. A paradigm shift is now occurring enabling a sharper focus on the molecular epidemiology and mechanisms underlying the causal chain (Fig. 4).

Epidemiology has adapted to the progress made in sister disciplines in the laboratory and in clinical practice. Among the most significant developments,

Table 2. Anticipated gains in life expectancy from risk factor modification

Risk factor modification	Gains in life expectancy		Reference
	Men	Women	
Smoking cessation*	1.2–2.3 years	1.5–2.8 years	TSEVAT et al. 1991
Cholesterol control*	0.5–4.2 years	0.4–6.3 years	TSEVAT et al. 1991
Weight control*	0.7–1.7 years	0.5–1.1 years	TSEVAT et al. 1991
Exercise participation$	0.1–1.3 years	n/a	PAFFENBARGER et al. 1993
Exercise participation and smoking cessation$	1.3–3.7 years	n/a	PAFFENBARGER et al. 1993

* Gains in life expectancy for 35 year olds at risk.
$ Gains in life expectancy for men aged over 45 to 84 years free from coronary heart disease.

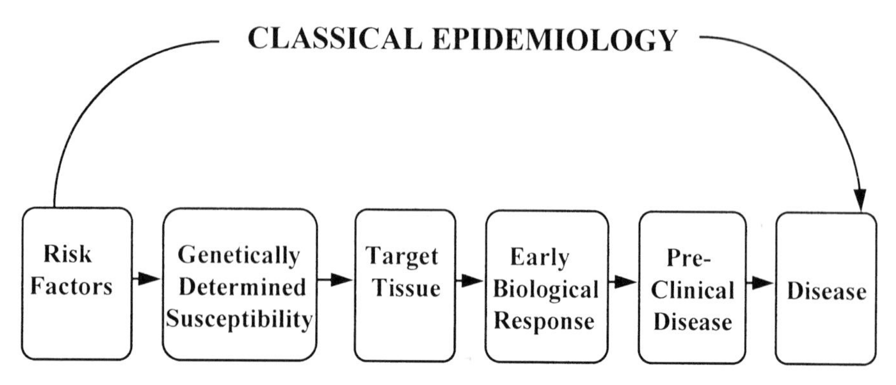

Fig. 4. Future directions in epidemiological research: Filling in the "pieces" (genetic factors, organ specificity and gene-environment interactions) between risk factors and disease (McMichael 1994)

perhaps, are the quantitative methods used in the genetic study of chronic diseases and in the synthesis of evidence (meta-analysis).

I. Genetic Epidemiology

Without a doubt, whether in the field of infectious or chronic disease, molecular epidemiology has gained a prominent place in medical science (McDADE and ANDERSON 1996; RISCH and MERIKANGAS 1996). For complex diseases, as for rare single gene defects, statistical analysis of family data can provide clues to the existence of a basic metabolic defect. However, the demonstration that a familial pattern of disease is consistent with Mendelian segregation ratios is considerably more difficult for a disease with a complex etiology than for a simple autosomal dominant or recessive disorder. Although susceptibility to thrombosis may be determined by numerous genetic loci, most may have only a minor effect. There may be one or a few which independently or in combination have a major effect. In addition, complex diseases such as coronary heart disease, may exhibit etiologic heterogeneity with one locus having a major impact in some families and another locus and or environmental variable having a bearing in others. Because of these complications, statistical analyses of family patterns of such diseases may sometimes yield ambiguous results.

The frequency of association between a gene marker and a "disease" gene will depend on variations in the frequency of the marker in different populations. If controls have a different ethnic background from cases, and if marker allele frequencies differ between ethnic backgrounds, spurious associations may emerge that are unrelated to a disease susceptibility gene. In addition, it is important to remember that these studies are based on a concept of statistical association rather than physical proximity of the relevant loci.

Another issue often not addressed in studies of genetic marker-disease associations is that of biological interaction between the genetic marker and other genetic and environmental factors. In the presence of gene-environment interaction, a truly causal association between a marker and disease may be diluted by the proportion of the population that are or are not exposed to environmental factors which interact with the susceptible genotype (SING and MOLL 1989).

Not surprisingly, analyses of data from different populations sometimes yield discrepant results. Disease susceptibility loci detectable in one population may not be discerned in another due to differences in environmental risk or other genetic or racial factors, or because of differing methods of ascertainment. A further problem resides in the fact that many genetic diseases exhibit etiological heterogeneity. While these potential shortcomings must always be borne in mind, more advanced statistical approaches beyond the scope of this section are continually being developed which can at least partially take account of such factors.

II. Evidence Synthesis

The Canadian epidemiologist David Naylor is of the view that the Cochrane Collaboration is an endeavor which will rival the Human Genome Project in its potential significance to clinical medicine. Set up originally in honor of the British epidemiologist Archie Cochrane, this is a worldwide network of biostatisticians, epidemiologists and clinicians whose primary purpose is to provide an unbiased synthesis of evidence on the efficacy of treatments or the etiology of diseases. A host of statistical issues in meta-analysis remain as much a matter of judgement as of science (COOK et al. 1995) as do the older debates between the traditional frequentist and Bayesian views of statistics. Frequentists deduce the probability of observing an outcome given the true underlying state while Bayesians induce the probability of the existence of the true but as yet unknown underlying state given the data. Probabilities of different states of nature can be combined with pre-specified benefits and costs that would result from taking a certain decision in the face of those "givens". The Bayesian approach would then be to choose the decision that maximizes the expected benefit. Such an approach, it is claimed, would have demonstrated that although it looks fairly likely that venous thromboembolism occurs somewhat more frequently with third generation OCPs, there is considerable doubt about which preparation is safer overall (LILFORD and BRAUNHOLTZ 1996). Faced with data presented in Bayesian terms, one would more readily appreciate how the probabilities attached to alternative states of nature vary according to different interpretations of the "starting" information and that the final decision (to initiate contraception) should take account of all the relevant personal trade-offs. A Bayesian approach can thus do justice to the disparate evidence that may impact upon prior beliefs and will allow for assumptions to be handled openly and explicitly. The reality is that degrees of belief are often continuous, not dichotomous, varying from one person to person or situation to situation, in the face of inconclusive evidence. The Bayesian approach seems certain to gain ground in the policy making arena (Freedman 1996).

References

Barker DJP (1994) Mothers, babies and disease in later life. BMJ Publishing Group, London

Barker DJP (1992) (ed) Fetal and infant origins of adult disease. BMJ Publishing Group, London

Barker DJ, Meade TW, Fall CH, Lee A, Osmond C, Phipps K et al (1992) Relation of fetal and infant growth to plasma fibrinogen and factor VII concentrations in adult life. BMJ 304:148–152

Bertina RM, Koeleman BP, Koster T, Rosendaal FR, Dirven RJ, de Ronde H et al (1994) Mutation in blood coagulation factor V associated with resistance to activated protein C. Nature 369:64–67

Bloemencamp KW, Rosendaal FR, Helmerhorst FM, Buller HR, Vandenbroucke JP (1995) Enhancement by factor V Leiden mutation of risk of deep-vein thrombosis associated with oral contraceptives containing a third-generation progestagen. Lancet 346:1593–1596

Boushey CJ, Beresford SA, Omenn GS, Motulsky AG (1995) A quantitative assessment of plasma homocysteine as a risk factor for vascular disease. Probable benefits of increasing folic acid intakes. JAMA 274:1049–1057

Brunner E, Smith GD, Marmot M, Canner R, Bekinska M, O'Brien J (1996) Childhood social circumstances and psychosocial and behavioural factors as determinants of plasma fibrinogen. Lancet 347:1008–1013

Carter CJ (1994) The natural history and epidemiology of venous thrombosis. Prog Cardiovasc Dis 36:423–438

Clarke R, Daly L, Robinson K, Naughten E, Cahalane S, Fowler B et al (1991) Hyperhomocysteinemia: an independent risk factor for vascular disease. N Engl J Med 324:1149–1155

Coon WW, Willis PW 3d, Keller JB (1973) Venous thromboembolism and other venous disease in the Tecumseh Community Health Study. Circulation 48:839–846

Cook DJ, Sackett DL, Spitzer WO (1995) Methodologic guidelines for systematic reviews of randomized control trials in healthcare from the Potsdam International Consultation on Meta-Analysis. J Clin Epidemiol 48:167–171

Dahlback B (1996) Are we ready for factor V Leiden screening? Lancet 347:1346–1347

Daly L, Robinson K, Tan KS, Graham IM (1996) Hyperhomocysteinaemia: a metabolic risk factor for coronary heart disease determined by both genetic and environmental influences? Q J Med 86:685–689

Department of Health, Central Health Monitoring Unit (1994) Coronary heart disease. An epidemiological overview. HMSO, London

Department of Health and Social Services (NI) (1990) Annual Report of the Registrar General for Northern Ireland. HMSO, Belfast

Doll R, Hill AB (1954) The mortality of doctors in relation to their smoking habits. A preliminary report. BMJ: I:1451–1455

Drill VA, Calhoun DW (1968) Oral contraceptives and thromboembolic disease. JAMA 206:77–84

Farmer RD, Lawrenson RA, Thompson CR, Kennedy JG, Hambleton IR (1997) Population based study of risk of venous thromboembolism associated with various oral contraceptives. Lancet 349:83–88

Fowkes FG, Connor JM, Smith FB, Wood J, Donnan PT, Lowe GD (1992) Fibrinogen genotype and risk of peripheral atherosclerosis. Lancet 339:693–696

Freedman L (1996) Bayesian statistical methods. BMJ 313:569–570
Fuster V (1994) Lewis A. Conner Memorial Lecture. Mechanisms leading to myocardial infarction: insights from studies of vascular biology. Circulation 90:2126–2145
General Register Office for Scotland (1990) Annual report of the Registrar General for Scotland. GRO for Scotland, Edinburgh
Goldhaber SZ, Morpurgo M (1992) Diagnosis, treatment and prevention of pulmonary embolism. Report of the WHO/International Society and Federation of Cardiology Task Force. JAMA 268:1727–1733
Green F, Humphries S (1994) Genetic determinants of arterial thrombosis. Balliere's Clin Haematol 7:675–692
Hamsten A, Iselius L, de Faire U, Blomback M (1987) Genetic and cultural inheritance of plasma fibrinogen concentration. Lancet II:988–991
Healthcare Management (1993) What price will you pay to prevent thrombosis? 1(3):40–42
Hill AB (1965) The environment and disease: association or causation? Proc R Soc Med 58:295–300
Isles CG, Hole DJ, Hawthorne VM, Lever AF (1992) Relation between coronary risk and coronary mortality in women of the Renfrew and Paisley survey: comparison with men. Lancet 339:702–706
Jick H, Jick SS, Gurewich V, Myers MW, Vasilakis C (1995) Risk of idiopathic cardiovascular death and nonfatal venous thromboembolism in women using oral contraceptives with differing progestagen components. Lancet 346:1589–1593
Juhan-Vague I, Alessi MC, Vague P (1991) Increased plasma plasminogen activator inhibitor 1 levels. A possible link between insulin resistance and atherothrombosis. Diabetologia 34:457-462
Kannel WB (1987) The Framingham Study: an epidemiological investigation of cardiovascular disease. NIH Publication No 87–2703. NIH, Springfield MA
Katerndahl DA, Realini JP, Cohen PA (1992) Oral contraceptive use and cardiovascular disease: is the relationship real or due to study bias? J Fam Pract 35:147–157
Kee F, Nicaud V, Tiret L, Evans A, O'Reilly D, De Backer G (1997) Short stature and heart disease: nature or nurture? The EARS Group. Int J Epidemiol 26:748–756
Lidegaard O, Milsom I (1996) Oral contraceptives and thrombotic diseases: impact of new epidemiological studies. Contraception 53:135–139
Lilford RJ, Braunholtz D (1996) The statistical basis of public policy: a paradigm shift is overdue. BMJ 313:603–607
Liu QZ, Knowler WC, Nelson RG, Saad MF, Charles MA, Leibow IM et al (1992) Insulin treatment, endogenous insulin concentration, and ECG abnormalities in diabetic Pima Indians. Cross-sectional and prospective analyses. Diabetes 41:1141–1150
Logan RL, Riemersma RA, Thomson M, Oliver MF, Olsson AG, Walldius G et al (1978) Risk factors for ischaemic heart disease in normal men aged 40. Edinburgh-Stockholm Study. Lancet I:949–954
Lolin YI, Sanderson JE, Cheng SK, Chan CF, Pang CP, Woo KS et al (1996) Hyperhomocysteinaemia and premature coronary artery disease in the Chinese. Heart 76:117–122
McCully KS, Wilson RB (1975) Homocysteine theory of arteriosclerosis. Atherosclerosis 22:215–227
McDade JE, Anderson BE (1996) Molecular epidemiology: applications of nucleic acid amplification and sequence analysis. Epidemiol Rev 18:90–97
McKeigue PM, Shah B, Marmot MG (1991) Relation of central obesity and insulin resistance with high diabetes prevalence and cardiovascular risk in South Asians. Lancet 337:382–386
McMichael AJ (1994) Invited commentary "molecular epidemiology": new pathway or new travelling companion? Am J Epidemiol 140:1–11

Meade TW (1995) Fibrinogen in ischaemic heart disease. Eur Heart J 16 (Suppl A): 31–35

Meade TW, Ruddock V, Stirling Y, Chakrabarti R, Miller GJ (1993) Fibrinolytic activity, clotting factors, and long-term incidence of ischaemic heart disease in the Northwick Park Heart Study. Lancet 342:1076–1079

Modan M, Or J, Karasik A, Drory Y, Fuchs Z, Lusky A et al (1991) Hyperinsulinaemia, sex, and risk of atherosclerotic cardiovascular disease. Circulation 84:1165–1175

Nicolaides AN, Irving D (1975) Clinical factors and the risk of deep venous thrombosis. In: Nicolaides AN (ed) Thromboembolism: etiology, advances in prevention, and management. University Park Press, Baltimore, pp 193–204

Office of Population Censuses and Surveys (1990) Mortality statistics: area review of the Registrar General on deaths by area of usual residence in England and Wales, 1990. HMSO, London

Paffenbarger RS Jr, Hyde RT, Wing AL, Lee IM, Jung DL, Kampert JB (1993) The association of changes in physical-activity level and other lifestyle characteristics with mortality among men. N Engl J Med 328:538–545

Paneth N, Susser M (1995) Early origin of coronary heart disease (the "Barker hypothesis"). Hypotheses, no matter how intriguing, need rigorous attempts at refutation. BMJ 310:411–412

Reaven GM (1988) Banting lecture 1988. Role of insulin resistance in human disease. Diabetes 37:1595–1607

Risch N, Merikangas K (1996) The future of genetic studies of complex human diseases. Science 273:1516–1517

Scarabin PY, Bara L, Ricard S, Poirier O, Cambou JP, Arveiler D et al (1993) Genetic variation at the β-fibrinogen locus in relation to plasma fibrinogen concentrations and risk of myocardial infarction. The ECTIM study. Arterioscler Thromb 13: 886–891

Sheu WH, Jeng CY, Shieh SM, Fuh MM, Shen DD, Chen YD et al (1992) Insulin resistance and abnormal electrocardiograms in patients with high blood pressure. Am J Hypertens 5:444–448

Sing CF, Moll PP (1989) Genetics of variability of CHD risk. Int J Epidemiol 18:S183–S195

Spitzer WO, Lewis MA, Heinemann LA, Thorogood M, MacRae KD (1996) Third generation oral contraceptives and risk of venous thromboembolic disorders: an international case-control study. BMJ 312:83–88

Stadel BV (1981) Oral contraceptives and cardiovascular disease. N Engl J Med 305:612–618, 672–677

Stampfer MJ, Malinow MR, Willet WC, Newcomer LM, Upson B, Ullmann D et al (1992) A prospective study of plasma homocysteine and risk of myocardial infarction in US physicians. JAMA 268:877–881

Stout RW, Vallance-Owen J (1969) Insulin and atheroma. Lancet I:1078–1080

Thromboembolic Risk Factors (THRIFT) Consensus Group (1992) Risk of and prophylaxis for venous thromboembolism in hospital patients. BMJ 305:567–574

Tsevat J, Weinstein MC, Williams LW, Tosteson AN, Goldman L (1991) Expected gains in life expectancy from various coronary heart disease risk factor modifications. Circulation 83:1194–1201

Tunstall-Pedoe H, Kuulasmaa K, Amouyel P, Arveiler D, Rajakangas AM, Pajak A (1994) Myocardial infarction and coronary deaths in the World Health Organization MONICA Project. Registration procedures, event rates, and case-fatality rates in 38 populations from 21 countries in four continents. Circulation 90:583–612

Turnbull A, Tindall VR, Beard RW, Robson G, Dawson IM, Cloake EP et al (1989) Report on confidential enquiries into maternal deaths in England and Wales 1982–1984. Rep Health Soc Subj (London) 34:1–166

Vagero D, Leon D (1994) Ischaemic heart disease and low birth weight: a test of the fetal-origins hypothesis from the Swedish Twin Registry. Lancet 343:260–263

Vandenbroucke JP, van der Meer FJ, Helmerhorst FM, Rosendaal FR (1996) Factor V Leiden: should we screen oral contraceptive users and pregnant women? BMJ 313:1127–1130

Vaughan CJ, Murphy MB, Buckley BM (1996) Statins do more than just lower cholesterol. Lancet 348:1079–1082

Wade DT (1994) Stroke. In: Stevens A, Raftery J (eds) Health care needs assessment of the epidemiologically based needs assessment reviews, vol 1. Radcliffe Medical Press, Oxford

Wilson TW, Kaplan GA, Kauhaanen J, Cohen RD, Wu M, Salonen R et al (1993) Association between plasma fibrinogen concentration and five socio-economic indices in the Kupio ischaemic heart disease risk factor study. Am J Epidemiol 137:292–300

World Health Organisation (1980–1995) World Health Statistics Annual. WHO, Geneva, Switzerland

World Health Organisation Collaborative Study of Cardiovascular Disease and Steroid Hormone Contraception (1995) Venous thromboembolic disease and combined oral contraceptives: results of an international multi-centre case-control study. Lancet 346:1575–1582

Wu LL, Wu J, Hunt SC, James BC, Vincent GM, Williams RR et al (1994) Plasma homocysteine as a risk factor for early familial coronary artery disease. Clin Chem 40:552–561

CHAPTER 4
In Vivo Models of Thrombosis

L. CHI, S. REBELLO, and B.R. LUCCHESI

A. Introduction

Under certain pathological conditions, thrombosis occurs in the arterial or venous system and may cause myocardial infarction, cerebral infarction, transient ischemic attack, or deep vein thrombosis (DVT). These thrombotic diseases are most common in the Western industrialized countries and affect millions of people every year (HIRSH and HOAK 1996). Over the last several decades, a variety of experimental models of thrombosis have been developed and utilized to elucidate the etiology and pathogenesis of thrombosis and to search for effective antithrombotic agents.

Thrombus formation is a complicated process which involves the interplay of several mechanisms. Any type of stimulus which can activate the platelets and/or coagulation system will in turn initiate platelet aggregation, convert fibrinogen to insoluble fibrin, and subsequently lead to intravascular clot formation. Over a century ago, VIRCHOW (1856) proposed three causative factors leading to thrombosis, now known as the classical Virchow's triad: obstruction to blood flow, changes in the property of the blood contents and alterations in the vessel wall. These principles have been applied by many investigators in the development of in vivo models of thrombosis. In this chapter, the commonly used models for both arterial and venous thrombosis will be discussed relative to the mechanisms by which the thrombi develop. These models include vessel wall injury-induced thrombosis (endothelial damage), stasis/hypercoagulable models, foreign surface-induced thrombosis, and transgenic animal models. Although in many cases, combinations of more than one thrombogenic stimulus may be applied to form a thrombus, the models will be summarized based on the primary factor which contributes to thrombosis. The general properties, specific characteristics, advantages and disadvantages of the representative models from each class will be discussed. The intent is to provide broad and general information in this area. We have elected, therefore, not to discuss each individual experimental technique in detail. The readers may refer to original articles cited here or recent reviews on the subject (BUSH and SHEBUSKI 1990; ANDERSON and WILLERSON 1994) for additional details.

B. Vessel Wall Injury-Induced Model of Thrombosis

Endothelial injury or changes in the properties of endothelial cells are among the most common causes of thrombosis in coronary artery disease. Endothelial cells are capable of synthesizing and secreting a number of antithrombotic substances such as prostacyclin (PGI_2), tissue factor pathway inhibitor (TFPI), thrombomodulin, and tissue plasminogen activator (t-PA). Atherosclerotic arteries with damaged endothelium not only lose some of these antithrombotic mechanisms, but may also expose the blood to the subendothelium, which is a highly reactive surface for the initiation of thrombosis. Platelet adhesion, activation, and initiation of both the intrinsic and extrinsic pathways of blood coagulation occur upon exposure of tissue factor and medial collagen (types I and III) to the blood. A mural thrombus subsequently develops at the site of vessel wall injury. Atherosclerotic plaque-induced narrowing of the vessel creates a turbulent blood flow pattern and higher shear rate, which is another factor facilitating platelet activation. Numerous animal models have been developed simulating these pathophysiological conditions.

I. Photochemical Reaction

Focal thrombosis of small vessels can be initiated by irradiating them with a high intensity light. ROSENBLUM and EL-SABBAN (1977) reported that platelet aggregation could be induced in the cerebral arterioles of mice by exposing the vessels to a filtered light in the presence of an intraluminal fluorescent dye. Subsequently, SATO and OHSHIMA (1984a,b) conducted detailed studies in the venules and arterioles of the rat mesenteric artery by applying this method. They quantitatively evaluated how the experimental conditions influenced the kinetics of the microthrombus formation. Arterioles 20–50 μm in diameter or venules 20–80 μm in diameter were selected to produce a microthrombus. Filtered light (400–500 nm wavelength), generated from a mercury lamp in an epi-illumination system, is used to irradiate a microvessel over an area approximately 130 μm in diameter. A solution of fluorescein sodium is then injected through the external jugular vein. Throughout the experiments, the dynamic course of thrombus formation is monitored continuously and recorded via an intravital microscope-television system. After testing at several levels of light intensity (20.7, 15.9, 9.2, 6.1 and 2.4 mW/mm^2) and different dye concentrations (10, 25 and 50 μg/kg), it was found that when the light intensity or the dye concentration increased, the initiation time of thrombus formation and time to occlusion of the lumen correspondingly shortened. The growth rate of the thrombus volume also depended on the dye concentration (SATO and OHSHIMA 1984b).

It has been suggested that free radicals (O_2--, HO•) and/or singlet molecular oxygen ($^1O_2$) which are generated from the photochemical reaction may be responsible for the endothelial damage (SANIABADI 1994; SANIABADI et al. 1995; JOURDAN et al. 1995). The singlet oxygen scavenger aminothiol (DL-cysteine) inhibits thrombotic occlusion of the guinea pig femoral artery (Saniabadi et al. 1995) and the hydroxyl radical scavengers dimethyl sulfoxide and glycerol prevent platelet aggregation in the pial arteriole of a mouse in the photochemical model (Rosenblum and El-Sabban 1982). It is not clear, however, whether the endothelial injury is the sole factor contributing to thrombus formation. In a study by Saniabadi et al. (1995), the green light irradiation was continued until the vessel occluded. If the light was intermittently inactivated, the growth of the platelet thrombus immediately ceased, with aggregation reinitiated by reactivated light irradiation (Sato and Ohshima 1984b). This suggested that light irradiation may have caused a change in aggregability of the platelets. In a recent report, however, Inamo et al. (1996) demonstrated that oxygen free radicals produced by photo-activated rose bengal exerted only a weak or no effect on platelet activation which was dependent upon the concentration of ADP used. Endothelial damage was nevertheless evident in this model. In a guinea pig femoral artery, electron microscopy revealed that following a photochemical reaction, endothelial cells first contract, then become detached from, the vessel wall; the cell membrane is destroyed at the irradiated site, where an occlusive, platelet-rich thrombus forms (Saniabadi et al. 1995).

The photochemical reaction has been applied in many studies using different animal species including rat mesenteric microvessels and middle cerebral arteries (Kawai et al. 1995; Jourdan et al. 1995; Sato and Ohshima 1984b), guinea pig (Saniabadi et al. 1995) and rabbit (Kawai and Tamao 1995) femoral arteries, as well as squirrel monkey mesenteric microvessels (Kaku et al. 1996). The filtered green light for irradiation can be generated from a mercury lamp through an epi-illumination system, or from a xenon lamp with a heat absorbing filter and a green filter. Commonly used sensitizing dyes are rose bengal and sodium fluorescein. Resulting thrombi are mainly composed of platelets (Sato and Ohshima 1984b; Saniabadi et al. 1995). The adhesion and aggregation of platelets in this model probably do not involve arachidonic acid metabolism, since neither aspirin (Jourdan et al. 1995) nor a thromboxane A_2 synthetase inhibitor, sodium ozagrel, can prevent thrombus formation while an anti-GPIIb/IIIa monoclonal antibody can (Kaku et al. 1996).

Although photochemically induced vessel damage can be achieved in large or small animals, clearly it is an advantage to use this technique in the latter, especially in the microvessels, where it is difficult to apply electrical or mechanical injury. The severity of injury or intensity of insults can be precisely controlled by adjusting the level of light intensity and/or duration of the irradiation, as well as the concentration of fluorescent dye injected. Although circulating platelets and blood cells may be involved, it is the endothelial cell which is the primary target in this model, which has been refined by the concomitant use of an intravital microscope-television system to permit evaluation of several end points. Using such a system, the initiation time of thrombus formation, time to occlusion, thrombus volume (Sato and Ohshima 1984b), number of emboli removed by the blood stream, and the number of emboli per minute (Jourdan et al. 1995) can all be quantified, allowing the dynamic course of intravascular thrombus formation and dose-response for an antithrombotic compound to be studied.

II. Laser

The basic characteristics of laser-induced endothelial injury and thrombus formation are similar to those of photochemically induced thrombosis. In the 1960s, Grand and Becker (1965) were able to induce a small injury in the venule of the rabbit ear chamber using a laser beam in the presence of carbon (Pelikan ink) in the blood stream. The microthrombus attached to the site of lesion and grew over a period of hours. They hypothesized that carbon particles absorbed the energy and generated heat, causing a microburn on the luminal side of the vessel wall. Arfors et al. (1968) also employed a laser to damage the endothelium in rabbit ear chamber preparations and in mesenteric vessels. A pulsed laser energy input of 180–200 J caused endothelial injury in a area approximately 6 μm in diameter. The thrombus grew rapidly by aggregation of platelets and was followed by repeated embolization and formation of microthrombi. Higher laser energy input resulted in more severe endothelial damage and subsequent formation of an occlusive thrombus which did not embolize (Arfors et al. 1968).

Different types of laser beams have been used by several research groups, including He-Ne gas laser (Kovacs et al. 1975; Yamamoto et al. 1989; Yamashita et al. 1993), argon laser (Imbault et al. 1996; Belougne et al. 1996), and ruby impulse laser (Weichert et al. 1983), with or without energy-absorbing materials. Although they all damage the tissue via the same mechanism, some differences do exist. For example, in rat mesenteric vessels, the ruby laser produces an intravascular heat precipitate, consisting of proteins, heat-damaged erythrocytes, and some platelets. This precipitate then acts as a strong stimulus for platelet adhesion and aggregation. Alternatively, the argon laser produces direct damage to the vessel wall; microburns rarely occur (Weichert et al. 1983). The He-Ne laser is often combined with Evans blue, which seems to be an ideal energy absorbing material for this particular laser. Its plasma concentration remains constant for at least 30 min after intravenous administration and it does not influence platelet function (Kovacs et al. 1975). The rat carotid artery irradiated by laser beam after administration of Evans blue exhibits severe intimal damage and disrupted internal elastic membrane. In contrast, without Evans blue, the artery displays normal structure after the same laser exposure (Kovacs et al. 1975).

It is possible to achieve thrombus formation without the use of any energy-absorbing materials, but this usually requires higher laser intensity. In experiments devoid of energy-absorbing materials (Arfors et al. 1968), only slight endothelial impairment is observed. The damaged red blood cells and platelets may play a role in the development of a platelet thrombus by releasing factors, such as ADP, but the thrombus is relatively "loose" and easily embolized, making it difficult to achieve vessel occlusion (IMBAULT et al. 1996). In contrast, when a laser is combined with Evans blue, the thrombus adheres firmly to the vessel wall and occludes the lumen after repeated irradiation (KOVACS et al. 1975; YAMASHITA et al. 1993). In light and electron microscopy studies, KOVACS et al. (1975) have confirmed severe endothelial damage with exposure of the subendothelium after laser irradiation in a number of animal models.

Platelets play an important role in this model. In fawn hooded bleeder rats, which have a platelet defect, the formation of microthrombi in small mesenteric vessels after laser is seen to be markedly reduced compared with normal rats (WEICHERT et al. 1983). Ticlopidine and dipyridamole produce a significant antithrombotic effect in this model, as reflected by an increase in the number of laser injuries required to induce a thrombus (WEICHERT et al. 1983). Platelets may even dominate in the formation of venous thrombi in this method. IMBAULT et al. (1996) reported that aspirin was as effective as low molecular weight heparin in preventing thrombus formation in rat mesenteric venules. It should be noted, however, that the thrombin inhibitors argatroban and PPACK dose dependently inhibited He-Ne laser-induced thrombosis in rat mesenteric microvessels (YAMASHITA et al. 1993), suggesting that this model may also be suitable for the evaluation of other antithrombotic strategies.

III. Mechanically Induced Injury

As early as the 1920s, experimental thrombosis was induced in damaged cortical and mesenteric arteries by pinching them with needle forceps (FLOREY 1925; HONOUR and RUSSELL 1962). The endothelial damage can also be created by balloon catheters, endarterectomy, or perfusing a segment of vessel with air or saline. These techniques are usually performed in large-diameter vessels, such as the coronary arteries, and the model closely simulates the consequences of coronary disease, i.e., acute coronary occlusion, ischemia, and sudden death. One of the commonly used thrombosis models in this category is the well-known "Folts coronary thrombosis model" (FOLTS et al. 1976; FOLTS 1991).

1. Pinching or Crushing

In anesthetized, open chest dogs, a segment of coronary artery is clamped with a vascular clamp to produce endothelial and medial damage. A plastic cylinder is placed externally around the injured artery to produce a critical stenosis (60%–80% diameter reduction). The coronary flow is gradually reduced to zero then is rapidly restored spontaneously or by manipulation of the plastic cylinder. This phenomenon occurs repetitively, causing the so-called "cyclic flow variations or reductions" (CFVs or CFRs) (FOLTS et al. 1976, 1982). Histological examination and coronary arteriography have demonstrated that the CFVs are caused by the repetitive accumulation and dislodgment of platelet aggregates at the site of injury/stenosis (FOLTS et al. 1982; BUSH et al. 1984). It has been suggested that in this model, the damaged or denuded endothelium provokes platelet aggregation while the turbulent flow generated through the narrowed lumen causes the aggregates to be dislodged. The basic features of a deep arterial injury observed clinically are reflected in this in vivo model of thrombosis.

The antiplatelet, antithrombotic efficacy of various compounds has been tested using the FOLTS model. Aspirin (35 mg/kg, i.v.) can abolish the CFRs (FOLTS et al. 1976). DMP 728 (0.01 mg/kg i.v. or 0.6 mg/kg p.o.), a platelet GPIIb/IIIa antagonist, prevented CFRs in the left circumflex coronary artery of dogs (MOUSA et al. 1996). LIN and YOUNG (1995) reported that epinephrine increased, but norepinephrine reduced, the frequency of CFRs in a canine study. Thus, the model is suitable for studying mechanisms and interventions that either enhance or inhibit platelet aggregation. In the experimental preparation, one can control the coronary arterial blood pressure via a servocontrol device to eliminate any influence on the CFRs due to changes in perfusion pressure (LIN and YOUNG 1995). The FOLTS model can be used not only for short-term, but also for long-term, studies (GOLINO et al. 1993) and in conscious dogs as well (GALLAGHER et al. 1985; Fig. 1). BN52021, a potent and

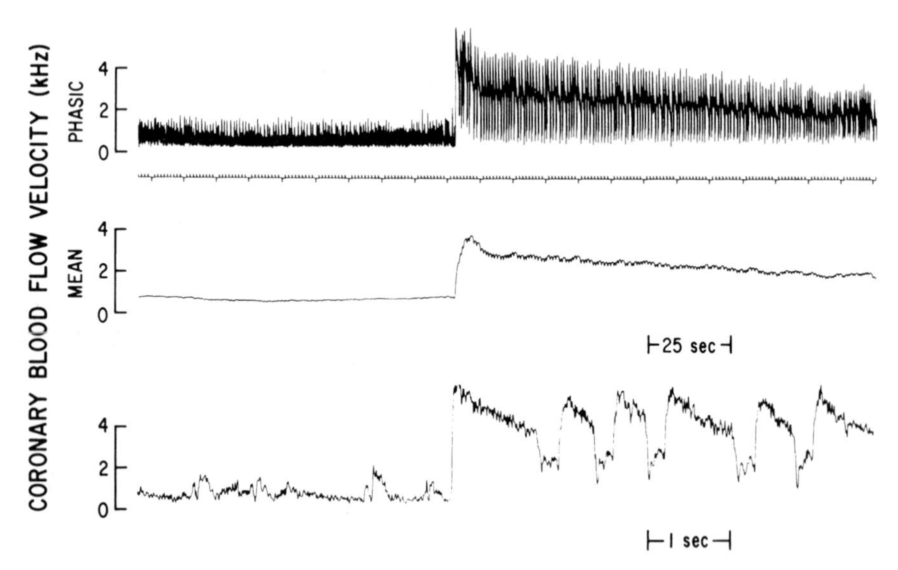

Fig. 1. Example of coronary blood flow velocity tracings to illustrate cyclical coronary flow variations (CFVs) in conscious dogs. The abruptness of flow change is demonstrated here. In the upper two panels, phasic and mean tracings of coronary flow velocity are shown at low paper speed. The lower panel, at higher paper speed, shows the exact cardiac cycle during which the abrupt flow change occurred. From GALLAGHER et al. (1985)

selective platelet activating factor (PAF) antagonist, acutely inhibited CFRs in rabbit carotid arteries, whereas it was relatively ineffective in abolishing CFRs in the canine left anterior descending coronary artery (GOLINO et al. 1993). To test the hypothesis that PAF could be an important mediator of CFRs in dogs in the subacute phase after endothelial injury, dogs were observed for an additional 8-h period. At the end of 8 h, a second dose of BN52021 was given, and CFRs were abolished. These results allowed the investigators to conclude that in this model, PAF becomes more important in platelet aggregation several hours after endothelial injury (GOLINO et al. 1993).

CFRs can also be produced in carotid arteries of the rabbit and monkey with a method similar to that described for canine coronary arteries (HILL et al. 1987; COLLER et al. 1989; GOLINO et al. 1992). After an external constrictor was placed around carotid arteries with endothelial injury, CFRs developed in 14 of 14 rabbits with a mean frequency of 16.5 ± 2.3 cycles/h. In this model, aspirin (10 mg/kg i.v.) had a minimal effect, whereas R68070 (TXA_2/PGH_2 receptor and TXA_2 synthase inhibitor) and ketanserin (serotonin S_2 receptor antagonist) abolished the CFRs in all rabbits (GOLINO et al. 1992). RAGNI et al. (1996) studied adjunctive therapies for thrombolysis by using the rabbit CFR model. In their study, AP-1, a monoclonal antibody against tissue factor, significantly shortened lysis time with t-PA and reduced the incidence of reocclusion (RAGNI et al. 1996). In guinea pigs, a platelet-rich carotid thrombus and CFRs could be induced even without a stenotic device (ROUX et al. 1994; CARTEAUX et al. 1995). The occurrence of CFRs, however, was not as reproducible as the original FOLTS model or in rabbit carotid arteries, and often required repeated "pinches" over a period of time. For this reason, the thrombosis index (the ratio of the number of CFRs to the number of pinches) was applied to quantify the process of thrombus formation (CARTEAUX et al. 1995). Because of the lower shear rate compared to the preparations with a critical stenosis, it has been suggested that the guinea pig CFRs model may be more suitable for mimicking the human pathophysiology of transient ischemic attacks (TIAs) (ROUX et al. 1994).

The extension of the original FOLTS model to the peripheral arteries of smaller animals offers some advantage. The rabbit and guinea pig are small, relatively inexpensive and practical for most investigators, and provide sufficient data when testing drugs that are available only in limited amounts. Access to the carotid artery is easy, and both carotid arteries can be used in the same animal. Unlike coronary artery thrombosis, in which lethal ventricular arrhythmias often occur, occlusive thrombi in carotid arteries cause no premature animal death, thereby minimizing loss of data and total animal usage.

Pinching-induced endothelial injury has been used in other types of in vivo models of thrombosis as well, in which different experimental parameters were assessed to analyze thrombus formation and growth. In a segment of porcine carotid artery deeply injured by hemostat crushes, platelet accretion was measured as ^{111}In radioactivity by a KI scintillation probe placed over the vessel (McBane et al. 1995). Occlusive thrombus could also be induced in the rat carotid artery after 5 min of clamp occlusion. The time to thrombotic occlusion was determined by measuring the decrease in temperature when occlusion occurred using a thermometer in contact with the arterial surface (Bagdy et al. 1992).

Pierangeli et al. (1994, 1995) reported an innovative model in mice, a species in which it is difficult to induce and monitor dynamic mural thrombus formation. They modified the method of Stockmans and coworkers (1991). A 1-cm segment of the mouse femoral vein was dissected free and a standardized "pinch" technique was performed to produce two small thrombogenic injury sites on the superior surface of the vein. The vessel was transilluminated using acrylic optical fibers, while the size and dynamics of thrombus formation were visualized and recorded through a surgical microscope equipped with a video camera, video recorder, and computer assisted digitized planimetry. The image of the thrombus appeared bright yellow white through the microscope due to the aggregating platelets, and displaced the darker flowing red blood cells. Therefore, the degree of brightness of the thrombus was used to indicate its size. In a typical preparation, a thrombus matures rapidly to reach its maximum size in the first 1–5 min and then decreases in area over 30 min. In mice which were passively immunized with IgG-APS (IgG from a patient with antiphospholipid syndrome) or the mice which received purified IgG anticardiolipin antibodies, the thrombi were significantly larger and persisted for a longer period of time than in control animals (Pierangeli et al. 1994, 1995). It is notable that the detailed kinetics of thrombus formation and dissolution can be quantified and analyzed in such small animals in this model. It enables investigators to obtain substantial information without using large amounts of investigative compounds. The application of this technique in transgenic mice may be useful for studying the mechanisms of thrombotic disorders in the future (see Chap. 6, this volume).

2. Perfusion with Saline or Air

Although the basic technique is mechanically to damage the vessel wall, differences in specific methods may give rise to variability in the nature or the severity of the injury. These in turn will influence the characteristics of thrombus formation and its composition, and eventually effect the response to antithrombotic agents. Millet et al. (1987) developed a model in rats using saline flushing and subsequent partial stasis in an isolated segment of the vena cava. It was noted that flushing of saline at the rate of 10 ml/min for 15 s induces only discrete endothelial lesions without evidence of widespread endothelial disruption as assessed using transmission electron microscopy. This is in contrast to the vessel, which is deeply damaged from crushing or pinching. In the model of Millet et al. (1987), lesions could be the result of a ballooning effect or flow turbulence created by flushing. The thrombus grows to its maximum weight 15 min after restoring the blood flow. Apparently platelets do not play an important role in this model, because thrombosis could be induced in thrombocytopenic rats. Using this technique, heparin and related compounds, but not aspirin, can prevent thrombus formation (MILLET et al. 1987).

A similar method has been applied by VAN RYN-MCKENNA et al. (1993) in the jugular veins of rabbits. A 2-cm segment of jugular vein was perfused with air, instead of saline, at 150 ml/min for 10 min. Scanning electron micrographs showed that the endothelium was almost completely removed, and the subendothelium was exposed after injury. There was marked accumulation of platelets and fibrin, interspersed with erythrocytes 4 h after restoring the blood flow. ^{125}I-fibrin accretion onto the injured jugular veins was enhanced 2.4-fold compared to that in sham-operated animals. The thrombogenicity of the injury was not inhibited by prophylactic doses of heparin, but it was attenuated by annexin V, which inhibits thrombin generation by preventing the assembly of prothrombinase complex on phospholipid surfaces (VAN RYN-MCKENNA et al. 1993). The results from these studies also suggest that a slight or mild vessel wall injury with or without partial stasis is sufficient to induce thrombosis in veins.

3. Endarterectomy and Balloon Angioplasty

Surgical endarterectomy and balloon angioplasty are commonly used in treating patients with atherosclerotic vascular disease. Because of the direct vascular injury and frequent acute thrombus formation, these interventional approaches have been adapted in animals to develop experimental models of thrombosis (SUAREZ and JACOBSON 1961; STEELE et al. 1985; LUMSDEN et al. 1993; KRUPSKI et al. 1991; LAM et al. 1991; MRUK et al. 1996). The morphology at the site of injury, platelet deposition, and mural thrombus formation in the damaged arteries after these procedures are well documented in several types of animal species. Instead of being a general model of thrombosis for evaluation of antithrombotic drugs, the results from these experiments may be more relevant to the rethrombotic complications associated with interventional therapies currently used in the management of arterial disease in humans.

IV. Electrical Current-Induced Injury

LUTZ et al. (1951) and SAWYER and PATE (1953a,b) reported that abnormal or reversed electrical potentials with sufficient current could cause blood vessel wall injury and thrombosis in the hamster and dogs, respectively. In 1980, ROMSON et al. developed a modified electrolytic model of canine coronary artery thrombosis which did not require fluoroscopic control. Since then, the basic technique of this model has been utilized with some modifications by many investigators in thrombosis research (SCHUMACHER et al. 1985; MICKELSON et al. 1990; BENEDICT et al. 1993; LYNCH et al. 1995; ROUX et al. 1996).

In the electrolytic injury model, an electrode is inserted into the lumen of the left circumflex (LCX) coronary artery of dogs. The electrode can be constructed from a 25- or 26-gauge stainless steel hypodermic needle tip attached to a Teflon-insulated, silver-coated, copper wire. LCX coronary artery injury is induced by the application of a continuous anodal direct current to the electrode that is in firm contact with the endothelium. The current delivered can range from 50 to 300 μA, though 150 μA is most commonly used. The current can be delivered using either a 9-V battery (Romson et al. 1980; Schumacher et al. 1985) or a constant current unit powered by a stimulator (Lynch et al. 1995).

The duration of electrical current applied to the artery varies among different laboratories. The current is maintained until LCX blood flow decreases to zero and remains at zero for a 30-min period in the preparations that develop occlusive thrombi, or for several hours in vessels that remain patent (Lynch et al. 1995). In the studies by Roux et al. (1996) and Shetler et al. (1996), for example, the current was applied for a total of 60 min and then switched off irrespective of the state of vessel patency. In the control dogs, nine out of ten vessels occluded; four vessels occluded before the current was switched off and five vessels occluded within 1 h of the current being switched off. The mean occlusion time was 67 ± 7 min (30–100 min) (Roux et al. 1996). In the experiments described by Benedict et al. (1993), the current was stopped when 50% occlusion of the vessel developed, indicated by a 50% increase in blood flow velocity, after which pharmacological interventions were initiated. In most studies, an adjustable mechanical occluder is placed near the needle electrode to create a critical stenosis, which greatly minimizes or eliminates the reactive hyperemic response. The electrolytic injury produces endothelial damage and formation of a platelet-rich thrombus (Sudo et al. 1995; Romson et al. 1980). Figure 2 illustrates the placement of the electrode, mechanical constrictor and flow probe on a artery of the anesthetized dog.

Numerous antiplatelet and antithrombin compounds have been evaluated using the electrolytic injury model of coronary artery thrombosis. The recombinant tick anticoagulant peptide (rTAP), a factor Xa inhibitor, significantly delayed time to thrombosis and reduced the residual thrombus mass (Lynch et al. 1995). Heparin and napsagatran, a synthetic direct thrombin inhibitor, were equally effective in preventing occlusive intracoronary thrombus formation, though nonocclusive CFRs were present in many dogs treated with high doses of these compounds (Roux et al. 1996). Benedict et al. (1993) reported that bovine glutamyl-glycinyl-arginyl-factor Xa (Xai), a competitive inhibitor of factor Xa assembly into the prothrombinase complex, dose dependently prolonged time to occlusion, and decreased the deposition of ^{125}I-fibrin and ^{111}In platelets at the site of the coronary thrombosis.

The application of this model also has been extended to study thrombolysis, rethrombosis, and adjunctive therapy with antithrombotic agents (Schumacher et al. 1985; Mickelson et al. 1990; Sitko et al. 1992; Jackson et

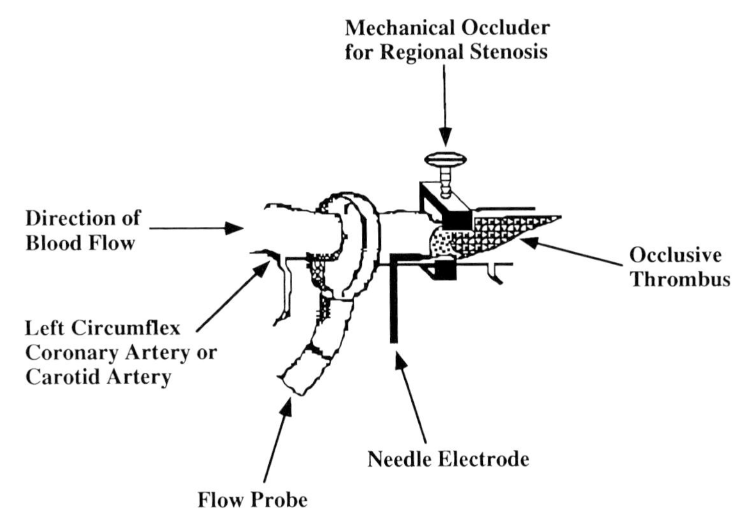

Fig. 2. Schematic illustration of the electrical current-induced thrombosis model in the dog. An anodal current is delivered to the intimal surface of an artery. An adjustable, mechanical occluder is placed near the needle electrode to create a regional stenosis. Occlusive platelet-rich thrombus develops at the site of injured vessel wall. The blood flow is measured using a flow probe. At the end of experiment, thrombus weight can be determined

al. 1993). In these experiments, intracoronary thrombi are lysed with tissue-type plasminogen activator (t-PA) or streptokinase. Antiplatelet antibody [7E3 $F(ab')_2$] (MICKELSON et al. 1990), rTAP, hirudin (SITKO et al. 1992) and tripeptide thrombin inhibitors (JACKSON et al. 1993) were administered along with t-PA. The incidence and time to thrombolytic reperfusion and reocclusion were determined. Thus, the effects of adjunctive treatment with these antithrombotic agents could be evaluated and compared.

Electrically induced endothelial injury and thrombosis have recently been applied to peripheral vessels, such as carotid arteries and jugular veins in dogs (ROTE et al. 1993; SUDO et al. 1995; SUDO and LUCCHESI 1996). Thrombosis was sequentially initiated in the contralateral carotid arteries in the same dog, using one artery as a control vessel (ROTE et al. 1993). The current can also be applied simultaneously to the carotid artery and jugular vein and the effect of a compound on both arterial and venous thrombosis and/or rethrombosis can be studied in a single animal (SUDO et al. 1995; SUDO and LUCCHESI 1996; REBELLO et al. 1997).

In summary, the technique of electrical current-induced vessel damage and thrombosis is relatively easy to perform and has been used in the carotid artery, the jugular vein, and the coronary artery. The extent of injury can be controlled precisely by adjusting the current. In addition, the coronary thrombosis model in a conscious, ambulatory animal provides an advantage over most of the other models for chronic studies (COUSINS et al. 1996). In fact, electrically induced coronary artery thrombosis in conscious dogs has long been used for myocardial infarction and arrhythmia studies (Uprichard et al. 1989; Chi et al. 1991).

C. Stasis/Hypercoagulability-Induced Models of Thrombosis

I. Wessler Test and Its Variants

Intravascular injection of procoagulable or proaggregatory substances is probably one of the oldest techniques of inducing thrombosis in animals. Various materials such as serum (Flexner 1902), chemicals (Yatsushiro 1913), thromboplastin (Copley and Stefko 1947), Russell viper venom (McLetchie 1952), thrombin (Wright et al. 1952), ADP (Jorgensen et al. 1970), and collagen (Nishizawa et al. 1972) have been used for some time. In the 1950s, Stanford Wessler (1952, 1955) combined local venous stasis with hypercoagulability and established an in vivo model for venous thrombosis that was subsequently known as the "Wessler model" or "Wessler test." In this model, serum from humans or dogs is infused into the systemic circulation of dogs or rabbits. After the infused serum has dispersed throughout the circulation (15 s to 1 min), the jugular vein segment is occluded with clamps. A large red clot forms in the isolated venous segment within a few minutes of clamping (Wessler 1955; Wessler et al. 1959). The serum or other substances used in this technique induce systemic hypercoagulability, and it has been shown that there is an inverse correlation between the duration of stasis and the amount of hypercoagulating agents required to produce a thrombus (Aronson and Thomas 1985).

The Wessler test has been a fully validated, standardized animal model for the study of venous thrombosis for over 40 years and is still being used with some modifications (Thomas 1996). More site specific (in the coagulation cascade) thrombogenic agents such as thromboplastin, activated prothrombin complex concentrates, thrombin, and factor Xa have been employed (Aronson and Thomas 1985; Fareed et al. 1985; Walenga et al. 1987). Thus, the site of action of test agents can be predicted by selectively activating a specific component of the coagulation system. Shortly after the systemic injection of the procoagulant substance, a segment (1–2 cm) of vein, commonly the inferior vena cava in rats (Tapparelli et al. 1995; Millet et al. 1996) and external jugular vein in rabbits (Walenga et al. 1987; Mauray et al. 1995), is isolated with two ligatures or clamps. The total stasis is maintained for 10 min after which the venous segment is opened and the thrombus is weighed or visually graded. Carter et al. (1982) and Buchanan et al. (1994) modified the model by injecting ^{125}I-labeled fibrinogen into the rabbits before infusing thrombin. The amount of fibrinogen converted to fibrin clot was then calculated by counting the specific radioactivity in the thrombus. This modification improves the sensitivity and accuracy of the measurements compared to the visually graded score or wet weight of the clot.

The Wessler model and its variants are used almost exclusively for venous thrombosis. Clinical and pathological observations indicate that vessel wall damage is absent in most cases of venous thrombosis, while circulating procoagulants play an important role in the pathogenesis of the disease, especially when local stasis is superimposed (THOMAS 1988). Experimental studies have shown that vessel wall damage alone is relatively ineffective in producing venous thrombosis (ARONSON and THOMAS 1985). In rabbit jugular veins, mechanical crushing caused endothelial disruption and platelet adhesion, but no thrombus formed over a period of 30 min of stasis. On the other hand, the Wessler test and its variants have clearly demonstrated the venous thrombogenicity of the combination of stasis and hypercoagulability.

There are number of limitations, however, in the stasis/hypercoagulability animal model. First, it is a static model which fails to take account of dynamic factors present in the clinical situations, such as shear stress and interactions between flowing blood components and the vessel wall. Second, the infusion of procoagulants often induces significant changes in the hemodynamic and coagulation status (TAPPARELLI et al. 1995). This makes it difficult to interpret the effects on coagulation solely caused by the test compounds. Heterologous serum or some other substances may also cause massive hemolysis, formation of immune complexes, and changes in the blood chemistry profile which may contribute to thrombogenesis in animals (FAREED at al. 1985; MILLET et al. 1996). Third, the experimental results are relative to the type of procoagulant used. The antithrombotic potency rank for some drugs could be different if the procoagulants differ (FAREED at al. 1985; WALENGA et al. 1987).

D. Foreign Surface-Induced Thrombosis

When foreign substances are introduced into the circulation, local activation of coagulation and platelets occurs. Thrombus gradually forms on the foreign surface, and may become occlusive over a period of time. Various thrombogenic materials and techniques have been used to develop experimental animal models. In contrast to the models described above, endothelial damage is not a prerequisite for the initiation of thrombus formation in these experiments.

I. Eversion Graft

HERGRUETER et al. (1988) developed an eversion graft model for producing thrombosis in the rabbit artery. This was later modified by JANG et al. (1989, 1990) and GOLD et al. (1991) using an everted segment of vessel in the canine coronary artery and rabbit femoral artery. A 4- to 6-mm segment of the right femoral artery is excised, everted, and then interposed by end-to-end anastomosis into the left femoral artery. A similar method is used in left circumflex coronary (LCX) artery of dogs. After the restoration of blood flow, the perfusion status can be monitored by a flow probe and can serve as the primary end point. A platelet-rich occlusive thrombus forms in approximately 5 min in the canine LCX artery and in less than 1 h in rabbit femoral artery (Jang et al. 1989, 1990; Gold et al. 1991).

The eversion graft model is reliably thrombogenic, although technically quite difficult and time consuming. The adventitial surface is a nonendothelial tissue containing tissue factor and collagen; thus both the coagulation system and blood platelets are activated. Thrombosis that develops does so in free-flowing blood, resulting in fibrin and platelet deposition.

II. Wire Coils

The insertion of wire coils into the lumen of blood vessels has long been used as a method to produce thrombosis (Stone and Lord 1951; Blair et al. 1964; Kordenat et al. 1972). The wire coils are made of magnesium, magnesium-aluminum (Stone and Lord 1951), copper (Kordenat et al. 1972), or stainless steel (Kumada et al. 1980), with different lengths of coil and a selected number of turns. In an early study by Kordenat et al. (1972), the coils were attached to a guide catheter and advanced into the coronary artery through the carotid artery of dogs using fluoroscopic control. An obstructing thrombus developed 1 h to several days after wire placement, depending on the composition and shape of the wire coil used and its location within the vessel (Kordenat et al. 1972). In the rat inferior vena cava, a stainless steel wire coil was directly inserted into the lumen by puncturing the vein wall (Kumada et al. 1980; Herbert et al. 1996). The thrombus size was determined by total protein content in the isolated thrombus (Kumada et al. 1980). The copper coil model has also been used in the femoral artery (Mellott et al. 1993) and carotid artery (Rubsamen and Hornberger 1996) of anesthetized dogs.

The formation of thrombus around the coil is highly reproducible and dose-response studies can be performed by quantifying the weight of thrombi. It is an obvious limitation of the model, however, that the anatomic and pathophysiologic environment of coronary thrombosis in this model do not simulate very precisely that of acute arterial thrombosis in man.

III. Preformed Thrombi

In an effort to study the effect of LMWH in preventing the extension of established thrombi, Boneu et al. (1985) developed a venous thrombosis model in the jugular veins of rabbits. In the anesthetized rabbit, a 2-cm segment of both jugular veins was isolated and blood was emptied from each segment. One milliliter of whole blood, collected from a carotid cannula, was mixed with 50 μl thrombin (2 U/ml) in a syringe and 150 μl clotted blood was immediately injected into each isolated venous segment via a 23-g needle. A suture was passed through the clot and vessel wall to prevent the thrombus from embolizing. Thrombi were allowed to age for 30 min, after which the blood flow was restored. ^{125}I-fibrinogen can be injected systemically and its accretion onto the preformed thrombus can be determined (BONEU et al. 1985). $CaCl_2$ can also be added into the mixture of blood and thrombin to enhance clot formation (AGNELLI et al. 1990; BIEMOND et al. 1996). Standard heparin (BUCHANAN et al. 1994), r-Hirudin (AGNELLI et al. 1990), rTAP, LMWH (Fraxiparin), and CVS#995, a direct inhibitor of thrombin (BIEMOND et al. 1996), have all reduced the accretion of ^{125}I-labeled fibrinogen onto preformed nonradioactive thrombi and subsequent thrombus growth in this model.

According to our own experience and that of FIORAVANTI et al. (1993), this model is associated with considerable variation in thrombus size even under controlled conditions. Improper mixing of thrombin, $CaCl_2$ and blood, or their leakage through the needle hole after injection, may cause variation in the size of preformed clot, which in turn may influence the thrombus growth. The gross morphology of the preformed thrombi inside the venous segment could also influence the final thrombus size. Another source of variability results from the embolization of thrombus fragments during the course of study (FIORAVANTI et al. 1993). This model was established mainly for thrombus growth or extension from already existing thrombi, and not for thrombus initiation.

IV. Hollenbach's Deep Venous Thrombosis Model

HOLLENBACH et al. (1994) developed a rabbit model of venous thrombosis by introducing cotton threads into the abdominal vena cava. Eight strands of cotton thread, 3 cm in length, were tied to a 14-cm-long looped copper wire. The wire was inserted into a polyethylene tubing fitted with a hub adaptor. The apparatus was then advanced into the abdominal vena cava via the femoral vein, and the cotton threads were pushed out of the tubing and exposed to the venous blood flow by advancing the copper wire. The cotton threads serve as a thrombogenic surface, and a thrombus forms around it growing to a maximum mass after 2–3 h. As with the preformed thrombus model described above, this model focuses on progression of thrombus formation rather than initiation. The infusion of antithrombotic agents is not started until 30 min after advancement of the cotton threads into the vena cava. Therefore, during the latter phase of the experiment, the effects of interventions are measured by the extent thrombus size increases above and beyond the thrombus that developed during the 30-min "incubation" period. The average thrombus growth in control experiments (lasting a total of 120 min) is 53 mg. Experiments with prothrombinase complex inhibitors (rXai and EGR-Xai), a thrombin inhibitor (PPACK), a factor Xa inhibitor (DX-9065a), and LMWH (Lovenox) were all associated with a dose-dependent inhibition of thrombus formation (HOLLENBACH et al. 1994, 1995), demonstrating the utility of this model.

It seems to be more difficult to develop a thrombosis model in the venous circulation than in the arterial system. Reproducibility and variability have been a challenge for many researchers in designing an experimental model for venous thrombosis in animals, particularly in larger animals. This might explain why complete stasis of a segment of vein is used by so many investigators to measure the effects of interventions on venous thrombosis. In the Hollenbach model, thrombus is formed under conditions which may more closely resemble pathophysiology in some humans because blood continues to flow throughout most of the experiment. The primary end point in their studies, however, is clot weight, which is determined at the conclusion of the experiment. The rate or kinetics of thrombus formation in this model is usually not evaluated.

V. A Novel Veno-Venous Shunt Model in Rabbit

To evaluate new antithrombotic agents, especially for their indications in prevention and/or treatment of deep venous thrombosis (DVT) in patients, our laboratory has developed a novel model of venous thrombosis in rabbits. In anesthetized New Zealand white rabbits, a midline laparotomy is performed and a segment of the abdominal vena cava is isolated. The segment is clamped at both distal and proximal ends and two incisions are made in the vessel wall, approximately 2 cm apart. A plastic tube is placed in the vessel through the incisions to serve as a venovenous shunt (Fig. 3). The shunt consists of a PE-280 tubing (3 cm long), the inside surface of which is siliconized. Six strands of cotton threads are passed inside the tube and anchored with a ligature on the outside of the tube.

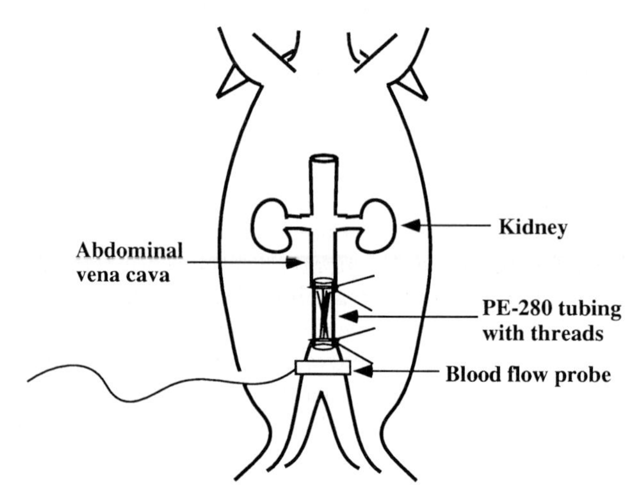

Fig. 3. Schematic illustration of the veno-venous shunt model in the vena cava of a rabbit. A 3 cm long PE-280 tubing containing 6 strands of cotton threads is cannulated into the middle of abdominal vena cava. A flow probe is placed at the distal end of the vessel and blood flow is continuously recorded with a flow meter. The shunt can be removed and replaced in the course of a single experiment, enabling use of each animal as its own control

The blood flow can be recorded continuously via a perivascular flow probe placed at the distal end of the vena cava. When the blood flow is restored, the thrombus develops inside the shunt indicated by the gradual reduction of vena caval blood flow. The cotton threads within the shunt to provide a thrombogenic surface on which the clot forms. Ten minutes after occlusive thrombus formation, the shunt is removed and the weight of the clot is determined. After administration of the test compound or vehicle, the second shunt is placed into the vena cava of the same animal.

Thus, the present preparation has two advantages compared to HOLLENBACH's model. First, in addition to the determination of thrombus size, the kinetics of thrombus formation over time can be analyzed by recording the blood flow. For example, the time to occlusion, incidence of occlusion, and integrated blood flow over the course of the entire experiment can be quantified. Second, two shunts can be tested consecutively in the same rabbit using the first shunt as a control. Data suggest that the variability of results within different animals or between the shunts in the same animal is minimal. In seven control rabbits, occlusive thrombi were formed in all of the first and second shunts and the mean times to occlusion were 20.6 ± 5.2 min and 20.2 ± 5.7 min, respectively. The net thrombus weights, calculated by subtracting the weight of cotton threads from the total clot weight, were 49.0 ± 3.5 mg and 47.0 ± 3.3 mg for the first and second shunts, respectively. The change in blood flow was similar in the first and second shunts (Fig. 4). This model has been characterized further and validated using an LMWH (Lovenox) and a direct thrombin inhibitor (inogatran). Both compounds showed dose-dependent inhibitory effects on thrombus formation, defined in terms of time to occlusion and thrombus weight.

The conventional arteriovenous (A-V) shunt models have been used widely in many different species including rat (HARA et al. 1995), rabbit (KNABB et al. 1992), pig (SCOTT et al. 1994), and nonhuman primates (YOKOYAMA et al. 1995). Because of the high pressure and shear rate inside the A-V shunts, the thrombi tend to be more arterial in character and have been utilized primarily for arterial thrombosis studies. In the venovenous shunt model we use, the composition of thrombi is fibrin-rich, containing red blood cells and platelets trapped in the fibrin meshwork (Fig. 5). One of the limitations of this model is that the initiation of thrombosis is probably tissue factor independent. We are currently examining, however, the utility of coating the cotton threads with a preformed plasma clot which can be induced by incubating the threads in plasma containing thrombin or thromboplastin.

E. Transgenic Animal Models

The technique to alter a single gene in vivo has been used to create animal models to study the pathophysiology and therapy for diseases in humans (BARRETT and MULLINS 1992). In the thrombosis area, both overexpression

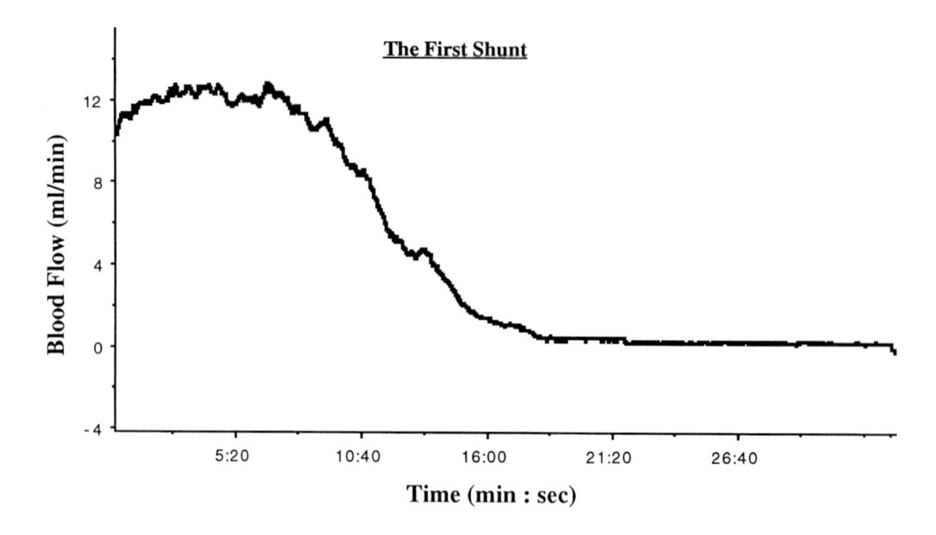

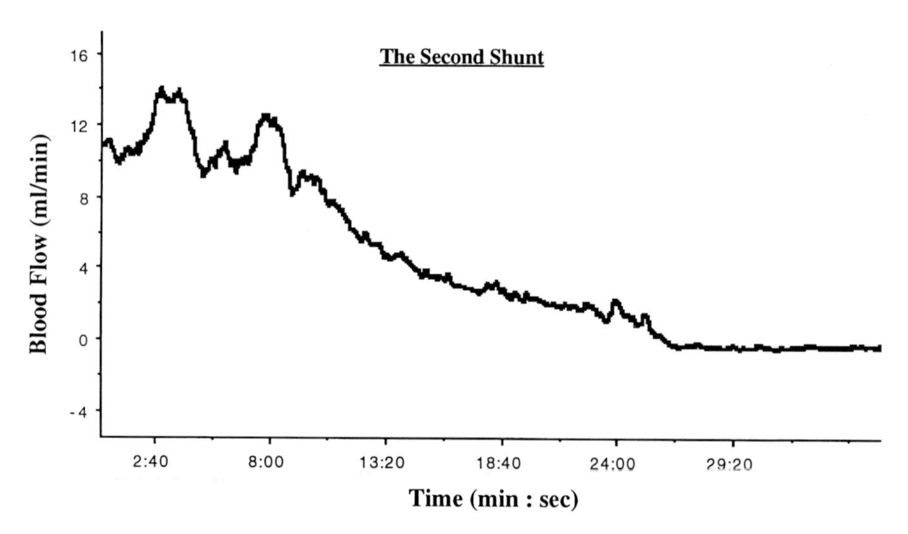

Fig. 4. Examples of mean blood flow tracings recorded from the first and second veno-venous shunts in an anesthetized rabbit under control (no treatment) conditions to illustrate the reproducibility of the technique. The blood flow measurements were digitized continuously with a MacLab/8e data acquisition system. Time to occlusion, indicated by zero flow, was 23 and 26 min in the first and second shunts, respectively

and knockout of a specific gene can be achieved in transgenic mice. ERICKSON et al. (1990) examined the contribution of PAI-1 to thrombus formation by using the metallothionine promoter to overexpress PAI-1 in transgenic mice. The mice with an increased level of PAI-1 developed venous, but not arterial, thrombotic occlusions (ERICKSON et al. 1990). Mice with single or combined

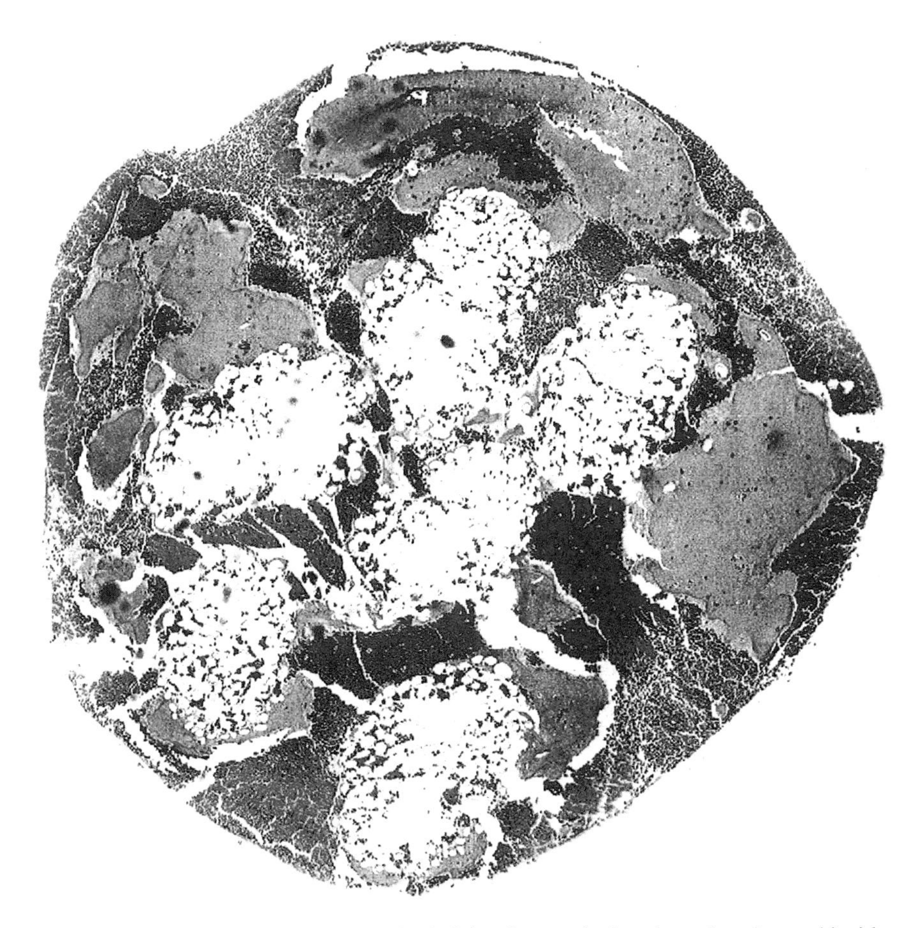

Fig. 5. Digitized image of a histological slide of an occlusive thrombus formed inside a veno-venous shunt. The thrombus was fixed in 1:10 buffered formaldehyde and the cross-sectional histological section was stained with Carstairs stain. Fibrin is indicated by the patches colored bright blue and red blood cells trapped in the fibrin meshwork appear red. The six round light (or open) areas in the center of the figure indicate the location of the six strands of cotton thread which were surrounded by a mass of fibrin and red blood cells

deficiency of t-PA and/or u-PA have been generated (CARMELIET et al. 1995). There was a higher incidence and a larger extent of venous thrombosis in mice with inactivated genes for t-PA or u-PA after local injection of endotoxin in the footpad (CARMELIET et al. 1994).

Although the development of transgenic animal models was intended to identify the role of a specific gene in thrombosis and thrombolysis, they can be utilized as models for evaluation of antithrombotic agents as well. In contrast to the conventional models, one of the advantages of the transgenic animal model is its ability to pinpoint a specific target for effective antithrombotic therapies. Detailed discussion on the transgenic model of thrombosis is presented in Chap. 6, this volume.

F. Conclusions

Animal models are useful experimental tools that can allow us to generate reliable, reproducible, and quantitative data that can be analyzed to predict the efficacy of a antithrombotic agent in the clinical setting. An important requirement in developing in vivo models is to simulate clinical situations as accurately as possible. Differences do exist in the pathogenesis of arterial and venous thrombotic disorders even though the basic elements of thrombus formation are the same. The factors that incite formation of a thrombus in the coronary artery are related to vessel wall changes due to atherosclerosis. The risk factors for deep vein thrombosis are substantially different. Accordingly, vessel wall injury is often applied in arterial thrombosis models whereas stasis, hypercoagulability, and thrombogenic surfaces are used frequently in venous thrombosis models.

Another major difference is the composition of the clots. Arterial thrombi which form in high shear settings are commonly platelet-rich (BUSH and SHEBUSKI 1990; ANDERSON and WILLERSON 1994), whereas venous thrombi are composed of mainly fibrin and red cells (THOMAS 1994; HOLLENBACH et al. 1994). This difference may contribute to the different sensitivity of the two systems in response to pharmacological interventions. Antiplatelet agents, such as GPIIb/IIIa antagonist (SUDO et al. 1995), are more efficacious in the inhibition of arterial than venous thrombosis. On the other hand, venous thrombosis tends to be more sensitive to the inhibition of fibrin generation which is achieved by heparin, factor Xa and thrombin inhibitors. Thus, depending on the molecular target of inhibition and the mechanism of inhibition, different results may be obtained in different experimental models.

We have learned a great deal about the mechanisms of thrombosis and the actions of antithrombotic agents from animal studies. It is still difficult, however, to directly extrapolate the experimental results to the clinical situations. First, like all other models of diseases, species difference must be taken into account. Drugs that show efficacy in an animal model may not be efficacious for a human disease. Second, most of the in vivo models of thrombosis are acute or subacute in nature. While it is true that arterial thrombi can develop very rapidly, venous thrombi may develop in humans over days or weeks. The risk factors and pathology which promote thrombosis in patients are far more complicated and subtle than the experimental conditions applied in the animal models. Last, it is not uncommon that the results of a particular antithrombotic drug being tested are model dependent. Therefore caution should be exercised before extrapolating the results of animal studies to the human condition.

References

Agnelli G, Pascucci C, Cosmi B, Nenci GG (1990) The comparative effects of recombinant hirudin (CGP 39393) and standard heparin on thrombus growth in rabbits. Thromb Haemost 63:204–207

Anderson HV, Willerson JT (1994) Experimental models of thrombosis. In: Loscalzo J, Schafer AI (eds) Thrombosis and hemorrhage. Blackwell Scientific, Boston, pp 385–393

Arfors KE, Dhall DP, Engeset J, Hint H, Matheson NA, Tangen O (1968) Biolaser endothelial trauma as a means of quantifying platelet activity in vivo. Nature 218:887–888

Aronson DL, Thomas DP (1985) Experimental studies on venous thrombosis: effect of coagulants, procoagulants and vessel contusion. Thromb Haemost 54:866–870

Bagdy D, Szabo G, Barabas E, Bajusz S (1992) Inhibition by D-MePhe-Pro-Arg-H (GYKI-14766) of thrombus growth in experimental models of thrombosis. Thromb Haemost 68:125–129

Barrett G, Mullins JJ (1992) Transgenic animal models of cardiovascular disease. Curr Opin Biotechnol 3:637–640

Belougne E, Aguejouf O, Imbault P, Azougagh-Oualane F, Doutremepuich F, Droy-Lefaix MT et al (1996) Experimental thrombosis model induced by laser beam. Application of aspirin and an extract of Ginkgo biloba: EGb 761. Thromb Res 82:453–458

Benedict CR, Ryan J, Todd J, Kuwabara K, Tijburg P, Cartwright J et al (1993) Active site-blocked factor Xa prevents thrombus formation in the coronary vasculature in parallel with inhibition of extravascular coagulation in a canine thrombosis model. Blood 81:2059–2066

Biemond BJ, Friederich PW, Levi M, Vlasuk GP, Buller HR, ten Cate JW (1996) Comparison of sustained antithrombotic effects of inhibitors of thrombin and factor Xa in experimental thrombosis. Circulation 93:153–160

Blair E, Nygren E, Cowley RA (1964) A spiral wire technique for producing gradually occlusive coronary thrombosis. J Thorac Cardiovasc Surg 48:476–485

Boneu B, Buchanan MR, Cade JF, Van Ryn J, Fernandez FF, Ofosu FA et al (1985) Effects of heparin, its low molecular weight fractions and other glycosaminoglycans on thrombus growth in vivo. Thromb Res 40:81–89

Buchanan MR, Liao P, Smith LJ, Ofosu FA (1994) Prevention of thrombus formation and growth by antithrombin III and heparin cofactor II-dependent thrombin inhibitors: importance of heparin cofactor II. Thromb Res 74:463–475

Bush LR, Shebuski RJ (1990) In vivo models of arterial thrombosis and thrombolysis. FASEB J 4:3087–3098

Bush LR, Campbell WB, Buja LM, Tilton GD, Willerson JT (1984) Effects of the selective thromboxane synthetase inhibitor dazoxiben on variations in cyclic blood flow in stenosed canine coronary arteries. Circulation 69:1161–1170

Carmeliet P, Schoonjans L, Kieckens L, Ream B, Degen J, Bronson R, DeVos R, van den Oord JJ, Collen D, Mulligan RC (1994) Physiological consequences of loss of plasminogen activator gene function in mice. Nature 368:419–424

Carmeliet P, Bouche A, de Clerco C, Janssen S, Pollefeyt S, Wyns S et al (1995) Biological effects of disruption of the tissue-type plasminogen activator, urokinase-type plasminogen activator, and plasminogen activator inhibitor-1 genes in mice. Ann NY Acad Sci 748:367–381

Carteaux JP, Gast A, Tschopp TB, Roux S (1995) Activated clotting time as an appropriate test to compare heparin and direct thrombin inhibitors such as hirudin or Ro 46–6240 in experimental arterial thrombosis. Circulation 91:1568–1574

Carter CJ, Kelton JG, Hirsh J, Cerskus A, Santos AV, Gent M (1982) The relationship between the hemorrhagic and antithrombotic properties of low molecular weight heparin in rabbits. Blood 59:1239–1245

Chi LG, Mu DX, Lucchesi BR (1991) Electrophysiology and antiarrhythmic actions of E-4031 in the experimental animal model of sudden coronary death. J Cardiovasc Pharmacol 17:285–295

Coller BS, Folts JD, Smith SR, Scudder LE, Jordan R (1989) Abolition of in vivo platelet thrombus formation in primates with monoclonal antibodies to the platelet GPIIb/IIIa receptor: correlation with bleeding time, platelet aggregation and blockade of GPIIb/IIIa receptors. Circulation 80:1766–1774

Copley AL, Stefko PL (1947) Coagulation thrombi in segments of artery and vein in dogs and genesis of thromboembolism. Surg Gynecol Obstet 84:451–459

Cousins GR, Friedrichs GS, Sudo Y, Rebello SS, Rote WE, Vlasuk GP et al (1996) Orally effective CVS-1123 prevents coronary artery thrombosis in the conscious dog. Circulation 94:1705–1712

Erickson LA, Fici GJ, Lund JE, Boyle TP, Polites HG, Marotti KR (1990) Development of venous occlusions in mice transgenic for the plasminogen activator inhibitor-1 gene. Nature 346:74–76

Fareed J, Walenga JM, Kumar A, Rock A (1985) A modified stasis thrombosis model to study the antithrombotic actions of heparin and its fractions. Semin Thromb Hemost 11:155–175

Fioravanti C, Burkholder D, Francis B, Siegl PK, Gibson RE (1993) Antithrombotic activity of recombinant tick anticoagulant peptide and heparin in a rabbit model of venous thrombosis. Thromb Res 71:317–324

Flexner S (1902) On thrombi composed of agglutinated red blood corpuscles. J Med Res 8:316–321

Florey H (1925) Microscopical observations on the circulation of the blood in the cerebral cortex. Brain 48:43–64

Folts JD (1991) An in vivo model of experimental arterial stenosis, intimal damage, and periodic thrombosis. Circulation 83:IV3–IV14

Folts JD, Crowell EB Jr, Rowe GG (1976) Platelet aggregation in partially obstructed vessels and its elimination with aspirin. Circulation 54:365–370

Folts JD, Gallagher KP, Rowe GG (1982) Blood flow reductions in stenosed canine coronary arteries: vasospasm or platelet aggregation? Circulation 65:248–255

Gallagher KP, Osakada G, Kemper WS, Ross J Jr (1985) Cyclical coronary flow reductions in conscious dogs equipped with ameroid constrictors to produce severe coronary narrowing. Basic Res Cardiol 80:100–106

Gold HK, Yasuda T, Jang IK, Guerrero JL, Fallon JT, Leinbach RD et al (1991) Animal models for arterial thrombolysis and prevention of reocclusion. Erythrocyte-rich versus platelet-rich thrombus. Circulation 83:IV26–IV40

Golino P, Ambrosio G, Pascucci I, Ragni M, Russolillo E, Chiariello M (1992) Experimental carotid stenosis and endothelial injury in the rabbit: an in vivo model to study intravascular platelet aggregation. Thromb Haemost 67:302–305

Golino P, Ambrosio G, Ragni M, Pascucci I, Triggiani M, Oriente A et al (1993) Short-term and long-term role of platelet activating factor as a mediator of in vivo platelet aggregation. Circulation 88:1205–1214

Grand L, Becker FF (1965) Mechanisms of inflammation. I. Laser-induced thrombosis, a morphologic analysis. Proc Soc Exp Biol Med 119:1123–1129

Hara T, Yokoyama A, Tanabe K, Ishihara H, Iwamoto M (1995) DX-9065à, an orally active, specific inhibitor of factor Xa, inhibits thrombosis without affecting bleeding time in rats. Thromb Haemost 74:635–639

Herbert JM, Bernat A, Samama M, Maffrand JP (1996) The antiaggregating and antithrombotic activity of ticlopidine is potentiated by aspirin in the rat. Thromb Haemost 76:94–98

Hergrueter CA, Handren J, Kersh R, May JW Jr (1988) Human recombinant tissue-type plasminogen activator and its effect on microvascular thrombosis in the rabbit. Plast Reconstr Surg 81:418–424

Hill DS, Smith SR, Folts JD (1987) The rabbit as a model for carotid artery stenosis and periodic acute thrombosis. Fed Proc 46:421

Hirsh J, Hoak J (1996) Management of deep vein thrombosis and pulmonary embolism. A statement for healthcare professionals. Council on thrombosis (in consultation with the Council on Cardivascular Radiology). Circulation 93:2212–2245
Hollenbach S, Sinha U, Lin PH, Needham K, Frey L, Hancock T et al (1994) A comparative study of prothrombinase and thrombin inhibitors in a novel rabbit model of non-occlusive deep vein thrombosis. Thromb Haemost 71:357–362
Hollenbach SJ, Wong AG, Ku P, Needham KM, Lin PH, Sinha U (1995) Efficacy of "fXa inhibitors" in a rabbit model of venous thrombosis. Circulation 92:I486–I487
Honour AJ, Ross Russell RW (1962) Experimental platelet embolism. Br J Exp Pathol 43:350–362
Imbault P, Doutremepuich F, Aguejouf O, Doutremepuich C (1996) Antithrombotic effects of aspirin and LMWH in a laser-induced model of arterials and venous thrombosis. Thromb Res 82:469–478
Inamo J, Belougne E, Doutremepuich C (1996) Importance of photo activation of rose bengal for platelet activation in experimental models of photochemically induced thrombosis. Thromb Res 83:229–235
Jackson CV, Wilson HC, Growe VG, Shuman RT, Gesellchen PD (1993) Reversible tripeptide thrombin inhibitors as adjunctive agents to coronary thrombolysis: a comparison with heparin in a canine model of coronary artery thrombosis. J Cardiovasc Pharmacol 21:587–94
Jang IK, Gold HK, Ziskind AA, Fallon JT, Holt RE, Leinbach RC et al (1989) Differential sensitivity of erythrocyte-rich and platelet-rich arterial thrombi to lysis with recombinant tissue-type plasminogen activator. A possible explanation for resistance to coronary thrombolysis. Circulation 79:920–928
Jang IK, Gold HK, Ziskind AA, Leinbach RC, Fallon JT, Collen D (1990) Prevention of platelet-rich arterial thrombosis by selective thrombin inhibition. Circulation 81:219–225
Jorgensen L, Hovig T, Rowsell HC, Mustard JF (1970) Adenosine diphosphate-induced platelet aggregation and vascular injury in swine and rabbits. Am J Pathol 61:161–176
Jourdan A, Aguejouf O, Imbault P, Doutremepuich F, Inamo J, Doutremepuich C (1995) Experimental thrombosis model induced by free radicals. Application to aspirin and other different substances. Thromb Res 79:109–123
Kaku S, Kawasaki T, Hisamichi N, Sakai Y, Taniuchi Y, Inagaki O et al (1996) Antiplatelet and antithrombotic effects of YM337, the Fab fragment of a humanized anti-GPIIb-IIIa monoclonal antibody in monkeys. Thromb Haemost 75:679–684
Kawai H, Tamao Y (1995) The combination of thrombin inhibitor and thromboxane synthase inhibitor on experimental thrombosis and bleeding. Thromb Res 80:429–434
Kawai H, Umemura K, Nakashima M (1995) Effect of argatroban on microthrombi formation and brain damage in the rat middle cerebral artery thrombosis model. Jpn J Pharmacol 69:143–148
Knabb RM, Kettner CA, Timmermans PB, Reilly TM (1992) In vivo characterization of a new synthetic thrombin inhibitor. Thromb Haemost 67:56–59
Kordenat RK, Kezdi P, Stanley EL (1972) A new catheter technique for producing experimental coronary thrombosis and selective coronary visualization. Am Heart J 83:360–364
Kovacs IB, Tigyi-Sebes A, Trombitas K, Gorog P (1975) Evans blue: an ideal energy-absorbing material to produce intravascular microinjury by HE-NE gas laser. Microvasc Res 10:107–124
Krupski WC, Bass A, Kelly AB, Hanson SR, Harker LA (1991) Reduction in thrombus formation by placement of endovascular stents at endarterectomy sites in baboon carotid arteries. Circulation 84:1749–1757

Kumada T, Ishihara M, Ogawa H, Abiko Y (1980) Experimental model of venous thrombosis in rats and effect of some agents. Thromb Res 18:189–203
Lam JY, Chesebro JH, Steele PM, Heras M, Webster MW, Badimon L et al (1991) Antithrombotic therapy for deep arterial injury by angioplasty. Efficacy of common platelet inhibition compared with thrombin inhibition in pigs. Circulation 84:814–820
Lin H, Young DB (1995) Opposing effects of plasma epinephrine and norepinephrine on coronary thrombosis in vivo. Circulation 91:1135–1142
Lumsden AB, Kelly AB, Schneider PA, Krupski WC, Dodson T, Hanson SR et al (1993) Lasting safe interruption of endarterectomy thrombosis by transiently infused antithrombin peptide D-Phe-Pro-ArgCH_2Cl in baboons. Blood 81:1762–1770
Lutz BR, Fulton GP, Akers RP (1951) White thromboembolism in the hamster cheek pouch after trauma, infection and neoplasia. Circulation III:339–351
Lynch JJ Jr, Sitko GR, Lehman ED, Vlasuk GP (1995) Primary prevention of coronary arterial thrombosis with the factor Xa inhibitor rTAP in a canine electrolytic injury model. Thromb Haemost 74:640–645
Mauray S, Sternberg C, Theveniaux J, Millet J, Sinquin C, Tapon-Bretaudiere J et al (1995) Venous antithrombotic and anticoagulant activities of a fucoidan fraction. Thromb Haemost 74:1280–1285
McBane RD, Wysokinski WE, Chesebro JH, Owen WG (1995) Antithrombotic action of endogenous porcine protein C activated with a latent porcine thrombin preparation. Thromb Haemost 74:879–885
McLetchie NGB (1952) The pathogenesis of atheroma. Am J Pathol 128:413–436
Mellott MJ, Stranieri MT, Sitko GR, Stabilito II, Lynch JJ Jr, Vlasuk GP (1993) Enhancement of recombinant tissue plasminogen activator-induced reperfusion by recombinant tick anticoagulant peptide, a selective factor Xa inhibitor, in a canine model of femoral arterial thrombosis. Fibrinolysis 7:195–202
Mickelson JK, Simpson PJ, Cronin M, Homeister JW, Laywell E, Kitzen J et al (1990) Antiplatelet antibody [7E3 $F(ab')_2$] prevents rethrombosis after recombinant tissue-type plasminogen activator-induced coronary artery thrombolysis in a canine model. Circulation 81:617–627
Millet J, Theveniaux J, Pascal M (1987) A new experimental model of venous thrombosis in rats involving partial stasis and slight endothelium alterations. Thromb Res 45:123–133
Millet J, Vaillot M, Theveniaux J, Brown NL (1996) Experimental venous thrombosis induced by homologous serum in the rat. Thromb Res 81:497–502
Mousa SA, DeGrado WF, Mu DX, Kapil RP, Lucchesi BR, Reilly TM (1996) Oral antiplatelet, antithrombotic efficacy of DMP 728, a novel platelet GPIIb/IIIa antagonist. Circulation 93:537–543
Mruk JS, Zoldhelyi P, Webster MW, Heras M, Grill DE, Holmes DR et al (1996) Does antithrombotic therapy influence residual thrombus after thrombolysis of platelet-rich thrombus? Effects of recombinant hirudin, heparin, or aspirin. Circulation 93:792–799
Nishizawa EE, Wynalda DJ, Suydam DE, Sawa TR, Schultz JR (1972) Collagen-induced pulmonary thromboembolism in mice. Thromb Res 1:233–242
Pierangeli SS, Barker JH, Stikovac D, Ackerman D, Anderson G, Barquinero J et al (1994) Effect of human IgG antiphospholipid antibodies on an in vivo thrombosis model in mice. Thromb Haemost 71:670–674
Pierangeli SS, Liu XW, Barker JH, Anderson G, Harris EN (1995) Induction of thrombosis in a mouse model by IgG, IgM and IgA immunoglobulins from patients with the antiphospholipid syndrome. Thromb Haemost 74:1361–1367
Ragni M, Cirillo P, Pascucci I, Scognamiglio A, D'Andrea D, Eramo N et al (1996) Monoclonal antibody against tissue factor shortens tissue plasminogen activator lysis time and prevents reocclusion in a rabbit model of carotid artery thrombosis. Circulation 93:1913–1918

Rebello SS, Miller BV, Basler GC, Lucchesi BR (1997) CVS-1123, a direct thrombin inhibitor, prevents occlusive arterial and venous thrombosis in a canine model of vascular injury. J Cardiovas Pharmacol 29:240–249
Romson JL, Haack DW, Lucchesi BR (1980) Electrical induction of coronary artery thrombosis in the ambulatory canine: a model for in vivo evaluation of anti-thrombotic agents. Thromb Res 17:841–853
Rosenblum WI, EL-Sabban F (1977) Platelet aggregation in the cerebral microcirculation: effect of aspirin and other agents. Circ Res 40:320–328
Rosenblum WI, EL-Sabban F (1982) Dimethyl sulfoxide (DMSO) and glycerol, hydroxyl radical scavengers, impair platelet aggregation within and eliminate the accompanying vasodilation of, injured mouse pial arterioles. Stroke 13:35–39
Rote WE, Mu DX, Roncinske RA, Frelinger AL III, Lucchesi BR (1993) Prevention of experimental carotid artery thrombosis by applaggin. J Pharmacol Exp Ther 267:809–814
Roux S, Carteaux JP, Hess P, Falivene L, Clozel JP (1994) Experimental carotid thrombosis in the guinea pig. Thromb Haemost 71:252–256
Roux S, Tschopp T, Baumgartner HR (1996) Effects of napsagatran (Ro 46–6240), a new synthetic thrombin inhibitor and of heparin in a canine model of coronary artery thrombosis: comparison with an ex vivo annular perfusion chamber model. J Pharmacol Exp Ther 277:71–78
Rubsamen K, Hornberger W (1996) Prevention of early reocclusion after thrombolysis of copper coil-induced thrombi in the canine carotid artery: comparison of PEG-hirudin and unfractionated heparin. Thromb Haemost 76:105–110
Saniabadi AR (1994) Photosensitisers and photochemical reactions. In: Nakashima M (ed) A novel photochemical model for thrombosis research and evaluation of antithrombotic and thrombolytic agents. Churchill Livingstone, New York, pp 1–19
Saniabadi AR, Umemura K, Matsumoto N, Sakuma S, Nakashima M (1995) Vessel wall injury and arterial thrombosis induced by a photochemical reaction. Thromb Haemost 73:868–872
Sato M, Ohshima N (1984a) A new experimental model for inducing platelet thrombus in the microvasculature: basic study on its feasibility. Blood Vessel 15:36–44
Sato M, Ohshima N (1984b) Platelet thrombus induced in vivo by filtered light and fluorescent dye in mesenteric microvessels of the rat. Thromb Res 35:319–334
Sawyer PN, Pate JW (1953a) Bio-electric phenomena as an etiologic factor in intravascular thrombosis. Am J Physiol 175:103–107
Sawyer PN, Pate JW (1953b) Electrical potential differences across the normal aorta and aortic grafts of dogs. Am J Physiol 175:113–117
Schumacher WA, Lee EC, Lucchesi BR (1985) Augmentation of streptokinase-induced thrombolysis by heparin and prostacyclin. J Cardiovasc Pharmacol 7:739–746
Scott NA, Nunes GL, King SB III, Harker LA, Hanson SR (1994) Local delivery of an antithrombin inhibits platelet-dependent thrombosis. Circulation 90:1951–1955
Shetler TJ, Crowe VG, Bailey BD, Jackson CV (1996) Antithrombotic assessment of the effects of combination therapy with the anticoagulants efegatran and heparin and the glycoprotein IIb-IIIa platelet receptor antagonist 7E3 in a canine model of coronary artery thrombosis. Circulation 94:1719–1725
Sitko GR, Ramjit DR, Stabilito II, Lehman D, Lynch JJ, Vlasuk GP (1992) Conjunctive enhancement of enzymatic thrombolysis and prevention of thrombotic reocclusion with the selective factor Xa inhibitor, tick anticoagulant peptide. Comparison to hirudin and heparin in a canine model of acute coronary artery thrombosis. Circulation 85:805–815
Steele PM, Chesebro JH, Stanson AW, Holmes DR Jr, Dewanjee MK, Badimon L et al (1985) Balloon angioplasty. Natural history of the pathophysiological response to injury in a pig model. Circ Res 57:105–112

Stockmans F, Deckmyn H, Gruwez J, Vermylen J, Acland R (1991) Continuous quantitative monitoring of mural, platelet-dependent, thrombus kinetics in the crushed rat femoral vein. Thromb Haemost 65:425–431

Stone P, Lord JW Jr (1951) An experimental study of the thrombogenic properties of magnesium and magnesium-aluminum wire in the dog's aorta. Surgery 30:987–993

Suarez EL, Jacobson JH (1961) Results of small artery endarterectomy-microsurgical technique. Surg Forum 12:256–257

Sudo Y, Lucchesi BR (1996) Antithrombotic effect of GYKI-14766 in a canine model of arterial and venous rethrombosis: a comparison with heparin. J Cardiovasc Pharmacol 27:545–555

Sudo Y, Kilgore KS, Lucchesi BR (1995) Monoclonal antibody [7E3 F(ab')$_2$] prevents arterial but not venous rethrombosis. J Cardiovasc Pharmacol 26:241–250

Tapparelli C, Metternich R, Gfeller P, Gafner B, Powling M (1995) Antithrombotic activity in vivo of SDZ 217–766, a low-molecular weight thrombin inhibitor in comparison to heparin. Thromb Haemost 73:641–647

Thomas DP (1988) Overview of venous thrombogenesis. Semin Thromb Hemost 14:1–8

Thomas DP (1994) Pathogenesis of venous thrombosis. In: Bloom AL, Forbes CD, Thomas DP, Tuddenham EGD (eds) Haemostasis and thrombosis, vol 2, 3rd edn. Churchill Livingstone, Edinburgh, pp 1335–1347

Thomas DP (1996) Venous thrombosis and the "Wessler test". Thromb Haemost 76:1–4

Uprichard AC, Chi LG, Kitzen JM, Lynch JJ, Frye JW, Lucchesi BR (1989) Celiprolol does not protect against ventricular tachycardia or sudden death in the conscious canine: a comparison with pindolol in assessing the role of intrinsic sympathomimetic activity. J Pharmacol Exp Ther 251:571–577

Van Ryn-McKenna J, Merk H, Muller TH, Buchanan MR, Eisert WG (1993) The effects of heparin and annexin V on fibrin accretion after injury in the jugular veins of rabbits. Thromb Haemost 69:227–230

Virchow R (1856) I. Über die Verstopfung der Lungenarterie. In: Gesammelte Abhandlungen zur wissenschaftlichen Medicin. Meidinger Sohn, Frankfurt, p 221

Walenga JM, Petitou M, Lormeau JC, Samama M, Fareed J, Choay J (1987) Antithrombotic activity of a synthetic heparin pentasaccharide in a rabbit stasis thrombosis model using different thrombogenic challenges. Thromb Res 46:187–198

Weichert W, Pauliks V, Breddin HK (1983) Laser-induced thrombi in rat mesenteric vessels and antithrombotic drugs. Haemostasis 13:61–71

Wessler S (1952) Studies in intravascular coagulation. I. Coagulation changes in isolated venous segments. J Clin Invest 31:1011–1014

Wessler S (1955) Studies in intravascular coagulation. III. The pathogenesis of serum-induced venous thrombosis. J Clin Invest 34:647–651

Wessler S, Reimer SM, Sheps MC (1959) Biological assay of a thrombosis-inducing activity in human serum. J Appl Physiol 14:943–946

Wright HP, Kubik MM, Hayden M (1952) Influence of anticoagulant administration on the rate of recanalization of experimentally thrombosed veins. Br J Surg 40:163–166

Yamamoto J, Iizumi H, Hirota R, Shimonaka K, Nagamatsu Y, Horie N et al (1989) Effect of physical training on thrombotic tendency in rats: decrease in thrombotic tendency measured by the He-Ne laser-induced thrombus formation method. Haemostasis 19:260–265

Yamashita T, Yamamoto J, Sasaki Y, Matsuoka A (1993) The antithrombotic effect of low molecular weight synthetic thrombin inhibitors, argatroban and PPACK, on He-Ne laser-induced thrombosis in rat mesenteric microvessels. Thromb Res 69:93–100

Yatsushiro T (1913) Experimentelle Untersuchungen über die Thrombosenfrage, nebst Angabe einer einfachen Methode zur Koagulationsbestimmung des Blutes. Dtsch Z Chir 125:559–612

Yokoyama T, Kelly AB, Marzec UM, Hanson SR, Kunitada S, Harker LA (1995) Antithrombotic effects of orally active synthetic antagonist of activated factor X in nonhuman primates. Circulation 92:485–491

CHAPTER 5

Monitoring Antithrombotic Therapy

R.J.G. Cuthbert

A. Introduction

I. Balancing Antithrombotic Efficacy Against the Risk of Bleeding

In health, the haemostatic system is in a constant state of low grade activation, and anticoagulant and antifibrinolytic pathways regulate the system to prevent spontaneous thrombosis. In evolutionary terms, a haemostatic system which is partially activated favours a successful response to injury. Agents such as warfarin and heparin which inhibit the procoagulant effect of the haemostatic system inevitably have anticoagulant effects which favour bleeding. An ideal agent would have specific antithrombotic activity to prevent pathological activation of haemostasis yet would produce negligible global anticoagulant activity so that the haemostatic system could respond appropriately to physiological procoagulant stimuli. During antithrombotic therapy precise dose adjustment is required to establish a clinically suitable effect with a minimal risk of bleeding. Because it is difficult to separate the antithrombotic effects from the global anticoagulant effects of currently available agents, the therapeutic window of such agents tends to be narrow, and dose adjustment requires regular therapeutic monitoring.

Clinical monitoring of antithrombotic therapy usually involves tests which assess some aspect of haemostatic function, e.g. prothrombin time (PT), activated partial thromboplastin time (APTT), etc. and rarely utilise direct pharmacological assays of the plasma concentration of the particular drug. In theory, it is usually possible to determine a therapeutic range for any particular antithrombotic agent within which therapeutic efficacy is established (usually in clinical trials) and the risk of bleeding is minimised.

B. Warfarin

I. Mechanism of Action

1. Effect on Vitamin K-Dependent Clotting Factors

Warfarin and other coumarins are structural analogues of vitamin K. They exert their anticoagulant effect by inhibition of vitamin K in the post-translational modification of the procoagulant proteins factors II, VII, IX and X (Paul et al. 1987; Xi et al. 1989). Glutamate residues at the N-terminal of

Table 1. Plasma half-life of vitamin K-dependent factors

Factor II	72 h
Factor VII	4–6 h
Factor IX	18–30 h
Factor X	48 h
Protein C	6–9 h

the proteins are normally converted to carboxyglumate (gla). Repeat gla sequences, usually of 10–13 residues, are involved in Ca^{2+} binding and conformational changes in the proteins that facilitate their interaction and activation on phospholipid surfaces (NELSESTUEN 1976). Reduction in the number of gla residues to 9 causes an approximate 30% reduction in procoagulant activity; reduction to 6 or less virtually abolishes procoagulant activity (MALHOTRA 1990). Warfarin also inhibits the naturally occurring anticoagulants protein C and protein S (WEISS et al. 1987). In the steady state, however, warfarin therapy causes a net anticoagulant effect on the haemostatic system. The effect of warfarin on each of the vitamin K-dependent factors is variable depending on the plasma half-life of each protein (THIJSSEN et al. 1988). (See Table 1.)

2. Kinetics of Vitamin K-Dependent Clotting Factors During Warfarin Therapy

After initiation of warfarin therapy, hepatic synthesis of functional vitamin K-dependent clotting factors is inhibited. There is a rapid reduction in factor VII since its half-life is relatively short and it is rapidly cleared from plasma. The protein C level also falls rapidly at the onset of warfarin therapy (WEISS et al. 1987). There is a delay of 48–72 h before a significant decrease in factors II, IX and X occurs. In the first 24–48 h, therefore, there is a potential change in haemostatic balance to favour fibrin deposition and thrombosis. This effect explains the rare phenomenon of warfarin-induced skin necrosis, most commonly associated with protein C deficiency (SALLAH et al. 1997). Traditionally, the antithrombotic effect of warfarin had been attributed to its global anticoagulant effects but there is evidence now to suggest that the antithrombotic effect of warfarin is predominantly due to depletion of factor II activity, and to a lesser extent factor X activity (ZIVELIN et al. 1993). For the management of acute thrombotic conditions, particularly venous thromboembolism, it is necessary to overlap warfarin therapy with 4–5 days of heparin therapy before a therapeutic warfarin-induced antithrombotic state is achieved (BRANDJES et al. 1992).

3. Variation in Pharmacological Response

There is wide inter-individual variation in the dose-response relationship to warfarin administration (THIJSSEN et al. 1988). This is due mainly to variation

Table 2. Typical activities of vitamin K-dependent procoagulant proteins during steady state warfarin therapy

Factor II	20%–30%
Factor X	10%
Factors VII, IX	40%–50%

in the rate of warfarin metabolism, and to a lesser degree to variable warfarin absorption and bioavailability in the body (HIRSH et al. 1995a). The plasma half-life of warfarin may vary from 15 to 50h. In the steady state, there is a balance between the new rate of synthesis of the vitamin K-dependent proteins and their rate of plasma clearance adjusted to lower plasma levels as outlined in Table 2. Increasing the intensity of warfarin therapy by increasing the daily dose causes a further reduction in the rate of synthesis of the vitamin K-dependent procoagulant proteins and a new steady state with lower factor levels is established (BERTINA 1984). The rate of warfarin metabolism declines with increasing age. Therefore, elderly patients tend to have a lower dose requirement to maintain therapeutic efficacy (GURWITZ et al. 1991; WYNNE et al. 1996).

II. Laboratory Monitoring

1. Prothrombin Time for Monitoring Warfarin Therapy

The prothrombin time (PT) is sensitive to changes in factor II, VII and X and may be utilised to monitor the anticoagulant effect of warfarin therapy. The PT assay utilises a thromboplastin preparation containing phospholipid and tissue factor required to activate factor VII. Formerly thromboplastins were prepared commercially from human brain extracts. For health and safety reasons currently available commercial thromboplastins are prepared from animal origins, usually rabbit brain extracts. The prothrombin time result is dependent on the thromboplastin reagent used in the assay. Traditionally the test PT is compared with the PT of a pool of normal plasma, and is reported as the prothrombin time ratio (PTR), where PTR = test PT/normal plasma pool PT.

In 1948, the American Heart Association suggested that the therapeutic range for warfarin therapy should be equivalent to a range of PTR 2.0–2.5. This was dependent on highly sensitive thromboplastin reagents used in the PT assay at the time. No change in the recommended therapeutic range was suggested when less sensitive thromboplastins were introduced in North America in the 1970s. Thus, in retrospect, patients received more intensive warfarin therapy (POLLER and TABERNER 1982; HIRSH 1991). Furthermore, wide variation in PTR values were obtained from different laboratories using different thromboplastin preparations (POLLER 1987; HIRSH and LEVINE 1988).

2. Standardisation of Thromboplastin Reagents

In the early 1980s, a model was developed for calibration of the PT assay for use with different thromboplastins. A linear relationship is observed between log PT values using a reference thromboplastin and log PT values using a comparison thromboplastin, when a range of plasmas from patients stabilised on warfarin and normal controls are assayed (KIRKWOOD 1983). The slope of the line gives a measure of the sensitivity of the test thromboplastin in comparison to the reference thromboplastin, and is called its international sensitivity index (ISI) (HERMANS et al. 1983). By 1985, many laboratories began to report PTRs as the international normalised ratio (INR) where $INR = PTR^{ISI}$ or $INR = -\log PTR \times ISI$ (POLLER 1988). Since the model is empirical (the original reference thromboplastin is assigned an ISI = 1.0), thromboplastins may deviate from the model slightly more than predicted from their assigned ISI (MORIARTY et al. 1990). Thromboplastins which are highly sensitive have ISI values close to 1.0 whereas less sensitive thromboplastins have ISI values of 1.5–2.8. These less sensitive thromboplastins produce more rapid factor VII/tissue factor-mediated activation of factor X. Thus, less prolongation of the PT is produced by a given warfarin dose compared with the PT result obtained for a more sensitive thromboplastin. By employing the ISI as a correction factor for the degree of sensitivity of a thromboplastin, calculation of the INR provides a more reliable method of monitoring the dose-response effect of warfarin therapy (BUSSEY et al. 1992; HIRSH et al. 1995a).

The original calibration model is based on a manual PT assay with a visual method to determine the end-point of clot formation. Most laboratories, however, now use automated instruments to perform PT assays, and have optical/electronic methods to determine the end-point. There is up to 10% variation in end-points between manual and automated PT determinations with wide variation between instruments (RAY and SMITH 1990; THOMSON et al. 1990). Manufacturers of thromboplastins should report ISI values for particular thromboplastins for use on specified automated instruments (POLLER et al. 1994). The inter-laboratory variation in calculation of the ISI for any given thromboplastin may be 2%–6%. Variation in estimation of the mean normal plasma PT (MNPT) may also occur. The MNPT should be estimated by measuring PT values from a pool of at least 20 fresh plasma samples from healthy adults assayed by the same technique and instrumentation as the test PT (VAN DEN BESSELAAR et al. 1993).

3. Choice of Thromboplastins for Clinical Monitoring

Clinical trials have shown that low intensity warfarin therapy – INR 2.0–3.0 (2.5–3.5 for mechanical valve prosthesis) – is safe and effective (HIRSH et al. 1995a). All such published studies have reported INR values assayed using highly sensitive thromboplastins with ISI values close to 1.0. There is concern that the efficacy of low intensity warfarin therapy is compromised when poorly responsive thromboplastins (i.e., with ISI values of 1.5–2.8) are used in the

INR estimation. In North America, increasing numbers of laboratories are reporting the INR rather than the PTR, and transferring to the use of highly sensitive thromboplastins (HIRSH et al. 1995a). The practice is well established in western Europe. Despite the limitations in estimation of the INR as outlined above, the degree of laboratory error in its calculation is considerably less than that experienced with the uncontrolled PTR reporting system. By reporting the INR using a highly sensitive thromboplastin on an instrument calibrated for its use, the analytical error is minimised (POLLER et al. 1995) and physicians can be confident that an accurate assessment of warfarin therapy is achieved. Despite this, some variation in the INR is observed over time for any particular patient. This is due to biological fluctuation in the patient's procoagulant and anticoagulant proteins. Fluctuation may be exacerbated by variation in dietary vitamin K intake (BOOTH et al. 1997) and interactions with concominant drug therapy (HIRSH et al. 1995a).

III. Determinants of Bleeding Risk

Bleeding is the most important complication of warfarin therapy. Bleeding is defined as major when any of the following occur: intracranial, intraspinal, or retroperitoneal bleeding, bleeding resulting in the need for transfusion, bleeding resulting in the need for hospital admission, or bleeding leading directly to death (LEVINE et al. 1995). Minor bleeding is defined as all other less serious bleeding episodes, such as epistaxis, menorrhagia, haematuria, and gastrointestinal bleeding not requiring transfusion or hospitalisation. Analysis of data from large series give estimates of annual major bleeding rates of up to 5% and fatal bleeding of up to 2%. The risk of bleeding is related to the intensity of warfarin therapy, the duration of warfarin therapy, characteristics of the individual patient, and the occurrence of adverse drug interactions (LEVINE et al. 1995).

Both major and minor bleeding occur at higher frequency when patients are maintained at INR levels above 2.5. Clinical trials have demonstrated that reducing the intensity of warfarin therapy from a target therapeutic INR range of 3.0–4.5 to 2.0–3.0 or 2.5–3.5 significantly reduces the rate of major bleeding with no significant reduction in antithrombotic efficacy in the management of DVT (HULL et al. 1982), tissue valve prosthesis (TURPIE et al. 1988), and mechanical valve prosthesis (SAOUR et al. 1990).

Clinical characteristics of patients may be an even more important determinant of the risk of bleeding than the intensity of therapy. The risk of major bleeding increases significantly at ages above 65 years, in patients with a history of previous gastrointestinal haemorrhage, severe hypertension, renal dysfunction, anaemia, cerebrovascular disease, serious cardiac disease, unreported additional drug ingestion (particularly aspirin) and binge alcohol drinking (FIHN et al. 1993; VAN DER MEER et al. 1993; BRIGDEN 1996). When bleeding occurs at INR values of 3.0 or less, investigation of an underlying secondary cause should be conducted (LANDEFELD et al. 1989). Gastrointestinal bleeding should always be investigated. The diagnostic yield is likely to be high at lower INR ranges. Since elderly individuals have a lower rate of warfarin clearance causing an exaggerated anticoagulant response, warfarin therapy should be initiated at lower doses in elderly patients.

IV. Practical Aspects of Warfarin Dosing

Before initiation of warfarin therapy, baseline evaluations of platelet count, PT, APTT, and liver and renal function should be conducted (PHILLIPS et al. 1997). Patients should be assessed clinically for risk factors for bleeding such as hypertension, active gastrointestinal bleeding, alcohol abuse, etc. A bleeding risk analysis index is valuable to stratify bleeding risk amongst patient populations. This can be used as a guide to employ lower initiation doses of warfarin for patients at higher risk of bleeding (LANDEFELD et al. 1990).

It should be noted that the INR is unreliable during initiation of warfarin therapy, being a poor predictor of both antithrombotic efficacy and bleeding risk (JOHNSTON et al. 1996). During the first 2–5 days of treatment, the PT (and therefore the INR) becomes prolonged due to the rapid fall in factor VII. The full antithrombotic effect, however, is more dependent on reduction in factor II and, to a lesser extent, factor X. Nonetheless, the INR remains more predictable than the uncorrected PTR for monitoring the early stages of warfarin therapy (JOHNSTON et al. 1996)

Loading doses of 9–10 mg on days 1 and 2 (reduced in elderly and high risk patients) are traditionally administered to initiate warfarin therapy. A lower loading dose of 5 mg, however (estimated to be closer to the eventual maintenance dose of warfarin therapy), should be considered (HARRISON et al. 1997). It is associated with less over-anticoagulation and consequent bleeding risk, and potentially lower risk of development of a transient hypercoagulable state due to the rapid fall in protein C levels during the early days of warfarin therapy. The time taken to reach the target maintenance therapeutic INR range is not significantly increased. Subsequent doses are adjusted according to the INR. For patients requiring heparin therapy, it should not be discontinued until the INR is greater than 2.0 for at least 48 h, to ensure an adequate reduction in factor II levels and an adequate antithrombotic effect. The INR should be repeated about 6 h after heparin is discontinued and the warfarin dose adjusted accordingly, as heparin may contribute to prolongation of the PT.

V. Maintenance Treatment

1. Anticoagulant Clinics

Long-term warfarin therapy requires meticulous monitoring to ensure efficacy and safety. Consistent levels of patient care are best provided by establishing a routine monitoring and dose adjustment system. Dedicated anticoagulant clinics may be employed for this purpose and have been shown to lead to improved control (CORTELAZZO et al. 1993; ANSELL and HUGHES 1996; ANSELL et al. 1997). Initially patients will require INR monitoring twice weekly or more frequently. Once stabilized on a suitable maintenance dose, patients are reviewed at longer intervals to a maximum of 6–8 weeks. More frequent follow-up will be required during dose changes and during the initiation or discontinuation of other drugs which may potentially interact with warfarin.

2. Computer-Assisted Monitoring and Patient Self-Monitoring

Computer based algorithms are available on a commercial basis to assist with anticoagulant monitoring, dose adjustment, patient records and date handling. Studies have demonstrated that adequate control of warfarin therapy can be achieved with computer algorithms, and these compare favourably with levels of control achieved by experienced practitioners (POLLER et al. 1993; VADHER et al. 1997). Recently, mobile automated assay systems have been developed to estimate INR values during warfarin therapy. These can be employed by practitioners allowing near-patient testing and monitoring, and can be employed to assist patient self-monitoring (ANSELL et al. 1995; ANSELL and HUGHES 1996).

VI. Alternative Methods of Monitoring Warfarin Therapy

1. Functional Prothrombin Assay

The antithrombotic effect of warfarin correlates well with inhibition of factor II activity. By employing a functional prothrombin assay it is possible to develop an efficacious and safe therapeutic range equivalent to a factor II activity of 10%–25%, leading to minimal bleeding risk and adequate antithrombotic effect. The advantage of this method is that it is easy to standardise. The disadvantages are that it is more time consuming to perform and more costly both in terms of labour and reagent costs. Automated immunoassays of prothrombin antigen may overcome these disadvantages (KORNBERG et al. 1993).

2. Prothrombin Fragment F1.2

This peptide is released during the activation of factor II to thrombin. The circulating half-life is approximately 90 min. Sensitive and specific immunoassays have been developed. Commercial systems include synthetic thrombin inhibitors to counteract the anticoagulant in the assay tube. Since warfarin suppresses prothrombin activation, the levels of prothrombin fragment F1.2 fall during warfarin therapy and thus treatment can be adjusted to maintain the prothrombin fragment F1.2 level below the normal range. The disadvantage is that the system does not provide adequate prediction of bleeding risk (FEINBERG et al. 1997; NAKAMURA et al. 1997).

3. Prothrombin-Proconvertin Ratio

The prothrombin-proconvertin ratio is a modified PT that is sensitive to warfarin-induced changes in factors II, VII, and X, which can be used to analyse stored samples kept at room temperature. To perform a PT assay, adequate amounts of factor V and fibrinogen are supplied by the addition of adsorbed bovine plasma to the thromboplastin reagent. The assay results correlate well with the INR conducted on fresh samples and with prothrombin antigen levels in warfarin treated patients (HARALDSSON et al. 1997).

C. Heparin

I. Heparin Structure

Heparin is a glycosaminoglycan consisting of a polypeptide matrix linked to a sulphated polysaccharide (see Chap. 9). Commercial heparins are prepared from bovine lung or porcine intestinal mucosa. The chain length is variable producing molecules of molecular weights between 5 and 30kDa. The mean molecular weight is about 15 kDa, with average chain lengths of 50 polysaccharide residues.

II. Mechanism of Action

Heparin exerts its anticoagulant effect by forming a complex with antithrombin (formerly identified as antithrombin III) (BJORK and LINDAHL 1982). This augments 1000–2000 fold antithrombin's activity to inhibit thrombin, factor Xa, and to a lesser extent factor XIa (OFOSU et al. 1989). The antithrombin binding region on the heparin molecule consists of a unique pentasaccharide sequence. This is present on only about 30% of the heparin chains. Heparin molecules consisting of less than 18 polysaccharide residues are unable to bind thrombin and antithrombin simultaneously and, therefore, cannot augment the inhibitory effect on thrombin. They do, however, have the ability to augment the antithrombin-mediated inactivation of factor Xa. At high doses, heparin also accelerates the inactivation of thrombin by a second thrombin inhibitor called heparin co-factor II. The reaction does not require binding to the unique pentasaccharide sequence. Heparin co-factor II does not influence heparin-mediated inactivation of factor Xa or factor XIa (MAIMONE and TOLLEFSEN 1988).

Low molecular weight heparin (LMWH) is prepared by chemical or enzymatic fractionation of heparin. Commercial LMWH preparations have an average molecular weight of 5–6kDa. Since they have a relatively lower proportion of molecules with 18 or more polysaccharide residues, their anticoagulant effect favours factor Xa inactivation over thrombin inactivation (BEGUIN et al. 1989; LANE and RYAN 1989).

III. Unfractionated Heparin

1. Pharmacokinetics

Heparin must be administered parenterally since it is not absorbed from the gut. Intramuscular administration may cause haematoma formation. Therefore, heparin is administered by intravenous or subcutaneous injection. Intermittent intravenous heparin administration is associated with an increased bleeding risk so continuous intravenous infusion or intermittent subcutaneous injection are the preferred methods of administration (HIRSH et al. 1995b).

The plasma bioavailability of unfractionated heparin is highly variable. Heparin is inactivated by circulating heparin-binding proteins including von Willebrand factor, platelet factor 4, fibronectin and histidine-rich glycoprotein (LANE et al. 1986; DE ROMEUF and MAZUIER 1993). Binding to such plasma proteins causes unpredictable bioavailability in various clinical states (YOUNG et al. 1997). Therefore, variable anticoagulant responses are observed when heparin is administered in fixed doses to different patients (HIRSH et al. 1995b). Binding of heparin to von Willebrand factor may cause some degree of inhibition of platelet adhesion (SOBEL et al. 1991) and potentially contribute to the bleeding risk of high dose heparin therapy. Furthermore, heparin may also cause increased vascular permeability (BLAJCHMAN et al. 1989).

High molecular weight heparin chains are rapidly cleared from plasma by binding to endothelial cell and macrophage receptors (BARZU et al. 1985). The heparin-receptor complex is internalised and heparin is degraded within the cell cytoplasm. The rate of clearance is accelerated in acute clinical states such as febrile illnesses and acute major thrombosis; thus, higher doses may be required to achieve the desired anticoagulant effect. Heparin at the lower end of the molecular weight spectrum is excreted by renal clearance by a slower mechanism. The dose-response effect of heparin is not linear; as the intensity of dosing increases the plasma half-life is increasingly prolonged. The half-life may vary from 30 to 150 min over the range of clinical therapeutic dosing.

Following administration of heparin by the subcutaneous route there is a delay before full anticoagulant effect occurs due to the slower rate of saturation of heparin-binding proteins and endothelial and macrophage receptors (BARA et al. 1985). If a rapid anticoagulant effect is required, an intravenous heparin bolus should be administered.

2. Laboratory Monitoring by the Activated Partial Thromboplastin Time (APTT)

The anticoagulant effect of heparin is most commonly monitored by the APTT (HIRSH et al. 1995b). The APTT is sensitive to heparin-induced inhibition of thrombin, factor Xa, and factor XIa. A direct relationship is observed between the therapeutic APTT level and the clinical efficacy of heparin (BASU et al.

1972; Hull et al. 1986; Raschke et al. 1993; Levine et al. 1994). Traditionally, the heparin dose is adjusted to maintain the APTT ratio between 1.5 and 2.5 (APTT ratio = patient APTT/APTT of normal plasma pool). This therapeutic range is based on an animal study in which experimental thrombus extension was prevented by heparin administration in a dose which prolonged the APTT to a ratio above 1.5, equivalent to a plasma heparin concentration measured by a protamine neutralisation assay of 0.2 iu/ml (Chiu et al. 1977). There is no conclusive evidence that an APTT above any given ratio can predict an increased bleeding risk (Levine et al. 1995). However, higher doses of heparin may be associated with more bleeding than lower doses (Reilly et al. 1993). Thus, arbitrarily, an upper therapeutic APTT ratio of 2.5, equivalent to a plasma heparin level of 0.4 iu/ml by protamine neutralisation, has been established (Hirsh et al. 1995b).

Different commercial APTT reagents assayed on different instruments produce variable responses to heparin therapy. This is due to different phospholipid and contact activators present in commercial APTT reagents (Brill-Edwards et al. 1993; Kitchen et al. 1996) and different methods of clot detection (D'Angelo et al. 1990). Depending on the assay system, a therapeutic heparin level may not occur at an APTT ratio of 1.5 (Brill-Edwards et al. 1993). Each laboratory should establish a local therapeutic range for the APTT ratio which is equivalent to a plasma heparin concentration of 0.2–0.4 iu/ml by protamine neutralisation, or 0.3–0.7 iu/ml by anti-factor Xa activity (Hirsh et al. 1995b; Brill-Edwards et al. 1993). This can be achieved by comparing APTT ratios with plasma heparin levels for a range of patients established on heparin therapy, or by calibrating the APTT ratio to in vitro heparin concentrations added to a normal plasma pool.

Heparin regimens and therapeutic monitoring protocols have been extensively studied in prospective randomised trials for treatment of venous thromboembolism and acute coronary disease (Hull et al. 1986; Brandjes et al. 1992; Turpie et al. 1989; Raschke et al. 1993; Harenberg et al. 1997). Commonly used regimens employ the use of an intravenous bolus of 5000 iu heparin followed by a continuous intravenous infusion of 30,000–40,000 iu heparin per 24 h (Anand et al. 1996). The APTT is measured 4–6 h after initiation of therapy and the heparin dose adjusted to achieve an APTT ratio within the therapeutic range as outlined above. The APTT ratio correlates reasonably well with plasma heparin levels over the range of heparin doses required for these disorders (Levine et al. 1994). Additional boluses may be administered if the APTT remains sub-therapeutic. The risk of thromboembolic recurrence is minimised if a therapeutic heparin level is achieved within the first 24 h.

3. Heparin Resistance

In a subgroup of patients treated with therapeutic doses of heparin at 30,000–40,000 iu/24 h, an APTT ratio of 1.5 or more is not achieved. The increased dose requirements in these patients is due to increased heparin neutralisation by heparin binding proteins in the plasma. Since many such proteins are acute phase reactants, their plasma concentration may rise during acute illness (YOUNG et al. 1997). Furthermore, elevated factor VIII levels cause shortening of the APTT of normal and heparin treated plasmas. An APTT ratio of more than 1.5 may not, therefore, be achieved due to raised levels of heparin-binding proteins and/or raised FVIII levels despite adequate heparin levels (YOUNG et al. 1994; LEVINE et al. 1994). Evidence exists to suggest that such laboratory "heparin resistance" is not associated with reduced antithrombotic efficacy if plasma heparin levels of 0.3–0.7 iu/ml, measured by anti-factor Xa activity, are achieved (LEVINE et al. 1994).

4. Dose-Adjustment Nomograms

Calculation of the heparin loading dose and maintenance infusion rate based on body weight may be helpful in achieving rapid therapeutic heparin levels. Since dose adjustment of maintenance infusions can be difficult, and sub-optimal therapy is not uncommonly observed (FENNERTY et al. 1985; LE BRAS et al. 1992), several groups have developed dose-adjustment nomograms to try to improve the efficacy of heparin therapy (CRUIKSHANK et al. 1991; RASCHKE et al. 1993; RASCHKE and REILLY 1994; SHALANSKY et al. 1996). Nomograms provide a convenient and relatively accurate method of dose-adjustment (BROWN and DODEK 1997). Since APTT reagents and assay methods are variable, each centre using a published nomogram should adapt it to the APTT therapeutic range established for their local centre (RASCHKE and HERTEL 1991).

5. Subcutaneous Heparin Regimens

Since there is a delay in full therapeutic effect following subcutaneous heparin administration, an intravenous bolus of 5000 iu heparin is required to achieve a rapid anticoagulant effect (HULL et al. 1986). Initial maintenance doses of 15,000–17,500 iu subcutaneously are given every 12 h thereafter. The APTT should be estimated at least 6 h after a dose of heparin, and the daily dose adjusted to an APTT ratio equivalent to a plasma heparin concentration of 0.2–0.4 iu/ml by protamine neutralisation or 0.3–0.7 iu/ml by anti-factor Xa activity (HIRSH et al. 1995b). Prospective studies have shown that therapeutic heparin administration by the subcutaneous route is associated with somewhat lower antithrombotic efficacy compared with continuous intravenous infusion therapy; a greater total daily dose is required to achieve therapeutic efficacy (HULL et al. 1986; HOMMES et al. 1992).

Prophylactic subcutaneous heparin regimens using low-dose heparin 5000 iu every 12 h are effective in the prevention of venous thromboembolism in surgical and medical patients (GALLUS et al. 1973). At this dose the APTT is not prolonged and therapeutic monitoring is not required. In hip surgery the risk of venous thromboembolism is higher but dose-adjusted low-dose subcutaneous heparin regimens have been employed effectively (POLLER et al. 1982; TABERNER et al. 1989). The APTT is measured 6 h after a dose, and the heparin dose adjusted to achieve an APTT ratio of 1.1–1.2. This approach is not associated with excessive post-operative bleeding (COLLINS et al. 1988). The disadvantage is that frequent APTT monitoring is required. The efficacy is dependent on the use of sensitive APTT reagents and assay systems (HIRSH et al. 1995b).

6. Determinants of Bleeding Risk

Five percent or more of patients receiving therapeutic doses of heparin may experience major bleeding episodes (LEVINE et al. 1995), depending on a number of factors including the method of heparin administration, the heparin dose, co-existing clinical conditions, and the use of other antithrombotic agents such as aspirin or fibrinolytic therapy. Retrospective analysis of bleeding episodes occurring during prospective clinical trials of heparin therapy indicate a weak positive correlation between the 24 h heparin dose and the risk of major bleeding (MORABIA 1986; REILLY et al. 1993). Trials comparing continuous intravenous heparin infusion with intermittent intravenous heparin injection have reported a significantly higher risk of major bleeding associated with intermittent therapy (HIRSH 1991). The latter group, however, required larger total 24 h heparin doses to maintain therapeutic APTT ratios, so the increased bleeding risk may be due to a dose effect rather than the mode of administration.

Patient characteristics are important determinants of bleeding risk for therapeutic heparin administration. Recent surgery or trauma, age over 65 years, renal failure, recent thrombotic stroke, active peptic ulcer disease, and poor performance status are associated with a higher risk of major bleeding events. In patients receiving concomitant aspirin therapy, a higher risk of bleeding may be observed (SETHI et al. 1990) although patients have been treated safely with heparin and aspirin for acute coronary disease (LEVINE et al. 1995). Administration of heparin following fibrinolytic therapy for acute myocardial infarction is associated with a substantially higher risk of haemorrhagic stroke, compared with coronary patients receiving heparin without thrombolytic therapy (GUSTO Investigators 1993).

Although there is evidence that higher doses of heparin may increase bleeding risk, therapeutic monitoring by the APTT is a poor predictor of bleeding risk. Nevertheless, current clinical practice usually employs an upper therapeutic range for heparin therapy of the APTT ratio equivalent to 0.4 iu/ml by protamine neutralisation or 0.6–0.7 iu/ml by anti-factor Xa activity (HIRSH et al. 1995b).

7. The Activated Clotting Time for Monitoring High Dose Heparin Therapy

The APTT is sensitive to heparin in the range 0.1–1 iu/ml and is therefore the most convenient laboratory test for monitoring heparin in the treatment of

DVT/PE, acute coronary disease, peripheral arterial thrombosis, etc. At heparin levels above 1 iu/ml, the APTT becomes too prolonged to be readable for practical purposes. For heparin doses above this level, such as those required following coronary angioplasty and during cardiopulmonary bypass, the activated clotting time (ACT) has been employed successfully to monitor heparin therapy (NOUREDINE 1995, REECE et al. 1996; SIMKO et al. 1995). The ACT is a measure of the time for whole blood to clot following activation by particulate matter such as kaolin (DESPOTIS et al. 1996). The ACT produces a linear response to heparin in the range 1–5 iu/ml, and a linear relationship between the ACT and log APTT. There is, however, a wide inter-individual variation in the ACT response to heparin, and each department should establish the ACT-plasma heparin correlation for their particular reagents and instruments.

IV. Low Molecular Weight Heparin

1. Pharmacokinetics

Low molecular weight heparins (LMWHs), like unfractionated heparin, are not absorbed following oral administration and must be administered parenterally. LMWHs have a lower affinity for circulating heparin-binding proteins and for binding to endothelial cells and macrophages (BARZU et al. 1985; YOUNG et al. 1994). Consequently, LMWHs have improved bioavailability with recovery of anti-factor Xa activity close to 100% (BARA and SAMANA 1988). The lack of significant binding to heparin binding proteins reduces the variability in the dose-anticoagulant response relationship since LMWH plasma concentration is not adversely influenced by variability in plasma protein concentration which can occur in ill patients (WEITZ 1997). Thus, there is a predictable response to fixed doses of LMWH in different patients. LMWHs have a lower affinity for von Willebrand factor binding (SOBEL et al. 1991), and may cause less inhibition of platelet function (HORNE and CHAO 1990; SERRA et al. 1997). Furthermore, LMWHs do not cause increased vascular permeability (BLAJCHMAN et al. 1989). Thus LMWHs produce less bleeding both in experimental systems and clinical settings. The lack of LMWH binding to endothelial cells and macrophages results in a longer plasma half-life, allowing once daily dosing. Excretion is principally by renal clearance; the half-life is increased in patients with renal dysfunction (CADROY et al. 1991).

2. Laboratory Monitoring by Chromogenic Anti-Factor Xa Assays

At therapeutic doses LMWHs do not cause significant prolongation of the APTT assay. There is a linear relationship between LMWH dose and plasma heparin concentrations measured by a chromogenic anti-factor Xa assay (TEIEN and LIE 1977; HOMER 1985). Because of the predictable anticoagulant response to weight-adjusted doses, it is not necessary to monitor LMWH therapy for most clinical conditions. Monitoring may be necessary, however, in renal failure (Harenberg et al. 1997) and potentially to assess efficacy in pregnancy (Barbour et al. 1995). Anti-factor Xa assays are commonly used for this purpose. Chromogenic substrate assays provide a convenient method which can be applied to automated coagulation instruments.

3. Clinical Efficacy and Bleeding Risks

Low dose LMWHs (equivalent to 0.1–0.2 iu/ml anti-factor Xa activity) have equal efficacy to fixed-dose subcutaneous unfractionated heparin for venous thromboembolism prophylaxis in general surgery and are associated with less post-operative bleeding (Kakkar et al. 1993). During hip and knee surgery, LMWHs are superior to fixed dose heparin (Levine et al. 1991), and are at least as effective (GHAT Group 1992) or superior (Dechavanne et al. 1989) to adjusted low-dose heparin. Overall the bleeding risks are lower for LMWHs used for venous thromboembolism prophylaxis during surgery (Nurmohamed et al. 1992; Imperiale and Speroff 1994). For the treatment of acute venous thromboembolism, therapeutic doses of LMWHs are given on a weight-adjusted basis. LMWHs are at least as effective as continuous intravenous unfractionated heparin and are associated with less bleeding in this setting (Siragusa et al. 1996). Since laboratory monitoring is not required, outpatient treatment is feasible for some patients (Levine et al. 1996; Kooman et al. 1996). LMWHs have proven value in patients with acute coronary disease (Cohen et al. 1997) but the dose should not exceed 1.0 iu/ml of antifactor Xa activity as the bleeding risk is excessive above this dose (TIMI IIa Investigators 1997)

D. Direct-Acting Antithrombin Agents

I. Role of Thrombin in Thrombogenesis

Thrombin plays a central role both in physiological haemostasis and in pathological thrombogenesis. During activation of the haemostatic system, thrombin catalyses its own activation from factor II through the "amplification" (intrinsic) pathway (see Chap. 1), contributing to a positive feedback effect on fibrin generation. In addition to catalysing conversion of fibrinogen to fibrin, thrombin activates factors V, VIII and XIII. Thrombin is also a potent activator of platelet aggregation and release reactions. Fibrin-bound thrombin contributes to thrombus propagation during the thrombogenic process by amplifying its own activity several 100-fold.

II. Mechanism of Action and Clinical Studies

Direct acting antithrombin agents have theoretical advantages over heparin. They act independently of antithrombin activity and since they have different plasma protein binding properties, the dose-antithrombotic response relationship is potentially more predictable. More critically, they have the ability to inhibit fibrin-bound thrombin and may, therefore, contribute more to inhibition of the thrombogenic process than heparin (WEITZ et al. 1990; PHILIPPIDES and LOSCALZO 1996; VERSTRAETE 1997). Clinical experience with direct acting antithrombin agents is limited. Hirudin, the principal peptide anticoagulant of the medicinal leech, binds directly to thrombin at the fibrin(ogen) cleaveage site, preventing thrombin's catalytic action. Desirudin, produced by recombinant techniques, and related hirudin analogues such as bivalirudin (Hirulog), have been most widely studied.

Initial clinical studies in acute coronary disorders demonstrated that the therapeutic index was much narrower than expected (ANTMAN 1994; GUSTO IIa Investigators 1994; NEUHAUS et al. 1994). Excessive bleeding, including intracranial bleeding, was observed during high-dose therapy (200–300mg/24h). At moderate doses, clinical efficacy was retained but more prolonged therapy may be required (OASIS Investigators 1997). Low doses of 10–20mg/12h have been shown to be superior to low-dose unfractionated heparin (ERIKSSON et al. 1996) and LMWH (ERIKSSON et al. 1997) in prophylaxis of venous thromboembolism in hip surgery. Notably, excessive bleeding was not a problem.

III. Laboratory Monitoring

At intermediate doses, the APTT is sensitive to the anticoagulant effect of hirudin in a dose-dependent fashion (VERSTRAETE et al. 1993; MARBET et al. 1993). At higher doses, however, linearity in the dose-APTT response may be lost (NURMOHAMED et al. 1994). Recently a novel whole blood activated clotting time, using the snake venom ecarin, has been shown to be of value for monitoring hirudin therapy during cardiopulmonary bypass because it can be used in the presence of heparin (POTZSCH et al. 1997). What may prove to be more useful, however, is a chromogenic anti-factor IIa assay which demonstrated linearity in the dose-anticoagulant response over the whole range of therapeutic doses (ESSLINGER et al. 1997).

E. Thrombolytic Agents

I. Clinical Use

Fibrinolytic or thrombolytic agents directly or indirectly activate plasma plasminogen to form the proteolytic enzyme plasmin which has a lytic effect on fibrin and fibrinogen. Dissolution of thrombotic lesions may improve outcome in the treatment of acute myocardial infarction, thrombotic stroke, peripheral arterial thrombosis, and major acute venous thromboembolism. Agents currently available for clinical study include streptokinase, recombinant tissue-type plasminogen activator (alteplase), plasminogen-streptokinase activator complex, urokinase-type plasminogen activator, and recombinant staphylokinase (COLLEN and LIJNEN 1991). Alteplase and staphylokinase have theoretical advantages in that they preferentially activate plasminogen at the fibrin-clot surface (SUENSEN and PETERSEN 1986; SAKHAROV et al. 1996). Heparin and/or aspirin is usually administered to prevent recurrence following successful thrombolysis. Such combined therapies result in an increased risk of bleeding complications compared with the use of heparin, aspirin, or warfarin alone (CALIFF et al. 1988).

II. Monitoring Thrombolytic Therapy

In early clinical trials, a number of haemostatic tests were conducted to assess the potential clinical efficacy and to attempt to predict the risk of bleeding during thrombolytic therapy (BOVILL et al. 1992). Prolongation of global haemostatic tests including PT, APTT and thrombin clotting time occur virtually universally. Unfortunately none of these tests are of value to either predict the efficacy of thrombolytic therapy or to assess bleeding risk (COLLEN and LIJNEN 1991). Although the plasma levels of plasminogen decrease, there is no level at which clinical efficacy of thrombolytic therapy can be predictably evaluated. Depletion of fibrinogen on the other hand, may be helpful in predicting bleeding risk; fibrinogen levels below 1 g/l are associated with a higher risk of intracerebral, retroperitoneal and gastrointestinal bleeding in regimens in which therapy is extended to 48–72 h by continuous intravenous infusion. Baseline assessment of haemostatic function may be useful to identify unrecognised, pre-existing bleeding diatheses (for example, due to liver disease, DIC, etc. which may increase the patient's risk of bleeding). Assessment of fibrin(ogen) degradation products are helpful to ensure a lytic state is achieved.

Current thrombolytic regimens tend to be of fixed duration based on trials showing no benefit for prolonged therapy (MARDER 1997). Therapeutic monitoring for these regimens is not indicated apart from the routine monitoring of concomitant or subsequent heparin therapy.

F. Antiplatelet Agents

I. Aspirin

1. Clinical Effects

Antiplatelet therapy has an important role in the management of serious disorders including coronary artery disease, cerebrovascular disease, and peripheral vascular disease. Aspirin is long-established as a useful antiplatelet agent and it is generally well tolerated. At the therapeutic doses currently used, however, there is a moderate risk of bleeding complications including easy bruising, epistaxis and gastrointestinal bleeding (SCPHSRG 1989; CUSLANDI 1997).

2. Mechanism of Action

Aspirin inhibits platelet and endothelial cell cyclo-oxygenase. Within the platelet there is consequent inhibition of thromboxane A_2 synthesis. Thromboxane A_2 is a potent agonist for platelet aggregation. Within the endothelial cell, inhibition of cyclo-oxygenase blocks prostacyclin synthesis. Prostacyclin can interact with a platelet membrane receptor causing inhibition of platelet aggregation by an effect on cyclic AMP. Theoretically this effect of aspirin is prothrombotic. Dose-finding studies have shown that there is a degree of selection for inhibition of thromboxane A_2 versus prostacyclin at lower aspirin doses (20–80 mg/day) (FITZGERALD et al. 1983). However, the inhibitory effect of aspirin on platelets is irreversible, whereas endothelial cells can promptly re-synthesize prostacylin synthetase. Thus, there is no evidence of a clinical prothrombotic effect of aspirin at daily doses of 100–350 mg, such as those used in current clinical practice.

3. Laboratory Monitoring

Aspirin causes prolongation of the template bleeding time (MIELKE 1982) but the effect is variable among individuals. The bleeding time is an invasive test associated with potential scar and keloid formation. It is difficult to standardise and cumbersome to perform. It is of no value in establishing a suitable anithrombotic effect for any given dose of aspirin and does not predict bleeding risk (RODGERS and LEVIN 1990; LIND 1991).

Turbidometric platelet aggregation is potentially more suitable to assess the antithrombotic effect of aspirin therapy. Aspirin inhibits collagen-induced platelet aggregation (CERLETTI et al. 1986). Aspirin also inhibits the second-wave response to weak agonists such as ADP and epinephrine (CERLETTI et al. 1986). When a range of collagen concentrations are used to stimulate platelet aggregation in samples from individuals receiving aspirin therapy at different doses, maximum inhibition of collagen-induced platelet aggregation occurs only at very high aspirin doses which are unsuitable for clinical use (due to excessive gastro-intestinal haemorrhage) (O'BRIEN and ETHERINGTON 1990). Platelet aggregation in response to aspirin therapy can be studied on citrated whole blood by impedance platelet aggregometry (KUNDU et al. 1995). Resistance across a pair of electrodes increases as platelet aggregates form on the electrodes in response to the agonists. Unfortunately the technique is currently no more sensitive to assess inhibition of platelet aggregation by aspirin therapy than turbidometric platelet aggregometry on preparations of platelet-rich plasma.

With current laboratory monitoring, therefore, it is not possible to establish optimal dosing of aspirin to balance clinical efficacy in vascular disorders with the risk of side-effects, particularly gastrointestinal bleeding (TOHG et al. 1992). Establishing optimal therapy remains dependent on the evaluation of arbitrary aspirin doses in large clinical trials (FORSTER and PARRATT 1997; FUTTERMAN and LEMBERG 1997; SANDERCOCK 1997).

4. Other Antiplatelet Agents

Numerous agents have been evaluated for their potential to inhibit platelet function in a therapeutic setting. Ticlopidine inhibits ADP-mediated platelet aggregation and has potential value in coronary artery surgery (Schror 1993). Recently agents which block the GPIIb/IIIa platelet membrane receptor have been demonstrated to have potent, reversible antiplatelet activity (Van de Werf 1997). These agents inhibit platelet aggregation in response to all known agonists, since GPIIb/IIIa is involved in the final common pathway of platelet aggregation (Lefkovits et al. 1995). Intravenous and, more recently, oral GPIIb/IIIa antagonists have been evaluated in clinical trials. Despite the steepness of the GP IIb/IIIa dose-response relationship, bleeding complications have not been excessive and favourable results were noted in the settings of unstable angina and PTCA (Topol 1998). Monitoring these agents, however, poses the same question associated with aspirin. Dosing in clinical trials with GPIIb/IIIa antagonists is often targeted to produce 70%–80% inhibition of agonist-induced aggregation in vitro. It is not established, however, if this will lead to the best balance between risk and benefit in vivo, especially with chronic oral dosing or in combination with thrombolytics (Willerson 1996). Accurate, predictive monitoring of GPIIb/IIIa antagonist therapy remains a significant challenge.

G. Summary/Conclusion

Warfarin is the principal orally-active agent for long-term antithrombotic therapy. It inhibits vitamin K-dependent hepatic synthesis of factors II, VII, IX, and X. The antithrombotic efficacy of warfarin is mainly due to inhibition of factor II, and to a lesser extent factor X.

There is a wide inter-individual variation in the dose-antithrombotic response. The prothrombin time (PT) is used to monitor warfarin therapy to ensure a therapeutic response and to minimise bleeding risks. Standardisation of thromboplastin reagents in the PT assay, and reporting the results as the international normalised ratio (INR), improves the safety and efficacy of warfarin therapy. The INR is maintained in the range 2.0–3.0 for most indications, and 2.5–3.5 for patients with mechanical heart valve prosthesis.

Heparin is the anticoagulant agent of choice when a rapid antithrombotic response is required. It accelerates the activity of antithrombin in the inhibition of thrombin, factor Xa and to a lesser extent factor XIa. Heparin is administered either by continuous intravenous infusion or intermittent subcutaneous injection. The anticoagulant effect is monitored by the activated partial thromboplastin time (APTT). APTT reagents have variable sensitivities to heparin. Traditionally, therapeutic efficacy and optimal safety are achieved at an APTT ratio of 1.5–2.5. Each laboratory, however, should establish an APTT therapeutic range by titration against a heparin assay equivalent to plasma heparin levels of 0.2–0.4 iu/ml by protamine neutralisation, or

0.3–0.7 iu/ml by chromogenic anti-factor Xa activity. Heparin dose-adjustment nomograms offer a convenient, accurate method of dose adjustment provided the nomogram is adapted to apply to the local therapeutic APTT ratio range.

Low molecular weight heparins (LMWHs) exert their antithrombotic effect primarily by accelerating antithrombin-mediated inhibition of factor Xa. They have advantages over unfractionated heparin. As a result of improved pharmacokinetics, there is a predictable anticoagulant response to weight-adjusted dosing. Due to a lack of effects on platelets or vascular permeability, LMWHs are associated with lower bleeding risks than are observed with unfractionated heparin. Monitoring of LMWH therapy is required only in a few clinical states. A chromogenic anti-factor Xa assay provides convenient, reliable prediction of therapeutic efficacy and safety.

Direct acting antithrombin agents have theoretical advantages over heparin and LMWH. They act independently of antithrombin activity by direct antagonism of thrombin. Their pharmacokinetic properties are favourable for predictable dose-antithrombotic responses. Laboratory monitoring is achieved by the APTT or ACT at moderate doses, and by a novel ecarin ACT at higher doses.

Several thrombolytic agents are available for clinical use. They directly or indirectly activate plasminogen to the lytic enzyme plasmin. When other antithrombotic agents are administered during or after thrombolytic therapy, there is a substantial bleeding risk. During prolonged therapy, monitoring of fibrinogen levels appears useful to predict the risk of bleeding: fibrinogen less than 1.0 g/l is associated with a marked increase in bleeding risk. None of the currently available haemostatic tests, however, are of clinical value in predicting efficacy of thromboloytic therapy.

Aspirin is the most widely used and studied antiplatelet agent. It inhibits platelet thromboxane A_2 synthesis and inhibits platelet aggregation by several agonists. The bleeding time and platelet aggregation assays have been employed to assess the antiplatelet efficacy of aspirin, but no tests are currently available to adequately predict either bleeding risks or therapeutic efficacy. Establishing optimal therapy remains dependent on the evaluation of arbitrary aspirin doses in large clinical trials. Newer antiplatelet agents, including ticlopidine and GPIIb/IIIa platelet membrane receptor inhibitors, are currently under investigation. In general, dosing regimens have been designed to achieve pre-set levels of inhibition of agonist-induced platelet aggregation in vitro. It remains to be determined if this will enable precise definition of optimal dosing.

Over the years, we have seen the development of treatment guidelines for monitoring conventional anticoagulant therapy, based mainly on our experience with heparin and warfarin. In addition to bleeding times, automated tests of coagulation parameters and direct assays of enzyme activity provide useful and predictable information to guide therapy. Newer approaches, however, based on direct inhibition of factors in the coagulation cascade, promotion of thrombolysis, and antagonism of platelet receptors pose challenging problems for monitoring. The old "rules of thumb" may not apply and working out new ones will require a substantial effort. Although new technology and concepts may make the task easier, it appears likely that the primary source for future monitoring guidelines will be empirical findings generated in clinical trials.

References

Anand S, Ginsberg JS, Kearon C, Gent M, Hirsh J (1996) The relation between the activated partial thromboplastin time response and recurrence in patients with venous thrombosis treated with continuous intravenous heparin. Arch Intern Med 156:1677–1681

Ansell JE, Patel N, Ostrovsky D, Nozzolillo E, Peterson AM, Fish L (1995) Long-term patient self-management of oral anticoagulation. Arch Intern Med 155:2185–2189

Ansell JE, Hughes R (1996) Evolving models of warfarin management: anticoagulation clinics, patient self-monitoring, and patient self-management. Am Heart J 132:1095–1100

Ansell JE, Buttaro ML, Thomas OV, Knowlton CH (1997) Consensus guidelines for coordinated outpatient oral anticoagulant therapy management. Anticoagulation Guidelines Task Force. Ann Pharmacother 35:604–615

Antman EM (1994) Hirudin in acute myocardial infarction. Safety report from the Thrombosis and Thrombin Inhibition in Myocardial Infarction (TIMI) 9 A trial. Circulation 90:1624–1630

Bara L, Billaud E, Gramond G, Kher A, Samama M (1985) Comparative pharmacokinetics of a low molecular weight heparin (PK 1169) and unfractionated heparin after intravenous and subcutaneous administration. Thromb Res 39:631–636

Bara L, Samama M (1988) Pharmacokinetics of low molecular weight heparins. Acta Chir Scand Suppl 543:65–72

Barbour LA, Smith JM, Marlar RA (1995) Heparin levels to guide thromboembolism prophylaxis during pregnancy. Am J Obstet Gynecol 173:1869–1873

Basu D, Gallus A, Hirsh J, Cade J (1972) A prospective study of the value of monitoring heparin treatment with the activated partial thromboplastin time. N Engl J Med 287:324–327

Barzu T, Molho P, Tobelem G, Petitou M, Caen J (1985) Binding and endocytosis of heparin by human endothelial cells in culture. Biochim Biophys Acta 845:196–203

Beguin S, Mardiguian J, Lindhout T, Hemker HC (1989) The mode of action of low molecular weight heparin preparation (PK 10169) and two of its major components on thrombin generation in plasma. Thromb Haemost 61:30–34

Bertina RM (1984) The relationship between international normalised ratio and coumarin-induced coagulation defect. In: Van den Besselaar A, Gralnick H, Lewis S (eds): Thromboplastin calibration and oral anticoagulant control, Martinus Nijhoff, Boston

Bjork I, Lindahl U (1982) Mechanism of the anticoagulant action of heparin. Mol Cell Biochem 48:161–182

Blajchman MA, Young E, Ofosu FA (1989) Effects of unfractionated heparin, dermatan sulfate and low molecular weight heparin on vessel wall permeability in rabbits. Ann NY Acad Sci 556:245–254

Booth SL, Charnley JM, Sadowski JA, Saltzman E, Bovill EG, Cushan M (1997) Dietary vitamin K1 and stability of oral anticoagulation: proposal of a diet with constant vitamin KI content. Thromb Haemost 77:504–509

Bovill EG, Becker R, Tracy RP (1992) Monitoring thrombolytic therapy. Prog Cardiovasc Dis 34:279–294

Brandjes DP, Heijboer H, Buller HR, de Rij K, Jagt H, ten Cate JW (1992) Acenocounmarol and heparin compared with acenocoumarol alone in the initial treatment of proximal-vein thrombosis. N Engl J Med 327:1485–1489
Brigden ML (1996) Oral anticoagulant therapy: practical aspects of management. Postgrad Med 99:81–84, 87–89, 93–94
Brill-Edwards P, Ginsberg JS, Johnston M, Hirsh J (1993) Establishing a therapeutic range for heparin therapy. Ann Intern Med 119:104–109
Brown G, Dodek P (1997) An evaluation of empiric vs nomogram-based dosing of heparin in an intensive care unit. Crit Care Med 25:1534–1538
Bussey HI, Force RW, Bianco TM, Leonard AD (1992) Reliance on prothrombin time ratios causes significant errors in anticoagulation therapy. Arch Intern Med 152:278–282
Cadroy Y, Pourrat J, Baladre MF, Saivan S, Hovin G, Montastruc JL et al (1991) Delayed elimination of enoxaparin in patients with chronic renal insufficiency. Thromb Res 63:385–390
Califf RM, Topol EJ, George BS, Boswick JM, Abbottsmith C, Sigmon KN et al (1988) Hemorrhagic complications associated with the use of intravenous tissue plasminogen activator in treatment of acute myocardial infarction. Am J Med 85:353–359
Cerletti C, Carriero MR, de Gaetano G (1986) Platelet-aggregation response to single or paired aggregating stimuli after low-dose aspirin. N Engl J Med 314:316–319
Chiu HM, Hirsh J, Yung WL, Regoeczi E, Gent M (1977) Relationship between the anticoagulant and antithrombotic effects of heparin in experimental venous thrombosis. Blood 49:171–184
Cohen M, Demers C, Gurfinkel EP, Turpie AG, Fromell GJ, Goodman S et al (1997) A comparison of low-molecular-weight heparin with unfractionated heparin for unstable coronary artery disease. Efficacy and safety of Subcutaneous Enoxaparin in Non-Q-Wave Coronary Events Study Group. N Engl J Med 337:447–452
Collen D, Lijnen HR (1991) Basic and clinical aspects of fibrinolysis and thrombolysis. Blood 78:3114–3124
Collins R, Scrimgeour A, Yusuf S, Peto R (1988) Reduction in fatal pulmonary embolism and venous thrombosis by perioperative administration of subcutaneous heparin. Overview of results of randomized trials in general orthopedic,and urologic surgery. N Engl J Med 318:1162–1173
Cortelazzo S, Finazzi P, Viero P, Galli M, Remuzzi A, Parenzan L, et al (1993) Thrombotic and haemorrhagic complications in patients with mechanical heart valve prosthesis attending an anticoagulation clinic. Thromb Haemost 69:316–320
Cruickshank MK, Levine MN, Hirsh J, Roberts R, Siguenz M (1991) A standard heparin nomogram for the management of heparin therapy. Arch Intern Med 151:333–337
D'Angelo A, Seveso MP, D'Angelo SV, Gilardoni F, Dettori AG, Bonini P (1990) Effect of clot-detection methods and reagents on activated partial thromboplastin time (APTT). Implications in heparin monitoring by APTT. Am J Clin Pathol 94:297–306
Dechavanne M, Ville D, Berruyer, Trepo F, Dalery F, Clermont N et al (1989) Randomized trial of low-molecular-weight heparin (kabi 2165) versus adjusted-dose subcutaneous standard heparin in the prophylaxis of deep-vein thrombosis after elective hip surgery. Haemostasis 19:5–12
de Romeuf C, Mazurier C (1993) Heparin binding assay of von Willebrand factor (vWF) in plasma milieu-evidence of the importance of the multimerization degree of vWF. Thromb Haemost 69:436–440
Despotis GJ, Alsoufeiv AL, Spitznagel E, Goodnough LT, Lappas DG (1996) Response of kaolin ACT to heparin: Evaluation with an automated assay at higher heparin doses, Ann Thorac Surg 61:795–779
Eriksson BI, Ekman S, Kalebo P et al (1996) Prevention of deep-vein thrombosis after total hip replacement: direct thrombin inhibition with recombinant hirudin, CGP 39393. Lancet 347:635–639

Eriksson BI, Wille-Jorgensen P, Kalebo P, Mouret P, Rosencher N, Bosch P et al (1997) A comparison of recombinant hirudin with a low-molecular weight-heparin to prevent thromboembolic complications after total hip replacement. N Engl J Med 337:1329–1335
Esslinger HU, Haas S, Maurer R, Lassman A, Dubbers K, Muller-Peltzer H et al (1997) Pharmacodynamic and safety results of PEG-hirudin in healthy volunteers. Thromb Haemost 77:919–919
Feinberg WM, Cornell ES, Nightingale SD, Pearce LA, Tracy RP, Hart RG et al (1997) Relationship between prothrombin activation fragment F1.2 and international normalized ratio in patients with atrial fibrillation. Stroke Prevention in Atrial Fibrillation Investigators 28:1101–1106
Fennerty AG, Thomas P, Backhouse G, Bentley P, Campbell IA, Routledge PA (1985) Audit of control of heparin treatment. Br Med J. Clin Res Ed. 290:27–28
Fihn SD, McDonnel M, Martin D, Henikoff J, Vermes D, Kent D et al (1993) Risk factors for complications of chronic anticoagulation. A multi-centre study. Warfarin Optimized Outpatient Follow-up Study Group. Ann Intern Med 118:511–520
FitzGerald GA, Oates JA, Hawiger J, Maas RL, Roberts LJ 2d, Lawson JA et al (1983) Endogenous biosynthesis of prostacyclin and thromboxane and platelet function during chronic administration of aspirin in man. J Clin Invest 71:676–688
Forster W, Parratt JR (1997) The case of low-dose aspirin for the prevention of mycocardial infarction: but how low is low? Cardiovasc Drugs Ther 10:727–734
Futterman LG, Lemberg L (1997) Harnessing the platelet. Am J Crit Care 6:406–414
Gallus AS, Hirsh J, Tutle RJ, Trebilocock R, O'Brien SE, Carroll JJ et al (1973) Small subcutaneous doses of heparin in prevention of venous thrombosis. N Engl J Med 288:545–551
German Hip Arthroplasty (GHAT) Group (1992) Prevention of deep vein thrombosis with low-molecular-weight heparin in patients undergoing total hip replacement. A randomized trial, Arch Orthop Trauma Surg 111:110–120
Gurwitz JH, Avorn J, Ross-Degnan D, Ansell J (1991) Age-related changes in warfarin pharmacodynamics. Clin Pharmacol Ther 49:166
Guslandi M (1997) Gastric toxicity of antiplatelet therapy with low-dose aspirin. Drugs 53:1–5
GUSTO Investigators (1993) An international randomized trial comparing four thrombolytic strategies for acute myocardial infarction. N Engl J Med 329:673–682
GUSTO IIa Investigators (1994) Randomized trial of intravenous heparin versus recombinant hirudin for acute coronary syndromes. Circulation 90:1631–1637
Haraldsson HM, Onundarson PT, Einarsdolter KA, Guomundsdottir BR, Petursson MK, Palsson K et al (1997) Performance of prothrombin-proconvertin time as a monitoring test of oral anticoagulation therapy. Am J Clin Pathol 107:672–680
Harenberg J, Haaf B, Dempfle CE, Stehle G, Heene DL et al (1995) Monitoring of heparins in haemodialysis using an anti-factor Xa-specific whole-blood clotting assay. Nephrol Dial Transplant 10:217–222
Harenberg J, Stehle G, Blauth M, Huck K, Mall K, Heene DL et al (1997) Dosage, anticoagulant, and antithrombotic effects of heparin and low-molecular-weight heparin in the treatment of deep vein thrombosis. Semin Thromb Hemost 23:83–90
Harrison L, Johnston M, Massicotte MP, Crowther M, Moffat K, Hirsh J et al (1997) Comparison of 5-mg and 10-mg loading doses in initiation of warfarin therapy. Ann Intern Med 126:133–136
Hermans J, Van den Besselaar AM, Loeliger EA,vander Velde EA (1983) A collaborative calibration study of reference materials for thromboplastins. Thromb Haemost 50:712–717
Hirsh J (1991) Oral anticoagulant drugs. N Engl J Med 324:1865–1875

Hirsh J, Levine M (1988) Confusion over the therapeutic range for monitoring oral anticoagulant therapy in North America. Thromb Haemost 59:129–132
Hirsh J, Dalen JE, Deykin D, Poller L, Bussey H (1995a) Oral anticoagulants. Mechanism of action, clinical effectiveness, and optimal therapeutic range. Chest 108:231s–246s
Hirsh J, Raschke R, Warkentin TE, Dalen JE, Deykin D, Poller L et al (1995b), Heparin: mechanism of action, pharmcokinetics, dosing considerations, monitoring, efficacy and safety. Chest 108:258s–275s
Homer E (1985) A new, simple chromogenic substrate assay for heparin and heparin-like anti FXa activity in plasma. Thromb Haemost 54:29–31
Hommes DW, Bura A, Mazzolai L, Buller HR, ten Cate JW (1992) Subcutaneous heparin compared with continuous intravenous heparin administration in the initial treatment of deep venous thrombosis. A meta-analysis, Ann Intern Med 116:279–284
Horne MK, Chao ES (1990) The effect of molecular weight on heparin binding to platelets. Br J Haematol 74:306–312
Hull R, Hirsh J, Jay R, Cater C, England C, Gent M et al (1982) Different intensities of oral anticoagulant therapy in the treatment of proximal-vein thrombosis. N Engl J Med 307:1676–1681
Hull RD, Raskob GE, Hirsch J, Jay RM, Leclerc JR, Geerts WM et al (1986) Continuous intravenous heparin compared with intermittent subcutaneous heparin in the initial treatment of proximal-vein thrombombosis. N Engl J Med 315:1109–1114
Imperiale TF, Speroff T (1994) A meta-analysis of methods to prevent venous thromboembolism following total hip replacement. JAMA 271:1780–1785
Johnston M, Harrison L, Moffat K, Willian A, Hirsh J (1996) Reliability of the international normalized ratio for monitoring the induction phase of warfarin: comparison with the prothrombin time ratio. J Lab Clin Med 128:214–217
Kakkar VV, Cohen AT, Edmonson RA, Phillips MJ, Cooper DJ, Das Sk et al (1993) Low molecular weight versus standard heparin for prevention of venous thromboembolism after major abdominal surgery. The Thromboprophylaxis Collaborative Group. Lancet 341:259–265
Kirkwood TB (1983) Calibration of reference thromboplastins and standardisation of the prothrombin time ratio. Thromb Haemost 49:238–244
Kitchen S, Jennings I, Woods TA, Preston FE (1996) Wide variability in the sensitivity of APTT reagents for monitoring of heparin dosage. J Clin Pathol 49:10–14
Kooman MM, Prandon P, Piovella F, Ockelford PA, Brandjes DP, vander Meer J et al (1996) Treatment of venous thrombosis with intravenous unfractionated heparin administered in the hospital as compared with subcutaneous low-molecular-weight heparin administered at home. The Tasman Study Group. N Engl J Med 334:682–687
Kornberg A, Francis CW, Pellegrini VD Jr, Gabriel KR, Narder VJ (1993) Comparison of native prothrombin antigen with the prothrombin time for monitoring oral anticoagulant prophylaxis. Circulation 88:454–460
Kundu SK, Heilman EJ, Sio R Garcia C, Davidson RM, Ostgaard RA (1995) Description of an in vitro platelet function analyzer– PFA-100. Semin Thromb Hemost 21:106–112
Landefeld CS, Rosenblatt MW, Goldman L (1989) Bleeding in outpatients treated with warfarin: relation to the prothrombin time and important remediable lesions. Am J Med 87:153–159
Landefeld CS, McGuire E 3d, Rosenblatt MW (1990) A bleeding risk index for estimating the probability of major bleeding in hospitalized patients starting anticoagulant therapy. Am JMed 89:569–578
Lane DA, Ryan K (1989) Heparin and low molecular weight heparin: is anti-factor Xa activity important? J Lab Clin Med 114:331–333
Lane DA, Pejler G, Flynn AM, Thompson EA, Lindahl U (1986) Neutralization of heparin-related saccharides by histidine-rich glycoprotein and platelet factor 4. J Biol Chem 261:3980–3986

Le Bras P, Halfon P (1992) Standardization of heparin therapy improved efficacy. Arch Intern Med 152:2140–2143

Lefkovits J, Plow EF, Topol EJ (1995) Platelet glycoprotein IIb/IIIa receptors in cardiovascular medicine. N Engl J Med 332:1553–1559

Levine M, Hirsh J, Gent M, Turpie AG, Leclerc J, Poswers PJ et al (1991) Prevention of deep vein thrombosis after elective hip surgery. A randomized trial comparing low molecular weight heparin with standard unfractionated heparin. Ann Intern Med 114:545–551

Levine M, Hirsh J, Gent M, Turpie AG, Cruickshank M, Weitz J et al (1994) A randomized trial comparing activated thromboplastin time with heparin assay in patients with acute venous thromboembolism requiring large daily doses of heparin. Arch Intern Med 154:49–56

Levine MN, Raskob G, Landefeld S, Hirsh J (1995) Hemorrhagic complications of anticoagulant treatment. Chest 108:276s–290s

Levine M, Gent M, Hirsh J, Leclerc J, Anderson D, Weitz J et al (1996) A comparison of low-molecular-weight heparin administered primarily at home with unfractionated heparin administered in the hospital for proximal deep-vein thrombosis. N Engl J Med 334:677–681

Lind SE (1991) The bleeding time does not predict surgical bleeding. Blood 77:2547–2552

Maimone MM, Tollefsen DM (1988) Activation of heparin co-factor II by heparin oligosaccharides, Biochem Biops Res Commun 152:1056–1061

Malhotra OP (1990) Dicoumarol-induced 9-gamma-carboxyglutamic acid prothrombin: Isolation and comparison with 6-, 7-, 8-, and 10-gammacarboxyglutamic acid isomers. Biochem Cell Biol 68:705–15

Marbet GA, Verstraete M, Kienast J, Graf P, Hoet B, Tsakiris DA et al (1993) Clinical pharmacology of intravenously administered recombinant desulfatohirudin (CGP 39393) in healthy volunteers. J Cardiovasc Pharmacol 22:364–372

Marder VJ (1997) How have trials of thrombolytics influenced clinical management of patients with acute myocardial infarction. Thromb Haemost 78:548–552

Mielke CH Jr (1982) Aspirin prolongation of the template bleeding time, influence of venostasis and direction of incision. Blood 60:1132–1142

Morabia A (1986) Heparin doses and major bleedings. Lancet 1:1278–1279

Moriarty HT, Lam Po Tang PR, Anastas N (1990) Comparison of thromboplastins using the ISI and INR system. Pathology 22:71–76

Nakamura K, Toyohira H, Kariyazono H, Yamada K, Moriyama Y, Taira A (1997) Relationship between changes in F1+2 and TAT levels and blood coagulation early after prosthetic valve replacement. Thromb Res 86:161–171

Nelsestuen GL (1976) Role of gamma-carboxyglutamic acid. An unusual protein transition required for the calcium-dependent binding of prothrombin to phospholipid. J Biol Chem 251:5648–5656

Neuhaus KL, von Essen RV, Tebbe U, Jessel A, Heinrichs H, Maurer W et al (1994) Safety observations from the pilot phase of the randomized r-Hirudin for Improvement of Thrombolysis (HIT-III) study. A study of the ALKK. Circulation 90:1638–1642

Nouredine SN (1995) Research review: use of activated clotting time to monitor heparin therapy in coronary patients. Am J Crit Care 4:272–277

Nurmohamed MT, Rosendaal FR, Buller HR, Dekker E, Hommes DW, Vandenbrouck JP et al (1992) Low-molecular-weight heparin versus standard heparin in general and orthopaedic surgery: a meta-anaylsis. Lancet 340:152–156

Nurmohamed MT, Berckmans RJ, Morrien-Salomons WM, Berends F, Hommes DW, Rijnierse JJ et al (1994) Monitoring anticoagulant therapy by activated partial thromboplastin time: hirudin assessment. Thromb Haemost 72:685–692

OASIS Investigators (1997) Comparison of the effects of two doses of recombinant hirudin compared with heparin in patients with acute myocardial ischemia without ST elevation: a pilot study. Circulation 96:769–777

O'Brien JR, Etherington MD (1990) How much aspirin? Thromb Haemost 64: 486–491
Ofosu FA, Hirsh J, Esmon CT, Modi GJ, Smith LM, Anvari N et al (1989) Unfractionated heparin inhibits thrombin-catalysed amplification reactions of coagulation more efficiently than those catalysed by factor Xa. Biochem J 257: 143–150
Paul B, Oxley A, Brigham K, Cox T, Hamilton PJ (1987) Factor II, VII, IX and X concentrations in patients receiving long-term warfarin. J Clin Pathol 40:94–98
Philippides GJ, Loscalzo J (1996) Potential advantages of direct-acting thrombin inhibitors. Coron Artery Dis 7:497–507
Phillips WS, Smith J, Greaves M, Preston FE, Channer KS (1997) An evaluation and improvement program for inpatient anticoagulant control. Thromb Haemost 77:283–288
Poller L (1987) Progress in standardization in anticoagulant control. Hematol Rev 1:225–241
Poller L (1988) A simple nomogram for the derivation of international normalised ratios for the standardisation of prothrombin times. Thromb Haemost 60:18–20
Poller L, Taberner DA (1982) Dosage and control of oral anticoagulants: An international collaborative survey. Br J Haematol 51:479–485
Poller L, Taberner DA, Sandilands DG, Galasko CS (1982) An evaluation of APTT monitoring of low-dose heparin dosage in hip surgery. Thromb Haemost 47:50–53
Poller L, Wright D, Rowlands M (1993) Prospective comparative study of computer programs used for management of warfarin. J Clin Pathol 46:299–303
Poller L, Thomson JM, Taberner DA, Clarke DK (1994) The correction of coagulometer effects on international normalized ratios: a multicentre evaluation. Br J Haematol 86:112–117
Poller L, Triplett DA, Hirsh J, Carroll J, Clarke K (1995) The value of plasma calibrants in correcting coagulometer effects on international normalized ratios: An international multicenter study. Am J Clin Pathol 103:358–365
Potzsch B, Madlener K, Seelig C, Riess CF, Greinacher A, Muller-Berghaus G (1997) Monitoring of r-hirudin anticoagulation during cardiopulmonary bypass–assessment of the whole blood ecarin clotting time. Thromb Haemost 77:920–925
Raschke R, Hertel G (1991) Clinical use of the heparin nomogram. Arch Intern Med 151:2318, 2321
Raschke RA, Reilly BM, Guidry JR, Fontana JR, Srinivas S et al (1993) The weight-based heparin dosing nomogram compared with a "standard care" nomogram. A randomized controlled trial. Ann Intern Med 119:874–881
Raschke RA, Reilly B (1994) Monitoring heparin therapy. Ann Intern Med 120:169–170
Ray MJ, Smith IR (1990) The dependence of the International Sensitivity Index on the coagulometer used to perform the prothrombin time. Thromb Haemost 63:424–427
Reece IJ, Linley G, al Tareif H, DeBroege R, Tolia J, Sheth J (1996) The activated clotting time loading dose response ratio (ACTLORR) as an indicator of heparin demand during cardiopulmonary by-pass. Perfusion 11:125–130
Reilly BM, Raschke R, Srinivas S, Nieman T (1993) Intravenous heparin dosing: patterns and variations in internists' practices. J Gen Intern Med 8:536–542
Rodgers RP, Levin J (1990) A critical reappraisal of the bleeding time. Semin Thromb Hemost 16:1–20
Sakharov DV, Lijnen HR, Rijken DC (1996) Interactions between staphylokinase, plasminogen, and fibrin. Staphylokinase discriminates between free plasminogen and plasminogen bound to partially degraded fibrin. J Biol Chem 271:27912–27918

Sallah S, Thomas DP, Roberts HR (1997) Wafarin and heparin-induced skin necrosis and the purple toe syndrome: infrequent complications of anticoagulant treatment. Thromb Haemost 78:785–790

Saour JN, Sieck JO, Mamo LA, Gallus AS et al (1990) Trial of different intensities of anticoagulation in patients with prosthetic heart valves, N Engl J Med 322: 428–432

Sandercock P (1997) Antiplatelet therapy with aspirin in acute ischaemic stroke. Thromb Haemost 78:180–182

SCPHSRG (Steering Committee of the Physicians' Health Study Research Group) (1989) Final report on the aspirin component of the ongoing Physicians' Health Study. N Engl J Med 321:129–135

Schror K (1993) The basic pharmacology of ticlopidine and clopidogrel. Platelets 4:252–261

Serra A, Esteve J, Reverter JC, Lozano M, Escolar G, Ordinas A (1997) Differential effect of a low-molecular-weight heparin (dalteparin) and unfractionated heparin on platelet interaction with the subendothelium under flow conditions. Thromb Res 87:405–410

Sethi GK, Copeland JG, Goldman S, Moritz T, Zadina K, Henderson WG (1990) Implications of preoperative administration of aspirin in patients undergoing coronary artery bypass grafting. Departments of Veterans Affairs Cooperative Study on Antiplatelet therapy. J Am Coll Cardiol 15:15–20

Shalansky KF, FitzGerald JM, Sunderji R, Traboulay SJ, O'Malley B, McCarron BI (1996) Comparison of a weight-based heparin nomogram with traditional heparin dosing to achieve therapeutic anticoagulation. Pharmacotherapy 16:1076–1084

Siragosa S, Cosmi B, Piovella F, Hirsh J, Ginsberg JS (1996) Low-molecular-weight heparins and unfractionated heparin in the treatment of patients with acute venous thromboembolism: results of a meta-analysis. Am J Med 100:269–277

Simko RJ, Tsung FF, Stanek EJ (1995) Activated clotting time versus activated partial thromboplastin time for therapeutic monitoring of heparin, Ann Pharmacother 29:1015–1021

Sobel M, McNeill PM, Carson PL, Kermode JC, Adelman B, Conroy R et al (1991) Heparin inhibition of von Willebrand factor- dependent platelet function in vitro and in vivo. J Clin Invest 87:1787–1793

Suenson E, Petersen LC (1986) Fibrin and plasminogen structures essential to stimulation of plasmin formation by tissue-type plasminogen activator. Biochim Biophys Acta 870:510–519

Taberner DA, Poller L, Thomson JM, Lemon G, Weighill FJ (1989) Randomized study of adjusted versus fixed low dose heparin prophylaxis of deep vein thrombosis in hip surgery. Br J Surg 76:933–935

Teien AN, Lie M (1977) Evaluation of an amidolytic heparin assay method: Increased sensitivity by adding purified antithrombin III. Thromb Res 10:399–410

Thijssen HH, Hamulyak K, Willigers H (1988) 4-Hydroxycoumarin oral anticoagulants: pharmacokinetics-response relationship. Thromb Haemost 60:35–38

Thomson JM, Taberner DA, Poller L (1990) Automation and prothrombin time: a United Kingdom field study of two widely used coagulometers. J Clin Pathol 43:679–684

Thromboloysis In Myocardial Infarction (TIMI) IIA Trial Investigators (1997) Dose-ranging trial of enoxaparin for unstable angina: Results of TIMI IIA. J Am Coll Cardiol 29:1474–1482

Tohg H, Konno S, Tamura K, Kimura B, Kawano K (1992) Effects of low-to-high doses of aspirin on platelet aggregability and metabolites of thromboxane A_2 and prostacyclin. Stroke 23:1400–1403

Topol EJ (1998) Toward a new frontier in myocardial referfusion therapy. Circulation 97:211–218

Turpie AG, Gunstensen J, Hirsh J, Nelson H, Gent M (1988) Randomised comparison of two intensities of oral anticoagulant therapy after tissue heart valve replacement. Lancet 1:1242–1245

Turpie AG, Robinson JG, Doyle DJ, Mulji AS, Mishkel GJ, Sealey BJ et al (1989) Comparison of high-dose with low-dose subcutaneous heparin to prevent left ventricular mural thrombosis in patients with acute transmural anterior myocardial infarction. N Engl J Med 320:352–357
Vadher B, Patterson DL, Leaning M (1997) Evaluation of a decision support system for initiation and control of oral anticoagulation in a randomised trial. BMJ 314:1252–1256
Van den Besselaar AM, Lewis SM, Mannucci PM, Poller L (1993) Status of present and candidate international reference preparations (IRP) of thromboplastin for the prothrombin time. A Report of the Subcommittee for Control of Anticoagulation. Thromb Haemost 69:85
Van de Meer FJ, Rosendaal FR, Vandenbroucke JP, Briet E (1993) Bleeding complications in oral anticoagulant therapy. An analysis of risk factors. Arch Intern Med 153:1557–1562
Van de Werf (1997) Clinical trials with glycoprotein IIb/IIIa receptor antagonists in acute coronary syndromes. Thromb Haemost 78:210–213
Verstraete M (1997) Direct thrombin inhibitors: appraisal of the antithrombotic/hemorrhagic balance. Thromb Haemost 78:357–363
Verstraete M, Nurmohamed M, Kienast J, Siebeck M, Silling-Engelhardt G, Buller H et al (1993) Biologic effects of recombinant hirudin (CGP 39393) in human volunteers. European Hirudin in Thrombosis Group J Am Coll Cardiol 22:1080–1088
Weiss P, Soff GA, Halkin H, Seligsohn U (1987) Decline of proteins C and S and factors II, VII, IX and X during the initiation of warfarin therapy. Thromb Res 45:783–790
Weitz JI (1997) Low-molecular-weight heparins. N Engl JMED 337:688–698
Weitz JI, Hudoba M, Massel D, Maraganore J, Hirsh J (1990) Clot-bound thrombin is protected from inhibition by heparin-antithrombin III but is susceptible to inactivation by antithrombin III – independent inhibitors. J Clin Invest 86:385–391
Willerson JT (1996) Inhibitors of Platelet Glycoprotein IIb/IIIa Receptors. Will they be useful when given chronically? Circulation 94:866–868
Wynne HA, Kamali F, Edwards C, Long A, Kelly P (1996) Effect of ageing upon warfarin dose requirements: A longitudinal study. Age Ageing 25:429–431
Xi, M, Beguin S, Hemker HC (1989) The relative importance of the factors II, VII, IX and X for the prothrombinase activity in plasma of orally anticoagulated patients. Thromb Haemost 67:788–791
Young E, Wells P, Holloway S, Weitz J, Hirsh J (1994) Ex-vivo and in-vitro evidence that low molecular weight heparins exhibit less binding to plasma proteins than unfractionated heparin. Thromb Haemost 71:300–304
Young E, Podor TJ, Venner T, Hirsh J (1997) Induction of acute-phase reaction increases heparin-binding proteins in plasma. Artioscler Thromb Vasc Biol 17:1568–1574
Zivelin A, Rao LV, Rapaport SI (1993) Mechanism of the anticoagulant effect of warfarin as evaluated in rabbits by selective depression of individual procoagulant vitamin K-dependent clotting factors. J Clin Invest 92:2131–2140

CHAPTER 6

Use of Transgenic Mice in the Study of Thrombosis and Hemostasis

J.M. PEARSON and D. GINSBURG

A. Introduction

Hemostasis is a delicate balance between the formation and lysis of insoluble blood clots and is required for the maintenance of vessel integrity. A complex, highly regulated system has evolved for the formation of a platelet plug and fibrin clot at the site of vessel injury, involving interactions between blood platelets, the endothelium, and a cascade of specific proteases. Once hemostasis has been achieved, a second system known as the fibrinolytic pathway is required for lysis and resolution of the fibrin and blood clot, leading to the repair of the damaged tissue and reestablishment of vascular integrity. A shift in the balance between these pathways can result in fatal thrombosis or hemorrhage. Recently, considerable insight into the functions and interactions between these pathways has come from the characterization of genetically altered mice generated by powerful transgenic technologies. This chapter will first briefly review the coagulation and fibrinolytic systems as a framework for discussion of these transgenic animal models.

B. Overview of Coagulation and Fibrinolysis

The coagulation system is an ordered cascade of proteolytic cleavage of zymogens to result in the formation of active proteases. Each active protease subsequently acts upon the next in the cascade, as depicted in Fig. 1. Although historically divided into the intrinsic and extrinsic pathways, the extrinsic pathway is currently viewed as the primary mechanism for the activation of coagulation. Tissue factor, exposed upon injury of the endothelium or induced by proinflammatory mediators such as endotoxin, binds factor VIIa (FVIIa) to initiate the extrinsic clotting pathway. The tissue factor/FVIIa complex activates factor Xa (FXa), factor IXa (FIXa) or both. Factor VIII (FVIII) and factor V (FV) are homologous nonenzymatic proteins that serve as essential cofactors for FIXa and FXa, respectively. FVIII and FV also require proteolysis for activation. The FVIIIa/IXa complex, often called the "X-ase" complex, specifically cleaves FX to create its active form, FXa. In turn, the FXa/FVa, or prothrombinase complex, specifically cleaves prothrombin to form active thrombin. Thrombin is the terminal protease in the cascade and directly

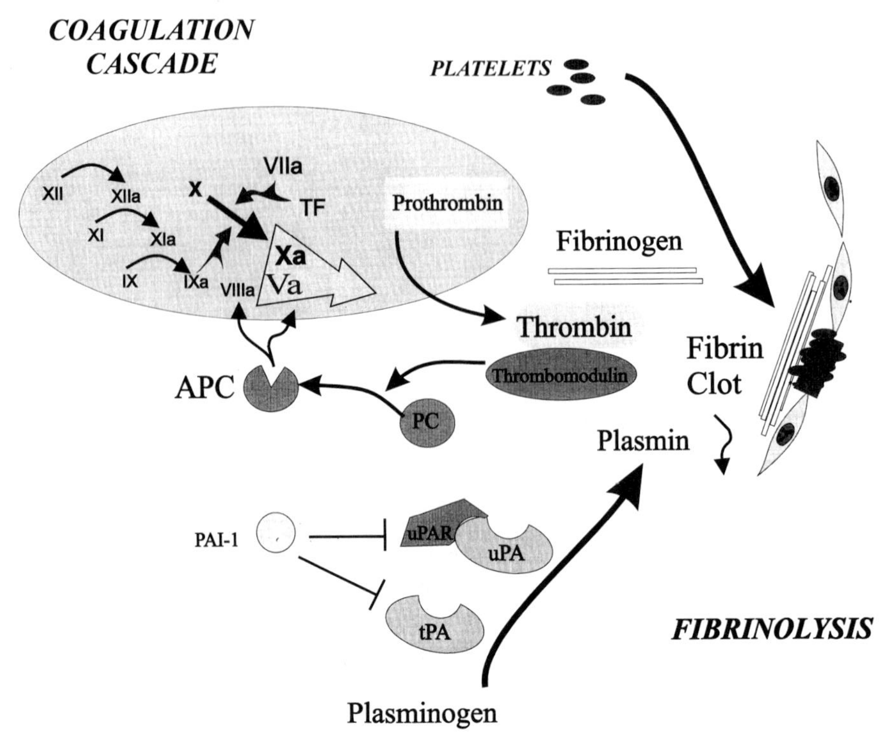

Fig. 1. The coagulation cascade is initiated by the intrinsic pathway (via Factor XII) or extrinsic pathway (via tissue factor (TF)) leading to the conversion of prothrombin to thrombin with subsequent cleavage of fibrinogen to form the fibrin clot. Plasmin-mediated fibrinolysis occurs following the conversion of plasminogen to plasmin by plasminogen activators, urokinase-type plasminogen activator (uPA) or tissue-type plasminogen activator (tPA). Both plasminogen activators are rapidly inhibited by plasminogen activator inhibitor-1 (PAI-1). [Modified from D. Ginsburg. 1997. Haemophilias and other disorders of heamostasis. In: Emery and Rimoin's Principles and Practice of Medical Genetics, Vol. II, 3rd Edition, RIMOIN DL, CONNOR JM, PYERITZ RE (eds), (New York: Churchill Livingstone), p. 1652]

cleaves fibrinogen to fibrin, providing the structural matrix of the blood clot (DAVIE et al. 1991).

This pathway is highly regulated by both positive and negative feedback mechanisms. For instance, thrombin has both procoagulant and anticoagulant properties. It promotes coagulation by directly activating FVIII, FV and FXI. In contrast, when thrombin binds the transmembrane protein thrombomodulin, its activity changes from procoagulant to anticoagulant. Through the formation of activated protein C (APC) and subsequent inactivation of FVa and FVIIIa by APC, thrombomodulin-bound thrombin dampens the coagulation process. The importance of these regulatory pathways is demonstrated by the thrombotic disorders in patients which result from deficiencies in protein C, protein S, and thrombomodulin and as well as the common

FV Leiden mutation which confers resistance to APC (ESMON 1989; MARLAR and NEUMANN 1990; GRIFFIN et al. 1981a; DAHLBACK 1995a–c).

Similar to the coagulation system, fibrinolysis also consists of a cascade of proteolytic enzymes. The terminal protease in this pathway is plasmin, a serine protease formed from its zymogen, plasminogen. Plasmin digests fibrin and converts the insoluble clot to soluble fibrin degradation products. Plasmin formation is regulated largely by two plasminogen activators (PAs), tissue-type and urokinase-type (t-PA and u-PA, respectively). Fibrinolysis is regulated through the inhibition of t-PA and u-PA by specific serine-protease inhibitors, plasminogen activator inhibitors 1 and 2 (PAI-1 and PAI-2), as well as direct inhibition of plasmin by α_2-antiplasmin. Deficiencies in plasminogen or its inhibitors can lead to thrombotic or bleeding disorders as a result of dysregulation of fibrinolysis.

C. Transgenic Technology

Our understanding of the biology of coagulation and fibrinolysis and the delicate balance between them has expanded dramatically in recent years through the generation and characterization of specific transgenic mouse models. The development of gene targeting techniques has allowed the generation of "designer" mice with precisely engineered deletions or subtle alterations in specific target genes. These mice permit analysis of the components of hemostasis within the context of the whole animal.

I. Generation of Standard Transgenic Mice by Zygote Injection

In 1981, five independent laboratories reported stable insertion of foreign DNA into the mouse germ line through microinjection of eggs and the term "transgenic" was used to describe such animals. The generation of transgenic mice is widespread and numerous applications have been developed. Two of the most common uses for transgenic mice are for the study of the regulation of gene expression and for the evaluation of the phenotype resulting from transgene expression. Generation of transgenic mice is schematically illustrated in Fig. 2. The transgenic DNA is directly microinjected into the pronucleus of a fertilized oocyte. In a subset of these cells, the injected DNA stably inserts at one or more random sites in the genome. The injected oocytes are implanted into the uterus of a pseudopregnant female and allowed to develop to term. The resulting mice are analyzed by Southern blotting or PCR for the presence of the transgene. The original mouse carrying the transgene is referred to as a "founder." The transgene integration sites differ among founders as does transgene copy number and level of transgene expression.

II. Generation of Knockout Mice

In contrast to the random insertion of DNA in the standard microinjection technique discussed above, the knockout approach targets a DNA modifica-

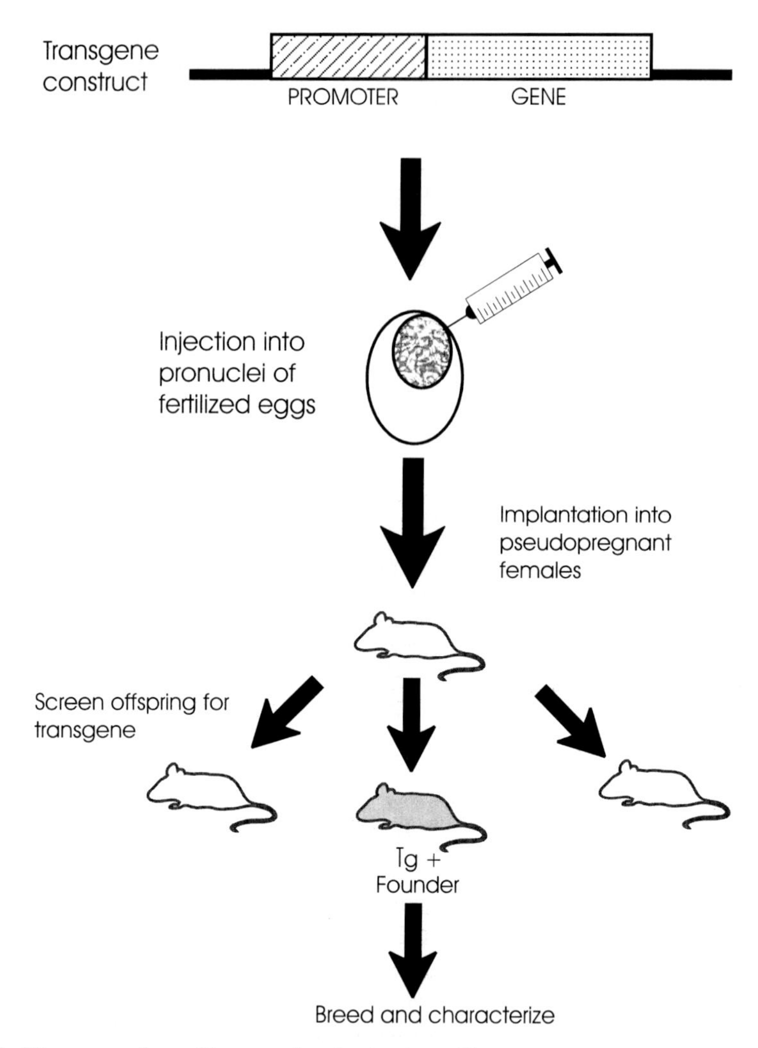

Fig. 2. The generation of transgenic mice begins with a transgene construct, containing the gene of interest driven by a ubiquitous or tissue-specific promoter. The construct is directly injected into the pronucleus of a fertilized oocyte which is then implanted into a pseudopregnant female mouse. The resulting progeny are screened for the presence of the transgene (founders). Founders and their offspring are analyzed further for transgene expression and phenotypic characterization

tion to a specific gene by taking advantage of homologous recombination. The refinement of methods to promote and select for homologous recombination, along with the development of embryonic stem cells that can be propagated in culture and subsequently contribute to the germ-line tissue of mice, opened the doorway for the application of gene-targeting or knockout technology.

The process of generating a knockout mouse is illustrated in Fig. 3. First, a targeting vector is constructed containing a selectable marker (such as the neomycin resistance gene) flanked by sequences homologous to portions of the gene being targeted. The targeting vector is introduced into embryonic stem (ES) cells by DNA transfection and grown in media which select for the presence of the neomycin resistance gene. ES cell colonies are screened by PCR or Southern blot analysis to identify those cells in which the desired homologous recombination event has occurred. Most commonly, the recombination experiment is designed to inactivate or "knockout" the target gene, though more subtle alterations can also be introduced. Successfully targeted ES cells are injected into blastocysts (3.5-day-old embryos) and implanted into foster mothers. Progeny derived from the modified ES cells carry the desired genetic alteration and can now be interbred to produce a stable line of mice, heterozygous and homozygous for this change.

D. Transgenic Mice Deficient in Coagulation Factors

Studies of human diseases have identified a number of clotting factors that are required to prevent spontaneous hemorrhage and to control bleeding following trauma. Hemophilias A and B (deficiencies in factors VIII and IX, respectively) are manifested clinically by hemorrhage into deep tissues, particularly into joints (hemarthroses) and soft tissues. Mice deficient in factors VIII and IX survive to term and display similar bleeding phenotypes (Bi et al. 1995; Lin et al. 1996). Both of these lines of mice should prove useful in the development of novel therapeutic agents for the treatment of the hemophilias, particularly studies evaluating the efficacy of gene therapy (Yao et al. 1991; Armentano et al. 1990).

Though FV deficiency in humans, also referred to as parahemophilia, causes a clinical disorder similar to classical hemophilia, FV knockout mice exhibit a much more severe phenotype (Cui et al. 1996a). Half of the homozygous FV deficient mice die between day 9 and 10 of gestation. The other half develop normally to term, only to die within hours of birth of massive abdominal hemorrhage. These findings suggest that the coagulation cascade plays a critical, previously unrecognized role in early mammalian development. Nearly all humans with parahemophilia are found to have detectable residual levels of FV activity. Taken together, these observations suggest that complete FV deficiency in humans may result in unrecognized early intrauterine death.

In contrast to the early embryonic lethality observed in the FV deficient mice, fibrinogen null mice survive to term (Suh et al. 1995). Approximately 30% of the mice develop spontaneous intra-abdominal hemorrhage at birth, yet remarkably more than 90% of the mice survive to adulthood. A striking observation in these mice is the frequent development of hematomas under the liver capsule. The hematomas become encapsulated by thick fibrous bands without fibroblast infiltration into the central portions of the hematoma.

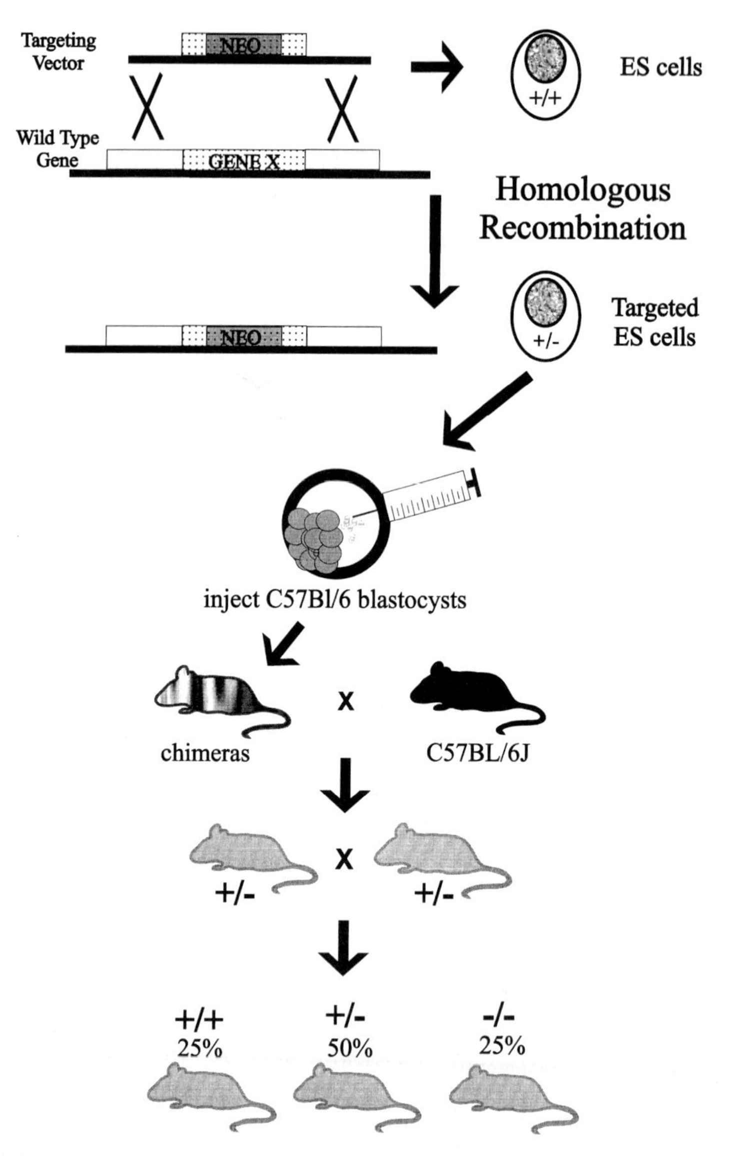

Fig. 3. The generation of knock-out mice begins with the construction of a targeting vector. This vector contains a selectable marker (usually neomycin resistance) flanked by sequences homologous to portions of the genomic target gene. The targeting vector is introduced into embryonic stem (ES) cells. Homologous recombination in these cells results in the replacement of one copy of the wild type gene with the mutant gene. Once the targeted cells have been identified by PCR and/or Southern blot analysis, the ES cells are injected into blastocysts and implanted into foster mothers. The resulting chimeras are interbred with wild type mice to generate mice heterozygous for the mutation. A cross between the heterozygous mice generates mice homozygous for the mutation, as well as animals heterozygous and homozygous for the wild type allele

This result suggests that fibrin may play a critical role as a provisional matrix for the migration of fibroblasts into a wound bed. These lesions resemble the pseudotumors which often develop in hemophiliacs following intramuscular hemorrhage, perhaps via a similar mechanism. In contrast to these visceral hematomas, fibrinogen deficient mice exhibit normal wound healing and keratinocyte migration in a cutaneous wound model (BUGGE et al. 1996b). Similar to reports of spontaneous abortions in afibrinogenemic humans (HAVERKATE and SAMAMA 1995), pregnancy in fibrinogen deficient mice results in consistent fetal loss, even when carrying heterozygous pups, presumably due to a critical requirement for fibrinogen for hemostasis in the gravid uterus (SUH et al. 1995).

The absence of thrombin-mediated platelet activation in addition to deficient fibrin clot formation may account for the more severe hemorrhage observed in FV null than in fibrinogen null mice. The combined deficiency of platelet and fibrin clot function might also explain the developmental defect in the FV null mice. Alternatively, thrombin generated in the early embryo may signal through another pathway that is critical for normal development. The latter hypothesis is supported by studies of thrombin receptor deficient mice (DARROW et al. 1996). Similar to the FV null mice, embryonic lethality is observed in half of the homozygous null mice. However, in contrast to the FV mice, the thrombin receptor deficient mice surviving to birth are viable and display no bleeding diathesis, even as adults. Thus, thrombin signaling through its receptor appears to be critical during embryogenesis but is not essential for postnatal survival. Of note, platelets from these mice can still be activated by thrombin, indicating the existence of a second platelet thrombin receptor in mice.

The exact role of thrombin during development remains to be determined. It is possible that thrombin plays a role not only in the developing embryo, but also at the placental interface in the maintenance of blood flow and vessel patency. Such a hypothesis could explain the striking lethal phenotype observed in thrombomodulin deficient embryos on day 9.5 of gestation (HEALY et al. 1995). The role of thrombomodulin in embryogenesis is not clear, yet in situ studies show that it is expressed early (E7.5) in the yolk sac, followed by expression in the vessels, lung buds, heart and central nervous system (E9.5–10.5). It has been proposed that thrombomodulin is needed to pacify maternal thrombin formed at the placental interface. Such a mechanism involving maternal coagulation factors could also be evoked to explain the spontaneous abortions observed in fibrinogen deficient women and mice (HAVERKATE and SAMAMA 1995; SUH et al. 1995).

In contrast to the FV and thrombin receptor deficient mice, deficiency in tissue factor (TF) is generally lethal between day 8.5 and 10.5 of gestation, though rare mice survive to term. Three laboratories have independently developed TF null mice, with each reporting similar embryonic lethal phenotypes (BUGGE et al. 1996c; CARMELIET et al. 1996; TOOMEY et al. 1997), though the proposed mechanism of lethality is controversial. CARMELIET et al (1996) suggest that TF is critical for vessel formation in the developing yolk sac, whereas BUGGE et al. (1996c) and TOOMEY et al. (1997) suggest that the absence of TF results in loss of hemostatic function and catastrophic hemorrhage. The embryonic lethal phenotype in the mouse is consistent with the absence of any reports of humans with complete tissue factor deficiency. Although the lethal phenotypes observed in a number of coagulation factor knockout mice renders these animals unavailable or difficult to study, others may prove of great value as models for human bleeding disorders and for the development of novel therapeutics.

In a recent preliminary report (CUI et al. 1996b), a "knock-in" transgenic strategy was used to generated a unique animal model for spontaneous thrombosis. These mice carry a specific mutation in FV, R506Q (referred to as FV Leiden), that was recently identified as an extremely common risk factor for venous thrombosis in humans (DAHLBÄCK and HILDEBRAND 1994; DAHLBÄCK 1995a–c; ZÖLLER et al. 1994; APARICIO and DAHLBÄCK 1996). The mutation confers APC resistance in carriers and accounts for >90% of individuals with APC resistance. Interestingly, R506Q has an allele frequency ranging from 3% to 7% in a number of European populations (REES 1996). To establish an appropriate animal model for the study of this mutation and its role in thrombosis, the R504Q (corresponding to the human R506Q) mutation was introduced into the murine FV cDNA which, in turn, was introduced into the mouse chromosome by homologous recombination in ES cells (CUI et al. 1996b). All of the newborn mice carrying the homozygous R504Q mutation develop spontaneous thrombosis in the liver and brain. Although 40%–50% of these homozygous mice appear sick at birth and die within 3 weeks of age, the remaining mice recover from the spontaneous thrombosis and survive to adulthood, with normal appearance and fertility. These mice may prove very useful for the development of novel anticoagulant therapeutic agents.

E. Transgenic Approaches to the Study of the Fibrinolytic System

Pathological alterations in the fibrinolytic system have long been postulated to play a critical role in the development of thrombosis. For instance, elevated plasma PAI-1 levels are correlated with increased risk of deep venous thrombosis and of thrombosis during surgery, trauma and sepsis (JORGENSEN and BONNEVIE-NIELSEN 1987; TABERNERO et al. 1989; ALMER and OHLIN 1987). Elevated plasma PAI-1 levels have been found in patients with recurrent myocardial infarction, angina pectoris and coronary artery disease (OSEROFF et al. 1989; OLOFSSON et al. 1989; PARAMO et al. 1985; AZNAR et al. 1988; HAMSTEN et al. 1987; BARBASH et al. 1989), although a causal relationship between elevated PAI-1 and the development of thrombosis has not been clearly established. In terms of plasminogen and the plasminogen activators, functional and quantitative differences in plasma plasminogen are associated with increased risk of thrombosis (ICHINOSE et al. 1991; AOKI et al. 1978); however,

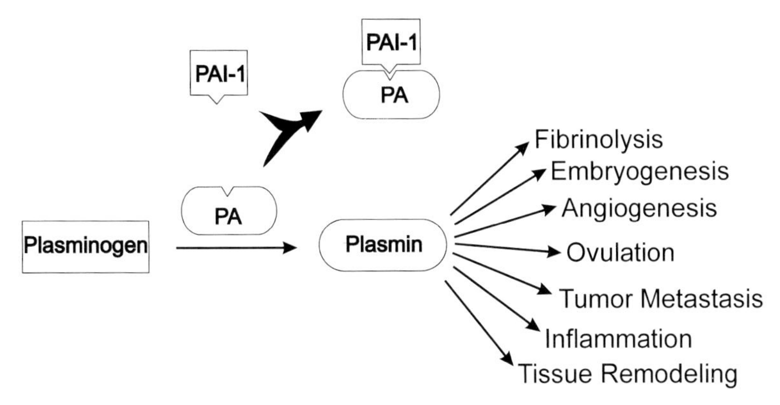

Fig. 4. Plasminogen is converted to plasmin by plasminogen activators (urokinase-type and tissue-type plasminogen activators (PA)) which are both rapidly inhibited by plasminogen activator inhibitor-1 (PAI-1). Plasmin has been implicated in a wide variety of physiological and pathological processes. [Adapted from ETIZMAN DT, FAY WP, and GINSBURG D. 1997. Plasminogen Activator Inhibitor-1. In: The Textbook of Coronary Thrombosis and Thrombolysis, BECKER RC (ed), (Boston: Kluwer Academic Publishers), p. 65, with permission]

no genetic deficiencies of u-PA and t-PA have been documented in humans to date.

In addition to its well established role in fibrinolysis, the PA system has also been proposed as a mediator of extracellular matrix degradation, tissue remodeling, cell migration and angiogenesis (SOFF et al. 1995; KWAAN 1992; SAPPINO et al. 1989; SHIRASUNA et al. 1993) (see Fig. 4). Mice with deficiencies in PA constituents, including plasminogen, u-PA, t-PA, u-PAR (u-PA receptor) and PAI-1, have been generated by targeted homologous recombination. These mice have provided powerful tools to explore and define the role of the PA system in thrombosis and fibrinolysis as well as other pathological conditions.

I. Plasminogen

The terminal step in the PA system is the proteolytic conversion of plasminogen to plasmin. Surprisingly, plasminogen deficient mice are grossly normal at birth, survive to adulthood and are fertile (BUGGE et al. 1995a; PLOPLIS et al. 1995). Thus, the PA system is not critical for development, growth or reproduction. However, plasminogen null mice exhibit high mortality and wasting after 3 months of age. Histologic evaluation of these mice revealed widespread spontaneous lesions, including gastrointestinal ulcers, rectal lesions and hepatic and alveolar fibrin deposits. In addition, when challenged in a cutaneous, full-thickness wound model, gross evaluation revealed dramatic inhibition of wound repair in these mice (ROMER et al. 1996). Histologic evaluation suggests the delayed wound healing may result from impaired keratinocyte migration across the wound bed.

II. Plasminogen/Fibrinogen

The primary mechanism for the spontaneous rectal, gastrointestinal and hepatic lesions, wasting and impaired wound healing in these mice is under investigation. In an interesting report, the plasminogen deficient and fibrinogen deficient mice were intercrossed to generate mice doubly deficient in both plasminogen and fibrinogen (BUGGE et al. 1996b). Fibrinogen deficiency completely rescues mice from the pleiotropic effects of plasminogen deficiency, including the development of rectal and gastrointestinal lesions, wasting, increased mortality and impaired wound healing. This finding strongly suggests that the pathologic consequences of plasminogen deficiency are primarily the result of impaired fibrinolysis and that the major functional role of plasminogen in vivo appears to be largely restricted to lysis of fibrin clots. This does not preclude the possibility that plasminogen plays other roles in specific pathological processes. Future studies employing disease models in these mice may identify contributions of plasminogen in other contexts such as infection, tumor metastasis, atherosclerosis and pulmonary fibrosis. However, the wasting, increased mortality and general poor health observed in adult plasminogen deficient mice could complicate the interpretation of results in long-term animal models required to evaluate chronic conditions such as atherosclerosis and tumor metastasis.

III. t-PA, u-PA and t-PA/u-PA

Other tools for evaluating the role of the PA system in various diseases include mice deficient in either t-PA or u-PA (CARMELIET et al. 1994, 1995). t-PA deficient mice exhibit no thrombotic phenotype whereas u-PA deficient mice display subtle thrombotic changes such as occasional rectal prolapse and ulcerations of eyelids and ears. Both types of mice, however, are more susceptible to the development of venous thrombosis induced by the injection of endotoxin in the footpad. As with spontaneous lesions, the development of venous thrombosis was more pronounced in u-PA than t-PA deficient mice. In order to evaluate clot lysis in vivo, one technique involves the injection of I^{125}-fibrinogen-labeled clots into the jugular vein, resulting in pulmonary emboli. Spontaneous clot lysis is measured by the loss of pulmonary radioactivity. Using this technique, there was no difference in spontaneous clot lysis between wild-type and u-PA deficient mice. In contrast, t-PA deficient mice lysed pulmonary clots at a significantly reduced rate (CARMELIET and COLLEN 1994; CARMELIET et al. 1994, 1995). These findings strongly support the hypothesis that t-PA is primarily responsible for fibrinolysis within the vasculature. Additional studies have also found that the t-PA deficient mice are resistant to neuronal degeneration and seizure after excitotoxin injection, thereby identifying an additional role for t-PA in neuronal activity (TSIRKA et al. 1995).

u-PA is proposed to play a role in cell-mediated fibrinolysis within tissues. Through its binding to its specific cell-surface receptor, u-PAR, u-PA is proposed to mediate proteolysis either directly or by the activation of other matrix degrading proteinases. Support for this hypothesis comes from studies on macrophages from u-PA and t-PA deficient mice. Thioglycollate elicited macrophages were analyzed ex vivo for their ability to lyse a radiolabeled fibrin matrix (Carmeliet et al. 1994, 1995). In comparison to wild type, macrophages from t-PA deficient mice do not differ in their ability to degrade the matrix, whereas macrophages from u-PA deficient mice were unable to lyse the matrix. This finding could explain the sporadic spontaneous lesions observed in the u-PA deficient mice as well as their susceptibility to endotoxin-induced venous thrombosis.

Mice doubly deficient in both u-PA and t-PA have a more pronounced thrombotic phenotype than mice with either single deficiency and exhibit a phenotype similar to that of plasminogen knockout mice. After 2–3 months of age, the t-PA:u-PA null mice develop spontaneous fibrin clots within the liver, gonads, lungs and intestines along with ulcerative lesions and occasional ischemic necrosis in the uterus and intestines. Grossly, the animals appear normal at birth, but after several months many develop rectal prolapse, nonhealing ulcerations, cachexia, wasting and increased mortality. These mice also display a pronounced impairment of pulmonary clot lysis, greater than that observed in t-PA deficient mice (Carmeliet and Collen 1994; Carmeliet et al. 1994, 1995; Bugge et al. 1996a).

IV. u-PAR and t-PA/u-PAR

Yet another tool to study the biology of plasminogen activation is u-PAR deficient mice (Bugge et al. 1995b; Dewerchin et al. 1996). Given the hypothesis that u-PA, upon binding to its receptor, mediates cell-mediated fibrinolysis, one would expect the u-PAR null mice to have a phenotype similar to that of the u-PA knockout mice and that mice deficient in both t-PA and u-PAR would display a phenotype like that of t-PA:u-PA deficient mice. Surprisingly, u-PAR null mice actually have a milder phenotype than u-PA deficient mice in that they do not exhibit spontaneous rectal prolapses or hepatic fibrin deposits. In addition, mice deficient in both t-PA and u-PAR exhibit a milder phenotype than t-PA:u-PA knockout mice as measured by incidence of rectal prolapse, hepatic and extrahepatic fibrin deposits and impairment of wound healing (Bugge et al. 1996a). Thus, evaluation of double deficient mice and comparisons among them suggest that u-PA effectively removes fibrin in the absence of either its receptor or t-PA. In addition, these studies imply that u-PA's action in cell-mediated fibrinolysis can involve either receptor-dependent or receptor-independent mechanisms.

The distinct phenotypes observed in mice with single or combined deficiencies of the plasminogen activators and cell-surface receptor suggest that u-PA and t-PA have distinct yet overlapping biological functions. These mice provide a unique tool to evaluate plasma, tissue and cell-mediated fibrinolysis independently or in combination. In addition, studies in these mice suggest that critical functions of these proteins may only become apparent when the mice are challenged.

V. PAI-1

Numerous studies conducted over the past decade have revealed a complex pattern of PAI-1 expression and regulation in a variety of tissues, as well as induction by cytokines and inflammatory mediators (PARAMO et al. 1988; MEDCALF et al. 1988; MEDINA et al. 1989; SAWDEY and LOSKUTOFF 1991). Due to its postulated role in the regulation of plasmin formation, PAI-1 has been proposed to participate in a number of physiologic processes in addition to fibrinolysis, such as embryo implantation and development, ovulation, tissue remodeling, tumor invasion and metastasis and wound healing (STRICKLAND and BEERS 1976; STRICKLAND et al. 1976; SAPPINO et al. 1989; DE VRIES et al. 1995; KWAAN 1992; ROMER et al. 1996). In humans, partial deficiency of PAI-1 has been identified in a few patients and is associated with clinical bleeding. Only one family with complete PAI-1 deficiency has been identified (FAY et al. 1992). The original patient had no apparent abnormalities in development, wound healing or response to infection despite several episodes of hemorrhage following minor trauma. The lack of other significant clinical findings in this patient suggests that the primary role of PAI-1 in vivo is regulation of fibrinolysis. The recent report of additional PAI-1 deficient patients identified in the same pedigree also supports these conclusions (FAY et al. 1997).

A similar phenotype is observed in mice deficient in PAI-1 (CARMELIET et al. 1993a,b, 1995), which exhibit normal fertility, survival and bleeding time. These mice, however, do not exhibit symptoms of hemorrhage, even after undergoing tail biopsy. Yet, when challenged, PAI-1 deficient mice lyse radiolabeled clots more rapidly and are resistant to endotoxin-induced venous thrombosis when compared to wild type mice.

Two laboratories have generated transgenic mice that overexpress PAI-1. One laboratory generated mice that overexpress human PAI-1 (ERICKSON et al. 1990) while the other laboratory created mice overexpressing murine PAI-1 (EITZMAN et al. 1996a,b). Mice transgenic for human PAI-1 develop venous occlusions evidenced by necrotic tails and swollen hind limbs. In these mice, the level of PAI-1 expression correlates with the presence of venous occlusions. In contrast, mice transgenic for murine PAI-1 do not develop venous occlusions and tails are normal, despite elevated plasma PAI-1 levels. The differences between these two transgenic mouse lines likely result from species differences of PAI-1 activity.

The development of both PAI-1 deficient and PAI-1 overexpressing mice provides unique tools to study the role of PAI-1 in disease processes. Studies in these mice support the hypothesis that PAI-1 plays a critical role in the regulation of fibrinolysis. In particular, PAI-1 deficient mice lyse radiolabeled clots at a higher rate and are resistant to endotoxin-induced venous thrombosis when compared to wild type mice (CARMELIET et al. 1993a,b, 1995; CARMELIET and COLLEN 1994).

In addition to evaluating the role of PAI-1 in fibrinolysis, PAI-1 transgenic mice have been utilized in the study of other disease processes such as tumor invasion and metastasis and pulmonary fibrosis. u-PA is expressed on a variety of tumor cells and is postulated to be a critical modulator of tumor cell invasiveness (KWAAN 1992; PYKE et al. 1991). Transgenic mice that overexpress human PAI-1 have been reported to experience significantly fewer spontaneous lung metastases in a Lewis lung carcinoma model (POGGI et al. 1993). In contrast, in another model employing u-PA expressing B16 melanoma cells, no differences in B16 melanoma growth, metastasis or animal survival were observed among mice lacking or overexpressing murine PAI-1 (EITZMAN et al. 1996a). Thus, the role of PAI-1 in tumor invasion and metastasis remains uncertain and appears to be dependent upon the animal model and neoplastic cell line.

Altered fibrinolytic activity has been observed in a number of pulmonary disease states. For instance, fibrinolytic activity is reduced in the bronchoalveolar lavage (BAL) fluid from patients with acquired respiratory distress syndrome (ARDS), idiopathic pulmonary fibrosis, bronchopulmonary dysplasia and sarcoidosis (HASDAY et al. 1988; KOTANI et al. 1995). Elevated BAL PAI-1 levels have been noted in ARDS patients and could contribute to the reduced fibrinolytic activity (MOALLI et al. 1989). In rodents, the intratracheal administration of bleomycin is used as a model of pulmonary fibrosis. Studies with the PAI-1 transgenic mice demonstrate that PAI-1 deficient mice are protected from bleomycin-induced pulmonary fibrosis whereas PAI-1 overexpressing mice develop more severe fibrosis (EITZMAN et al. 1996b), pointing to a critical role for PAI-1 in the pathogenesis of fibrosis. These findings are consistent with the hypothesis that the fibrin meshwork serves as a provisional matrix for cell migration and collagen deposition following tissue injury and hemorrhage. Prompt removal of the fibrin would minimize connective tissue deposition and hence reduce fibrosis. A similar scenario could apply to other disorders associated with fibrin deposition, including neointima formation and atherosclerosis. Several laboratories are currently pursuing these avenues with the transgenic reagents available.

F. Summary

Transgenic mice provide unique tools and reagents for studying the biology of particular genes and their role in diseases. However, these mice are only a model and it is likely that there are significant differences in the pathogenesis of disease between mice and humans. Clearly, when knockout mice are described as having a mild or no phenotype, this must be interpreted in the context of our limited ability to detect functional, sensory and behavioral changes in mice. In addition, phenotypes are described within the confines of a pathogen and predator-free environment, under conditions of controlled diet, temperature and light and within the limited life span of a mouse. In addition, the overall health and vitality of a given transgenic mouse model must be taken into consideration when conducting more chronic studies, such as wound healing, atherosclerosis and tumor metastasis. Clearly, cachexia, wasting and the development of inflammatory lesions may modify the pathogenic mechanisms associated with disease.

Nonetheless, the development of transgenic mice deficient in coagulation factors has provided powerful models for the study of human bleeding disorders and thrombosis. In addition, these studies have revealed a previously unrecognized role for the coagulation system in early mammalian development and have generated several exciting new areas of research.

Over the past several years, remarkable progress in molecular genetics has resulted in the generation of numerous transgenic mice over- or underexpressing most of the proteins of the fibrinolytic cascade. These valuable reagents have shed new light on previous studies and altered our understanding of the role of the PA system in biologic processes. Observations in these mice suggest that fibrinolysis is not as essential in reproduction, embryonic development and tissue remodeling as previously proposed. The generation of mice with single and combined deficiencies is invaluable for our understanding of the fibrinolytic system's role in pathologic conditions such as thrombosis, infection, fibrosis and metastasis. Current research is already utilizing these mice to evaluate fibrinolysis in more complex and diverse diseases such as atherosclerosis, vascular injury and neointima formation, restenosis, multiple inflammatory organ diseases and pathogenic infections. Future studies using these and other mouse models may also lead to the development of novel therapeutic approaches.

References

Almer LO, Ohlin H (1987) Elevated levels of the rapid inhibitor of plasminogen activator (t-PAI) in acute myocardial infarction. Thromb Res 47:335–339

Aoki N, Moroi M, Sakata Y, Yoshida N, Matsuda M (1978) Abnormal plasminogen. A hereditary molecular abnormality found in a patient with recurrent thrombosis. J Clin Invest 61:1186–1195

Aparicio C, Dahlbäck B (1996) Molecular mechanisms of activated protein C resistance – properties of factor V isolated from an individual with homozygosity for the Arg506 to Gln mutation in the factor V gene. Biochem J 313:467–472

Armentano D, Thompson AR, Darlington G, Woo SLC (1990) Expression of human factor IX in rabbit hepatocytes by retrovirus-mediated gene transfer: potential for gene therapy of hemophilia B. Proc Natl Acad Sci USA 87:6141–6145

Aznar J, Estelles A, Tormo G, Sapena P, Tormo V, Blanch S, Espana F (1988) Plasminogen activator inhibitor activity and other fibrinolytic variables in patients with coronary artery disease. Br Heart J 59:535–541

Barbash GI, Hod H, Roth A, Miller HI, Rath S, Zahav YH, Modan M, Zivelin A, Laniado S, Seligsohn U (1989) Correlation of baseline plasminogen activator inhibitor activity with patency of the infarct artery after thrombolytic therapy in acute myocardial infarction. Am J Cardiol 64:1231–1235

Bi L, Lawler AM, Antonarakis SE, High KA, Gearhart JD, Kazazian HH Jr (1995) Targeted disruption of the mouse factor VIII gene produces a model of haemophilia A. Nat Genet 10:119–121
Bugge TH, Flick MJ, Daugherty CC, Degen JL (1995a) Plasminogen deficiency causes severe thrombosis but is compatible with development and reproduction. Genes Dev 9:794–807
Bugge TH, Suh TT, Flick MJ, Daugherty CC, Romer J, Solberg H, Ellis V, Dano K, Degen JL (1995b) The receptor for urokinase-type plasminogen activator is not essential for mouse development or fertility. J Biol Chem 270:16886–16894
Bugge TH, Flick MJ, Danton MJS, Daugherty CC, Romer J, Dano K, Carmeliet P, Collen D, Degen JL (1996a) Urokinase-type plasminogen activator is effective in fibrin clearance in the absence of its receptor or tissue-type plasminogen activator. Proc Natl Acad Sci USA 93:5899–5904
Bugge TH, Kombrinck KW, Flick MJ, Daugherty CC, Danton MJS, Degen JL (1996b) Loss of fibrinogen rescues mice from the pleiotropic effects of plasminogen deficiency. Cell 87:709–719
Bugge TH, Xiao Q, Kombrinck KW, Flick MJ, Holmbäck K, Danton JS, Colbert MC, Witte DP, Fujikawa K, Davie EW, Degen JL (1996c) Fatal embryonic bleeding events in mice lacking tissue factor, the cell-associated initiator of blood coagulation. Proc Natl Acad Sci USA 93:6258–6263
Carmeliet P, Collen D (1994) Evaluation of the plasminogen/plasmin system in transgenic mice, Fibrinolysis 8:269–276
Carmeliet P, Kieckens L, Schoonjans L, Ream B, Van Nuffelen A, Prendergast GC, Cole MD, Bronson R, Collen D, Mulligan RC (1993a) Plasminogen activator inhibitor-1 gene-deficient mice. I. Generation by homologous recombination and characterization, J Clin Invest 92:2746–2755
Carmeliet P, Stassen JM, Schoonjans L, Ream B, van den Oord JJ, De Mol M, Mulligan RC, Collen D (1993b) Plasminogen activator inhibitor-1 gene-deficient mice. II. Effects on hemostasis, thrombosis, and thrombolysis. J Clin Invest 92:2756–2760
Carmeliet P, Schoonjans L, Kieckens L, Ream B, Degen JL, Bronson R, De Vos R, van den Oord JJ, Collen D, Mulligan RC (1994) Physiological consequences of loss of plasminogen activator gene function in mice. Nature 368:419–424
Carmeliet P, BouchÈ A, De Clercq C, Janssen S, Pollefeyt S, Wyns S, Mulligan RC, Collen D (1995) Biological effects of disruption of the tissue-type plasminogen activator urokinase-type plasminogen activator, and plasminogen activator inhibitor-1 genes in mice Ann N Y Acad Sci 748:367–382
Carmeliet P, Mackman N, Moons L, Luther T, Gressens P, Van Vlaenderen I, Demunck H, Kasper M, Breier G, Evrard P, Müller M, Risau W, Edgington T, Collen D (1996) Role of tissue factor in embryonic blood vessel development. Nature 383:73–75
Cui J, O'Shea KS, Purkayastha A, Saunders TL, Ginsburg D (1996a) Fatal haemorrhage and incomplete block to embryogenesis in mice lacking coagulation factor V. Nature 384:66–68
Cui J, Purkayastha A, Yang A, Yang T, Gallagher K, Metz A, Ginsburg D (1996b) Spontaneous thrombosis in mice carrying the APC resistance mutation (FV Leiden) introduced by gene targeting. Blood 88:440a (Abstract)
Dahlbäck B (1995a) Factor V gene mutation causing inherited resistance to activated protein C as a basis for venous thromboembolism. J Intern Med 237:221–227
Dahlbäck B (1995b) Resistance to activated protein C, the Arg506 to Gln mutation in the factor V gene, and venous thrombosis. Functional tests and DNA-based assays, pros and cons, Thromb Haemost 73:739–742
Dahlbäck B (1995c) Molecular genetics of thrombophilia: factor V gene mutation causing resistance to activated protein C as a basis of the hypercoagulable state. J Lab Clin Med 125:566–571
Dahlbäck B, Hildebrand (1994) Inherited resistance to activated protein C is corrected by anticoagulant cofactor activity found to be a property of factor V. Proc Natl Acad Sci USA 91:1396–1400

Darrow AL, Fung-Leung W-P, Ye RD, Santulli RJ, Cheung W-F, Derian CK, Burns CL, Damiano BP, Zhou L, Keenan CM, Peterson PA, Andrade-Gordan P (1996) Biological consequences of thrombin receptor deficiency in mice. Thromb Haemost 76:860–866

Davie EW, Fujikawa K, Kisiel W (1991) The coagulation cascade: initiation, maintenance, and regulation Biochem 30:10363–10370

De Vries TJ, Kitson JL, Silvers WK, Mintz B (1995) Expression of plasminogen activators and plasminogen activator inhibitors in cutaneous melanomas of transgenic melanoma-susceptible mice. Cancer Res 55:4681–4687

Dewerchin M, Van Nuffelen A, Wallays G, BouchÈ A, Moons L, Carmeliet P, Mulligan RC, Collen D (1996) Generation and characterization of urokinase receptor-deficient mice. J Clin Invest 97:870–878

Eitzman DT, Krauss JC, Shen T, Cui J, Ginsburg D (1996a) Lack of plasminogen activator inhibitor-1 effect in a transgenic mouse model of metastatic melanoma. Blood 87:4718–4722

Eitzman DT, McCoy RD, Zheng X, Fay WP, Shen T, Ginsburg D (1996b) Bleomycin-induced pulmonary fibrosis in transgenic mice that either lack or overexpress the murine plasminogen activator inhibitor-1 gene. J Clin Invest 97:232–237

Eitzman DT, Fay WP, Ginsburg D (1997) Plasminogen activator inhibitor-1. In: Becker RC (ed) The textbook of coronary thrombosis and thrombolysis. Kluwer Academic Publishers, Boston, p 65

Erickson LA, Fici GJ, Lund JE, Boyle TP, Polites HG, Marotti KR (1990) Development of venous occlusions in mice transgenic for the plasminogen activator inhibitor-1 gene. Nature 346:74–76

Esmon CT (1989) The roles of protein C and thrombomodulin in the regulation of blood coagulation. J Biol Chem 264:4743–4746

Fay WP, Shapiro AD, Shih JL, Schleef RR, Ginsburg D (1992) Complete deficiency of plasminogen-activator inhibitor type 1 due to a frame-shift mutation. N Engl J Med 327:1729–1733

Fay WP, Parker AC, Condrey LR, Shapiro AD (1997) Human plasminogen activator inhibitor-1 (PAI-1) deficiency: characterization of a large kindred with a null mutation in the PAI-1 gene. Blood 90:204–208

Ginsburg D (1997) Haemophilias and other disorders of haemostasis. In: Rimoin DL, Connor JM, Pyeritz RE (eds) Emery and Rimoin's principles and practice of medical genetics, vol. II, 3rd edn. Churchill Livingstone, New York, p 1652

Griffin JH, Evatt B, Zimmerman TS, Kleiss AJ (1981) Deficiency of protein C in congenital thrombotic disease. J Clin Invest 68:1370–1373

Hamsten A, de Faire U, Walldius G, Dahlen G, Szamosi A, Landou C, Blombäck M, Wiman B (1987) Plasminogen activator inhibitor in plasma: risk factor for recurrent myocardial infarction. Lancet 2:3–9

Hasday JL, Bachwich PR, Lynch JP, Sitrin RG (1988) Procoagulant and plasminogen activator activities of bronchoalveolar fluid in patients with pulmonary sarcoidosis. Exp Lung Res 14:261–278

Haverkate F, Samama M (1995) Familial dysfibrinogenemia and thrombophilia. Report on a study of the SSC subcommittee on fibrinogen. Thromb Haemost 73:151–161

Healy AM, Rayburn HB, Rosenberg RD, Weiler H (1995) Absence of the blood-clotting regulator thrombomodulin causes embryonic lethality in mice before development of a functional cardiovascular system. Proc Natl Acad Sci USA 92:850–854

Ichinose A, Espling ES, Takamatsu J, Saito H, Shinmyozu K, Maruyama I, Petersen TE, Davie EW (1991) Two types of abnormal genes for plasminogen in families with a predisposition for thrombosis. Proc Natl Acad Sci USA 88:115–119

Jorgensen M, Bonnevie-Nielsen V (1987) Increased concentration of the fast-acting plasminogen activator inhibitor in plasma associated with familial venous thrombosis. Br J Haematol 65:175–180

Kotani I, Sato A, Hayakawa H, Urano T, Takada Y, Takada A (1995) Increased procoagulant and antifibrinolytic activities in the lungs with idiopathic pulmonary fibrosis. Thromb Res 77:493–504

Kwaan HC (1992) The plasminogen-plasmin system in malignancy. Cancer Metastasis Rev 11:291–311

Lin HF, Maeda N, Smithies O, Straight DL, Stafford DW (1996) Coagulation factor IX-deficient mice developed by gene targeting. Blood 88:657a (Abstract)

Marlar RA, Neumann A (1990) Neonatal purpura fulminans due to homozygous protein C or protein S deficiencies. Semin Thromb Hemost 16:299–309

Medcalf RL, Kruithof EKO, Schleuning W-D (1988) Plasminogen activator inhibitor 1 and 2 are tumor necrosis factor/cachectin-responsive genes. J Exp Med 168: 751–759

Medina R, Socher SH, Han JH, Friedman PA (1989) Interleukin-1, endotoxin or tumor necrosis factor/cachectin enhance the level of plasminogen activator inhibitor messenger RNA in bovine aortic endothelial cells. Thromb Res 54:41–52

Moalli R, Doyle JM, Tahhan HR, Hasan FM, Braman SS, Saldeen T (1989) Fibrinolysis in critically ill patients. Am Rev Respir Dis 140:287–293

Olofsson BO, Dahlen G, Nilsson TK (1989) Evidence for increased levels of plasminogen activator inhibitor and tissue plasminogen activator in plasma of patients with angiographically verified coronary artery disease. Eur Heart J 10:77–82

Oseroff A, Krishnamurti C, Hassett A, Tang D, Alving B (1989) Plasminogen activator and plasminogen activator inhibitor activities in men with coronary artery disease. J Lab Clin Med 113:88–93

Paramo JA, Colucci M, Collen D, van de Werf F (1985) Plasminogen activator inhibitor in the blood of patients with coronary artery disease. BMJ (Clin Res Ed) 291:573–574

Paramo JA, Diaz FJ, Rocha E (1988) Plasminogen activator inhibitor activity in bacterial infection. Thromb Haemost 59:451–454

Ploplis VA, Carmeliet P, Vazirzadeh S, Van Vlaenderen I, Moons L, Plow EF, Collen D (1995) Effects of disruption of the plasminogen gene on thrombosis, growth, and health in mice. Circulation 92:2585–2593

Poggi A, Bellelli E, Castelli MP, Marinacci R, Rella C, Erickson LA, Donati MB, Bini A, Consorzio MNS, Santa Maria Imbaro I (1993) Reduced Lewis lung carcinoma (3LL) metastases in mice transgenic for human plasminogen activator inhibitor-1 (PAI-1). Proc Annu Meet Am Assoc Cancer Res 34: 72 (Abstract)

Pyke C, Kristensen P, Ralfkiaer E, Grondahl-Hansen J, Eriksen J, Blasi F, Dano K, Grndahl-Hansen J, Dan K (1991) Urokinase-type plasminogen activator is expressed in stromal cells and its receptor in cancer cells at invasive foci in human colon adenocarcinomas, Am J Pathol 138:1059–1067

Rees DC (1996) The population genetics of factor V Leiden (Arg 506 Gln). Br J Haematol 95:79–586

Romer J, Bugge TH, Pyke C, Lund LR, Flick MJ, Degen JL, Dano K (1996) Impaired wound healing in mice with a disrupted plasminogen gene. Nature Med 2:287–292

Sappino AP, Huarte J, Belin D, Vassalli JD (1989) Plasminogen activators in tissue remodeling and invasion: mRNA localization in mouse ovaries and implanting embryos. J Cell Biol 109:2471–2479

Sawdey MS, Loskutoff DJ (1991) Regulation of murine type 1 plasminogen activator inhibitor gene expression in vivo. Tissue specificity and induction by lipopolysaccharide, tumor necrosis factor-α, and transforming growth factor-α. J Clin Invest 88:1346–1353

Shirasuna K, Saka M, Hayashido Y, Yoshioka H, Sugiura T, Matsuya T (1993) Extracellular matrix production and degradation by adenoid cystic carcinoma cells: participation of plasminogen activator and its inhibitor in matrix degradation. Cancer Res 53:147–152

Soff GA, Sanderowitz J, Gately S, Verrusio E, Weiss I, Brem S, Kwaan HC (1995) Expression of plasminogen activator inhibitor type 1 by human prostate carcinoma

cells inhibits primary tumor growth, tumor-associated angiogenesis, and metastasis to lung and liver in an athymic mouse model. J Clin Invest 96:2593–2600
Strickland S, Beers WH (1976) Studies on the role of plasminogen activator in ovulation. In vitro response of granulosa cells to gonadotropins, cyclic nucleotides, and prostaglandins. J Biol Chem 251:5694–5702
Strickland S, Reich E, Sherman MI (1976) Plasminogen activator in early embryogenesis: enzyme production by trophoblast and parietal endoderm. Cell 9:231–240
Suh TT, Holmbäck K, Jensen NJ, Daugherty CC, Small K, Simon DI, Potter SS, Degen JL (1995) Resolution of spontaneous bleeding events but failure of pregnancy in fibrinogen-deficient mice. Genes Dev 9:2020–2033
Tabernero MD, Estells A, Vicente V, Alberca I, Aznar J (1989) Incidence of increased plasminogen activator inhibitor in patients with deep venous thrombosis and or pulmonary embolism. Thromb Res 56:565–570
Toomey JR, Kratzer KE, Lasky NM, Broze GJ Jr (1997) Effect of tissue factor deficiency on mouse and tumor development. Proc Natl Acad Sci USA 94:6922–6926
Tsirka SE, Gualandris A, Amaral DG, Strickland S (1995) Excitotoxin-induced neuronal degeneration and seizure are mediated by tissue plasminogen activator. Nature 377:340–344
Yao S-N, Wilson JM, Nabel EG, Kurachi S, Hachiya HL, Kurachi K (1991) Expression of human factor IX in rat capillary endothelial cells: toward somatic gene therapy for hemophilia B. Proc Natl Acad Sci USA 88:8101–8105
Z^ller B, Svensson PJ, He X, Dahlbäck B (1994) Identification of the same factor V gene mutation in 47 out of 50 thrombosis-prone families with inherited resistance to activated protein C. J Clin Invest 94:2521–2524

CHAPTER 7

Current Antiplatelet Therapy

J.A. JAKUBOWSKI, R.E. JORDAN, and H.F. WEISMAN

A. Introduction

Initially considered circulating cell fragments, the blood platelet is now recognized to play vital roles in both hemostasis and thrombosis. Recognition of its role in arterial thrombosis has, in part, been driven by the study of the effects of platelet inhibitory drugs on thrombus formation. There are many platelet inhibitory (antiplatelet) compounds described in the literature, with a wide range of mechanisms that mediate the platelet inhibitory effects. In order to cover current antiplatelet therapy in more detail, this review will deal primarily with currently available and commonly used antiplatelet agents that have regulatory approval. This will limit the review to the most widely used inhibitors, aspirin (a cyclo-oxygenase inhibitor), dipyridamole (a phosphodiesterase inhibitor), ticlopidine (an ADP antagonist) and the relatively recently introduced agent abciximab (a glycoprotein IIb/IIIa inhibitor). Given the novelty of abciximab, a substantial portion of this chapter will review the preclinical and clinical development of this monoclonal antibody as antiplatelet therapy for the treatment of ischemic complications associated with percutaneous coronary revascularization. A brief description of investigational agents that are in late stage clinical evaluation is also included.

B. Platelets: Physiological and Pathological Activities

Circulating blood platelets (also referred to as thrombocytes) are cell fragments (not containing nuclei they cannot accurately be referred to as cells) originating from megakaryocytes where their formation is modulated by thrombopoietin and other cytokines. The circulating, nonactivated, platelet has a disk-like shape, being 1–3 μm in diameter and approximately 0.5 μm in thickness. There is size heterogeneity within the platelet population with a normal volume range of 4.5–8.5 fl, with the larger platelets being more active (THOMPSON and JAKUBOWSKI 1988). The normal platelet count range is 150,000–400,000/ml in whole blood, with about one-third of the total platelet population residing, freely exchangeable, in the spleen. The half-life of a platelet is 4–5 days with a lifespan of 8–10 days. Examination of the platelet by transmission electron microscopy reveals the presence of numerous and diverse granules in the cytoplasm, the contents of which may be secreted upon platelet activation (the release reaction). For more thorough reviews of platelet morphology see LIND (1994) and WHITE (1994).

I. Physiological Activities

The primary physiological role of the platelet is the promotion of hemostasis. This function was described in 1882 by both HAYEM and BIZZOZERO, who observed platelet aggregates at sites of vascular injury. Platelets mediate primary hemostasis by accumulating (aggregating) at sites of vessel wall damage and providing a primary barrier to blood loss. Furthermore, adherent, activated platelets promote hemostasis indirectly by providing a localized surface upon which coagulation protein complexes assemble and are activated, notably the tenase complex (generating activated factor X) and the prothrombinase complex catalyzing the formation of thrombin (WALSH 1994). Thrombin subsequently cleaves fibrinogen to fibrin, which consolidates the hemostatic plug. In addition, thrombin interacts with receptors on platelets and promotes further platelet activation.

Platelet interactions and reactions constitute several discrete activities. Following vascular disruption and exposure to subendothelial collagen and von Willebrand factor, platelet adhesion occurs. Adhesion activates the platelets, which then undergo a shape change characterized by the platelet assuming a more spherical form and developing pseudopodia. Depending upon the strength of the stimulus, activation may also result in the release reaction during which a wide variety of prepackaged materials are secreted from platelet granules. The secreted substances include ADP and ATP (adenosine di- and triphosphate), serotonin, platelet derived growth factor, transforming growth factor (TGF)-β, plasminogen activator inhibitor-1 (PAI-1), β-thromboglobulin and platelet factor 4. In nonstatic environments, activation and release are typically accompanied by platelet aggregation, in which activated platelets bind to each other, generating platelet aggregates and forming a hemostatic plug. Upon activation, platelets also synthesize, de novo, various mediators which can promote or inhibit platelet aggregation and thus locally modulate the hemostatic response. These mediators include the activators thromboxane (TX) A_2 and platelet activating factor (PAF) and the inhibitors prostaglandin D_2 and nitric oxide (NO).

The platelet membrane has numerous receptors for ligands that can initiate or modulate platelet aggregation. Descriptions of platelet receptor types, G-protein systems, phospholipases, second messengers such as inosital phosphates, diacyl glycerol and ionized calcium that mediate the platelet responses are presented in Chap. 8, this volume, and other reviews (BRASS et al. 1993; KROLL 1994). Following receptor activation these intracellular second messenger systems mediate induction of the "final common pathway" – the cross-bridging of activated platelets to each other via binding of fibrinogen, von Willebrand factor (vWF) and other adhesive proteins to activated glycoprotein (GP) IIb/IIIa platelet surface receptors (also known as $\alpha_{IIb}\beta_3$) (COLLER

1990). GPIIb/IIIa on resting platelets does not bind the circulating plasma proteins fibrinogen and von Willebrand factor (vWf). However, in response to platelet activation and second messenger signaling, conformational changes in GPIIb/IIIa occur. This "activated" form of GPIIb/IIIa is a functional high affinity receptor for fibrinogen and vWf which, by virtue of their multivalent nature, mediate interplatelet bridging by binding to GPIIb/IIIa on adjacent platelets (COLLER 1990; CALVETTE 1995).

II. Pathological Activities

As previously described, the primary physiological function of platelets is to halt the loss of blood from damaged blood vessels. The pathological counterpart of this is thrombosis, which has been described as "hemostasis in the wrong place" (MACFARLANE 1977). In the arterial circulation, especially in the coronary arteries, platelet initiated thrombosis reflects platelet adhesion and aggregation on ruptured atherosclerotic plaques with subsequent arterial occlusion and ischemic sequelae. Many lines of evidence support the involvement of platelets in coronary thrombosis. These include the observation of platelet aggregates and platelet-rich thrombi on ruptured atherosclerotic plaques within thrombosed coronary arteries at autopsy following myocardial infarction (FRIEDMAN and VAN DEN BOVENKAMP 1966), direct angioscopic observation of platelet-rich thrombi in patients with ischemic coronary disease (MIZUNO et al. 1992), and elevation of platelet activation markers in patients with coronary ischemia and following PTCA (BRADEN et al. 1991; FITZGERALD et al. 1986).

Based on this clear evidence of platelet involvement in arterial thrombosis, platelet aggregation inhibitors have become a common component of antithrombotic therapy in the arterial circulation. The success of such therapy in reducing primary and secondary ischemic events is further evidence for platelet involvement in arterial thrombosis. Aspirin is currently the most widely used of platelet inhibitory agents. However, several other agents, with differing mechanisms of action, are available to the physician. The following sections deal with the mode of action of these agents and the clinical indications.

C. Current Antiplatelet Therapy

I. Aspirin

Originally synthesized in 1853 and marketed for its anti-inflammatory, analgesic and antipyretic properties in 1899, aspirin (acetylsalicylic acid, ASA) is presently the most commonly used antiplatelet agent. Later, noting that aspirin had both analgesic and anticoagulant properties, GIBSON, in 1949, presented initial evidence that aspirin was useful in treating vascular disease. Subsequently, CRAVEN provided additional support for the use of aspirin to protect against coronary heart disease (CRAVEN 1950, 1953). Both of these early investigators felt that the large doses of aspirin they administered (grams/day) inhibited hemostasis and thrombus formation via an anticoagulant mechanism. In reality, the effects probably reflected both anticoagulant and antiplatelet effects since at the large doses employed, aspirin has both antiplatelet and anticoagulant activities, although it was not appreciated at the time that aspirin had antiplatelet properties. While the acetyl moiety of ASA (aspirin) is not obligatory for its anti-inflammatory activities, it plays a unique role in platelet inhibition. Aspirin's platelet inhibitory properties reflect its ability to inhibit the enzyme cyclo-oxygenase, a ubiquitous enzyme which in the platelet plays a central role in thromboxane A_2 (TXA_2) synthesis. Following platelet activation, free arachidonic acid is transformed sequentially by cyclo-oxygenase and thromboxane synthase to the prostaglandin (PG) endoperoxides PGG_2/PGH_2 and TXA_2, respectively (ROTH 1986). Thromboxane A_2 released from platelets binds to receptors on platelets and promotes the aggregation response. Elucidation of this metabolic pathway and aspirin's inhibitory effect on it provided clues to the biochemical basis of aspirin's platelet inhibitory effects. The biochemical and molecular pharmacological actions and consequences of aspirin on cyclo-oxygenase are quite unique. Aspirin covalently modifies cyclo-oxygenase by acetylating a serine residue adjacent to the enzyme's active site and thereby abolishes catalytic activity (ROTH 1986). Since the platelet is essentially unable to synthesize proteins, the irreversible covalent modification of the platelet enzyme is long-lived. Indeed, reestablishment of the full biochemical potential of circulating platelets requires approximately 10 days for the complete turnover of the platelet pool initially exposed to aspirin (PATRIGNANI et al. 1982).

Near maximal inhibition (>95%) of platelet TXA_2 synthesis is achieved by many different aspirin dosing regimens, including single doses of 80 mg or repeated daily doses of 40 mg (PATRIGNANI et al. 1982). Inhibition is also achieved with enteric-coated preparations and is maintained chronically with 80 mg every day or 325 mg every other day (JAKUBOWSKI et al. 1985; STAMPFER et al. 1986). For both standard and enteric-coated aspirin, low-dose and every-other-day dosing provides continuous inhibition of platelet TXA_2 synthesis, due to the cumulative and irreversible acetylation of platelet cyclo-oxygenase. Intravenous (WILSON et al. 1990) and dermal routes (KEIMOWITZ et al. 1993) have also been described for platelet inhibition, but oral preparations are by far the most commonly used.

While TXA_2 strongly promotes platelet aggregation, there are alternative pathways that support platelet aggregation. Accordingly, while aspirin is clinically effective and widely used in cerebral and myocardial ischemia (MORAN and FITZGERALD 1994), the platelet dysfunction resulting from aspirin is relatively mild. Aspirin's platelet inhibitory effect can be detected in in vitro platelet aggregation responses to certain activating agents such as arachidonic acid and low-dose ADP or collagen. Aspirin's effect on hemostasis is reflected in a mild prolongation of the template bleeding time, which is approximately doubled by aspirin. In the presence of otherwise normal hemostatic mechanisms, aspirin is a relatively safe agent.

II. Dipyridamole

With the realization that adenosine inhibited platelet involvement in animal models of thrombosis and that the investigational agent RA8 (Persantin, dipyridamole) increased plasma levels of adenosine, investigations into the in vitro and in vivo effects of dipyridamole on human platelet function were initiated (EMMONS et al. 1965). Since these early studies, several additional mechanisms by which dipyridamole inhibits platelet function have been reported. The compound, a dipiperidino-pyrimido-pyrimidine, is a weak inhibitor of platelet phosphodiesterase. It also inhibits the uptake of adenosine by erythrocytes and endothelial cells leading to locally elevated adenosine concentrations; it may also increase vascular prostacyclin production (SHEBUSKI 1994). All of these actions can ultimately increase intraplatelet concentrations of cyclic nucleotides. The specific mechanisms underlying its in vivo antiplatelet activity at therapeutic concentrations may reflect one or more of these activities. While elevation of platelet cyclic AMP and cyclic GMP can provide maximal inhibition of platelet function, the in vitro and ex vivo platelet dysfunction associated with dipyridamole is marginal. It is likely that in vivo platelet inhibition is greater as a result of inhibition secondary to interaction with other cells and tissues in vivo. Despite its weak antiplatelet effects, dipyridamole has been evaluated in numerous clinical trials and is a marketed agent. The recommended dose of this orally active agent is 150–400 mg daily, with vasodilation being a dose limiting side effect.

Dipyridamole has undergone several clinical studies for the primary and secondary prevention of myocardial infarction and other thrombotic disorders. However, it is not clear from the early studies whether dipyridamole alone or in combination with aspirin provides any benefit over aspirin alone (FITZGERALD 1987). More recently, extended release formulations of dipyridamole (Persantin Retard) have been produced and tested in the secondary prevention of stroke. The new formulations provided more uniform levels of dipyridamole than the original formulation and dosing regimen. The results of the studies, while controversial, do suggest that Persantin Retard, 200 mg twice daily, may be an effective antiplatelet agent for secondary prevention of stroke and TIA and that its benefit may be additive to that provided by aspirin alone (DIENER et al. 1996). Dipyridamole is commonly used in conjunction with oral anticoagulant agents such as coumarin or warfarin for the reduction of thromboembolic complications of cardiac valve replacement. It is for this indication that dipyridamole is registered in the United States.

III. Ticlopidine

Ticlopidine (Ticlid) is a first generation, thienopyridine oral platelet aggregation inhibitor. Inhibition of human platelet aggregation by ticlopidine was originally described in 1975, at which time the mechanism of action was not known but was thought to be distinct from that of aspirin (Thebault et al. 1975). Subsequent reports concluded that ticlopidine inhibited fibrinogen binding to GPIIb/IIIa; however, it is now accepted that ticlopidine's platelet inhibitory effects result from interruption of the platelet ADP receptor (Schror 1993). Of note, the parent compound has little activity in vitro, and a hepatic metabolite and/or activity at the megakaryocyte level is felt to be responsible for the antiplatelet/antithrombotic properties. In vivo activity only becomes apparent after 5–6 days of administration and remains for at least 3 days after the last dose (Thebault et al. 1975). A serious side effect of ticlopidine is agranulocytosis, with neutropenia being observed in approximately 2.3% of patients and severe neutropenia in approximately 1% (Flores-Runk and Raasch 1993). Ticlopidine (250 mg twice daily) has provided benefit in several large trials of secondary stroke prevention and overall provides a similar degree of protection as aspirin (Flores-Runk and Raasch 1993). In recent years ticlopidine has become widely used (without regulatory approval) in conjunction with aspirin to reduce acute closure of coronary arteries following stent placement. A recently completed study, the STent Anticoagulation Regimen Study (STARS), supports the view that a combination of aspirin plus ticlopidine is more effective than either aspirin alone or aspirin with warfarin (Ferguson and Fox 1997). A second generation thienopyridine orally active platelet aggregation inhibitor, clopidogrel (Plavix), will be briefly described later (see Sect. D, below).

IV. Abciximab

1. Preclinical Development

Abciximab (ReoPro, c7E3 Fab) is the Fab fragment of the human/murine chimeric antibody 7E3 and is the first of the class of antiplatelet agents that directly bind and block the GPIIb/IIIa integrin receptor (Knight et al. 1995). Abciximab also binds to the closely related $\alpha_v\beta_3$ receptor, an adhesive receptor present in small numbers on platelets and in larger numbers on activated vascular endothelial and smooth muscle cells (Coller 1997a). As discussed above, the importance of the GPIIb/IIIa receptor as a therapeutic target was established when it was shown to be the platelet binding site for the adhesive molecules fibrinogen and von Willebrand factor. Through this adhesive function, GPIIb/IIIa is the final common pathway leading to platelet aggregation and the formation of occlusive intravascular thrombi. Abciximab is the first rationally designed antiplatelet and anti-integrin receptor agent and is the prototype for other agents in this class that are under development (Coller 1997b).

The binding of abciximab to platelet surface GPIIB/IIIa receptors is high affinity, saturable and readily quantifiable (COLLER 1985). At saturation, approximately 80,000 molecules of abciximab bind per platelet. There is a two-fold higher level of binding of c7E3 Fab (abciximab) to platelets compared to 7E3 IgG or 7E3 $F(ab')_2$ (WAGNER et al. 1996; COLLER 1986; KUTOK and COLLER 1994). The lower levels of 7E3 IgG and $F(ab')_2$ binding are due to their simultaneous, bivalent interaction with two adjacent GPIIb/IIIa receptors, leading to one-half the numbers of bound molecules at saturation compared to abciximab (WAGNER et al. 1996).

The equilibrium dissociation constant for the binding of abciximab to human platelets is 5 nM and is similar to the affinity of the bivalent forms of the 7E3 molecule [7E3 IgG and 7E3 $F(ab')_2$]. The 7E3 antibody and its fragments bind to platelets of cynomolgus monkeys and baboons with equivalent affinities as to human platelets. The binding affinity of 7E3 to dog platelets is lower than to primate platelets although numerous preclinical canine studies have been conducted using appropriate doses of 7E3 $F(ab')_2$.

The rates of binding and dissociation from platelets of the smaller, monovalent forms of 7E3, including abciximab, are different from the larger, bivalent forms of 7E3 (COLLER 1986). Unlike the slow dissociation of the bivalent 7E3 $F(ab')_2$ fragments from platelets, there is an appreciable rate of dissociation of abciximab over several hours. Approximately 50% of bound abciximab dissociated from platelets within 5 h under conditions in which little or no 7E3 $F(ab')_2$ dissociated during the same time period (WAGNER et al. 1996).

Flow cytometry illustrates the reversibility of abciximab binding to platelets. Figure 1 demonstrates the results of an experiment in which untreated platelets were mixed at 37°C with an equivalent number of platelets that had been saturated with abciximab. The association of abciximab to individual platelets was monitored in samples that were removed from the stirred platelet suspension at various times. Immediately after mixing ($t = 0$), the untreated platelets (left histogram) and the abciximab-saturated platelets (right histogram) were separate, identifiable populations. By 90 min after mixing, the two distinct platelet populations began to merge, and by 3 h, the platelet peaks had fully merged, indicating that abciximab was distributed uniformly over the entire population. In another study, full in vitro redistribution of abciximab among platelets maintained at 22°C occurred in approximately 10 h (CHRISTOPOULOS et al. 1993). These studies suggest a dynamic and reversible binding of abciximab that leads to a continuous, redistribution of abciximab among available platelets.

The binding of abciximab to platelets has been quantified and expressed as the percentage of GPIIb/IIIa receptors that are occupied (blocked) following treatment. Abciximab receptor blockade studies were regularly performed in both animal and human studies. Receptor blockade measurements have been performed using ex vivo incubations with ^{125}I-labelled 7E3 to measure unblocked GPIIb/IIIa on platelet surfaces (COLLER et al. 1985). Receptor

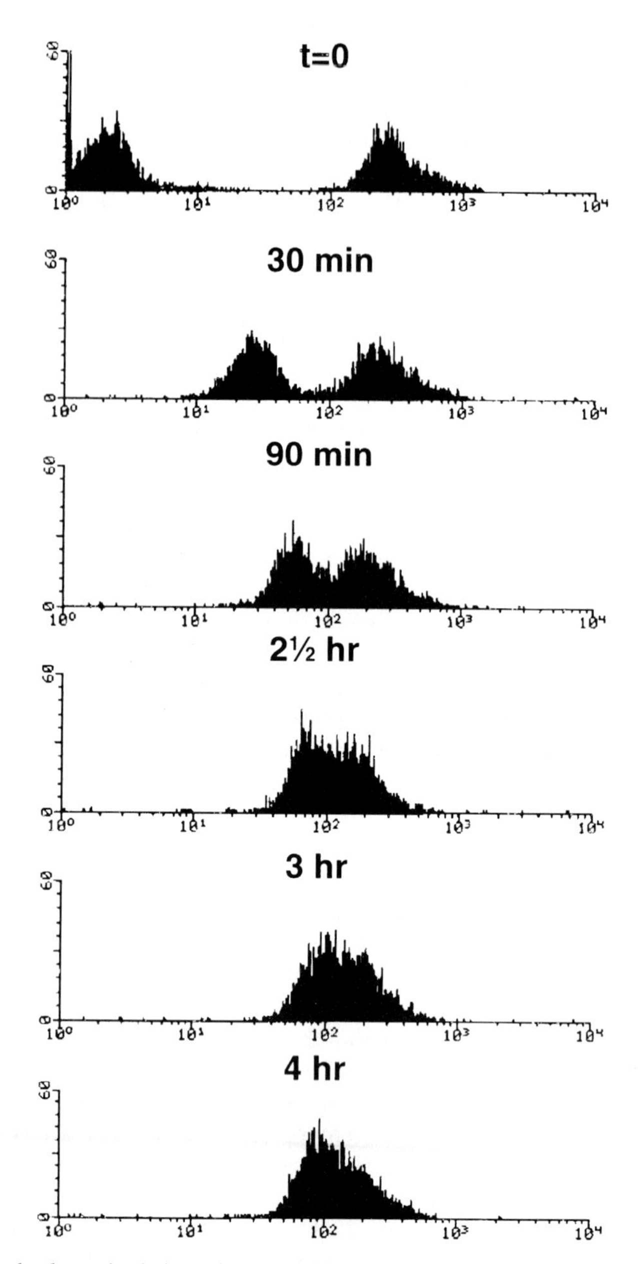

Fig. 1. Platelet-bound Abciximab Redistributes to Unlabeled Platelets under *In Vitro* Mixing Conditions. Equal volumes of washed abciximab-treated platelets and control (saline-treated) platelets were combined and continuously mixed at 37°C. Samples were periodically removed from the incubation and treated with a fluorescein-conjugated rabbit anti-abciximab antibody preparation and then fixed with 2% formalin. The samples were evaluated for the presence of platelet-bound anti-abciximab on a Becton-Dickinson FACScan flow cytometer. From each sample, 5000 events were analyzed in the forward-versus-side scatter gate that defined the platelet population. Individual platelet histograms are shown. The convergence of the two separate peaks into a unimodal pattern indicates that all platelets eventually acquired comparable amounts of abciximab (Jordan et al., 1996)

blockade measurements allowed a direct correlation to the inhibition of platelet aggregation to establish the in vitro and ex vivo dose response profiles of abciximab.

Figure 2a illustrates a typical in vitro dose response evaluation for abciximab binding and inhibition of platelets obtained from a cynomolgus monkey. Full inhibition of platelet aggregation was achieved at an abciximab concentration of 1.75 μg/ml at a corresponding receptor blockade of 82%. A second, in vivo phase of the cynomolgus monkey experiment comprised sequential intravenous administrations of weight-adjusted, 0.05-mg/kg doses of abciximab at 15-min intervals. Following each incremental dose, a blood sample was drawn for the evaluation of the effects on platelets (Fig. 2b). With increasing cumulative doses, there was a stepwise increase in the degree of receptor blockade and inhibition of platelet aggregation. Nearly complete inhibition of platelet aggregation was achieved at the cumulative dose of 0.25 mg/kg, corresponding to 82% receptor blockade. The results of these preclinical studies in monkeys were highly predictive of the subsequent studies in humans.

2. Clinical Pharmacology

The dose response of abciximab for inhibition of platelet function was investigated in volunteers and patients with coronary artery disease (Bhattacharya et al. 1995; Tcheng et al. 1994). Figure 3 depicts the abciximab dose relationship in human patients of receptor blockade and inhibition of platelet aggregation at 2 h after bolus doses that ranged from 0.15 to 0.30 mg/kg. Receptor blockade increased progressively at increasing abciximab doses and 80% blockade was achieved at 0.20 or 0.25 mg/kg in most individuals. Ex vivo platelet aggregation to 5 and 20 μM ADP was markedly inhibited at the 0.20-mg/kg dose and essentially abolished at 0.25 mg/kg. The high level of receptor blockade and inhibition of platelet aggregation at 2 h was followed, over the ensuing 12–24 h, by a gradual reduction of GPIIb/IIIa receptor blockade and a substantial return of platelet function. The results for patients receiving the 0.25-mg/kg bolus are shown in Fig. 4. There was a partial recovery of platelet aggregation at 4 and 6 h that coincided with the fall of receptor blockade to <80% during this period. The subsequent recovery of receptor blockade and platelet aggregation remained gradual. At 24 h, the receptor blockade had decreased to 50%–60% and platelet aggregation was ~40% of baseline.

The rapid recovery of partial platelet function following the 0.25-mg/kg bolus dose suggested that this duration of protection would be inadequate for the prevention of arterial thrombosis at sites of vessel wall injury such as occurs during balloon angioplasty. Thus, the bolus dose of abciximab was combined with subsequent continuous low level intravenous infusions to maintain the profound inhibition of platelets. Both non-weight-adjusted infusions of 10 μg/min, as well as weight-adjusted infusions of 0.125 μg/kg-min,

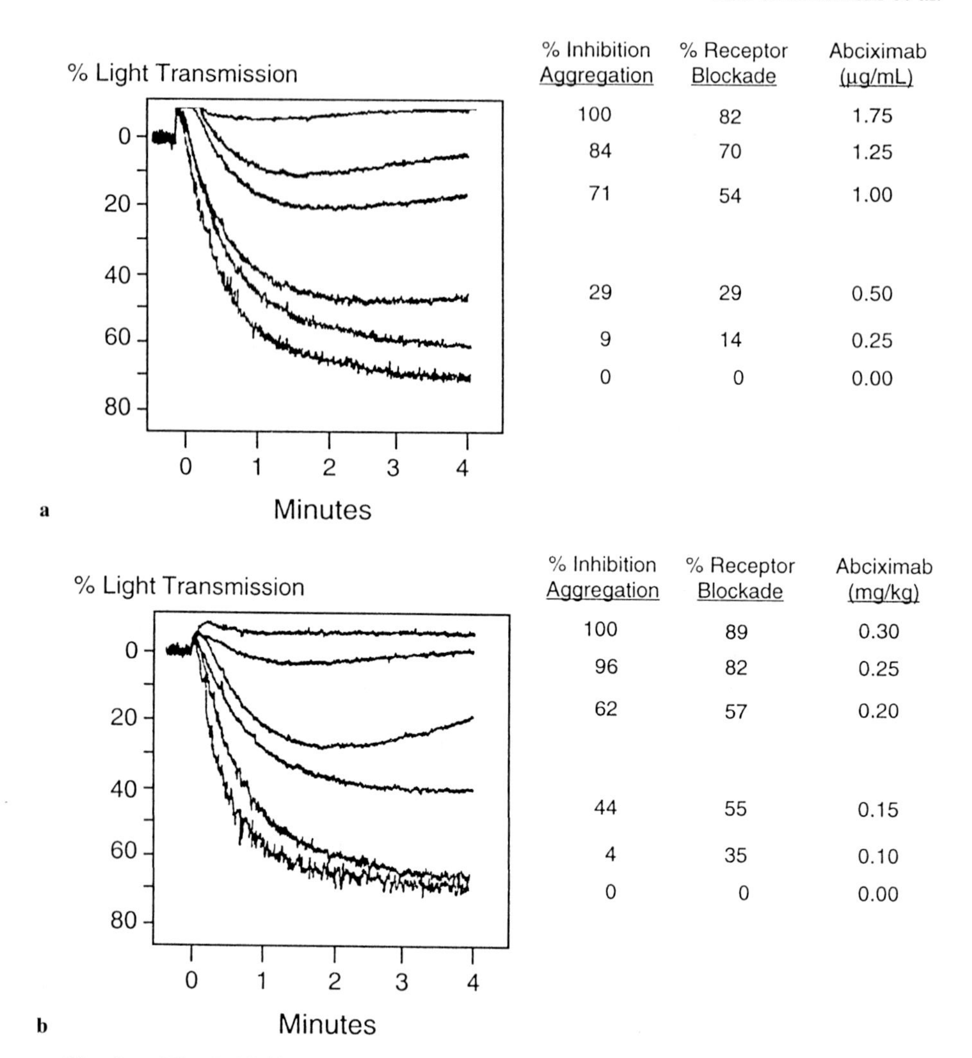

Fig. 2. **a** The Inhibition of *In Vitro* Monkey Platelet Aggregation by Abciximab in Relation to GPIIb/IIIa Receptor Blockade. Aliquots of PRP, were incubated with the indicated concentrations of abciximab for 15 min at 37°C prior to initiating the aggregation reaction by addition of 20 μM ADP. Individual tracings were overlayed to produce the composite figure. The calculated levels of GPIIb/IIIa receptor blockade are also indicated next to the corresponding aggregation tracing. **b** Parallel Measurements of Platelet Aggregation and GPIIb/IIIa Blockade in a Monkey Receiving Incremental Doses of Abciximab. Aggregation was initiated with 20 μM ADP. The abciximab column summarizes the total cumulative dose (JORDAN et al., 1996)

have been used after a 0.25-mg/kg bolus of abciximab. Both regimens maintained GPIIb/IIIa receptor blockade at >80% and platelet aggregation at less than 20% of baseline. Figure 5 depicts the sustained inhibition of platelet aggregation and GPIIb/IIIa receptor blockade in patients receiving a 24-h

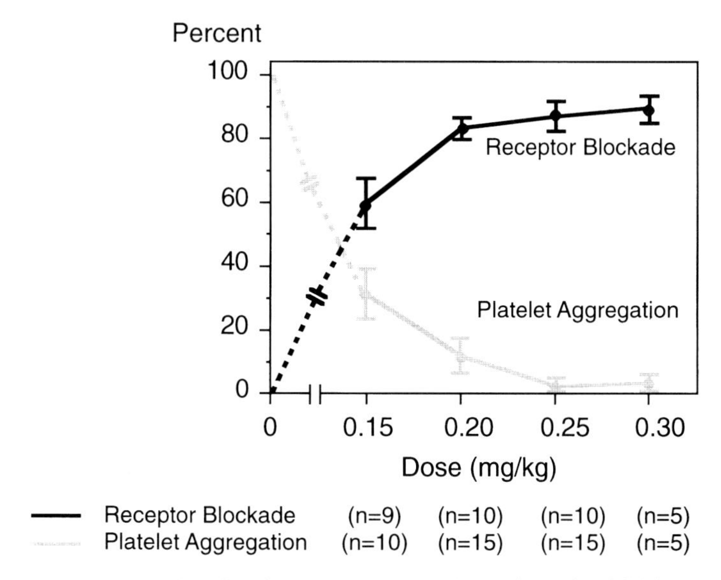

Fig. 3. Dose Response for GPIIb/IIIa Receptor Blockade and Inhibition of Platelet Aggregation in Patients Receiving Bolus Doses of Abciximab. Data are plotted as mean ± SEM. The number of subjects analyzed for each parameter is indicated. This composite figure was based on patient data from Centocor clinical trials

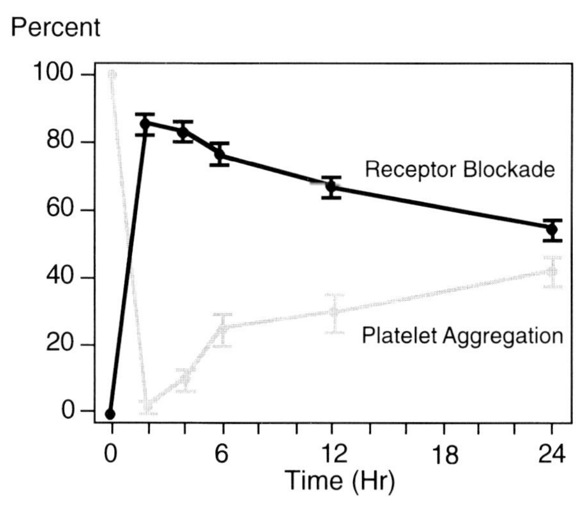

Fig. 4. Duration of Receptor Blockade and Inhibition of Platelet Aggregation (5 μM ADP) in Patients (n = 5) Receiving a 0.25 mg/kg Bolus of Abciximab. Data are plotted as mean ± SEM

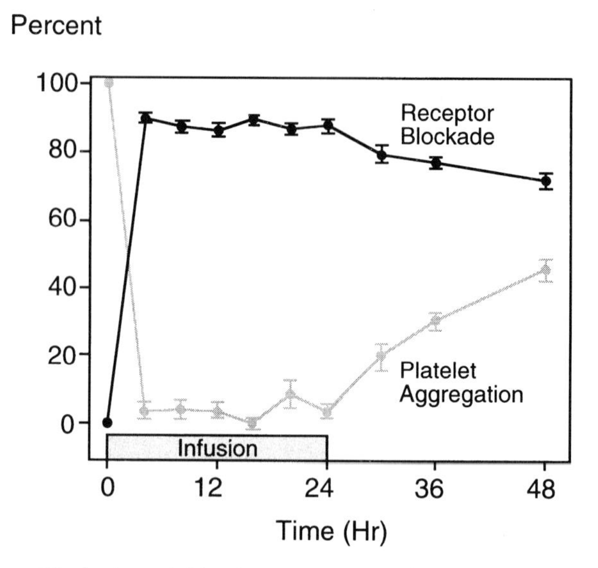

Fig. 5. Receptor Blockade and Platelet Aggregation (5 μM ADP) Results in Patients (n = 5) Receiving a 0.25 mg/kg Bolus and 24 Hour 10 mg/min Infusion of Abciximab. The bar indicates the duration of the infusion

abciximab infusion. A regimen of a 0.25-mg/kg bolus of abciximab followed by a 12 h infusion of 10 μg/min was adopted as the standard clinical therapy for the prevention of thrombotic ischemic events following coronary balloon angioplasty procedures (Weisman et al. 1995).

The apparent rate of recovery of ex vivo platelet function following abciximab bolus plus infusion therapy is dependent to some degree on the agonist or testing method that is used to test platelet aggregation. The use of different ADP concentrations to stimulate platelet aggregation revealed different profiles for the kinetics of platelet function recovery. Figure 6 presents the recovery of platelet aggregation in an individual who received a 0.25-mg/kg abciximab bolus. Recovery of 20 *μM* ADP-induced aggregation was >50% in 6 h whereas 50% recovery of 2 *μM* ADP-induced aggregation occurred at >48 h. A differential inhibition of platelet aggregation induced by ADP or thrombin receptor activation peptides was also noted in patients treated with abciximab (Kleiman et al. 1995).

Turbulent blood flow can cause pathological stress forces on flowing platelets in diseased, stenotic arteries. High shear can also be used to induce in vitro platelet aggregation. Konstantopoulos et al. (1995) employed a cone and plate viscometer system to induce ex vivo platelet aggregation in platelet samples obtained from patients treated with abciximab. High shear forces of 140–185 dynes/cm^2 for 30 s were employed. Following administration of a 0.25-mg/kg bolus of abciximab, shear-induced platelet aggregation at 2 h was

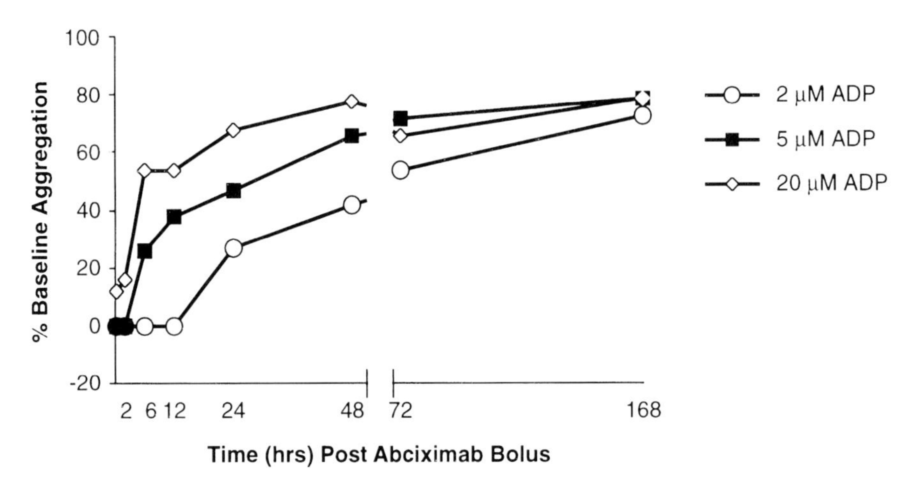

Fig. 6. Recovery of Platelet Aggregation in a Patient Treated with a 0.25 mg/kg Bolus of Abciximab. Aggregations determined at the indicated times were compared to the predose determinations for each ADP concentration

reduced by approximately 50%. The inhibitory effects of abciximab were even more pronounced when expressed in terms of the formation of large platelet aggregates defined as particles $>10 \mu M$ diameter. The administration of abciximab completely abolished the formation of large platelet aggregates during treatment and a reduction in the number of shear-induced large platelet aggregates was still evident at 1 week after treatment.

The antithrombotic benefits of sustained partial inhibition of platelet aggregation for several days after PTCA are not known. However, it is likely that the healing, or passivation, of the vessel wall occurs gradually for some period following injury. The pharmacodynamic profile of abciximab, characterized by gradual recovery of platelet function and tapered recovery from high level receptor blockade, may confer additional benefits during the period when vessel wall thrombogenicity decreases as arterial healing takes place.

The gradual recovery from receptor blockade by abciximab is a function of all of the platelets in circulation and is not due to an averaging effect of new platelets that have entered the circulation after cessation of abciximab treatment. Figure 7 shows a series of flow cytograms of platelets from an individual who received a bolus plus 12-h infusion of abciximab. The platelets obtained prior to abciximab treatment (upper panel) are nonfluorescent and indicate the flow cytometric profile of platelets without bound abciximab. At 30 min after treatment, all circulating platelets are coated with large numbers of abciximab as indicated by the substantial right-shifted position of the histogram. Also indicated is the number of abciximab molecules detected per platelet (102,000/platelet at 30 min). At 24 h after the bolus (12 h postcessation of the 12-h infusion), a single unimodal histogram is still evident although less

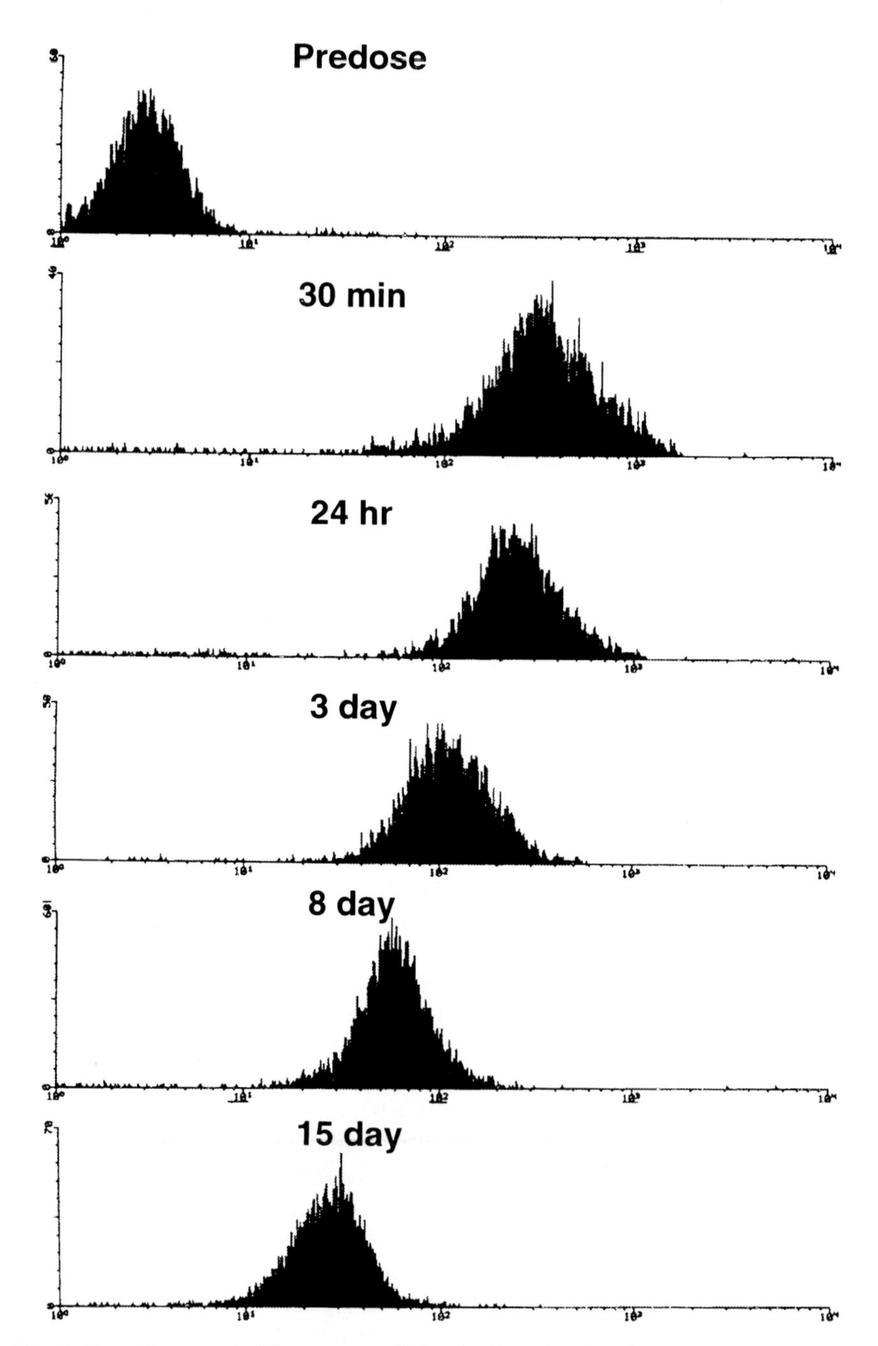

Fig. 7. Flow Cytometric Histograms of Platelet Samples Obtained from a Patient Who Received a 0.25 mg/kg Bolus plus a 12-hour Infusion of 10 mg/min of Abciximab. Platelet samples were treated with a fluorescein-conjugated rabbit anti-abciximab antibody preparation and then fixed with 1% formalin. Samples were evaluated on a Becton-Dickinson FACScan flow cytometer. The predose sample exhibits a non-fluorescent peak of platelets corresponding to the absence of abciximab. Post-treatment samples demonstrate a single, unimodal fluorescent platelet peak without evidence of a second population of platelet without abciximab

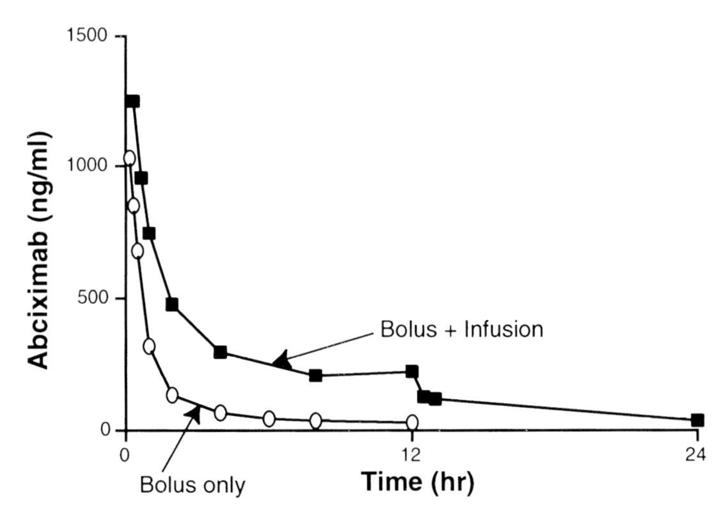

Fig. 8. Abciximab Plasma Concentrations Determined by an ELISA Assay Specific for the Variable Region of the Fab Molecule. Plasma samples were prepared from citrate anti-coagulated blood and assayed to determine abciximab concentrations to a level of detection of 20 ng/ml. The results presented are the median values at each timepoint for the bolus plus infusion (n = 4, 0.25 mg/kg plus 10 μg/min for 12 hr, Trial C0116T17) treatment groups and bolus only group (n = 32, 0.25 mg/kg, Trial C0116T15)

fluorescent than at the earlier 30-min determination. The unimodal nature of the histograms indicates that all platelets contain bound abciximab and that there are no uncoated platelets equivalent to those at pretreatment. Unimodal platelet histograms were also found at 3 days, 8 days and 15 days post-treatment. Although the average fluorescence of the platelet peaks declined progressively, there was no point during this period when a separate population of uncoated platelets could be detected. The persistence of platelet-bound abciximab at prolonged times suggests that abciximab continuously redistributes among all circulating platelets including those newly entered into the system. The recovery of platelet function is also a property of the entire platelet population since receptor blockade decreases uniformly among all circulating platelets.

The free plasma pharmacokinetic profile of abciximab is markedly different than the platelet-bound profile (KLEIMAN et al. 1995). Pharmacokinetic data for groups of PTCA patients who received the bolus alone or the bolus plus infusion regimens are presented in Fig. 8. Free plasma abciximab disappeared rapidly from plasma following the bolus dose. The bolus plus infusion regimen maintained a median plasma concentration of abciximab of >100 ng/ml for the duration of the infusion. Following the cessation of the 12-h infusion, abciximab disappeared rapidly from the plasma. The rapid disappearance of free plasma abciximab is a consistent finding in all patient groups and contrasts with the prolonged circulation of platelet-bound abciximab. The rapid plasma disappearance of abciximab is likely due to several factors including the significant proportion of the total dose that binds to platelets (Coller et al. 1991a), distribution to the extravascular space, and clearance by the kidney as previously shown for immunoglobulin Fab fragments (Yokota et al. 1993).

The continuous infusion provides a source of abciximab to prevent the gradual decline of receptor blockade to below the 80% level. The continuous infusion of abciximab typically maintains a plasma concentration >100 ng/ml (2 n*M*) which is in range of the calculated equilibrium dissociation constant for binding to GPIIb/IIIa (5 n*M*). Abciximab has equivalent binding affinity to $\alpha_v\beta_3$ receptors on platelets and other cells and the continuous infusion of abciximab is likely to result in an equivalent blockade of any available $\alpha_v\beta_3$ receptors. The potential significance of abciximab blockade of $\alpha_v\beta_3$ function is discussed in a later section.

The rapid pharmacokinetic clearance of abciximab suggested that platelet transfusion might successfully reverse the pharmacologic effects since little free plasma abciximab would be available to inhibit the newly transfused platelets. Also, the redistribution of abciximab among platelets should result in a gradual decline of receptor blockade to <80% in the host platelets that were blocked by the initial treatment due to dilutional redistribution onto the new platelets. An investigation of the effects of platelet transfusion was conducted in monkeys (Wagner et al. 1995). A group of cynomolgus monkeys received a 0.25-mg/kg bolus dose of abciximab that resulted in >80% receptor blockade and nearly complete abolition of platelet aggregation. One hour after the administration of abciximab, the monkeys received cross-matched platelets from donor monkeys in a quantity scaled to be equivalent to a human receiving 5–10 human units of packed platelets. At 30 min following platelet transfusion, all parameters of platelet function showed a significant level of restoration to baseline values. Partial recovery of platelet aggregation to a mean of ~25% of the baseline level was accompanied by a decrease of mean receptor blockade from 84% to 58%. All circulating platelets were observed to possess bound abciximab, indicating that redistribution from treated platelets to transfused platelets had occurred. The bleeding times that were initially prolonged to >15 min were shortened to ~6 min after transfusion of platelets. In contrast, control abciximab-treated animals who did not receive a platelet transfusion exhibited a substantially longer period for recovery of the bleeding times. Importantly, the restoration of platelet function following the transfusion of platelets did not revert and recovery to baseline values occurred progressively thereafter. Data obtained in a limited number of abciximab-treated patients indicate that hemostatic function can be effectively and rapidly restored and that patients can successfully and safely undergo coronary artery bypass surgery following abciximab treatment (Boehrer et al. 1994).

3. Additional Consequences

The activation of platelets results not only in an upregulation of the number and functional activity of surface GPIIb/IIIa receptors, but also leads to the release of the contents of internal granules. Platelets contain two morphologically distinct storage granules in which intracellular substances are selectively segregated: dense granules that contain platelet agonists [i.e., adenosine diphosphate (ADP), serotonin]; and α-granules that contain numerous proteins including adhesive proteins (fibrinogen, von Willebrand factor, P-selectin), mitogens [platelet derived growth factor (PDGF), thrombospondin] and inhibitors of thrombolytic and anticoagulant mechanisms [plasminogen activator inhibitor-1 (PAI-1) and platelet factor 4].

a) Inhibition of Platelet Release

In vitro experiments demonstrated a direct correspondence between the inhibition of platelet aggregation by abciximab and the inhibition of granule release (M.A. Mascelli, unpublished observations). Abciximab was found to exert a dose-dependent inhibitory effect on the release of a number of platelet granule components that paralleled the dose-dependent inhibition of platelet aggregation. Thus, abciximab inhibition of platelet granule release may provide additional antithrombotic benefit, which adds to its blockade of GPIIb/IIIa-mediated adhesion and aggregation. In addition, long-term outcome may be enhanced by decreasing the release of proliferative mediators from platelets to the vessel wall.

b) Inhibition of Mac-1 Upregulation

Activated, but not unactivated platelets, can activate leukocytes through adhesive interactions. Leukocyte activation results in the deleterious expression of mitogens, cytokines and procoagulants such as tissue factor and also causes activation of the leukocyte integrin Mac-1. Blocking Mac-1 upregulation interrupts the adhesive and migratory capability of leukocytes and reduces tissue injury in models of inflammation.

Abciximab has been shown to prevent activation-dependent upregulation of Mac-1 by leukocytes (Mickelson et al. 1996). Patients undergoing coronary angioplasty who were treated with abciximab showed decreased Mac-1 expression and had fewer detectable platelet-leukocyte complexes. P-selectin on activated platelets activates monocytes to express tissue factor (Celi et al. 1994) and abciximab inhibits P-selectin expression by activated platelets (Dalesandro et al. 1996; Bihour et al. 1996). Thus, the inhibition of platelet activation may provide the additional benefit of preventing platelet and P-selectin-mediated leukocyte activation. This type of indirect inhibition of Mac-dependent function by abciximab (Mickelson et al. 1996) may jointly contribute to the regulation of vascular repair and to the sustained clinical benefits observed with abciximab after coronary angioplasty.

c) *Inhibition of Platelet-Mediated Thrombin Generation*

Activated platelets participate in blood coagulation by providing a catalytic surface on which coagulation reactions can occur. Specifically, the surface membrane of activated platelets facilitates a rapid, explosive formation of thrombin which produces fibrin clots and is also a potent stimulus for platelet aggregation. The potentiation of thrombin generation on platelets is mediated, in part, by the surface expression of the coagulation cofactor, factor V. One anticoagulant effect of abciximab is the reduction of the mass of available platelets that provide coagulant active membrane. It also prevents the local expression of platelet factor V. In an analysis of the EPIC trial, MOLITERNO et al. (1995) noted that patients receiving abciximab had longer activated clotting times than placebo patients receiving similar doses of heparin but no abciximab. Since the activated clotting time is performed in whole blood with platelets present, the EPIC result suggested that abciximab was exerting an anticoagulant effect. This phenomenon has been confirmed in recent in vitro analyses of the prolongation of activated clotting times by abciximab (AMMAR et al. 1997).

A direct role for platelets in the generation of thrombin was shown by REVERTER et al. (1996). In an in vitro study, thrombin generation in the presence of platelets was initiated by the addition of tissue factor and the resulting burst of thrombin activity was quantified. Abciximab inhibited tissue factor-induced thrombin generation by approximately 48%. Another equally potent anti-GPIIb/IIIa antibody (10E5) inhibited thrombin generation by only ~23%. A major distinction between abciximab and 10E5 antibodies is the ability of abciximab to bind to the platelet $\alpha_v\beta_3$ and GPIIb/IIIa receptor with equal affinity. Thus, the different potencies of abciximab and 10E5 on platelet-mediated thrombin formation suggested that the $\alpha_v\beta_3$ receptor on platelets might be important in this phenomenon.

The $\alpha_v\beta_3$ receptor is structurally related to the GPIIb/IIIa receptor and is present on platelet surfaces at numbers lower (~200/platelet) than GPIIb/IIIa (~80,000–100,000/platelet) (COLLER et al. 1991b). REVERTER et al. (1996) directly tested the contribution of platelet $\alpha_v\beta_3$ to thrombin generation by using a specific, potent anti-$\alpha_v\beta_3$ antibody, LM609. Interestingly, LM609 inhibited thrombin generation by only ~5%, suggesting that the specific blockade of only the $\alpha_v\beta_3$ receptor had minimal impact on thrombin generation. However, the combination of the anti-$\alpha_v\beta_3$ antibody, LM609, with the anti-GPIIb/IIIa antibody, 10E5, began to approach the magnitudes of effect produced by abciximab, with ~30% inhibition of thrombin generation seen with the combination, suggesting that the blockade of both $\alpha_v\beta_3$ and GPIIb/IIIa was needed to minimize the platelet contribution to coagulation. The results of this in vitro study are consistent with ex vivo observations of prolonged clotting times in abciximab-treated patients and point to a potentially important attribute of abciximab in inhibiting the propagation of coagulation on platelet surfaces.

d) Characterization of Abciximab Binding to $\alpha_v\beta_3$

The intact 7E3 IgG molecule binds to $\alpha_v\beta_3$ and blocks certain in vitro functions of this receptor such as the attachment and spreading of cells to surfaces coated with adhesive proteins (CHARO et al. 1987). Additionally, 7E3 IgG inhibits $\alpha_v\beta_3$-mediated human umbilical vein endothelial cell (HUVEC) attachment to fibrinogen coated plates (MARTINEZ et al. 1989), $\alpha_v\beta_3$-mediated sickle red blood cell adherence to HUVEC cells (SUGIHARA et al. 1992), vitronectin-coated bead binding to endothelial cells (ZANETTI et al. 1994), and fibroblast adhesion to fibronectin and vitronectin-coated plates (GAILIT and CLARK 1996).

Equilibrium binding experiments have demonstrated that the affinity constant (K_a) for ^{125}I-abciximab binding to HUVEC cells is $10^8 M^{-1}$ (dissociation constant $K_d = 0.37 \mu$g/ml). The affinity of 7E3 IgG for HUVEC cells, $K_a = 2.40 \times 10^8 M^{-1}$ ($K_d = 0.62 \mu$g/ml), is similar to that seen for the Fab fragment and also similar to the affinity of abciximab binding to GPIIb/IIIa on platelets, $K_a = 1.94 \times 10^8$ ($K_d = 0.26 \mu$g/ml). Abciximab bound to purified $\alpha_v\beta_3$ with an affinity that was similar to that seen for its binding to GPIIb/IIIa. Abciximab bound to a variety of human $\alpha_v\beta_3$-expressing cell lines, including endothelial cells (CONFORTI et al. 1992), fibroblasts (CONFORTI et al. 1992), and smooth muscle cells (HOSHIGA et al. 1995), with similar affinity as for purified $\alpha_v\beta_3$. Abciximab binding was consistent with the previously reported cell type distribution for $\alpha_v\beta_3$.

As will be described below, the clinical benefits of a single treatment with abciximab were found to be sustained for prolonged periods of up to 3 years in many patients. The surprisingly durable benefit of abciximab in the Phase III EPIC trial suggested other therapeutic modes of action in addition to the acute inhibition of platelet thrombosis via blockade of platelet GPIIb/IIIa. Specifically, the ability of abciximab to bind to the $\alpha_v\beta_3$ receptor and the well-recognized involvement of the $\alpha_v\beta_3$ receptor in smooth muscle cell migration and proliferation suggested that abciximab might also be blocking restenosis. The blockade of $\alpha_v\beta_3$ receptors by monoclonal antibodies and peptides had been independently shown to result in an inhibition of the extent of intimal hyperplasia after vascular injury in animal models (MATSUNO et al. 1994; VAN DER ZEE et al. 1996).

Support for a potential anti-$\alpha_v\beta_3$ action of abciximab was gained in independent experiments in a SCID mouse model of tumor angiogenesis. In this model, in which the development of new blood vessels was shown to be dependent on an $\alpha_v\beta_3$-mediated proliferation of endothelial cells, 7E3 IgG demonstrated a potent inhibitory effect (VARNER et al. 1997). This study was the first in vivo demonstration of inhibition by 7E3 of an $\alpha_v\beta_3$-dependent process. These results point to the possibility that abciximab may confer additional clinical benefit by inhibiting vascular $\alpha_v\beta_3$ receptors and intimal hyperplasia after vascular injury. However, additional studies will be required to prove this mechanism and to establish the appropriate dose-response profile for abciximab.

4. Clinical Experience

a) Early Human Efficacy Studies

Several open-label and nonrandomized phase I and phase II clinical studies were performed to obtain preliminary assessments of the antithrombotic potential of abciximab. The ability of abciximab to prevent platelet thrombus formation at sites of coronary injury was investigated by measuring coronary flow velocity in patients undergoing percutaneous coronary interventions (PCI) (ANDERSON et al. 1992, 1994). Cyclic flow variations within the coronary arteries are related to transient formation of platelet thrombi at the sites of vascular injury. Complete thrombotic occlusions that stop all blood flow can occur in such circumstances and lead to severe ischemia. In the study by ANDERSON et al. (1994), abciximab stabilized blood flow in all patients who had developed cyclic flow variations after PCI despite concomitant treatment with aspirin, high-dose intravenous heparin and nitroglycerin. In a phase II study in high-risk patients undergoing PCI, abciximab-treated patients were found to incur fewer clinical events related to recurrent ischemia (death, myocardial infarction and urgent coronary intervention) than a nonrandomized control group (TCHENG et al. 1994). Additional evidence that abciximab could inhibit platelet thrombus formation was gained in the setting of acute myocardial infarction (MI) in an early study employing the murine Fab fragment of 7E3. The Thrombolysis and Angioplasty in Myocardial Infarction (TAMI) study group conducted a 60-patient study demonstrating that patients who were treated with murine 7E3 Fab more often achieved coronary artery patency after coronary thrombolysis with recombinant tissue plasminogen activator (rt-PA) than control subjects treated only with rt-PA (KLEIMAN et al. 1993). Although these phase II studies were not large enough or designed to achieve statistical efficacy, they served to establish that the potent inhibition of platelet function with abciximab was safe and well tolerated in these clinical settings of acute MI and PCI (TOPOL and PLOW 1993).

A phase II randomized, placebo-controlled trial was conducted in 60 patients with unstable angina who were characterized as refractory to treatment with heparin, aspirin and nitrates (SIMOONS et al. 1994). Patients randomized to abciximab were treated with a 0.25-mg/kg bolus and 10-μg/min infusion beginning 18 h before PTCA and continuing until 1 h after PTCA. Compared to the control group receiving conventional therapy, abciximab reduced recurrent myocardial ischemia by 50% from the time the treatment was started until PTCA was performed 18-24 h later. Seven of 30 patients (23.3%) in the placebo treatment group exhibited a major clinical event whereas 1 of 30 patients (3.3%) in the abciximab treatment group had an event ($p = 0.052$, two-sided). The results of this pilot study indicated that effective blockade of the platelet GPIIb/IIIa receptor could reduce myocardial infarction and facilitate successful PTCA in patients with refractory unstable angina.

b) The Phase III EPIC Trial

The EPIC trial (*E*valuation of c7E3 Fab for the *P*revention of *I*schemic *C*omplications) was a multicenter, randomized, double-blind, placebo-controlled, three-arm trial in patients who were at high risk for ischemic complications following coronary artery procedures. Patients in EPIC underwent coronary balloon angioplasty (PTCA) or directional coronary atherectomy procedures that were high risk as defined by the criteria established by the American Heart Association/American College of Cardiology Task Force on coronary angioplasty (ACC/AHA Task Force 1988). A total of 2099 patents were randomized into one of three treatment arms: (1) a bolus of placebo followed by a 12-h infusion of placebo (placebo), (2) a bolus of abciximab (0.25 mg/kg) followed by a 12-h infusion of placebo (bolus) or (3) the abciximab bolus (0.25 mg/kg) followed by a 12-h infusion of abciximab (10 μg/min)(bolus plus infusion). All patients received standard therapy with aspirin and high-dose intravenous heparin. The primary efficacy end point was a composite comprising any of the following within 30 days of randomization: death, MI or urgent intervention for recurrent ischemia [PTCA, coronary artery bypass graft (CABG) surgery, intracoronary stent placement or in patients unable to undergo repeat PTCA, intraaortic balloon pump (IABP) placements] (The Epic Investigators 1994). The "intention-to-treat" method was used to evaluate the 30-day primary end point event rates for each treatment group (Weisman 1996). In addition to the primary end point at 30 days, a secondary end point investigated 6-month benefit. The 6-month efficacy end point was defined as death, MI or revascularization procedures (CABG or PTCA) (Topol et al. 1994). In contrast to the primary end point, the 6-month revascularization end point included all CABG and both urgent and nonurgent PTCA. In addition, a long-term analysis of the EPIC patients was conducted approximately 3 years after the initial treatment using the same end points that were employed at 6 months (Topol et al. 1997). All potential efficacy and major safety events (bleeding and stroke) were reviewed and confirmed by an independent, blinded, clinical end point committee.

In the EPIC trial, the benefit of abciximab treatment was apparent by the 1st day after randomization and was maintained for the entire 30-day period that comprised the primary end point. Table 1 lists the efficacy results of the EPIC trial at the primary 30-day end point as well as results at the secondary 6-month end point and at 3-year follow-up. There was a 34.8% reduction in 30-day end points in the bolus plus infusion group compared to the placebo group (8.3% vs 12.8%, $p = 0.009$). There was a 10% reduction in end points in the abciximab bolus group compared to placebo (11.5% vs 12.8%), which was not statistically significant ($p = 0.428$). The difference in efficacy achieved in the abciximab bolus plus infusion group compared to the bolus group indicates that the more prolonged duration of treatment is necessary for therapeutic benefit.

At the 30-day primary end point analysis, the most marked dose–response effects were seen in the incidence of MI ($p = 0.013$) and urgent intervention ($p = 0.003$) (THE EPIC INVESTIGATORS 1994). In the bolus plus infusion treatment group, a 39.4% reduction in the incidence of MI (5.2% in the bolus plus infusion treatment group versus 8.6% in the placebo treatment group, $p = 0.014$, pairwise) and a 49.1% reduction in the incidence of urgent intervention (4.0% in the bolus plus infusion treatment group versus 7.8% in the placebo treatment group, $p = 0.003$, pairwise) were observed. Urgent interventions most often included urgent PTCA and urgent CABG; urgent stent or intra-aortic balloon pump (IABP) placement was rare. The few deaths during the 30-day follow-up period included 12 in the placebo group, 9 in the abciximab bolus group and 12 in the abciximab bolus plus infusion group (however, 3 patients in the bolus plus infusion arm who died never received abciximab).

The 6-month analyses of efficacy demonstrated that the 30-day benefit of ReoPro bolus plus infusion treatment was maintained for 6 months and suggested that additional benefits continued to accrue beyond 30 days (TOPOL et al. 1994). Clinical end points that occurred within 30 days of balloon angioplasty procedures have been attributed primarily to platelet thrombosis. Beyond 30 days, the reclosure of treated arteries may also be due to nonthrombotic events such as vessel wall remodeling and the restenosis caused by cellular proliferation. In EPIC, the clinical benefits of abciximab in patients receiving the bolus plus infusion were sustained even if all the early 30-day thrombotic events were excluded and only clinical end points that occurred thereafter were counted.

Table 1 summarizes the 6-month end point event rates for the 6-month period after study entry. Among all randomized patients, fewer clinical events occurred during the entire 6-month follow-up period in the bolus plus infusion treatment group (27.0%) than in the placebo treatment group (35.1%), representing a 22.9% reduction ($p = 0.001$). Even with the exclusion of the initial 2 days of platelet blockade provided by the abciximab bolus plus infusion treatment regimen, the time when most acute events occur, there was a 24.5% reduction in later clinical events ($p = 0.007$). Late benefit of bolus plus infusion treatment beyond 30 days was also suggested by a strong trend ($p = 0.071$) for fewer clinical events (20.6% less) after 30 days until 6 months in patients who did not have end point events during the initial 30-day follow-up. The 6-month benefit of abciximab bolus plus infusion treatment versus placebo was strongest for the end points of MI (34.2% reduction, $p = 0.018$) and repeat PTCA (31% reduction, $p = 0.01$). As shown in Table 1, treatment with only a bolus of abciximab did not confer comparable long-term benefits. The sustained 6-month results following abciximab bolus plus infusion treatment provided evidence that short-term, potent antiplatelet therapy could confer

Table 1. 30-day, 6-month and 3-year outcomes by treatment group in the EPIC trial

	Placebo	Bolus	Bolus + infusion
Patients randomized	696	695	708
Patients with death, MI or urgent intervention within 30 days	89 (12.8%)	79 (11.5%)	59 (8.3%)
% change vs placebo		−10.4%	−34.8%
P-value vs placebo		0.428	0.008
Patients with death, MI or repeat revascularization within 6 months	241 (35.1%)	224 (32.6%)	189 (27.0%)
% change vs placebo		−7.1%	−22.9%
P-value vs placebo		0.276	0.001
Patients with death, MI, or repeat revascularization within 3 years	319 (47.2%)	321 (47.4%)	283 (41.1%)
% change vs placebo		0.3%	−13.0%
P-value vs placebo		0.779	0.009

clinical benefits much longer than expected, based on a short (12-h) therapeutic regimen.

At 3 years, the reduction in the composite event rate of death, myocardial infarction, or the need for coronary revascularization for patients who received the abciximab bolus plus infusion treatment continued to be maintained (TOPOL et al. 1997). As also listed in Table 1, 47.2% of placebo patients incurred a composite end point at 3 years compared to 41.1% of patients receiving abciximab bolus plus infusion treatment ($p = 0.009$). At 3 years there was a nonsignificant reduction in mortality (8.6% for placebo; 6.8% for bolus plus infusion therapy) ($p = 0.20$), a reduction in myocardial infarction (13.6% vs 10.7%, respectively; $p = 0.075$) and a reduced need for repeat revascularization (40.1% vs 34.8%, respectively; $p = 0.02$). These data provided evidence that the 12-h periprocedural treatment with abciximab, which was primarily aimed at improving the safety of the index procedure, favorably affected long-term outcome.

When initiated, EPIC was the largest prospective trial yet conducted in patients undergoing PTCA. This trial provided strong confirmation for a causative role of platelet thrombosis in the most frequent major complications of PTCA (COLLER et al. 1995). The ability of abciximab to prevent these complications was consistent across all of the patient subgroups within the EPIC trial. Importantly, the benefit of abciximab treatment observed at 30 days was maintained for 3 years. Efforts are underway to understand and better characterize the mechanism by which abciximab achieves such durable benefits. These efforts include investigations into the binding of abciximab to $\alpha_v\beta_3$ integrin on platelets, endothelial cells and smooth muscle cells. In addition, the prolonged presence of abciximab on circulating platelets and the gradual manner in which platelet function is restored after therapy may confer both antithrombotic and antirestenotic advantages.

The major safety concern that was revealed in the EPIC trial was an increased bleeding rate in patients receiving abciximab. In EPIC, patients received high doses of intravenous heparin while undergoing coronary intervention whether or not they received abciximab. In retrospect, the increased rates of bleeding that occurred as a consequence of abciximab treatment were largely attributable to the potent platelet inhibition that was superimposed on a high level of heparin anticoagulation (Aguirre et al. 1995). Nevertheless, abciximab treatment did not increase the risk of intracerebral hemorrhage or the need for surgical intervention to correct the bleeding. Patients who received abciximab and underwent urgent coronary artery bypass graft surgery (CABG) experienced similar blood loss, transfusion rates and complication rates to patients in the placebo group who underwent CABG, even though most CABG procedures occurred early after onset of abciximab treatment, at which time substantial inhibition of platelet function would still be expected (Boehrer et al. 1994). In EPIC, a standard bolus heparin dose was used and this dose was usually not adjusted for patients of different body weight. Thus, lighter patients received a proportionately higher heparin dose. This higher heparin dose was associated with higher levels of anticoagulation as evidenced by longer activated clotting times (ACT) in lighter patients. Lighter patients demonstrated more blood loss in all treatment groups, including patients in the placebo arm not treated with abciximab. Interestingly, lighter patients who received proportionately higher heparin doses did not exhibit an apparent efficacy benefit from these higher heparin doses. These observations suggested that the standard heparin dose was inappropriately high when combined with abciximab and prompted the adoption of a modified heparin regimen for the subsequent clinical trials.

Before embarking on a second, large phase III trial with abciximab in PTCA, a pilot study (PROLOG) was conducted to investigate strategies for reducing the risks of bleeding without decreasing the antithrombotic efficacy (Lincoff et al. 1995). The strategies included a number of procedural aspects relating to device removal and access site care as well as a reduction in the heparin dose in one of the test groups. The results of the PROLOG trial indicated that the rates of bleeding complications in these patients who received abciximab were reduced to rates similar to those observed in the placebo group in the previous EPIC trial. Importantly, the PROLOG trial indicated that excessive bleeding could be minimized by adjustments in the heparin dose and suggested that these gains in safety were not offset by decreases in the efficacy of abciximab.

c) The EPILOG Trial

The second large phase III trial of abciximab in PTCA patients was the EPILOG trial *(E*valuation of *P*TCA to *I*mprove *L*ong-term *O*utcome by abciximab *G*PIIb/IIIa Receptor Blockade). This 4800-patient trial in patients who were undergoing either high-risk or low-risk PTCA procedures incorporated the procedural features of the PROLOG pilot trial. Patients with acute myocardial infarction or refractory unstable angina were excluded from EPILOG because these patients had statistically significant reductions in death and myocardial infarction in the EPIC trial and it was felt to be unethical to randomize these patients to placebo treatment. Patients enrolled in EPILOG were randomized to one of three treatment groups: (1) placebo coupled with a standard weight-adjusted heparin regimen designed to achieve an ACT >300 s, (2) abciximab (0.25-mg/kg bolus followed by a 12-h infusion at 0.125 μg/kg-min not to exceed the EPIC infusion of 10 μg/min and the standard weight-adjusted heparin regimen), and (3) the above abciximab dose coupled with a "low-dose"-weight-adjusted heparin regimen incorporating a 70-unit/kg initial bolus to achieve an ACT >200 s.

During the EPILOG trial, an interim analysis of the primary efficacy end point of myocardial infarction or death was performed on the first 1500 patients. The statistical analysis of the efficacy in the experimental arms of the trial indicated that the efficacy compared to the placebo group exceeded the predetermined stopping level ($p < 0.0001$) (Ferguson 1996). Based on this overwhelming efficacy result, the safety and efficacy monitoring committee (SEMC) recommended that enrollment into the trial be terminated. Importantly, the SEMC noted that the rate of major bleeding events in patients receiving abciximab was equivalent to that in patients receiving placebo. The interim results were fully confirmed when all of the 2792 patients who were enrolled at the time the EPILOG trial enrollment was stopped were analyzed (The Epilog Investigators 1997).

Table 2 details the results of the final efficacy analysis of the primary end points of death or MI at 30 days, death, MI, or urgent intervention at 30 days, and death, MI or any intervention at 6 months. There was a 58.8% reduction in the composite of death or MI at 30 days in the abciximab plus low-dose heparin group ($p < 0.0001$) and a 54.4% reduction in the abciximab plus standard-dose heparin group ($p < 0.0001$). Similar large reductions are evident for the 30-day composite end point including urgent revascularization. At 6 months, the event rate for death, MI or repeat revascularization (urgent and nonurgent) was 25.8% in the placebo group, 22.8% in the abciximab plus low dose heparin group ($p = 0.034$), and 22.3% in the abciximab plus standard dose heparin group ($p = 0.020$).

A major aim of the EPILOG study was to test if the benefit of abciximab in reducing serious complications in high risk patients could be extended to lower risk patients. An analysis of primary end points by risk classification showed that the composite of death and MI was reduced at 30 days by 64.1% ($p < 0.001$) in the lower risk group and by 53.8% ($p < 0.001$) in the high risk group. The composite of death, MI and urgent revascularization was reduced by 52.9% ($p < 0.001$) and 60.3% ($p < 0.001$) in the low and high risk groups, respectively. The reductions in these events were maintained at 6 months for both high and lower risk patients. The clinical benefits in CAPTURE applied

Table 2. 30-day and 6-month outcomes by treatment group in the EPILOG trial

	Total	Placebo	ReoPro + Low-Dose Heparin	ReoPro + Std.-Dose Heparin	Combined ReoPro
Patients randomized	2792	939	935	918	1853
Patients with death or MI within 30 days	158 (5.7%)	85 (9.1%)	35 (3.8%)	38 (4.2%)	73 (4.0%)
% change vs placebo			−58.8%	−54.4%	−56.6%
P-value vs placebo			<0.001	<0.001	<0.001
Patients with death, MI or urgent revascularization within 30 days	206 (7.4%)	109 (11.7%)	48 (5.2%)	49 (5.4%)	97 (5.3%)
% change vs placebo			−55.8%	−54.1%	−54.9%
P-value vs placebo			<0.001	<0.001	<0.001
Patients with death, MI, or repeat revascularization within 6 months	656 (23.6%)	241 (25.8%)	212 (22.8%)	203 (22.3%)	415 (22.5%)
% change vs placebo			−11.7%	−13.7%	−12.6%
P-value vs placebo			0.034	0.020	0.011

regardless of the type of intervention including balloons, stents or atherectomy.

With regard to bleeding complications, 2.9% of all patients in EPILOG experienced a major bleeding event (including those associated with coronary artery bypass grafts). In the placebo group, the major bleeding rate was 3.1% compared to 2.0% in the abciximab plus low-dose heparin group and 3.5% in the abciximab plus standard-dose heparin group. These rates are all markedly lower than the 14.0% major bleeding rate that had been observed with the abciximab bolus plus infusion regimen in the earlier EPIC trial. These results established that weight adjustment of heparin, in combination with other strategies such as strong recommendation for early sheath removal, specific vascular access site and patient management guidelines, and weight adjustment of the abciximab infusion, reduces the rate of major bleeding to that of the rate of patients receiving the placebo.

d) The CAPTURE Trial

CAPTURE denotes *C*himeric 7E3 *A*ntiplatelet *T*herapy in *U*nstable Angina *RE*fractory to Standard Treatment. This trial was designed as a double-blind, randomized, placebo-controlled study of abciximab in 1400 patients with refractory unstable angina scheduled to undergo PTCA. This trial was a test of

Table 3. CAPTURE trial primary end point event rates for all randomized patients[a]

	Total (n = 1265)	Placebo (n = 635)	ReoPro (n = 630)	% Change vs Placebo	P-value
Patients who died, had MI or urgent intervention	172 (13.6%)	101 (15.9%)	71 (11.3%)	−28.9%	0.012
Death	14 (1.1%)	8 (1.3%)	6 (1.0%)	−25.2%	0.603
MI	78 (6.2%)	52 (8.2%)	26 (4.1%)	−49.6%	0.002
Urgent intervention	118 (9.4%)	69 (10.9%)	49 (7.8%)	−28.0%	0.054

[a] Patients were counted once within a component, but could have been counted in more than one component.

the hypothesis that potent platelet inhibition could stabilize patients with severe unstable angina and reduce postcoronary intervention thrombotic complications. In the CAPTURE trial, patients received either a placebo or a bolus (0.25 mg/kg) plus continuous infusion (10 μg/kg) of abciximab beginning 18–24 h before the PTCA procedure and extending to 1 h after the PTCA. Patients in the CAPTURE trial received a standard regimen of heparin anticoagulation with a recommendation of weight adjusted heparin dosing during PTCA (100 units/kg heparin bolus dose). The composite clinical end point comprised death, myocardial infarction or the need for urgent intervention.

An interim analysis was conducted on the initial 1050 patients (Ferguson 1996). An SEMC review of the primary end points found that the efficacy had exceeded a predetermined stopping level ($p < 0.0072$). Based on this interim analysis, the SEMC recommended that the CAPTURE trial be stopped due to positive results. A full analysis of all 1267 enrolled patients in CAPTURE confirmed the results of the interim analysis (The Capture Investigators 1997). Table 3 shows the intention-to-treat analysis of the CAPTURE primary composite end point of death, MI or urgent intervention within 30 days of randomization. Abciximab treatment resulted in a 28.9% reduction in primary end point events from 15.9% in the placebo treatment group to 11.3% in the abciximab treatment group ($p = 0.012$). Although the occurrence of each of the components of the composite end point was reduced by abciximab, there was a particularly strong reduction of MI by 49.6% from 8.2% in the placebo groups to 4.1% in the abciximab treatment group ($p = 0.002$). In addition to the 30-day results, it was found that abciximab reduced the rate of myocardial infarctions that occurred before, during and after percutaneous intervention. The clinical benefits in CAPTURE applied regardless of the type of intervention including balloons, stents or atherectomy.

In CAPTURE, the rates of major bleeding were found to be low: 1.7% in the placebo group and 2.9% in the abciximab group. Although these bleeding rates were considerably less than previously observed in the EPIC trial, they were somewhat greater than in EPILOG, likely reflecting the higher heparin dose and the extended duration of sheath placement. The results of the CAPTURE trial demonstrated that a 24-h infusion regimen of abciximab could be used to stabilize unstable angina patients who are unresponsive to standard medical therapy and reduce the incidence of acute ischemic events prior to, during, and after PTCA.

e) Clinical Summary of Abciximab

Abciximab was developed during a period of rapidly expanding knowledge on the role of platelets in coronary artery disease. The overwhelmingly positive efficacy results obtained with abciximab in three large phase III trials confirmed the rationale that potent inhibition of platelets via the GPIIb/IIIa receptor would confer marked clinical and antithrombotic benefit. In the future, abciximab may find similar beneficial use in other settings of platelet-mediated thrombosis and ischemic disease. Stroke, unstable angina and myocardial infarction, with or without percutaneous interventions or thrombolytic agents, are indications currently being investigated to test whether patients with these conditions benefit from the potent antithrombotic effect of abciximab.

D. Investigational Agents

This section will deal briefly with investigational agents that are in late stage clinical evaluation, the assumption being that either the specific agent described, or a member of its class, is likely to obtain regulatory approval.

I. GPIIb/IIIa Antagonists

As described above, abciximab is a chimeric monoclonal antibody Fab fragment directed against GPIIb/IIIa. In part encouraged by the success of abciximab both preclinically and clinically, many other GPIIb/IIIa antagonists have been designed and are under clinical evaluation (Lefkovits et al. 1995). These agents fulfill the expectation of GPIIb/IIIa antagonists in that they all inhibit platelet aggregation irrespective of the platelet activator. In a large variety of animal models of arterial thrombosis, all types of GPIIb-IIIa antagonist show superiority over other classes of antiplatelet agents (Weller et al. 1994; Samanen 1996). The large majority of these investigational agents are based on the amino acid sequence Arg-Gly-Asp (R-G-D) found in fibrinogen and that is thought, in part, to mediate the binding of fibrinogen to GPIIb/IIIa. This sequence has been mimicked in cyclic peptides such as Integrilin (eptifibatide), a cyclic heptapeptide containing a modified lysine-glycine-aspartic acid (K-G-D) sequence (Scarborough et al. 1993). These modifications conferred specificity to the agent for GPIIb/IIIa over $\alpha_v\beta_3$ (Phillips and Scarborough 1997). As discussed earlier, $\alpha_v\beta_3$ is an integrin related to GPIIb/

IIIa and which also recognizes the R-G-D sequence in its natural ligands (vitronectin, fibronectin, etc.). Eptifibatide inhibits the binding of fibrinogen to GPIIb/IIIa and thereby effectively inhibits platelet aggregation. Eptifibatide's peptide nature currently limits its delivery to the parenteral route and it has undergone extensive clinical evaluation via the intravenous route. Several phase II and III clinical trials have been completed, the preliminary results of which support the hypothesis that eptifibatide provides superior antiplatelet and antithrombotic protection to that afforded by aspirin (Tcheng 1997; Harrington 1997). Other parenteral agents that have undergone advanced clinical testing include tirofiban and lamifiban (Theroux et al. 1996). Both of these agents are small molecule mimetics of the R-G-D sequence and like eptifibatide are highly specific for GPIIb/IIIa and limited to parenteral delivery (Samanen 1996). The clinical data with these agents lend further support to the superiority of GPIIb/IIIa antagonists over conventional therapy (aspirin plus heparin) for the treatment of acute coronary syndromes (Tcheng 1996). In addition to the parenteral agents, a large number of orally active small molecule GPIIb/IIIa antagonists are under clinical evaluation. These agents, which include Xemlofiban, Fradofiban and Ro 44-9883, are all specific GPIIb/IIIa antagonists (Samanen 1996). To improve the oral bioavailability of these zwitterionic molecules, many are prodrugs that are hydrolyzed to the active moiety following absorption. This class of agents has the potential to allow treatment to extend beyond the acute hospital period.

II. Clopidogrel

Clopidogrel (Plavix) is a second generation thienopyridene, i.e., an orally active ticlopidine-like platelet aggregation inhibitor. Like ticlopidine, clopidogrel's precise mechanism of action is unknown but appears to indirectly influence the platelet ADP receptor (Schror 1993). The compound appears to have greater efficacy in animal models of thrombosis than ticlopidine (Herbert et al. 1993). Clopidogrel has additional advantages over ticlopidine, perhaps the most important being that it does not appear to cause neutropenia. It also appears to be somewhat faster acting, although oral dosing for 3 days is still required for the full antiplatelet effect, but it only requires once-daily dosing. A recently completed phase III trial (CAPRIE) of clopidogrel (75 mg once daily) in patients with recent ischemic stroke, MI or peripheral arterial disease suggested some improvement over aspirin in reducing the composite end point of ischemic stroke, MI and vascular death (Caprie Steering Committee 1996). [Note added in proof: clopidogrel has gained regulatory approval for the above indications.]

References

ACC/AHA Task Force (1988) Guidelines for percutaneous transluminal coronary angioplasty. A report of the American College of Cardiology/American Heart

Association Task Force on Assessment of Diagnostic and Therapeutic Cardiovascular Procedures (Subcommittee on Percutaneous Transluminal Coronary Angioplasty). J Am Coll Cardiol 12:529–545
Aguirre FV, Topol EJ, Ferguson JJ, Anderson K, Blankenship JC, Heuser RR et al (1995) Bleeding complications with the chimeric antibody to platelet glycoprotein IIb/IIIa integrin in patients undergoing percutaneous coronary intervention. EPIC Investigators. Circulation 91:2882–2890
Ammar T, Scudder LE, Coller BS (1997) In vitro effects of the platelet glycoprotein IIb/IIIa receptor antagonist c7E3 Fab on the activated clotting time. Circulation 95:614–617
Anderson HV, Revana M, Rosales O, Brannigan L, Stuart Y, Weisman H et al (1992) Intravenous administration of monoclonal antibody to the platelet GPIIb/IIIa receptor to treat abrupt closure during coronary angioplasty. Am J Cardiol 69:1373–1376
Anderson HV, Kirkeeide RL, Krishnaswami A, Weigelt LA, Revana M, Weisman HF et al (1994) Cyclic flow variations after coronary angioplasty in humans: clinical and angiographic characteristics and elimination with 7E3 monoclonal antiplatelet antibody. J Am Coll Cardiol 23:1031–1037
Bhattacharya S, Jordan R, Machin S, Senior R, Mackie I, Smith CR et al (1995) Blockade of the human platelet GPIIb/IIIa receptor by a murine monoclonal antibody Fab fragment (7E3): potent dose-dependent inhibition of platelet function. Cardiovasc Drugs Ther 9:665–675
Bihour C, Durrieu-Jais C, Macchi L, Coste P, Besse P, Nurden P et al (1996) Circulating activated platelets are not seen following ReoPro infusion in patients with refractory unstable angina with total inhibition of ADP-induced platelet aggregation. Haemostasis 26: Abstract 430
Bizzozero J (1882) Ueber einen neuen Formbestandteil des Blutes und dessen Rolle bei der Thrombose und der Blutgerinnung. Virch Arch [A] 90:261–267
Boehrer JD, Kereiakes DJ, Navetta FI, Califf RM, Topol EJ, (1994) Effects of profound platelet inhibition with c7E3 before coronary angioplasty on complications of coronary bypass surgery. EPIC Investigators. Evaluation Prevention of Ischemic Complications. Am J Cardiol 74:1166–1170
Braden GA, Knapp HR, FitzGerald GA (1991) Suppression of eicosanoid biosynthesis during coronary angioplasty by fish oil and aspirin. Circulation 84:679–685
Brass LF, Hoxie JA, Manning DR (1993) Signaling through G proteins and G protein-coupled receptors during platelet activation. Thromb Haemost 70:217–223
Calvete JJ (1995) On the structure and function of platelet integrin alpha IIb beta 3, the fibrinogen receptor. Proc Soc Exp Biol Med 208:346–360
CAPRIE Steering Committee (1996) A randomised, blinded, trial of clopidogrel versus aspirin in patients at risk of ischaemic events (CAPRIE). Lancet 348:1329–1339
Celi A, Pellegrini G, Lorenzet R, De Blasi A, Ready N, Furie BC et al (1994) P-selectin induces the expression of tissue factor on monocytes. Proc Natl Acad Sci USA 91:8767–8771
Charo IF, Bekeart LS, Phillips DR (1987) Platelet glycoprotein IIb-IIa-like proteins mediate endothelial cell attachment to adhesive proteins and the extracellular matrix. J Biol Chem 262:9935–9938
Christopoulos C, Macklie IJ, Lahiri A, Machin S (1993) Flow cytometric observations on the in vivo use of a chimaeric monoclonal antibody to platelet glycoprotein IIb/IIIa. Blood Coagul Fibrinolysis 4:729–737
Coller BS (1985) A new murine monoclonal antibody reports an activation-dependent change in the conformation and/or microenvironment of the platelet glycoprotein IIb/IIa complex. J Clin Invest 76:101–108
Coller BS (1986) Activation affects access to the platelet receptor for adhesive glycoproteins. JCell Biol 103:451–456
Coller BS (1990) Platelets and thrombolytic therapy. N Engl J Med 332:33–42

Coller BS (1997a) GPIIb/IIIa antagonists: pathophysiologic and therapeutic insights from studies of c7E3 Fab. Thromb Haemost 78:730–735
Coller BS (1997b) Platelet GPIIb/IIIa antagonist: the first anti-integrin receptors therapeutics. J Clin Invest 99:1467–1471
Coller BS, Cheresh DA, Asch E, Seligsohn U (1991a) Platelet vitronectin receptor expression differentiates Iraqi–Jewish from Arab patients with Glanzmann thrombasthenia in Israel. Blood 77:75
Coller BS, Scudder LE, Beer J, Gold HK, Folts JD, Cavagnaro J et al (1991b) Monoclonal antibodies to platelet glycoprotein IIb/IIIa as antithrombotic agents. Ann N Y Acad Sci 614:193–213
Coller BS, Anderson K, Weisman HF (1995) New antiplatelet agents: platelet GPIIb/IIIa antagonists. Thromb Haemost 74:302–308
Conforti G, Dominguez-Jimenez C, Zanetti A, Gimbrone MA Jr, Cremona O, Marchisio PC et al (1992) Human endothelial cells express integrin receptors on the luminal aspect of their membrane. Blood 80:437
Craven LL (1950) Acetylsalicylic acid, possible preventive of coronary thrombosis. Ann West Med Surg 4:95–99
Craven LL (1953) Experiences with aspirin (acetylsalicylic acid) in the nonspecific prophylaxis of coronary thrombosis. Miss Vall Med J 75:38–40
Dalesandro MR, Frederick B, Gumbs CI, Deutsch E, Fowler AB, Mascelli MA et al (1996) Defining the effectiveness of anti-platelet therapy by measurement of platelet and plasma P-selectin levels. Blood 88 Supp 1:51b
Diener HC, Cunha L, Forbes C, Sivenius J, Smets P, Lowenthal A (1996) European Stroke Prevention Study 2. Dipyridamole and acetylsalicylic acid in the secondary prevention of stroke. J Neurol Sci 143:1–13
Emmons PR, Harrison MJ, Honour AJ, Mitchell JR (1965) Effect of dipyridamole on human platelet behaviour. Lancet 2:603–606
Ferguson JJ (1996) EPILOG and CAPTURE trials halted because of positive interim results. Circulation 93:637
Ferguson JJ, Fox R (1997) Meeting highlights. The 69th Scientific sessions of the American Heart Association in New Orleans, LA, November 10–13, 1996. Circulation 95:761–764
FitzGerald GA (1987) Dipyridamole. N Engl J Med 316:1247–1257
Fitzgerald J, Roy L, Catella F, FitzGerald GA (1986) Platelet activation in unstable coronary disease. N Engl J Med 315:983–989
Flores-Runk P, Raasch RH (1993) Ticlopidine and antiplatelet therapy. Ann Pharmacother 27:1090–1098
Friedman M, Van den Bovenkamp GJ (1966) The pathogenesis of a coronary thrombus. Am J Pathol 48:19–44
Gailit J, Clark RA (1996) Studies in vitro on the role of α_v and β_1 integrins in the adhesion of human dermal fibroblasts to provisional matrix proteins fibronectin, vitronectin, and fibrinogen. J Invest Dermatol 106:102
Gibson PC (1949) Aspirin in the treatment of vascular diseases. Lancet 2:1172–1174
Harrington RA (1997) Design and methodology of the PURSUIT trial: evaluating eptifibatide for acute ischemic coronary syndromes. Platelet Glycoprotein IIb/IIIa in Unstable Angina: Receptor Suppression Using Integrilin Therapy. Am J Cardiol 80:34B-38B
Hayem G (1882) Sur le mÈcanisme de l'arrÍt des hÈmorrhagies. C R Acad Sci Paris 95:18–25
Herbert J, Frehel D, Vallee E, Kieffer G, Gouy D, Berger Y et al (1993) Clopidogrel, a novel antiplatelet agent. Cardiovas Drug Revs 11:180–198
Hoshiga M, Alpers CE, Smith LL, Giachelli CM, Schwartz SM (1995) $\alpha_v\beta_3$ integrin expression in normal and atherosclerotic artery. Circ Res 77:1129
Jakubowski JA, Stampfer MJ, Vaillancourt R, Deykin D (1985) Cumulative antiplatelet effect of low-dose enteric coated aspirin. Br J Haematol 60:635–642
Keimowitz RM, Pulvermacher G, Mayo G, Fitzgerald DJ (1993) Transdermal modification of platelet function. A dermal aspirin preparation selectively inhibits

platelet cyclooxygenase and preserves prostacyclin biosynthesis. Circulation 88:556–561

Kleiman NS, Ohman EM, Califf RM, George BS, Kereiakes D, Aguirre FV et al (1993) Profound inhibition of platelet aggregation with monoclonal antibody 7E3 Fab after thrombolytic therapy. Results of the Thrombolysis and Angioplasty in Myocardial Infarction (TAMI) 8 Pilot Study. J Am Coll Cardiol 22:381–389

Kleiman NS, Raizner AE, Jordan R, Wang AL, Norton D, Mace KF et al (1995) Differential inhibition of platelet aggregation induced by adenosine diphosphate or a thrombin receptor-activating peptide in patients treated with bolus chimeric 7E3 Fab: implications for inhibition of the internal pool of GPIIb/IIIa receptors. J Am Coll Cardiol 26:1665–1671

Knight DM, Wagner C, Jordan R, McAleer MF, DeRita R, Fass DN et al (1995) The immunogenicity of the 7E3 murine monoclonal Fab antibody fragment variable region is dramatically reduced in humans by substitution of human for murine constant regions. Mol Immunol 32:1271–1281

Konstantopoulos K, Kamat SG, Schafer AI, Banez EI, Jordan R, Kleiman NS et al (1995) Shear-induced platelet aggregation is inhibited by in vivo infusion of an anti-glycoprotein IIb/IIIa antibody fragment, c7E3 Fab, in patients undergoing coronary angioplasty. Circulation 91:1427–1431

Kroll MH (1994) Mechanisms of platelet activation. In: Loscalzo J, Schafer AI (eds) Thrombosis and hemorrhage. Blackwell Scientific Publications, Oxford, pp 247–277

Kutok JL, Coller BS (1994) Partial inhibition of platelet aggregation and fibrinogen binding by a murine monoclonal antibody to GPIIIa: requirement for antibody bivalency. Thromb Haemost 72:964–972

Lefkovits J, Plow EF, Topol EJ (1995) Platelet glycoprotein IIb/IIIa receptors in cardiovascular medicine. N Engl J Med 332:1553–1559

Lincoff AM, Tcheng JE, Bass TA, Popma JJ, Teirstein PS, Kleiman NS (1995) A multicenter, randomized, double-blind pilot trial of standard versus low dose weight-adjusted heparin in patients treated with the platelet GP IIb/IIIa receptor antibody c7E3 during percutaneous coronary revascularization. J Am Coll Cardiol Spec Issue 80A-81 A:711–713

Lind SE (1994) Platelet morphology. In: Loscalzo J, Schafer AI (eds) Thrombosis and hemorrhage. Blackwell Scientific Publications, Oxford, pp 201–208

Macfarlane RG (1977) Haemostasis: introduction. Br Med Bull 33:183–184

Martinez J, Rich E, Barsigian C (1989) Transglutaminase-mediated cross-linking of fibrinogen by human umbilical vein endothelial cells. J Biol Chem 264:20502

Matsuno H, Stassen JM, Vermylen J, Deckmyn H (1994) Inhibition of integrin function by a cyclic RGD-containing peptide prevents neointima formation. Circulation 90:2203

Mickelson JK, Kleiman NS, Lakkis NM, Chow TW, Hughes BJ, Smith CW (1996) Chimeric 7E3 Fab (ReoPro®), decreases detectable CD11b on neutrophils from patients undergoing coronary angioplasty. Circulation 94:I42

Mizuno K, Satomura K, Miyamoto A, Arakawa K, Shibuya T, Arai T et al (1992) Angioscopic evaluation of coronary-artery thrombi in acute coronary syndromes. N Engl J Med 326:287–291

Moliterno DJ, Califf RM, Aguirre FV, Anderson K, Sigmon KN, Weisman HF et al (1995) Effect of platelet glycoprotein IIb/IIIa integrin blockade on activated clotting time during percutaneous transluminal coronary angioplasty or directional atherectomy (the EPIC trial). Am J Cardiol 75:559–562

Moran A, FitzGerald GA (1994) Mechanisms of action of antiplatelet drugs. In: Colman RW, Hirsh J, Marder VJ, Saltman EW (eds) Hemostasis and thrombosis: basic principles and clinical practice. J.P. Lippincott Company, Philadelphia, pp 1623–1637

Patrignani P, Filabozzi P, Patrono C (1982) Selective cumulative inhibition of platelet thromboxane production by low-dose aspirin in healthy subjects. J Clin Invest 69:1366–1372

Phillips DR, Scarborough RM (1997) Clinical pharmacology of eptifibatide. Am J Cardiol 80:11B–20B

Reverter JC, Beguin S, Kessels H, Kumar R, Hemker HC, Coller BS (1996) Inhibition of platelet-mediated, tissue factor-induced thrombin generation by the mouse/human chimeric 7E3 antibody. Potential implications for the effect of c7Ee Fab treatment on acute thrombosis and "clinical restenosis". J Clin Invest 98:863

Roth AA (1986) Platelet arachidonate metabolism and platelet-activating factor. In: Phillips DR, Schuman MA (eds) Biochemistry of platelets. Academic Press, New York, pp 69–113

Samanen J (1996) GPIIb/IIIa antagonists. Ann Rep Med Chem 31:91–100

Scarborough RM, Naughton MA, Teng W, Rose JW, Phillips DR, Nannizzi L et al (1993) Design of potent and specific integrin antagonists. Peptide antagonists with high specificity for glycoprotein IIB/IIIa. J Biol Chem 268:1066–1073

Schror K (1993) The basic pharmacology of ticlopidine and clopidogrel. Platelets 4:252–261

Shebuski RJ (1994) Pharmacology of antiplatelet agents. In: Loscalzo J, Schafer AI (eds) Thrombosis and hemorrhage. Blackwell Scientific Publications, Oxford, pp 1139–1154

Simoons ML, de Boer MJ, van den Brand MJ, van Miltenburg AJ, Hoorntje JC, Heyndrickx GR et al (1994) Randomized trial of a GPIIb/IIIa platelet receptor blocker in refractory unstable angina. European Cooperative Study Group. Circulation 89:596–603

Sugihara K, Sugihara T, Mohandas N, Hebbel RP (1992) Thrombospondin mediates adherence of CD36+ sickle reticulocytes to endothelial cells. Blood 80(10):2634

Stampfer MJ, Jakubowski JA, Deykin D, Schafer AI, Willett WC, Hennekens CH (1986) Effect of alternate-day regular and enteric-coated aspirin on platelet aggregation, bleeding time, and thromboxane A2 levels in bleeding-time blood. Am J Med 81:400–404

Tcheng JE (1996) Glycoprotein IIb/IIIa receptor inhibitors: putting the EPIC, IMPACT II, RESTORE and EPILOG trials into perspective. Am J Cardiol 783A:35–40

Tcheng JE (1997) Impact of eptifibatide on early ischemic events in acute ischemic coronary syndromes: a review of the impact II trial. Integrilin to Minimize Platelet Aggregation and Coronary Thrombosis. Am J Cardiol 804A:21B–28B

Tcheng JE, Ellis SG, George BS, Kereiakes DJ, Kleiman NS, Talley JD et al (1994) Pharmacodynamics of chimeric glycoprotein IIb/IIIa integrin antiplatelet antibody Fab 7E3 in high-risk coronary angioplasty. Circulation 90:1757–1764

Thebault JJ, Blatrix CE, Blanchard JF, Panak EA (1975) Effects of ticlopidine, a new platelet aggregation inhibitor in man. Clin Pharmacol Ther 18:485–490

Theroux P, Kouz S, Roy L, Knudtson ML, Diodati JG, Marquis JF et al (1996) Platelet membrane receptor glycoprotein IIb/IIIa antagonism in unstable angina. The Canadian Lamifiban Study. Circulation 94:899–905

The CAPTURE Investigators (1997) Randomised placebo-controlled trial of abciximab before and during coronary intervention in refractory unstable angina: the CAPTURE study. Lancet 349:1429–1435

The EPIC Investigators (1994) Use of a monoclonal antibody directed against the platelet glycoprotein IIb/IIIa receptor in high-risk coronary angioplasty. N Engl J Med 330:956–961

The EPILOG Investigators (1997) Platelet glycoprotein IIb/IIIa receptor blockade and low-dose heparin during percutaneous coronary revascularization. N Engl J Med 336:1689–1696

Thompson CB, Jakubowski JA (1988) The pathophysiology and clinical relevance of platelet heterogeneity. Blood 72:1–8

Topol EJ, Plow EF (1993) Clinical trials of platelet receptor inhibitors. Thromb Haemost 70:94–98

Topol EJ, Califf RM, Weisman HF, Ellis SG, Tcheng JE, Worley S et al (1994) Randomised trial of coronary intervention with antibody against platelet IIb/IIIa integrin for reduction of clinical restenosis: results at six months. The EPIC Investigators. Lancet 343:881–886

Topol EJ, Ferguson JJ, Weisman HF, Tcheng JE, Ellis SG, Kleiman NS et al (1997) Long-term protection from myocardial ischemic events in a randomized trial of brief integrin β_3 blockade with percutaneous coronary intervention. EPIC Investigator Group. JAMA 278:479–484

van der Zee R, Passeri J, Barry JJ, Cheresh DA, Isner JM (1996) A neutralizing antibody to the alpha v beta 3 integrin reduces neointimal thickening in a balloon-injured rabbit iliac artery. Circulation 94:I–257

Varner JA, Nakada M, Jordan R, Coller B (1997) Anti-angiogenic properties of 7E3, an integrin β_3 subunit antagonist, in the SCID mouse-human skin model of human angiogenesis. Thromb Haemost Suppl:158

Wagner CL, Cunningham MR, Wyand MS, Weisman HF, Coller BS, Jordan RE (1995) Reversal of the anti-platelet effects of chimeric 7E3 Fab treatment by platelet transfusion in cynomolgus monkeys. Thromb Haemostasis 73(6):899–1534

Wagner CL, Mascelli MA, Neblock DS, Weisman HF, Coller BS, Jordan RE (1996) Analysis of GPIIb/IIIa receptor number by quantification of 7E3 binding to human platelets. Blood 88:907–914

Walsh PN (1994) Platelet-coagulant protein interactions. In: Colman RW, Hirsh J, Marder VJ, Salzman EW (eds) Hemostasis and thrombosis: basic principles and clinical practice. J.P. Lippincott Company, Philadelphia, pp 629–651

Weller T, Alig L, Muller MH, Kouns WC, Steiner B (1994) Fibrinogen receptor antagonists – a novel class of promising antithrombotics. Drugs Fut 19:461–476

Weisman HF (1996) ReoPro clinical development. Future directions and therapeutic approaches. J Invasive Cardiol 8:51B–61B

Weisman HF, Schaible TF, Jordan RE, Cabot CF, Anderson KM (1995) Anti-platelet monoclonal antibodies for the prevention of arterial thrombosis: experience with ReoPro, a monoclonal antibody directed against the platelet GPIIb/IIIa receptor. Biochem Soc Trans 23:1051–1057

White JG (1994) Anatomy and structural organization of the platelet. In: Colman RW, Hirsh J, Marder VJ, Salzman EW (eds) Hemostasis and thrombosis: basic principles and clinical practice. J.P. Lippincott Company, Philadelphia, pp 397–413

Wilson KM, Siebert DM, Duncan EM, Somoygi AA, Lloyd JV, Bochner F (1990) Effect of aspirin infusions on platelet function in humans. Clin Sci Colch 79:37–42

Yokota T, Milenic DE, Whitlow M, Wood JF, Hubert SL, Schlom J (1993) Microautoradiographic analysis of the normal organ distribution of radioiodinated single-chain Fv and other immunoglobulin forms. Cancer Res 53:3776–3783

Zanetti A, Conforti G, Hess S, Martin-Padura I, Ghibaudi E, Preissner KT et al (1994) Clustering of vitronectin and RGD peptides on microspheres leads to engagement of integrins on the luminal aspect of endothelial cell membrane. Blood 84(4):1116

CHAPTER 8

Platelet Membrane Receptors and Signalling Pathways: New Therapeutic Targets

S.P. WATSON, D. KEELING, and M.D. HOLLENBERG

A. Introduction

Platelets circulate in the blood stream in a quiescent state but undergo explosive activation at sites of damage to the blood vessel wall, leading to formation of a haemostatic plug and cessation of bleeding. The quiescent state of the platelet is maintained in part by the inhibitory action of constitutively released nitric oxide from endothelial cells lining the blood vessels. Activation is brought about by contact with extracellular matrix proteins exposed at the site of tissue damage, notably collagen fibres, and is reinforced by release and generation of a large number of substances such as thrombin and thromboxanes. The remarkable number and diversity of these agents facilitates rapid formation of the platelet aggregate leading to arrest of bleeding. Activation of platelets within intact blood vessels, however, can lead to vascular occlusion with subsequent ischaemic tissue damage. In the most severe cases, this leads to myocardial infarction or stroke, two of the major causes of death in the Western world. The platelet, therefore, is a major target for therapeutic intervention in the prevention of thrombotic disease.

I. Platelet Activation

Platelet activation can be subdivided into a number of individual responses. The initial interaction of platelets with the vessel wall is termed adhesion and serves to trap the cells at the site of damage, to provide a platform for the haemostatic plug and to initiate platelet activation. The major protein interactions which support adhesion vary with shear force. Binding of the platelet glycoprotein (GP) GPIb-IX-V to von Willebrand factor (vWF) is essential for adhesion at the high shear forces found within small or diseased arteries and arterioles (RUGGERI 1997; CLEMETSON 1997). The bound vWF attaches to the vascular wall through interaction with exposed collagen fibres bringing the platelet surface into contact with other subendothelial proteins. Collagen fibres also play a direct role in adhesion through the integrin GPIa-IIa ($\alpha_2\beta_1$) and other surface proteins (see below) reinforcing adhesion to vWF at high shear and supporting adhesion independently of vWF at low shear (MOROI and JUNG 1997; SIXMA et al. 1995, 1997). Platelets also adhere to subendothelial, immobilized fibrinogen which serves to reinforce adhesion to vWf at high shear and to independently support adhesion at low shear (RUGGERI 1997). Although initial adhesion does not involve signal transduction events, these play important early roles in reinforcement through formation of filipodia and platelet spreading which increases the number of sites of contact; exposure of the RGD binding site on GPIIb-IIIa ($\alpha_{IIIb}\beta_3$) enables further interactions with extracellular matrix proteins including vWF, fibronectin and fibrinogen (RUGGERI 1997). Adhesion is therefore a complex process involving several subendothelial proteins and intracellular signals, the importance of which vary with the shear force in the vessel and with time.

Adhesion leads to a number of functional responses which serve to recruit further platelets into the platelet monolayer/developing haemostatic plug. The principal responses are aggregation, secretion, generation of thromboxanes, cytoskeletal rearrangement and expression of procoagulant activity. Aggregation requires a conformational change in the platelet integrin GPIIb-IIIa leading to exposure of the RGD binding site, a process known as inside-out signalling. Binding of circulating fibrinogen molecules to exposed RGD sites on GPIIb-IIIa enable platelet–platelet interaction (i.e. aggregation) to occur. A variety of proaggregatory substances are released from dense granules including ADP, which in concert with thromboxanes generated downstream of phospholipase A_2 (PLA_2), reinforce development of the aggregate through activation of circulating platelets. Alpha (α) granules secrete a number of substances which also support haemostasis, either directly or indirectly, such as chemokines (which attract monocytes and neutrophils,) adhesion molecules (e.g. thrombospondin, fibronectin and fibrinogen), and platelet derived growth factor (which facilitates long term repair through stimulation of smooth muscle proliferation.) Exposure of aminophospholipids on the outer leaflet of the platelet membrane facilitates generation of thrombin through binding of intermediates in the coagulation cascade (procoagulant activity). Thrombin is one of the most powerful platelet stimuli and also plays a key role in the coagulation cascade. Over a period of 15–40 min the loose platelet aggregate is strengthened through clot retraction which stabilizes the link between the platelet cytoskeleton and extracellular matrix. Stimulation of clot retraction is mediated through activation of GPIIb-IIIa by fibrinogen, a process known as outside-in signalling. Activation of GPIIb-IIIa also regulates the size of the aggregate, converting small aggregates into larger ones.

II. Platelet Inhibition

Unwanted activation of platelets within the vasculature is life-threatening and can occur in regions of high shear force, reduced blood flow and damaged atherosclerotic plaques. It is therefore essential for platelets to have well developed inhibitory pathways to counteract activation under these conditions. These inhibitory effects are mediated by the cyclic nucleotides cAMP and cGMP. Nitric oxide is constitutively released from endothelial cells lining the vessel walls giving rise to elevation of platelet cGMP. This provides tonic control over platelet activity. Prostacyclin is formed in endothelial cells from arachidonic acid released from activated platelets. Together with nitric oxide, it plays a vital role in limiting platelet activation at the edge of sites of damage to the vasculature. Prostacyclin binds to the IP prostanoid receptor on the platelet surface, stimulating formation of cAMP.

III. Regulation of Platelet Activation

The last 15 years have seen a remarkable advance in knowledge of signalling events that initiate the responses described above. The intracellular signals generated by the majority of surface receptors have been identified and significant progress has been made in linking their formation to the regulation of functional responses. There is still a great deal to be achieved in this area, though, because the increase in understanding has brought increased realization of the complexity of signalling cascades, such as the existence of large superfamilies of proteins with close sequence homology. A further problem is that we know relatively little about the final events regulating secretion, aggregation, clot retraction and exposure of aminophospholipids, although it seems likely that there are multiple control mechanisms in each case.

In this short review, we shall focus on the intracellular events that give rise to platelet activation and describe how these are generated by the diversity of receptors found on the platelet surface. There are several examples of convergent evolution which have enabled receptors of distinct structure to activate the same pathway within the cell, although there may be important temporal and spatial differences. An understanding of the molecular basis of receptor signalling provides a molecular framework as to why agonists synergise with each other and alter the sensitivity of platelets to other stimuli. This information will ultimately provide a molecular explanation for bleeding disorders caused by lesions in intracellular signalling events and identify important new targets for pharmaceutical intervention.

B. Signalling by Cell Surface Receptors

Most endogenous agents that modulate platelet function act via a variety of membrane-localized cell surface receptors. Pharmacologically, the term "receptor" is used to designate a macromole that performs two distinct, but related functions. First, a receptor must bind its specific ligand with (usually) high affinity and stereospecific chemical specificity. Secondly, the receptor must respond to the binding of its specific ligand by a conformational change that leads to an intracellular signal. It is this dual recognition-signalling property that distinguishes pharmacologic receptors from other membrane constituents, such as nutrient transporters and some but not all adhesion receptors (see below).

For plasma membrane receptors, there are three general paradigms that lead to the initiation of a cellular signal (SEVERSON and HOLLENBERG 1997):

(A) the receptor may comprise a ligand-gated ion channel; (B) the receptor may be a ligand-regulated transmembrane enzyme, like the tyrosine kinase receptors for platelet-derived growth factor; and (C) the receptor may not of itself possess either intrinsic enzymatic or transporter activity, but may upon activation interact specifically and directly with other membrane-localized signal adaptor molecules (or effectors). This third type of paradigm is dramatically represented by the family of G protein-coupled receptors (see Sect. D). Well over 300 of these putative 7-transmembrane-spanning (so-called serpentine) G protein receptors have been cloned and sequenced (STROSBERG 1991; SAVARESE and FRASER 1992; WATSON and ARKINSTALL 1994; GUDERMAN and NURNBERG 1995). Because of the importance of G protein-coupled receptors in regulating platelet function, the signal transduction mechanisms for the serpentine receptors will be dealt with in some detail (Sect. C and D).

Both receptor paradigms B and C can be seen as a variant of the mobile or floating receptor model of agonist action, as developed some time ago (BOEYNAEMS and DUMONT 1977; DE HAEN 1976; CUATRECASAS and HOLLENBERG 1976). This receptor model of hormone action, as depicted in Fig. 1 led to the hypothesis that receptor dimerization (or multimerization) was an essential event in the activation of growth factor receptors (SCHLESSINGER and ULLRICH 1992). The essential tenet of this model of receptor activation is that ligand binding initiates a number of membrane-localized receptor-effector interactions within the plane of the membrane. During the course of these mobile reactions, in theory, multiple membrane effector moieties might be activated, generating multiple transmembrane signals from single ligand-receptor interactions (Fig. 1). The receptor domains that may be involved (1) in ligand binding and (2) in receptor-effector interactions (e.g. via

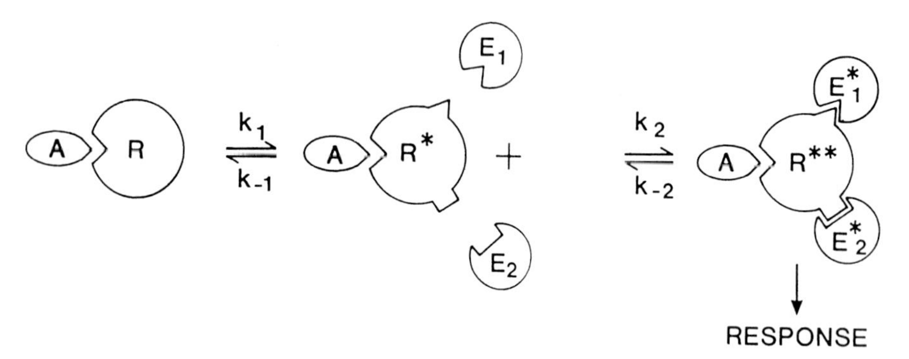

Fig. 1. The mobile receptor model. The binding of a single ligand (A) to a specific receptor (R) induces conformational changes (R*) thereby enabling the ligand activated receptor to interact in the plane of the membrane with multiple effectors (E1, E2). Effector activation (E*1, E*2) may be caused concurrently, as suggested by the supramolecular complex (A, R**, E*, E2*) and as suggested in Fig. 2 for the regulation of both adenylyl cyclase (negatively) and phospholipase C (positively) via a single agonist-receptor interaction (A2/R_i)

SH2 domains – see below) can each be thought of as related, but independent sites on the receptor. The implications of this model for platelet receptor function are at least twofold: First, the model points to two types of domains on the receptor (ligand binding domain and receptor-effector domain) that may be targets for drug development (see Sect. H). Second, the model suggests that even in the absence of a receptor's specific ligand, receptor clustering (e.g., via anti-receptor antibodies or interactions with the extracellular matrix) could potentially modulate platelet function.

I. G Protein-Coupled Receptor Signalling

1. Guanine Nucleotide Binding Proteins and Effector Regulation

Receptor-mediated G protein activation represents one of the most commonly used mechanisms for generating intracellular signals. It was the recognition of the essential role of GTP in the regulation of membrane-associated adenylyl cyclase that led RODBELL and colleagues to propose the existence of an intermediary signal transducer that acted as a coupling device between the receptor and the cyclase (RODBELL 1995). The subsequent identification of the heterotrimeric GTP-binding protein responsible for the stimulation of adenylyl cyclase (G_s) and concurrent isolation of both the cyclase-inhibitory protein (G_i) and the retinal cyclic AMP phosphodiesterase-regulating counterpart, transducin (G_t), set the stage for the cloning of multiple members of the G protein superfamily (summarized by SIMON et al. 1991 and by GILMAN 1995). In the resting state, the G protein complex exists as a heterotrimer, composed of a GDP-bound α-subunit complexed with a tightly associated $\beta\gamma$ dimer. It is the catalytic interaction between the ligand-occupied receptor and the oligomeric G protein that triggers the exchange of GTP for GDP on the α-subunit and that promotes the interaction of αGTP and $\beta\gamma$ with downstream effectors, such as adenylyl cyclases and phospholipases.

There are four subclasses of α-subunits comprising 20 or more family members including a number of mRNA splice product variants: (1) the G_s subgroup, known for the ability to activate adenylyl cyclase, (2) the G_i subgroup, originally identified as substrates for pertussis toxin (except for G_z) and known to be involved in attenuating adenylyl cyclase and in modulating ion channel function, (3) the G_q subgroup involved in the activation of phospholipase C (PLC) β isoforms and (4) the G_{12} subgroup, for which specific target effectors have yet to be identified unequivocally. Multiple forms of the β (6 isoforms) and γ (11 isoforms) subunits are also known. In the platelet, in addition to α_S, α-subunit members of the G_i subclass (α_{i1}, α_{i2}, α_{i3}, α_z), the Gq subclass (α_q), and the G_{12} subclass (α_{12}, α_{13}) have been detected (reviewed by Brass et al. 1993). The identities of the $\beta\gamma$-subunit combinations present in platelets have yet to be established with precision.

A simplified overview of G protein-coupled effector regulation is shown in Fig. 2. Regulating production of a principal intracellular platelet signal

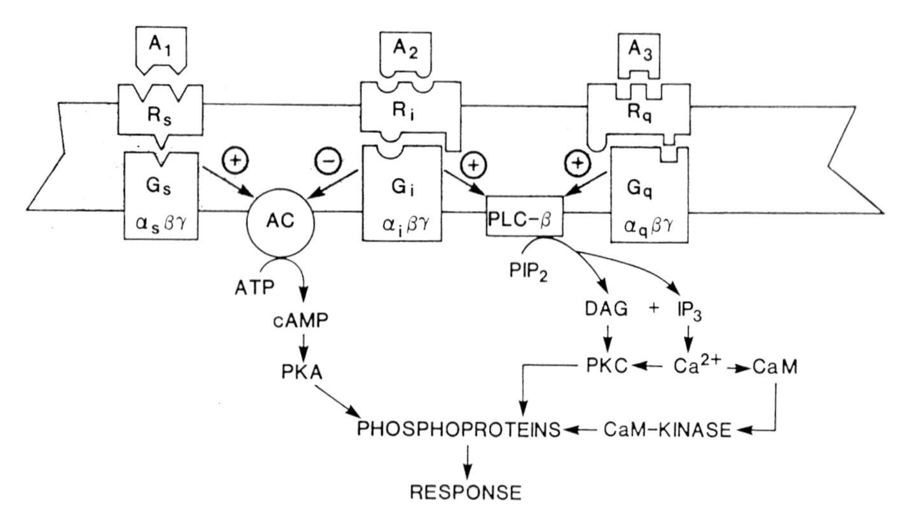

Fig. 2. Receptor-mediated regulation of G protein-effector interactions and the generation of signal mediators. Three distinct agonists (A1, A2, A3) are shown to bind to three distinct receptors (R_s, R_i, R_q) that preferentially interact with three distinct heterotrimeric G protein family members represented by G_s ($\alpha_s\beta\gamma$), G; ($\alpha_i\beta\gamma$) and G_q ($\alpha_q\beta\gamma$). The shapes of receptors R_i and R_q are shown as potentially interactive with both G_i and G_q, although in reality this is not the case for all heptahelical receptors. In turn, the G protein transducers (or effectors) are shown to regulate membrane-localized effector enzymes: adenylyl cyclase (AC) and phospholipase C isoforms (PLCβ_{1-3}). The activation of effector enzymes by the $\beta\gamma$ subunit is not shown. In this scheme, an individual receptor, like the one for thrombin (R_i), is shown to regulate both adenylyl cyclase (AC: negative regulation) and phospholipase Cβ_3, (PLCβ_3: positive regulation), resulting in the hydrolysis of phosphatidyl inositol 4, 5 bisphosphate, PIP2). A receptor like the one for PGI_2 (R_s) is shown to stimulate adenylyl cyclase resulting in the formation of cyclic AMP from ATP. Effector activation results in the generation of intracellular signal mediators such as cyclic AMP (cAMP), diacylglycerol (DG) and inositol 1, 4, 5 trisphosphate (IP3). In turn, these mediators activate protein kinase signal pathways via protein kinase A (PKA) protein kinase C (PKC) or, via elevated intracellular Ca^{2+} that stimulates Ca^{2+}-calmodulin (CaM)-activated kinases (CaM kinase)

mediator, cAMP, is independently under both positive (e.g., for prostacyclin, via G_s) and negative (e.g., for thrombin, via G_i) control. Although a given receptor may display a preferential affinity for one class of G proteins (e.g. most G_s-targeted receptors, like the one for prostacyclin, do not interact appreciably with G_i or G_q), it is also the case that receptors such as the one for thrombin can interact with two sub-classes of G proteins. The relative abundance of the interacting species (i.e. the receptor and the various G proteins) and the protein-protein affinities would, by mass action, govern the predominance of the effect of an individual agonist in a particular setting, such as the platelet. The entire process of G protein-effector activation is started by the receptor-mediated GTP/GDP exchange reaction.

2. The G Protein GTP/GDP Cycle and Effector Modulation

As indicated above, when the G protein is activated by the receptor-catalyzed GTP/GDP exchange reaction, both the α and $\beta\gamma$-subunits can go on to regulate a number of effectors, including adenylyl cyclases, phospholipases and ion channels. The time frame of these G protein effector interactions is governed by the rate of hydrolysis of GTP bound to the α-subunit; this hydrolysis returns αGTP to its ground state, αGDP, permitting a downregulation of effector activity and a reassociation of the α-subunit with the $\beta\gamma$ dimer. The domains that govern the mutual interactions between the G proteins and their target effectors could, in theory, provide targets for the development of drugs, although this is unlikely to be exploited because of a lack of specificity. The cycling of the G protein heterotrimer between the GDP-bound form and the receptor-catalyzed GTP-activated state is illustrated in Fig. 3.

Although a popular model of G protein function envisions the α-subunit shuttling freely between the ligand-occupied receptor and the effector (e.g. adenylyl cyclase; GILMAN 1995), it also possible that the G protein and effector may be precoupled in the membrane (LEVITZKI and BAR-SINAI 1991). Based on kinetic considerations, this situation appears to be the case for the G_s-

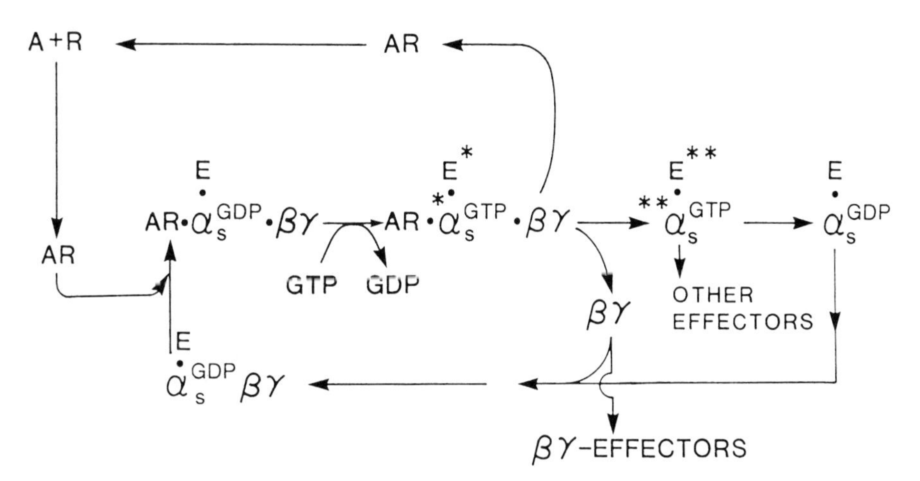

Fig. 3. The G Protein GDP/GTP cycle. The binding of an agonist (A) to its receptor (R) is shown, in catalytic manner, to trigger the exchange of GTP for GDP on the G protein α-subunit, depicted as the cyclase-activating subunit, α_s-GDP. The heterotrimeric protein is shown to be pre-coupled to an effector, E, as suggested by the kinetic data of LEVITZKI et al. (1993) for adenylyl cyclase. Upon exchanging GTP for GDP, the activated α-subunit (*αGTP) can stimulate its effector target (E*) which can proceed to a fully activated state (**αGTPE**) as the agonist receptor complex and the $\beta\gamma$ subunit dissociates. The $\beta\gamma$ subunit can go on to regulate other effectors (not shown), whilst the hydrolysis of GTP bound to α_s returns the α_s-subunit and the associated effector to a ground state that is capable of recombining with free $\beta\gamma$ subunit. This re-association completes the GDP/GTP cycle

regulated cyclase system (LEVITZKI and BAR-SINAI 1991); it can also be postulated that other effectors may exist in the membrane already precoupled with a ground-state G protein. Such a model is shown in Fig. 3, where the "effector" E is seen as being already coupled to the G protein (LEVITZKI et al. 1993). This situation might rationalize the very rapid time course of G protein-coupled receptor signalling, which would be limited only by a receptor diffusion process, and not by a concurrent diffusion of the activated G protein to the effector. Further, this possibility would provide for the compartmentalization of G protein-effector systems at discrete membrane loci in tissues such as the platelet. The compartmentalization of G proteins and their interactions with effectors may be governed also by a number of post-translational covalent modifications of the G protein subunits, of which the addition of lipid moieties may be of particular significance (WEDEGAERTNER et al. 1995).

3. G Protein-Regulated Effectors

a) Adenylyl Cyclase

The effector first shown to be regulated by G proteins was adenylyl cyclase. All nine isoforms of this enzyme now known to exist are activated by the α_s-subunit (SUNAHARA et al. 1996); subtype VII is thought to represent the predominant form in platelets. Other effectors that are targets for α_S-regulation include voltage-gated Ca^{2+} channels and certain Na^+ channels, but neither is present in the platelet. Early on it was recognized that a number of agonists cause inhibition of adenylyl cyclase acting via a pertussis toxin-sensitive G protein (the toxin ADP-ribosylates the α_i-subunit and thereby abrogates receptor-triggered $\alpha_i\beta\gamma$ dissociation). In the platelet, this G_i-mediated inhibitory regulation of cyclase activity can be seen upon activation by thrombin and adrenaline. The mechanism of negative regulation of adenylyl cyclase is complex and determined by the isoform of the enzyme which is present (both α_i and $\beta\gamma$-subunits can negatively regulate the enzyme). The inhibition can be either via a direct interaction between the α_i-subunit and the cyclase (e.g., isoforms V and VI) or indirect (e.g., platelet cyclase isoform VII), due to the sequestration of active α_s by an excess of $\beta\gamma$, made available, for example, by activation of the quantitatively predominant G_i oligomer. This latter $\beta\gamma$-mediated α_s sequestration mechanism most likely explains the cyclase-inhibitory action of thrombin and adrenaline in platelets. To complicate matters further, the $\beta\gamma$-subunit itself can stimulate rather than inhibit selected cyclase isoforms (types II and IV), one of which (IV) may be present along with type VII in platelets. The $G\alpha_i$ family isoform α_z, present in comparatively high amounts in platelets, is a substrate for protein kinase C (PKC); phosphorylation could provide for signal amplification, since this blocks the interaction of α_z with the $\beta\gamma$-subunit (KOZASA and GILMAN 1996; FIELDS and CASEY 1995). Direct regulation of adenylyl cyclases by the $\beta\gamma$ dimer must be kept in mind when interpreting measurements of platelet cyclase regulated by agonists that act via either G_i or G_q.

b) Phospholipase C

The effector that has one of the most dramatic impacts on platelet function is phosphoinositide-specific PLC (see Sect. C.I). The β- and γ-isoforms of this enzyme lead to agonist-stimulated hydrolysis of platelet membrane phospholipids, so as to yield the intracellular mediators, diacylglycerol (DG) and inositol 1,4,5-trisphosphate (IP3) (see Sect. C and Fig. 2). The β-isoforms of the enzyme can be stimulated by members of the α_q-family to hydrolyse phosphatidylinositol bisphosphate. The G protein $\beta\gamma$-subunit can also independently activate PLC β-isoforms, with a preference for activating PLC $\beta 3$ (PARK et al. 1993; summary by FIELDS and CASEY 1997). Thus, in the platelet, one could envision activation of PLC β both via activation of G_q family members and via activation of the G_i members. In the case of the thrombin receptor (that couples to members of both the G_i and G_q family members) there could be concurrent activation of PLC β (via α_q as well as via released $\beta\gamma$) along with a parallel $\beta\gamma$-mediated inhibition of adenylyl cyclase (it is only the $\beta\gamma$ from G_i that is thought to play a signalling role in this way because it is present at a much higher level). Stimulation of PLC β isoforms via G protein-coupled processes would complement activation of PLC γ isoforms via tyrosine kinase-mediated pathways (see Sects. B.II, C.I). The concurrent activation of both groups of PLC isoforms would yield an abundance of the dual intracellular mediators, IP3 and DG.

c) Other $\beta\gamma$-Regulated Effectors: src Family Kinases and PI 3-Kinase

In addition to the ability of the $\beta\gamma$-subunits to modulate adenylyl cyclase and PLC β, this G protein-derived dimer is now known to regulate other effectors. Several of these such as the N-and P/Q-type Ca^{2+} channels have little relevance to platelet function. Other interactions, however, have uncertain function in platelets such as $\beta\gamma$ activation of tyrosine kinase pathways involving c-src family members and possibly pleckstrin homology (PH)-domain-containing tyrosine kinases (LUTTRELL et al. 1995; VAN BIESEN et al. 1996). G protein $\beta\gamma$-subunits have also been observed to stimulate an isoform of PI 3-kinase present in human platelets, PI 3-kinase γ (THOMASON et al. 1994; ABRAMS et al. 1996). The γ-isoform of PI 3-kinase present in platelets, upon activation by released $\beta\gamma$, might play a role in recruiting PH-domain-containing signal proteins such as the serine threonine kinase AKT/protein kinase B and the phosphoinositide-binding protein, Grpl, to the platelet plasma membrane (LEMMON et al. 1997). Such PI 3-kinase-promoted membrane interactions could play a role in modulating platelet secretion and in regulating integrin-dependent 'inside-out' adhesion processes (see below). The modulation of src-family kinases via a $\beta\gamma$-mediated process provides for crosstalk between the serpentine receptor systems and the receptor systems mediated via those tyrosine kinase pathways described in (Sect. B.II).

II. Tyrosine Kinase-Linked Receptors

Tyrosine phosphorylation was described in 1979 (ECKHART et al. 1979) and is estimated to account for approximately 1% of phosphorylation of intracellular proteins. From early on it was recognized that many oncogenes encode for tyrosine kinases leading to the speculation that tyrosine phosphorylation was involved in gene regulation. The first description of tyrosine phosphorylation in platelets was not made until 1983 (TUY et al. 1983), with the high kinase level in this cell challenging the idea that tyrosine kinases participate solely in gene expression.

A key breakthrough in understanding the role of protein tyrosine phosphorylation was made in the early 1990s when it was recognized that phosphorylated tyrosine residues bind with high affinity to sequences within proteins of approximately 100 amino acids in length known as src homology 2 or SH2 domains (KOCH et al. 1991; PAWSON 1995). Binding of phosphorylated tyrosine residues to SH2 domains enables protein-protein interaction thereby localizing proteins in the cell.

Tyrosine kinases can be divided into receptor and non-receptor classes. Receptor tyrosine kinases have a single tyrosine kinase domain in their intracellular region and most are receptors for growth factors. They are activated by crosslinking which often leads to autophosphorylation (more strictly, transphosphorylation). This enables association with SH2 domain-containing proteins thereby recruiting enzymes or adaptor molecules to the receptor (e.g., PLC γ, p85-subunit of PI 3-kinase, etc.). Adaptors are docking proteins, themselves devoid of intrinsic enzymatic activity, but which play pivotal roles in cell activation. They may contain SH2 domains and/or other domains such as SH3 domains, which bind to proline rich regions, and PH domains, which bind to lipids and G protein $\beta\gamma$-subunits.

Three receptor tyrosine kinases have been described in the platelet, the β chain of the insulin receptor, the c-kit receptor (also known as stem cell factor receptor) and the platelet derived growth factor α-receptor, but all are present in extremely low levels and there is little evidence that they are of importance physiologically. Insulin has been reported to weakly inhibit aggregation through potentiation of PGE_1-stimulated formation of cAMP despite the absence of the receptor α chain (JAP et al. 1988). Activation of the c-kit receptor by stem cell factor has no effect on its own but potentiates the second phase of aggregation induced by adrenaline and ADP but not by the thromboxane mimetic U46619 (GRABAREK et al. 1994). Platelet derived growth factor caused weak inhibition of aggregation and secretion by several agonists including thrombin (VASSBOTN et al. 1994).

In contrast, at least 15 non-receptor tyrosine kinases have been identified in the platelet, some of which are present in extremely high levels, notably src and syk which each make up 1%–2% of platelet protein (for review see PRESEK and MARTINSON 1997). This diverse family of tyrosine kinases play important roles in intracellular signalling by a variety of surface receptors including those for cytokines, antigens, integrins and heptahelical receptors. As with transmembrane tyrosine kinases, many of these non-receptor kinases undergo autophosphorylation enabling them to recruit SH2 domain-containing proteins and stimulate the events described above.

Platelet receptors that signal through non-receptor tyrosine kinases include the receptor for the cytokine thrombopoietin (known as c-mpl), the collagen receptor, GPVI, and the low affinity receptor for immune complexes, FcγRIIA. Platelets also express other receptors that stimulate tyrosine kinase-dependent pathways although much less is known about these and their importance. These include integrins, which are activated by extracellular matrix proteins, and several single transmembrane proteins that can be activated by crosslinking. In many of these cases, it is unclear whether the initial event is activation of a tyrosine kinase or another intracellular signalling enzyme.

III. Ion Channels and Their Receptors

Platelets are considered as non-excitable cells because their resting membrane potential falls within a similar range to that of other cells but undergoes only a small depolarization on activation. Platelets also lack voltage-operated Ca^{2+} channels and so the significance of this depolarisation is not known.

The $P2x_1$ purinoceptor on the platelet surface is a voltage-independent channel which permits entry of Ca^{2+} and other cations (MACKENZIE et al. 1996). It is activated by purine nucleotides such as ADP and ATP and is present at a density of approximately 30 sites per platelet (MACKENZIE et al. 1996). This is considered too low to make a significant impact on platelet activation by ADP and its functional significance is not known.

The only other class of ion channels on the platelet surface which permit Ca^{2+} entry are those activated by depletion of intracellular stores. Store depletion, mediated physiologically by the action of IP3 and possibly other mechanisms, leads to entry of Ca^{2+} through a channel known as I_{CRAC}. This is discussed further in Sect. C.I.

C. Signal Enzymes and Mediators

Despite the complexity of signalling pathways that exist within platelets and other cells, Ca^{2+} ions can be considered to be the major intracellular messenger controlling cellular activity. Ca^{2+} ionophores are powerful platelet agonists and the major inhibitory action of cyclic nucleotides is to prevent Ca^{2+} elevation and its actions (see below). There is considerable interest in the regulation of this cation within the cell from the viewpoint of development of novel antithrombotics. Surprisingly, however, we still know little about the mecha-

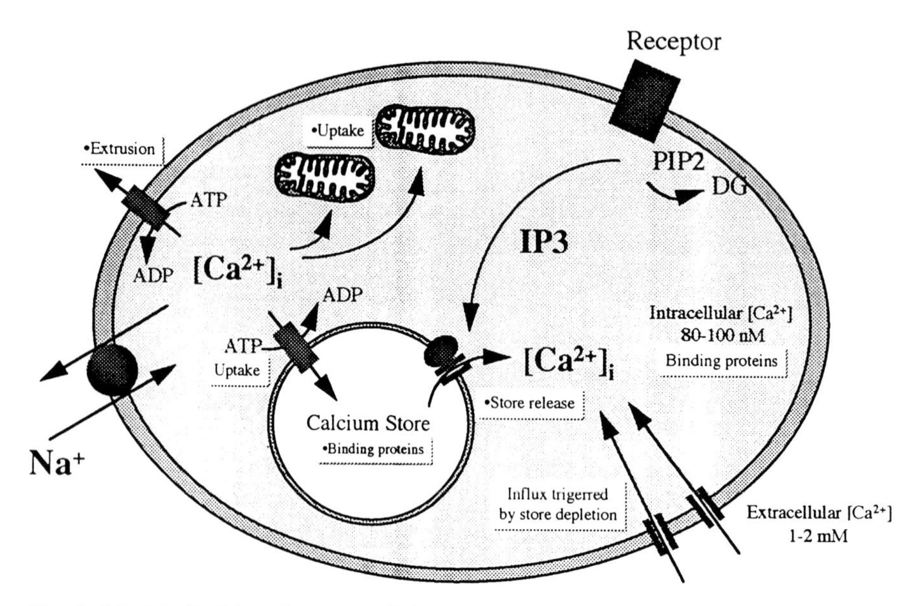

Fig. 4. Model of calcium homeostasis in non-excitable cells. The model depicts mechanisms for the regulation of intracellular calcium $[Ca^{2+}]$i both at rest and in response to agonist challenge

nisms controlling entry of Ca^{2+} and its release from intracellular pools through IP3-independent mechanisms. Although depletion of intracellular Ca^{2+} stores gives rise to Ca^{2+} entry through the non-specific cation channel known as I_{CRAC}, controversy surrounds the molecular mechanism that opens this channel, with several theories having been proposed. The confusion is likely to remain until the sequence of I_{CRAC} is obtained and it is established whether it exists in a superfamily of channels subject to distinct modes of regulation. Figure 4 summarizes current understanding of Ca^{2+} regulation.

I. Phosphoinositide Metabolism

The majority of platelet agonists induce activation through the phosphoinositide second messenger pathway which liberates the intracellular messengers IP3 and DG. IP3 mobilizes Ca^{2+} from intracellular stores and DG activates conventional forms of PKC, of which PKCα and PKCβ are expressed in platelets. Platelets also express several DG-independent forms of PKC which are regulated in other ways (e.g. PI 3-kinase pathway) (see Sect. C.III).

The three major subclasses of PLC are regulated by distinct pathways: PLCβ isoforms are regulated by G proteins, PLCγ isoforms by tyrosine phosphorylation while the mechanism of regulation of PLCδ isoforms is not established. PLCβ2 and PLCγ2 are expressed in platelets at substantially higher levels than other isoforms of PLC (LEE et al. 1996). Evidence that G protein-coupled and tyrosine kinase-linked receptors activate platelets through different isoforms of PLC can be demonstrated through the use of inhibitors of tyrosine kinases. Thus the mixed kinase inhibitor staurosporine and the tyrphostin ST271 completely block formation of inositol phosphates in response to collagen and Fcγ-IIA receptor activation but have marginal effects on the response to the G protein coupled receptor agonists thrombin and thromboxane (BLAKE et al. 1994a; YANAGA et al. 1995).

The role of IP3/Ca^{2+} and DG/PKC in the activation of platelets varies according to the functional response in question. The earliest response that platelets undergo following activation of PLC is a change in shape from discoid to spherical with formation of pseudopodia. Shape change can be mediated by Ca^{2+}-activation of myosin light chain kinase and is independent of PKC. This is illustrated by the ability of the Ca^{2+}-chelator BAPTA-AM but not the PKC inhibitor Ro 31-8220 to prevent shape change by Ca^{2+} ionophores. Shape change by G protein-coupled aganists however is independent of Ca^{2+}, and is probably mediated by G12 stimulation of rho proteins. Aggregation is stimulated by both Ca^{2+}-dependent and PKC-dependent pathways. Ca^{2+} ionophores stimulate rapid aggregation in the absence of PLC activation whereas the response to phorbol esters or membrane-permeable DGs is much slower (WALKER and WATSON 1993). When combined, the two arms of the messenger pathway undergo synergistic interaction giving rise to an increased rate of aggregation (WALKER and WATSON 1993). Together, PKC and Ca^{2+} exhibit remarkable synergy in stimulating secretion from dense granules whereas either agent on its own stimulates only a small or negligible response (WALKER and WATSON 1993). This result demonstrates that it is the degree of synergy between the two messengers which regulates the level of secretion. Activation of the phosphoinositide pathway gives rise to other responses such as secretion from α granules and expression of procoagulant activity although the relative importance of DG/PKC and IP3/Ca^{2+} is not established.

Despite the ability of Ca^{2+} and DG to cause these responses, it is likely that agonist-specific pathways underlie many of these effects. For example, although collagen is in general a far weaker agonist than thrombin, it is a more powerful stimulant of procoagulant activity (DACHERY-PRIGENT et al. 1993) suggesting regulation by receptor-specific pathways.

II. Phospholipase A_2

The 85 kDa cytosolic PLA_2 ($cPLA_2$) is probably the only form of PLA_2 responsible for release of arachidonic acid following receptor activation. The liberated arachidonic acid is converted to cyclooxygenase and lipoxygenase products and may also play a role as a second messenger pathway through potentiation of PKC activation by DG. The most important of these events is the production of thromboxanes (TXA_2) which are released from the platelet and activate additional platelets through the G protein-coupled TXA_2

receptor. The importance of this pathway is demonstrated by the prophylactic action of the cyclooxygenase inhibitor aspirin in the prevention of myocardial infarction and stroke in individuals at risk of cardiovascular events (CATELLA-LAWSON and FITZGERALD 1997).

The significant breakthrough in understanding the regulation of $cPLA_2$ came from the sequencing of the enzyme in 1991. $cPLA_2$ has no structural homology to other forms of PLA_2 but important clues for its regulation are provided by the presence of a calcium-ligand-binding-domain (C2 domain) in its N-terminal region and a consensus phosphorylation site for p42/44 mapkinases at position 505. Recombinant $cPLA_2$ is a substrate for p42/44 mapkinase and phosphorylation brings about a threefold increase in intrinsic activity in vitro (LIN et al. 1993).

$cPLA_2$ is highly active in its non-phosphorylated state, but is unable to reach its substrates due to being localized in the cytosol. Elevation of Ca^{2+} to approximately 300 nM causes translocation to the membrane through the binding of its C2-domain to the phospholipid surface, causing activation. This can be illustrated in platelets by liberation of arachidonic acid through the action of Ca^{2+} ionophores in the absence of significant phosphorylation (BORSCH-HAUBOLD et al. 1995). $cPLA_2$ undergoes phosphorylation on serine residues in platelets following activation of PLC-coupled receptors or challenge with phorbol esters. Phosphorylation induced by phorbol esters is mediated through PKC-dependent activation of p42/44 mapkinases but fails to activate the phospholipase which remains in the cytosol; however, the response to the Ca^{2+} ionophore is potentiated (BORSCH-HAUBOLD et al. 1995).

Surprisingly, given the evidence for a role of p42/44 mapkinases in the regulation of $cPLA_2$ described above (LIN et al. 1993), inhibitors which prevent activation of mapkinase by acting on upstream kinases, namely Ro 31-8220, (which inhibits PKC) or PD98059 (which inhibits mapkinase kinase), have little effect on phosphorylation of $cPLA_2$ in platelets following stimulation by collagen or thrombin (BORSCH-HAUBOLD et al. 1995, 1996). On the other hand, two inhibitors of the closely related stress activated protein kinases (SAPK) partially inhibit phosphorylation of $cPLA_2$ and partially inhibit release of arachidonic acid (BORSCH-HAUBOLD et al. 1997; and unpublished). The two inhibitors, SB203580 and SB2021990, block activation of SAPK2 A and SAPK2B (also known as p38 mapkinase and p38 mapkinase β) isoforms, implicating one or both of these in phosphorylation of $cPLA_2$.

Together, the mechanism of regulation of $cPLA_2$ in platelets is beginning to emerge. Ca^{2+} is essential for activity as it enables recruitment of the enzyme to the membrane, the location of the lipid substrate. Phosphorylation is unable to cause activation on its own but undergoes synergy with Ca^{2+} to increase the activity of $cPLA_2$ by 20%–70%, dependent on agonist concentration. The enzymes responsible for phosphorylation are SAPK2 A and/or SAPK2B, and at least one other kinase. The relative importance of phosphorylation by the latter, unidentified kinase is not known.

III. PI 3-Kinase

In addition to being hydrolyzed to IP3 and DG by the action of PLC, phosphatidylinositol 4,5-bisphosphate can be converted to a further second messenger, phosphatidylinositol 3,4,5-trisphosphate (PIP3) by the action of PI 3-kinase. PIP3 or its metabolite phosphatidylinositol 3,4-bisphosphate bind to PH domains in a number of proteins causing translocation to the membrane. Mounting evidence suggests that PIP3 activates a number of pathways in the platelet. This work has been facilitated by the use of two inhibitors of PI 3-kinase, wortmannin and LY294002, although caution is required in that both compounds exert other actions. PI 3-kinase is required for aggregation by submaximal (but not maximal) concentrations of thrombin, possibly through activation of the atypical PKC isoform, PKCζ (Toker et al. 1995; Zhang et al. 1996). PI 3-kinase may also play a role in the sensitizing action of thrombopoietin to platelet activation by other agonists and for full activation of PLCγ2 by collagen. PIP3 has also recently been described to activate a kinase upstream of protein kinase B and this may underlie several of its actions in the cell.

IV. Cyclic Nucleotides

1. cAMP

The second messenger cAMP activates protein kinase A leading to phosphorylation of many proteins in the cell including a major substrate of 55 kDa known as VASP. cAMP is formed by activation of seven transmembrane receptors coupled to the G protein subunit, G_s as described above. A superfamily of phosphodiesterases break down cAMP (and cGMP), thereby terminating its action. Currently, seven homologous phosphodiesterase families are recognized and most contain multiple isoforms. Three distinct phosphodiesterases exist in the human platelet, cGMP-stimulated phosphodiesterase type II, cGMP-inhibited phosphodiesterase and cGMP-specific phosphodiesterase (for review see Butt and Walter 1997).

Inhibition by cAMP is mediated by a multitude of mechanisms. Elevation of cAMP is associated with marked inhibition of activation of PLCβ isoforms by G protein receptor-coupled agonists, an action mediated by inhibition of $\beta\gamma$-regulation of PLC. This process does not prevent activation of PLCγ2 by the tyrosine kinase-dependent receptor agonists collagen and immune complexes. cAMP also inhibits activation of the fibrinogen receptor GPIIb-IIIa, preventing aggregation by all extracellular agonists. Phosphorylation of VASP by protein kinase A, which is also a substrate for cGMP-dependent protein kinase (PKG), is implicated in the regulation of activation of GPIIb-IIIa by cAMP through binding to F-actin and profilin, and regulating actin polymerisation (Horstrup et al. 1994). Phosphorylation of myosin light chain kinase by protein kinase A would be expected to inhibit shape change under certain conditions. Other pathways are also implicated in the inhibition of platelet function by cAMP but many of these are secondary to inhibition of PLC (for a full review see KOESLING and NURNBERG 1997).

2. cGMP

cGMP levels are elevated in platelets by activation of soluble guanylyl cyclase through liberation of nitric oxide. The constitutive release of nitric oxide from the endothelial cells helps to prevent activation of platelets in the normal circulation. This inhibition occurs therapeutically following administration of nitrovasodilators such as sodium nitroprusside, used in the treatment of angina. The inhibitory action of NO/cGMP is mediated through activation of protein kinase G and shares many features with the action of cAMP, including inhibition of PLCβ isoforms and activation of GPIIb-IIIa, along with phosphorylation of a number of common substrates (KOESLING and NURNBERG 1997). Part of the action of cGMP is mediated through elevation of cAMP by inhibition of cGMP-inhibited phosphodiesterase (BUTT and WALTER 1997).

D. Platelet G Protein-Coupled Receptors

I. Thrombin (PAR_1)

1. Thrombin Binding Sites

The serine proteinase, thrombin, can be classified as a strong platelet agonist. Activation of platelets by thrombin can lead to multiple responses, including phosphoinositide hydrolysis, activation of $cPLA_2$, activation of PI 3-kinase, elevation of intracellular Ca^{2+}, an increase in intracellular protein phosphorylation on serine, threonine and tyrosine residues and a decrease in cAMP. These biochemical responses, in turn, lead to changes in platelet shape, exposure of receptors for fibrinogen, granule secretion and aggregation (for an excellent synopsis, see GRAND et al. 1996). All of these responses require that thrombin be esterolytically active, and many studies have implicated a key role for G proteins as mediators of these thrombin-stimulated platelet responses (BRASS et al. 1993). Ligand-binding studies aimed at characterizing the platelet receptor responsible for these actions of thrombin identified both high ($K_D <$ 10 nM; approx. 50 sites/platelet) and moderate (K_D 10 nM; about 2000 sites/platelet) affinity thrombin binding sites (HARMON and JAMIESON 1986b; DE MARCO et al. 1994). High affinity binding sites are thought to reside on the platelet GPIb-IX-V complex (HARMON and JAMIESON 1986a). It has not been possible to attribute pharmacological receptor status to these binding sites, which nonetheless may well play an important role in platelet/thrombin interactions, in addition to the G protein-coupled receptor for thrombin described in the following paragraphs. In this regard, the ability of thrombin to remain bound in an active form to extracellular matrix constituents (BAR-SHAVIT et al. 1989) and to regulate cell function via non-catalytic protein domains (BAR-SHAVIT et al. 1986a,b; GLENN et al. 1988) may also bear on the interactions between thrombin and platelets.

2. PAR$_1$, A G Protein-Coupled Receptor for Thrombin

Major insight concerning the action of thrombin resulted from expression cloning of the G protein-coupled human and hamster receptors that, when expressed in *Xenopus laevis* oocytes, led to thrombin-dependent Ca^{2+} mobilization (Vu et al. 199la; Rasmussen et al. 1991). The unique mechanism that leads to activation of the receptor comprises the proteolytic unmasking of an N-terminal receptor domain that becomes a 'tethered' or 'anchored' receptor-activating ligand (Fig. 5) (Vu et al 1991a; Coughlin et al. 1992b). Remarkably, synthetic peptides (SFLLR...) based on the sequence of the human receptor revealed by the cleavage at serine residue 41 (LDPR/S41FLLRNPNDKYEPF) were found in isolation to activate the human platelet receptor, mimicking the effects of thrombin (Vu et al 1991a). It was also demonstrated via site-directed mutagenesis that proteolysis of the N-terminal receptor domain to reveal the anchored neoligand was a sine qua non for receptor activation (Scarborough et al. 1992; Vu et al. 1991b). It was found

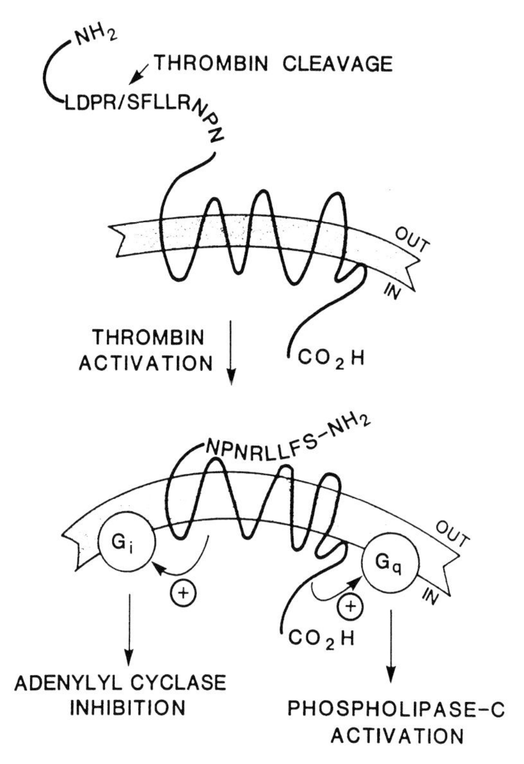

Fig. 5. Model of Thrombin Receptor Activation. The binding of thrombin is shown to result in proteolytic cleavage of the N-terminal arginine 41/serine 42 bond of human PAR$_1$ resulting in the exposure of a new N-terminal tethered sequence, NH_2-SFLLRNPN. The revealed neoligand is then shown to activate the receptor, promoting its interaction with G_q (thereby stimulating phospholipase Cβ) or with G_i (thereby inhibiting adenylyl cyclase)

that synthetic peptides as short as 5 amino acids in length (e.g. SFLLR-NH2), used as receptor-activating probes, could mimic the actions of thrombin in a wide variety of tissues ranging from platelets to smooth muscle (VU et al. 1991b; VASSALLO et al. 1992; CHAO et al. 1992; SCARBOROUGH et al. 1992; SABO et al. 1992; HOLLENBERG et al. 1992). Based on the receptor-activating peptide sequence, it was possible to synthesize an antagonist capable of blocking thrombin-mediated platelet activation (COUGHLIN et al. 1992a, 1992b; see Sect. H). The receptor has been termed proteinase-activated receptor-1 or PAR_1 and is thought to underlie activation of human platelets by thrombin and trypsin.

3. Other Protease-Activated Receptors (PARs)

Although the thrombin receptor-activating peptides (TRAPs) could mimic the majority of the actions of thrombin in human platelets, it was soon apparent that platelets from other species (rabbit, rat: KINLOUGH-RATHBONE et al. 1993a; CONNOLLY et al. 1994), which were otherwise sensitive to thrombin, were not sensitive to the TRAPs. These data, along with the observed thrombin responsiveness of platelets from mice in which the initially cloned thrombin receptor had been eliminated (Connolly et al. 1996) pointed to the existence of a receptor for thrombin other than the one that had been originally described. During that time period, a receptor that was related in sequence to the originally cloned thrombin receptor, but that was activated by trypsin rather than by thrombin, was discovered (NYSTEDT et al. 1994, 1995). This second proteinase-activated receptor, PAR_2, was also activated by a tethered ligand mechanism and by synthetic peptides (SLIGRL-NH_2) based on the trypsin-revealed tethered neoligand (NYSTEDT et al. 1994). However, PAR_2 did not account for the thrombin-sensitive receptor in rabbit and rat platelets; and PAR_2-activating peptides were unable to activate human platelets, indicating an absence of PAR_2. In rabbit and rat platelets, activation by thrombin is probably mediated by a distinct thrombin-activated receptor (proteinase-activated receptor 3 or PAR_3) recently cloned from a human intestinal cDNA library (ISHIHARA et al. 1997). Like PAR_1, PAR_3 is proteolytically activated to reveal a presumed activating neoligand; unlike PAR_1, PAR_3 appears to be resistant to activation by short synthetic peptides derived from the presumed activating sequence (TFRAGPPNS) that would otherwise be capable of activating PAR_1 or PAR_2 (ISHIHARA et al. 1997; NATARAJAN et al. 1995; HOLLENBERG et al. 1997).

Thus in summary, three novel G protein-coupled proteinase-activated receptors have now been discovered, two of which are activated by thrombin (PAR_1 and PAR_3) and one of which (PAR_2) can be activated by trypsin but not by thrombin. All of the PARs, presumably via interactions with G_q, are able to modulate phosphoinositide hydrolysis, so as to mobilize intracellular Ca^{2+}.

4. PAR_1 and Human Platelet Activation

Taken together, the data obtained with human platelets, including studies with PAR-activating peptides (which do not activate PAR_3), and with fluorescently labelled anti-PAR antibodies to localize platelet membrane PAR_1 in numbers (about 2000) that correspond to the moderate affinity binding sites (BRASS et al. 1992), suggest that PAR_1 probably represents the major G protein-coupled receptor for thrombin in the human platelet. Further work with PAR_3-targeted antibodies will be required to substantiate this conclusion and to determine whether or not PAR_3 might also regulate human platelet function.

Given that PAR_1 may represent the principal functional thrombin receptor in human platelets, it is possible to rationalize most of the actions of thrombin in terms of the interactions between PAR_1 and members of either the G_i (especially, $G\alpha_z$) or G_q G protein family members (see Sect. B.II). Interactions with G_i would lead to a concurrent inhibition of adenylyl cyclase and an increase in platelet protein tyrosine phosphorylation via the released $\beta\gamma$-subunit. The released $\beta\gamma$-subunit could also activate PLC β, and the increase in tyrosine phosphorylation might underlie the small increase in tyrosine phosphorylation of PLC $\gamma 2$ by thrombin which is substantially lower than that stimulated by collagen (DANIEL et al. 1994; BLAKE et al. 1994b). PAR_1-mediated activation of G_q could also lead to increased phosphoinositide hydrolysis, with a resulting increase in protein serine/threonine phosphorylation caused either by DG-stimulated PKC or via Ca^{2+}-calmodulin-regulated kinases. Elevated intracellular Ca^{2+} might also play a role in increasing platelet protein tyrosine phosphorylation. These PAR_1-mediated actions of thrombin are illustrated in Fig. 5.

II. Thromboxane A_2 (TP Receptor)

The ability of arachidonic acid liberated by the action of $cPLA_2$ to stimulate platelet activation is mediated through metabolic conversion to the endoperoxides, prostaglandin G_2, prostaglandin H_2, and TXA_2. Endoperoxides and TXA_2 activate the thromboxane TP receptor with TXA_2 having the higher potency; TXA_2 is the major metabolite of arachidonic acid in the platelet. The TP receptor is a seven transmembrane protein present at a density of 2000 sites/platelet. It is coupled to G_q stimulating breakdown of phosphoinositides, to G_{i2} inhibiting adenylyl cyclase and to G_{12} and G_{13} (for a review see HALUSHKA et al. 1997).

The physiological importance of the TP receptor is demonstrated by the effectiveness of aspirin in reducing the rate of thrombotic disease (see Chap. 7). A wide range of potent and highly selective antagonists have therefore been developed for the TP receptor in an attempt to inhibit thromboxane action while preserving the vasodilatory and antiplatelet action of prostacyclin. TXA_2 synthase inhibitors have also been developed since they increase formation of prostacyclin from prostaglandin H_2, (the so-called "endoperoxide shunt"). Endoperoxides activate the TP receptor, however, limiting the antiplatelet effect of this class of drug. The combination of a PG synthase inhibitor and TP receptor antagonist has been shown to be synergistic. Molecules able to mediate both actions have been developed (e.g. Ridogrel), and have undergone limited clinical testing but with disappointing results (see CATELLA-LAWSON and FITZGERALD 1997).

III. ADP Receptors

ADP was the first agonist to be proposed as a physiological regulator of platelets; yet the surface receptors for ADP which underlie activation and their mechanism of action remains controversial (for review see GACHET et al. 1997). There are several factors that explain this. The activity of purine nucleotides is influenced markedly by interconversion through the action of extracellular enzymes such as ectonucleotidases. Further, there are at least two and almost certainly three subtypes of receptor on the platelet surface. These include the $P2X_1$ receptor which contains an intrinsic ion channel (see above), the $P2_T$ receptor which is expressed only on platelet and is activated by ADP (ATP is an antagonist) and a third receptor that may be $P2Y_1$. Until recently, there has been a lack of selective pharmacological agents to distinguish between these receptors, although the recent development of selective and potent $P2_T$ receptor antagonists has been a major advance. The $P2_T$ receptor has not been cloned and this has also hampered its characterization, in particular its mechanism of signalling.

The selective, potent $P2_T$ receptor antagonists developed by Astra are structural analogues of ATP itself. This series is represented by ARL 66096 and ARL 67085, both of which have nanomolar affinities at the $P2_T$ receptor and micromolar affinities or less at other purinoceptors. The clinical importance of the $P2_T$ receptor is indicated by the efficacy of these antagonists in arterial thrombosis models in which they have been demonstrated to have a greater effect against thrombus formation than bleeding time (HUMPHRIES et al. 1995). This selectivity represents an advantage over some other forms of anti-thrombotic treatment.

There are two ADP receptor antagonists currently in the clinic, clopidogrel and ticlopidine (both are thienopyridines). There is uncertainty regarding their mode of action as both are metabolized in vivo to active forms which, as yet, have not been identified. The nature of the P2 purinoceptor underlying their action is also not established. Ticlopidine causes a number of serious adverse effects making clopidigrel the preferred therapeutic agent. A recent randomized, blind trial of clopidigrel versus aspirin involving 19185 patients at risk of ischaemic events demonstrated that clopidogrel is slightly more effective than medium-dose aspirin in reducing risk (CAPRIE study 1996). It was also concluded that the overall safety profile of clopidogrel is comparable to that of medium-dose aspirin. Its slightly greater efficacy, however, may be offset by a higher cost.

It is not clear whether the inhibitory action of the ARL series of antagonists and the thienopyridines is through the same receptor. Neither antagonist blocks shape change suggesting selectivity within the population of platelet P2 receptors. The thienopyridines partially inhibit binding of [^{33}P]2MeSADP and have no effect against elevation of [Ca^{2+}]i, further emphasizing their selectivity within the population of platelet P2 receptors (GACHET et al. 1997).

A further area of uncertainty in regard to platelet ADP receptors is the signal transduction pathways underlying activation. Although the $P2_T$ receptor inhibits adenylyl cyclase, this is not enough to trigger platelet activation, at least on its own. ADP does stimulate entry of Ca^{2+} following depletion of intracellular stores but there are conflicting reports on the ability of ADP to stimulate metabolism of phosphoinositides in platelets. This uncertainty makes it unclear whether mobilisation of Ca^{2+} is mediated by IP3 or another messenger. ADP also stimulates Ca^{2+} entry through a receptor operated channel, $P2X_1$. The latter appears to be of little importance as the selective $P2X_1$-agonist $\alpha\beta$MeATP does not induce functional responses and platelet activation is maintained following removal of Ca^{2+}.

Although the platelet ADP receptor represents one of the most important targets for the development of new antithrombotics, research is hampered by the uncertainty concerning the number of subtypes of receptors on the platelet and their mode of action. Significant advances in these two areas are expected in the near future through cloning of the $P2_T$ receptor and development of more selective ligands.

Platelets also express receptors for adenosine which inhibit activation of platelets through elevation of cAMP. The physiological significance of this is likely to be small because of the modest size of the effect, but it may complicate interpretation of experiments involving the study of ADP and other nucleotides.

IV. 5-Hydroxytryptamine ($5HT_{2A}$ Receptor)

Platelet dense granules contain the highest levels of 5-HT of any cell in the body and release their contents at sites of damage to the vasculature. However, 5-HT is a very weak agonist giving rise to shape change and weak primary aggregation through activation of the phosphoinositide pathway. The ability of 5-HT to synergise with other stimuli may be of greater significance physiologically, although an even more important role may be to limit blood flow in damaged vessels through vasoconstriction.

V. Vasopressin (V_1 Receptor)

Vasopressin is a weak platelet agonist stimulating activation through the seven transmembrane V_1 receptor coupled to the phosphoinositide second messenger pathway. The weak action of vasopressin is explained by the low density of receptors on the surface, estimated as 95/platelet. Platelet activation by vasopressin is of little physiological relevance not only because of the weak nature of its action but also because the concentration of the peptide usually does not change at the site of injury.

VI. Platelet Activating Factor (PAF)

PAF (acetylglycerylether phosphorylcholine) takes its name from the fact that it is a very potent and powerful activator of rabbit platelets producing strong activation at concentrations of 1 n*M* or less. However, it is approximately two orders of magnitude less potent on human platelets and produces only weak aggregation. Despite its name, PAF is unlikely to have a physiological role in regulating platelet function because it is present only in very low levels within platelets and at sites of damage to the vasculature. There is little evidence for a role of PAF in human cardiovascular diseases (NATHAN and DENIZOT 1995).

There is only one subtype of PAF receptor. It has seven transmembrane regions and couples to the G_q family of G proteins stimulating phosphoinositide metabolism. Estimates of the number of receptors in humans vary widely with lower estimates in the order of 160 sites/platelet, comparable to estimated numbers of other G protein-coupled receptors and consistent with its weak activity.

VII. Adrenaline (α_2-Adrenoceptor)

Platelet aggregation induced by adrenaline is distinct from that by other agonists because it is not accompanied by shape change or elevation of Ca^{2+}. The seven transmembrane receptor which mediates activation, the α_2-adrenoceptor, is different from other receptors on the platelet surface in that it couples only to the G_i family and not to the G_q family. On the face of it, this could explain the inability of adrenaline to raise Ca^{2+}, because of the absence of α_q-dependent activation of PLC. On the other hand, adrenaline does release G protein $\beta\gamma$-subunits which, as discussed above, have been shown to activate PLCβ isoforms. The explanation for this paradox is not known. It could reflect compartmentalization, too low a concentration of liberated $\beta\gamma$-subunits or the fact that in vivo the action of $\beta\gamma$ requires co-release of α-subunits from the G_q family. The mechanism of platelet aggregation induced by adrenaline is not established. It is not mediated through inhibition of adenylyl cyclase but might occur through activation of PI 3-kinase (by G protein $\beta\gamma$-subunits) leading to PIP3-dependent activation of PKC.

Physiologically, the importance of the action of adrenaline on platelets is likely to be limited. Adrenaline is a weak agonist, consistent with the low number of α_2-adrenoceptors (estimated at 300/platelet). Adrenaline, however, is able to undergo synergistic interactions with other platelet agonists, notably ADP, an action that may be facilitated through the different signalling pathways used by these agonists.

Platelets also express β-adrenoceptors which are unlikely to have physiological significance given their very low density.

VIII. Prostacyclin (IP Receptor)

Prostacyclin (prostaglandin I_2) inhibits platelet aggregation. It is a major metabolite of arachidonic acid produced downstream of cyclooxygenase activity from the endoperoxide prostaglandin H_2. It is not formed within platelets but is made within endothelial cells lining the vessel wall from prostaglandin H_2 liberated by the activated platelet. Prostacyclin, together with nitric oxide, is thought to play an important role in maintaining a nonthrombogenic barrier between circulating platelets and the blood vessel wall at the site of damage.

The prostacyclin receptor (IP receptor) is coupled to the G protein subunit G_s giving rise to activation of adenylyl cyclase and elevation of cAMP. It is the major inhibitory G_s-coupled receptor on the platelet surface with a density of approximately 2700 sites/platelet. It is activated by other prostanoids, including PGE_1 and PGD_2, which are formed in small amounts in activated platelets but their affinity for the IP receptor is lower.

IX. Other Seven Transmembrane Receptors

Platelets undergo activation by the lipid phosphates lysophosphatidic acid and sphingosine 1-phosphate. Both compounds are released following platelet activation suggesting that they may have physiological roles in haemostasis (Yatomi et al. 1995, 1997). There is some evidence that they mediate their actions through a common cell surface receptor (Yatomi et al. 1997).

There are several reports that endothelin weakly potentiates primary aggregation but inhibits secondary aggregation, although it has no effect on its own (e.g. Dockrell et al. 1996). There is evidence for the presence of both ET_A and ET_B receptors on the platelet surface mediating these effects. The small size of these effects suggests that they are likely to have little physiological significance.

E. Tyrosine Kinase-Linked Receptors

I. Collagen

Collagen has a major role in supporting adhesion of platelets to the subendothelium (see Sect. A.I) and it has long been recognized that the integrin GPIa-IIa plays a pivotal role in this event (Sixma et al. 1995, 1997). The action of collagen is distinct from that of other adhesion proteins in that it stimulates significant activation of the phosphoinositide pathway in platelet suspensions giving rise to aggregation and secretion (Watson et al. 1985).

Moreover, evidence has continued to emerge since the 1980s that activation of this pathway by collagen is mediated through a receptor other than GPIa-IIa.

Activation of platelets in suspension by collagen, unlike that by G protein receptor-coupled agonists, is preceded by a lag of 15–40 s and is blocked completely by inhibitors of tyrosine kinases. The molecular basis of this was revealed when collagen was found to stimulate tyrosine phosphorylation of PLCγ2 (BLAKE et al. 1994b; DANIEL et al. 1994). The only other cell surface receptor found to stimulate a similar level of tyrosine phosphorylation of the phospholipase was the low affinity receptor for immune complexes, FcγRIIA (BLAKE et al. 1994b). This led the group of Watson to propose that collagen signals through the same pathway as that used by FcγRIIA and other immune (or antigen) receptors (BLAKE et al. 1994b). This family of receptors signal through sequential activation of two tyrosine kinases, a member of the src family and a member of the syk family (WEISS and LITTMAN 1994). The src family kinase phosphorylates a motif found in all antigen receptors or their associated chains known as an immunoreceptor tyrosine-based activation motif (ITAM). Phosphorylation on the two conserved tyrosines in the ITAM enables binding of the tandem SH2 domains in the tyrosine kinase syk leading to autophosphorylation and activation. In turn, syk phosphorylates a number of target proteins leading to tyrosine phosphorylation of PLCγ2 (it is not known whether this is a direct effect or whether it is mediated by a downstream kinase).

Further evidence that collagen signals through the same pathway as that used by immune receptors came from the demonstration that the tandem SH2 domains of syk, expressed as a GST fusion protein associate with a tyrosine phosphorylated protein of 14 kDa in lysates from cells stimulated by collagen (GIBBINS et al. 1996). This protein was identified as the Fc receptor chain (FcR γ-chain) by immunoblotting. The FcR γ-chain (which should not be confused with FcγRIIA) has one ITAM and is essential for signalling by several Fc receptors, including the mast cell receptor FceRI. An essential role for the FcR γ-chain and syk in signalling by collagen was confirmed by the absence of aggregation and dense granule secretion to collagen in platelets from mice genetically engineered to lack either protein (POOLE et al. 1997). In contrast, responses to the G protein receptor-coupled agonist thrombin were unaltered.

This work identifies collagen as the first example of a non-immune receptor stimulus to signal through a pathway that was previously thought to be used only by immune receptors. Collagen therefore activates platelets through a unique pathway during haemostasis, since FcγRIIA does not play a major role in this process. This mechanism has several important physiological and pathological implications. The different mechanism of signalling by collagen enables synergistic interactions with other stimuli as exemplified by the remarkable synergy that occurs between collagen and thromboxanes in the activation of platelets (NAKANO et al. 1989). This also raises the possibility that disorders which result from impairment in signalling by immune receptors in other haematopoietic cells such as immunodeficiencies and leukaemias may have associated mild bleeding problems because of a loss of response of platelets to collagen. The relatively mild phenotype of disorders associated with a loss of response to collagen can be explained by the large number of other agonists which support activation of platelets during haemostasis and by the fact that the importance of activation by collagen varies according to the vessel and conditions.

Increasing evidence suggests that the collagen receptor which underlies activation of the above pathway is GPVI. The group of Okuma provided the first lines of evidence that GPVI is the collagen signalling receptor (SUGIYAMA et al. 1987; MOROI et al. 1989). Okuma and coworkers identified several individuals, all Japanese, who presented with mild bleeding problems that could be traced to a deficiency or lack of GPVI. The antiserum from one patient with autoimmune thrombocytopenia recognizes GPVI and this has been used to confirm the absence of this glycoprotein in the other patients (SUGIYAMA et al. 1987; MOROI et al. 1989). Crosslinking GPVI with $f(ab)_2$ fragments from this antiserum mimics the pattern of platelet activation by collagen, including stimulation of tyrosine phosphorylation of FcR γ-chain, syk and PLCγ2 (GIBBINS et al. 1997; ICHINOHE et al. 1995). Moreover, collagen is unable to stimulate phosphorylation of these proteins in GPVI-deficient patients (ICHINOHE et al. 1997).

Given the importance of GPVI in the action of collagen, it is important to reconsider the role played by the integrin GPIa-IIa. The group of SIXMA and others has generated considerable evidence for an essential role of GPIa-IIa in supporting adhesion to collagen under flow conditions, strongly suggesting that this is the first point of attachment between collagen and the platelet surface (for reviews see SIXMA et al. 1995, 1997; MOROI and JUNG 1997). This serves to bring the collagen molecule into the vicinity of other receptors on the platelet surface including GPVI, a set of events that has been termed the two-site, two-step model of collagen receptor signalling (SANTORO et al. 1991). Variations of this model can be envisaged such that initial contact of platelets with collagen is mediated through binding to either receptor, the interaction with the other receptor reinforcing attachment (MOROI et al. 1996). Either model can account for the mild bleeding disorders in patients who have a low level of expression or absence of GPIa-IIa or GPVI. More severe cases of bleeding associated with a low level of expression of GPIa-IIa seem likely to be due to additional factors (for review see Watson and Gibbins 1998). There is no information on signalling events linked to activation of GPIa-IIa and this is an important area for investigation.

Work carried out by the group of Barnes and colleagues has shown that the interaction between collagen and its receptors, GPIa-IIa and GPVI, is mediated by distinct epitopes in the adhesion molecule. The first evidence for this came from separation of the actions of collagen on adhesion and aggregation by the generation of cyanogen bromide fragments (MORTON et al. 1989). Later, the same group showed that simple, triple-helical collagen-related peptides based on a glycine-proline-hydroxyproline repeat motif were able to

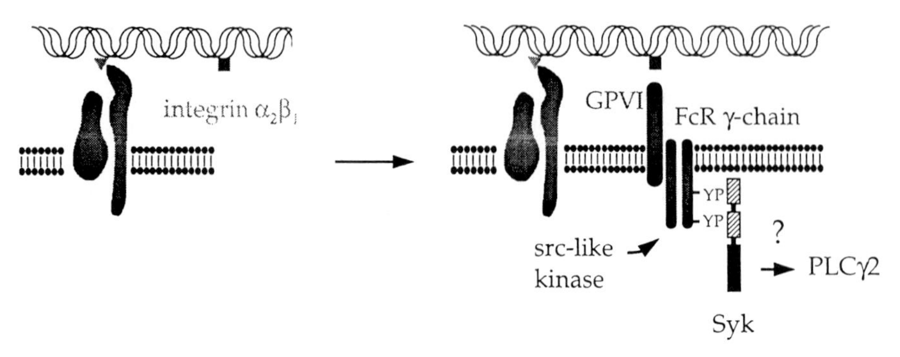

Fig. 6. Two-site, two-step model of platelet activation by collagen. The initial interaction is via the integrin $\alpha_2\beta_1$ which brings the collagen molecule into the vicinity of the second receptor, GPVI. Crosslinking of GPVI leads to tyrosine phosphorylation of the Fc receptor γ chain on the two conserved tyrosines in the ITAM by a src-like kinase, recruitment and phosphorylation of the tyrosine kinase syk and tyrosine phosphorylation and activation of phospholipase Cγ2

mimic platelet activation by collagen but were unable to support adhesion to the integrin GPIa-IIa (MORTON et al. 1995; ASSELIN et al. 1997). These observations suggest it is possible to develop antagonists which inhibit either adhesion or activation.

A model of the mechanism of platelet activation by collagen is shown in Fig. 6. The importance of GPVI identifies this glycoprotein as an important new target for development of antithrombotics, especially because collagen is a *primary* stimulus of platelets, acting upstream of the generation of thromboxanes (see Sect. H).

II. FcγRIIA

Platelets express a single Fc receptor, the 40 kDa FcγRIIA which is present at between 500 and 1500 copies per platelet and exists in two major polymorphic forms. Crosslinking of FcγRIIA leads to platelet activation through the phosphoinositide pathway (ANDERSON and ANDERSON 1990), and this is regulated through the same pathway as that used by collagen (BLAKE et al. 1994a; YANAGA et al. 1995). FcγRIIA is activated by clustering which can occur in vivo through formation of immune complexes or generation of certain antiplatelet antibodies. Antibodies to certain proteins have a much greater tendency than others to induce activation through the latter route, notably those raised against CD9. The FcγRIIA plays an important role in the pathophysiology of immune-mediated thrombocytopenia and thrombosis (ANDERSON et al. 1995). Examples of disorders in which it is implicated include heparin-induced thrombocytopenia, bacterial sepsis-associated thrombocytopenia and the antiphospholipid syndrome. A full disussion of this is beyond the scope of this review but clearly FcγRIIA is an important target for the prevention of such disorders.

III. Thrombopoietin

Thrombopoietin is essential for the growth and development of the platelet precursor cell, the megakaryocyte. Mice deficient in this cytokine have only 15% of the normal level of platelets in the circulation. The level of free thrombopoietin in the circulation is inversely related to the platelet density: it binds to the platelet with high affinity and this binding serves to lower its concentration as the platelet density increases.

Thrombopoietin does not have a direct effect on platelets on its own but potentiates activation by a wide range of agonists including thrombin, collagen and ADP (CHEN et al. 1995; RODRIGUEZ-LINARES and WATSON 1996). This effect is most striking at threshold concentrations of agonist where shape change or weak aggregation is converted into full aggregation. Thrombopoietin causes only a small increase in sensitivity to higher concentrations of agonists. The action of thrombopoietin is mediated through potentiation of PLC activity (RODRIGUEZ-LINARES and WATSON 1996), although other events may also play a role. The physiological significance of this action of thrombopoietin may be to increase the sensitivity of platelets during thrombocytopenia thereby helping to compensate for the reduction in cell number. Correspondingly, a high level of thrombopoietin may be a risk factor for thrombotic disease although the receptor is not a practical target for antithrombotics because of its role in megakaryocyte growth and differentiation.

F. Adhesion Receptors

Although the focus of this chapter is on surface receptors and their signalling pathways, it is no longer possible to separate receptor-mediated signals from those initiated by platelet adhesion. It is now recognized that adhesion receptors in many cell types give rise to intracellular signals controlling responses such as gene regulation, cytoskeletal rearrangement, kinase cascades, etc. Platelets express an abundance of proteins involved in adhesion although relatively little is known of their ability to generate intracellular signals with the exception of GPIIb-IIIa. This section will describe briefly the major adhesion receptors on the platelet surface and their signalling pathways.

I. Integrins

Integrins are heterodimers made up of one α and one β chain (see example in Fig. 7; for review see HYNES 1992). They are widely distributed and mediate cell-matrix and cell-cell interactions. Twelve α chains and seven β chains have

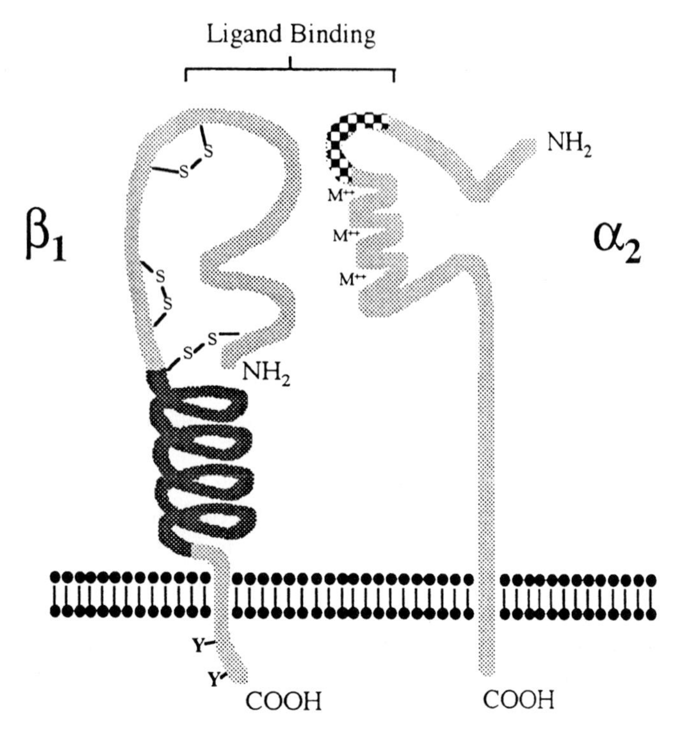

Fig. 7. Schematic representation of the integrin $\alpha_2\beta_1$. The darker region of the β subunit indicates the location of four internally folded cysteine repeat units. The relative location of tyrosine residues on the cytoplasmic tail of the β subunit are marked with Y. The shaded region of the α subunit indicates the relative position of the I-domain. M^{++} denotes the three putative divalent cation domains of the α subunit. The region thought to be important for ligand binding is enclosed by the bracket

been described in humans with the majority having molecular masses of between 100 and 150 kDa and short cytoplasmic sequences. Many α and β chains bind to several partners, although some associate only with one chain. Certain α/β combinations interact with several adhesion molecules (e.g., GPIIb-IIIa), while others bind only a single ligand (e.g., GPIa-IIa). In addition to supporting adhesion, integrins generate intracellular signals influencing responses such as cell growth, morphology and movement.

1. GPIIb-IIIa

The integrin GPIIb-IIIa is the major integrin on the platelet surface. It is present at a density of 80,000 copies/cell, representing 1%–2% of total cell protein, and is also found in membranes of the surface canalicular system and α-granules, and can be expressed from these pools following activation. The α chain, GPIIb, is expressed only in platelets whereas GPIIIa ($\beta3$)

is present in many other cells and combines with several α-subunits including α_v in platelets.

GPIIb-IIIa is the major protein supporting platelet-platelet interaction, i.e. aggregation, in activated cells. Aggregation is mediated by binding of fibrinogen (or vWF) to activated GPIIb-IIIa complexes on adjacent cells. Unstimulated platelets do not bind soluble fibrinogen as GPIIb-IIIa needs to undergo a conformational change leading to exposure of the RGD binding site, the site of interaction within the adhesion molecule; this process is known as "inside-out signalling". Other adhesion molecules containing the amino acid sequence RGD include fibronectin, vitronectin and thrombospondin, which can also bind to activated GPIIb-IIIa. The final event which regulates exposure of this site has not been identified. Binding of fibrinogen to GPIIb-IIIa generates intracellular signals, known as 'outside-in signalling', which support several responses including platelet spreading on a monolayer, the size of the aggregate in platelet suspensions and clot retraction; little is known about the signalling events which are directly linked to activation of the integrin.

There is strong evidence that the cytoplasmic tails of the integrin are essential for both inside-out and outside-in signalling (for review see SHATTIL et al. 1997). However, the cytoplasmic tails of GPIIb and GPIIIa are short, only 20 and 47 amino acids in length respectively, and provide few clues as to their regulation since they do not possess intrinsic enzymatic activity or recognized signalling motifs. There is mounting evidence for a role of the cytoskeleton in the regulation of the integrin, although the identity of the proteins which interact with the cytoplasmic tails is not known (for discussion of proteins implicated in binding to the integrin the reader is referred to the recent review by SHATTIL et al. 1997).

The biochemical events which signal activation of GPIIb-IIIa are not defined, although it seems likely that it is regulated by multiple pathways. The phosphoinositide-derived messengers Ca^{2+} and PKC are able to trigger activation independently of each other (WALKER and WATSON 1993), and there is evidence for undefined roles of the small molecular weight G protein RhoA, tyrosine kinases and PI 3-kinase in activation. The action of PI 3-kinase may be mediated by activation of novel forms of PKC (ZHANG et al. 1996). It should be borne in mind, however, that evidence for involvement of a number of these events is based on the use of pharmacological inhibitors which may exert other actions or induce intracellular changes which have little physiological counterpart. In addition to affinity changes in GPIIb-IIIa, clustering of the integrin can also bring an increase in ligand binding through an increase in "avidity", and this may be regulated by outside-in signalling (for discussion see SHATTIL et al. 1997).

Many downstream signalling responses linked to activation of GPIIb-IIIa have been documented. Aggregation is associated with a net increase in protein tyrosine phosphorylation (FERRELL and MARTIN 1989), although a limited number of proteins undergo dephosphorylation (for review see WATSON 1997).

The tyrosine kinase syk is thought to play an early role in mediating the increase in tyrosine phosphorylation, with focal adhesion kinase and src involved in later events (SHATTIL et al. 1997). Activation of GPIIb-IIIa is associated with migration of a number of proteins to the cytoskeleton, notably those which have undergone tyrosine phosphorylation. A significant number of these proteins concentrate at sites, known as focal contacts, where GPIIb-IIIa links the cytoskeleton to the bound fibrinogen molecule. Despite the emphasis on the role of tyrosine phosphorylation in signalling by GPIIb-IIIa, a remarkable number of other proteins such as lipid kinases, serine-threonine kinases, adaptors, protein phosphatases, and small molecular weight G proteins have been identified in focal adhesions and are implicated in signalling by the integrin (see SHATTIL et al. 1997). Further work is required to establish the role of these proteins in downstream events such as actin polymerization, cytoskeletal reorganization, generation of filopodia, and platelet spreading.

2. Other Platelet Integrins

The vitronectin receptor ($\alpha_v\beta_3$) is a relatively minor platelet integrin present at a level of between 1000 and 1400 sites/platelet. It binds a variety of RGD-containing adhesive proteins such as vitronectin, fibronectin, vWf and thrombospondin. GPIc-IIa ($\alpha_5\beta_1$) and $\alpha_6\beta_1$ are minor integrins on the platelet surface supporting adhesion to fibronectin and laminin, respectively. The other major integrin on the platelet surface, the collagen receptor GPIa-IIa ($\alpha_2\beta_1$), is discussed above. It is present at a density of between 1000 and 2000 copies/platelet. There is little information on the signalling events mediated by binding of ligands to these integrin receptors.

II. GP-IX-V (vWf Receptor)

The other major receptor for adhesion molecules on the platelet surface is the GPlb-IX-V complex. It is present at a density of 25,000 copies/cell and its major function is to bind vWf supporting adhesion at high shear (see Sect. A.I). The nature of the interaction between vWf and the platelet surface has been elucidated and recently reviewed (RUGGERI 1997). Briefly, platelet GPIb binds to the vWf A1 domain through an interaction characterized by fast association, thereby helping platelets attach to the vessel wall at high rates of flow, with attachment being favoured by the multimeric nature of vWf. The interaction is also characterized, however, by rapid dissociation, so platelets continue to move in the direction of flow. Slowing the velocity of flow allows other platelet receptors to interact with adhesion proteins exposed in the subendothelial matrix such as collagen and fibrinogen. In contrast to binding to vWf, adhesion to collagen and fibrinogen is instantaneous, thereby stopping the movement of cells. Binding to vWf generates intracellular signals leading to exposure of the RGD binding site on GPIIb-IIIa (DE MARCO et al. 1985; IKEDA et al. 1991; PETERSON et al. 1987). This reinforces binding of the platelet surface to vWf, via the carboxyl terminal C1 domain of the adhesion molecule, and also supports binding to other adhesion molecules. Evidence for this model can be shown experimentally using only purified vWf under conditions where the action of released agents is blocked. Under this condition, thrombus formation by wWf is slow and is observed only at high shear (SAVAGE et al. 1996); in vivo, however, adhesion to vWf via GPlb is reinforced by binding to other adhesion molecules and by release of mediators such as ADP.

The importance of the interaction of GPIb-IX-V is shown by the bleeding disorder, Bernard-Soulier syndrome, which, in general, is caused by the absence or a low level of expression of GPlb-IX-V. Individual mutations that lead to the disorder have been found in GPIb and GPIX but not GPV (CLEMETSON 1997). GPIb-IX-V is also a receptor for thrombin, although the significance of this interaction is uncertain.

Only in the last few years, has it been realized that binding of vWf generates intracellular signals, but little additional information is available beyond that. This is largely due to the fact that vWf is unable to bind to the GPIb-IX-V receptor complex when added to platelets in solution. Experimentally this can be overcome by use of the bacterial glycopeptide ristocetin or the snake protein botrocetin, although the physiological relevance of such experiments is uncertain. The ability of vWf to interact with GPIb under high flow conditions may reflect the ability of shear forces to act via the cytoskeleton to induce a conformational change in the complex (CLEMETSON 1997). There is evidence for involvement of non-receptor tyrosine kinases (ODA et al. 1995; RAZDAN et al. 1993) and Ca^{2+} (CHOW et al. 1992) in signalling by GPIb-IX-V. It is important further to understand the events mediated by the receptor as these may represent novel targets for therapeutic intervention.

III. PADGEM (P-Selectin)

The α-granule membrane protein, P-selectin (also known as GMP-140, PADGEM and CD62), is expressed on the surface of the activated platelet at a density of up to 13,000 sites/platelet following α-granule secretion. It supports adhesion of neutrophils and monocytes to activated platelets through binding to sugar residues. This binding localizes the cells at the site of injury and enables monocytes to remove activated platelets from the circulation. There are no reports of intracellular signalling by binding of P-selectin to its ligand.

IV. PECAM-1

Platelet endothelial cell adhesion molecule-1 (PECAM-1; also known as CD31) is the only member of the cell adhesion molecule superfamily of immunoglobulin-like proteins expressed on the platelet surface. It has six extracellular immunoglobulin-like homology units and undergoes homophilic interaction (i.e. it binds to itself), and this may play a minor role in platelet aggregation. It is also implicated in adhesion to monocytes and granulocytes and may bind to viruses supporting their entry into the cell. CD31 is present at a density of approximately 5000 copies/platelet.

Recently, evidence has been presented to suggest that CD31 is involved in intracellular signalling. It becomes tyrosine phosphorylated in mast cells following activation of the FceRI receptor (SAGAWA et al. 1997), possibly on a residue in a motif closely related to the ITAM (described above) known as an immunoreceptor tyrosine-based inhibitory motif or ITIM. Tyrosine phosphorylation of this motif enables recruitment of the tyrosine phosphatase, SHP-1 or the inositol 5-phosphatase, SHIP, to the membrane. Recruitment of these proteins has been shown to inhibit signals generated by immune receptors (e.g. ONO et al. 1997).

V. GPIV

GPIV (also known as CD36 or GPIIIb) is a major platelet glycoprotein unrelated to other proteins on the cell surface. It is present at a level of 12,000 copies/platelet. It is known to play a role in supporting attachment and spreading of endothelial and melanoma cells, and to generate intracellular signals in monocytes. It is a receptor for thrombospondin in platelets and this interaction may play a role in the stabilization of fibrinogen binding and development of irreversible aggregation. It is also a receptor for collagen. The physiological significance of GPIV in platelets is uncertain. It is absent in about 10% of the Japanese population yet this defect is not associated with recognized haemostatic problems.

GPIV is potentially involved in the generation of intracellular signals in platelets since it is associated with three members of the src family of tyrosine kinases under resting conditions: $pp60^{fyn}$, $pp62^{yes}$ and $pp54/58^{lyn}$ (Huang et al. 1991).

G. Clinical Settings for Antiplatelet Drugs

Platelets play a key role in normal haemostasis and defects lead to a characteristic bleeding tendency. However, in the modern industrial world thrombosis, particularly arterial thrombosis, has become a much more significant problem than bleeding. Platelets play a pivotal role in arterial thrombosis which occurs under conditions of high flow. Clinically, antiplatelet agents can be used to treat arterial disease or as primary or secondary preventive therapy.

I. The Platelet as a Target in Thrombotic Disease

That aspirin has an important role in the treatment of acute myocardial infarction was established by the ISIS-2 Trial (SECOND INTERNATIONAL STUDY OF INFARCT SURVIVAL 1988). Patients were randomized to intravenous streptokinase, oral aspirin, both or neither. Aspirin reduced vascular deaths by 23%

and aspirin plus streptokinase was significantly better than either agent alone. The RISC Group (Research Group on Instability in Coronary Artery Disease 1990) randomized 796 men with unstable angina (or non-Q-wave infarction) to treatment with a daily dose of 75 mg aspirin, intravenous heparin, both or neither. They concluded that aspirin reduces the risk of myocardial infarction by 50% at 3 months. Two recent large trials of the early use of aspirin in ischaemic stroke have shown a small but definite benefit. One death is prevented for every 100 patients treated during the first few weeks (ISTST, International Stroke Trial Collaborative Group 1997; CAST, Chinese Acute Stroke Trial Collaborative Group 1997).

The first report of the Antiplatelet Trialists' Collaboration (1994a) made the case for secondary prevention of occlusive vascular disease. They showed that antiplatelet therapy reduces the incidence of arterial vascular events (non-fatal myocardial infarctions, non-fatal strokes, or vascular deaths) by a quarter. They recommended that it should be considered for almost all patients with suspected acute myocardial infarction, unstable angina, stroke, transient ischaemic attack, arterial bypass surgery, or angioplasty. The relative risk reduction was similar in different patient groups so the largest absolute benefit was in those at highest risk. The trialists also concluded that there is no clear evidence that antiplatelet therapy is indicated for routine use in primary prevention in subjects at low risk of occlusive vascular events. This conclusion is supported by two studies which have addressed the use of aspirin in healthy individuals. In a trial in British male doctors (Peto et al. 1988), 6 years of treatment with 500 mg aspirin a day did not produce a significant effect compared with placebo. In the study of American doctors (Steering Committee of Physicians' Health Study Research Group 1989), 325 mg aspirin on alternate days reduced myocardial infarction by 44%. There was, however, a slight increase in haemorrhagic stroke and no reduction in mortality from all cardiovascular causes. The second report of the Antiplatelet Trialists' Collaboration (1994b) made the case for use of antiplatelet agents in the maintenance of vascular graft or arterial patency.

It had previously been supposed that antiplatelet therapy did not influence venous thromboembolism. The third report of the Antiplatelet Trialists' Collaboration (1994c) showed that venous thromboembolism was reduced by about a quarter (that is approximately the same as the reduction in arterial disease). Subcutaneous heparin (unfractionated and low molecular weight) and oral anticoagulants are more effective, however, and current antiplatelet drugs have a limited role in this context. In a patient with thrombophilia, in whom the benefits of oral anticoagulants do not outweigh the risks, then low dose aspirin is sometimes given to take advantage of its small risk reduction. It does not, however, replace more vigorous prophylaxis for high-risk situations.

In a recent pooled analysis of three randomized trials, aspirin was shown to reduce the risk of stroke in patients with atrial fibrillation by 21% (95% confidence interval 0%–38%). This compares with a 68% reduction with full-dose oral anticoagulant therapy (Atrial Fibrillations Investigators 1997).

Warfarin is preferable to aspirin except in patients with a low risk of stroke or a high risk of haemorrhage. Aspirin, therefore, is still used in patients with atrial fibrillation and no other risk factors for stroke (a previous transient ischaemic attack, or stroke, hypertension, diabetes, or heart failure).

II. Overview of Currently Used Agents

The reader is referred to the preceding chapter for a full discussion of this.

III. Possible Settings for New Antiplatelet Agents

Aspirin is extremely cheap and has proven efficacy. Can new antiplatelet drugs improve on aspirin and, if so, in which clinical settings would they be used? In general terms, we would like a new drug to be either more effective or to have less side effects or both, and the improvement would have to be sufficient to justify an increase in cost. If new drugs were shown to be an improvement, then they could replace aspirin as the treatment of choice for the conditions discussed above-myocardial infarction, unstable angina, ischaemic stroke and transient ischaemic attacks – and following arterial grafts and angioplasty. Of more interest is whether efficacy could reach levels where new agents could challenge warfarin as prophylaxis for recurrent venous thromboembolism or as prophylaxis for stroke in patients with atrial fibrillation and no additional risk factors. The most intriguing question is whether efficacy and safety could reach such levels as to justify primary prophylaxis.

The difficulty for pharmaceutical companies in improving on aspirin is exemplified by the CAPRIE STUDY (1996). This was a huge randomized trial involving 19,185 patients which compared long-term aspirin with clopidogrel in patients with atherosclerotic vascular disease. Clopidogrel reduced ischaemic stroke, myocardial infarction or vascular death by 8.7% compared with aspirin. If aspirin reduces this endpoint by a quarter (see above) then clopidogrel is estimated to reduce it by a third. These figures imply that if 200 such patients are treated for a year, aspirin will prevent four events and clopidogrel five. On this basis, it seems unlikely that clopidogrel will replace the very cheap aspirin, except in patients who cannot tolerate aspirin or perhaps in those that can be shown to be non-responsive to it. There is a suggestion that a third of patients are non-responsive to aspirin (GROTEMEYER et al. 1993; BUCHANAN and BRISTER 1995), a finding which does not seem to be dose-dependent. If this is so, then identification of these patients and treatment with alternative antiplatelet therapy may improve clinical outcomes.

H. New Targets for Drug Development

Since all new potential targets must be judged against aspirin, inhibitors of GPIIb-IIIa and, to a more limited extent, clopidigrel, the question that must be considered is *what advantages over and above these three classes must new drugs possess?* High on the list of considerations are whether the new drugs will have fewer side-effects or whether they are able treat the disease rather than the symptoms. For example, the mechanism of action of GPIIb-IIIa antagonists is to block activation of GPIIb-IIIa complexes rather than to inhibit the mechanisms which give rise to this response. Targeting cell surface receptors and signalling pathways which underlie activation of the integrin will not only prevent its activation but may also block other events which contribute to the haemostatic process, such as activation of additional platelets and participation of other cells including sub-endothelial smooth muscle cells and macrophages.

It is also important to emphasize the dramatic difference in efficacy between aspirin, clopidogrel and GPIIb-IIIa antagonists and to consider requirements of new inhibitors. Aspirin and clopidigrel target agonists, namely thromboxanes and ADP, respectively, that serve as positive feedback stimuli to activation by many stimuli, although neither inhibitor is able to block activation by intermediate concentrations of strong agonists such as thrombin or combinations of agonists (e.g., adrenaline and collagen). In contrast, GPIIb-IIIa antagonists block aggregation to all agonists accounting for their powerful antithrombotic action and necessitating care in selection of dose. All three of these drugs, however, target events which reinforce the haemostatic (or thrombotic) process rather than those which cause initiation. It is likely that targeting individual agonists will only succeed for those with major roles such as thrombin, or which play very early roles, such as collagen and vWf. As an alternative, it may be possible to target common intracellular signalling pathways, as these are used by all or subgroups of agonists.

I. Receptors as Targets

Cell surface receptors continue to represent the major targets for development of new drugs primarily because of access and the ability to test specificity. It is also easier to set up screens to identify new lead compounds. The challenge of targeting drugs at intracellular pathways is substantially greater (see Sect. H.II).

1. ADP Receptors

Although ADP receptor antagonists are already in the clinic, it is clear that a great deal of additional work needs to be done on ADP receptors. Their potential is illustrated by the efficacy of clopidogrel, which is comparable to that of aspirin, yet the active metabolite of clopidogrel has not been identified. The potential importance of the ADP receptor is also emphasized by the efficacy of the ARL series of $P2_T$ antagonists in arterial models of thrombosis relative to effects on bleeding times. It is vital to know whether these two classes of ADP receptor antagonists are working through the same receptor (if not, then a mixed inhibitor at both sites is likely to be even more effective) and to identify all of the ADP receptor subtypes on the surface on the platelet.

The reason for the disparity between the concentration of the $P2_T$ receptor antagonist required to prolong bleeding time and to inhibit arterial thrombosis should also be determined. There is a wealth of evidence demonstrating that ADP plays a critical role in the spreading of platelets and that ADP is essential for many of the initial events in haemostasis in vivo (WEISS et al. 1986). ADP may be released from non-activated platelets enabling it to synergise with signals generated by adhesion molecules. In this regard, ADP appears to play an early role in platelet activation, an effect which may contribute importantly to the efficacy of ADP antagonists. Agents which prevent the release of ADP or its ability to activate all of its receptors on the surface may offer many advantages over available inhibitors.

2. Adhesion Receptors

One of the most promising areas for the development of new drugs is interference with platelet adhesion, notably under high shear conditions. Shear stress is low in veins, high in arteries, and can be very high indeed in stenotic arteries. As discussed in Sect. A, at low shear, platelets can adhere to a type I collagen surface in the absence of vWF and agonist-induced aggregation can occur as a result of fibrinogen binding to GP IIb-IIIa. At high shear, platelet adhesion to type I collagen is mediated primarily by surface-bound vWF interacting with GP Ib-IX-V. Platelet aggregation can be initiated by high shear stress which induces soluble vWF to bind to GP Ib-IX-V, generating intracellular signals which cause a functional change in GP IIb-IIIa; binding of vWF to GPIIb-IIIa then mediates aggregation. Inhibition of cyclooxygenase by aspirin has little effect on shear stress-induced platelet aggregation, which may account for the lack of potency of aspirin in some cases of arterial thrombosis. It is possible that pharmacological interference with the binding of vWF to GP Ib-IX-V may result in reduced arterial thrombosis while preserving haemostasis at sites where shear stress is lower. Similarly, interference of platelet binding to collagen may result in reduced thrombosis in areas of lower shear force. Drugs aimed at these interactions may have less side effects than those aimed at the GPIIb-IIIa complex.

GPIb-IX-V and GPVI are expressed only on platelets, suggesting that antagonists are likely to have only limited side-effects; in contrast, GPIa-IIa is a ubiquitous receptor for collagen so does not represent a suitable target. The relative importance of these receptors in most cases of thrombotic disease is not clear, since thrombosis is not usually caused by rupture of the vessel wall. However, the high shear forces generated in stenosed arteries may give rise to activation of GPIb, whereas collagen fibres may become exposed following the rupture of atherosclerotic plaques. It seems reasonable to assume that antagonists of these receptors may only be effective in this subpopulation of patients with thrombotic disease.

3. Thrombin Receptor

Because of the profound impact of thrombin on platelet function, the receptor responsible for thrombin-mediated platelet activation has long been considered a target for drug development. Targeting the receptor for thrombin, rather than the enzyme itself, offers the theoretical advantage of blocking the direct actions of thrombin on the platelet, as well as on other target tissues, without affecting the ability of thrombin to catalyze the final stages of the clotting process so necessary for hemostasis. Thus, receptor inhibitors would, in principle, provide a safer alternative to thrombin-directed enzyme inhibitors, not only for the control of platelet function, but also for blocking the potentially adverse actions of thrombin on other target tissues, such as the endothelial and smooth muscle elements of the vasculature, blood monocytes and neuronal dendrites. The receptor antagonist might also provide a wider therapeutic index than antagonists of the GPIIb-IIIa integrin receptor. Another potential advantage of blocking the receptor is that its activation by proteases other than thrombin would also be inhibited.

In view of the theoretical advantages outlined in the paragraph above for receptor antagonists, news of the cloning of the G protein-coupled receptor for thrombin (PAR_1) kindled great excitement. The search for a suitable receptor antagonist appeared to be especially promising, since a peptide motif composed of only four to five amino acids, with few prominent side-chain pharmacophores, was found to be fully capable of activating the thrombin receptor in platelets (Scarborough et al. 1992; Vassallo et al. 1992, Hui et al. 1992; Chao et al. 1992) and in other targets such as the vasculature (Muramatsu et al. 1992; Simonet et al. 1992). It was determined in NMR studies that the receptor activating peptide, SFLLR-NH2 appears to adopt a cyclic conformation in a membrane-mimicking environment and it was observed that cyclic peptides having Phe and Arg pharmacophores were biologically active in a smooth muscle assay system (Matsoukas et al. 1996). Peptide structure-activity studies for a receptor antagonist quickly identified an interesting antagonist, 3-mercaptopropionyl-Phe–cyclohexyl-Ala-cyclohexyl-Ala-Arg-Leu-Pro-Asn-Asp-Lys-amide, that was able to block activation of platelets both by thrombin and by the PAR_1 receptor-activating peptide, Ser-Phe-Leu-Leu-Arg-Asn-Pro (Coughlin and Scarborough 1992; Seiler et al. 1995). Surprisingly, however, this 'antagonist' was found to be a potent contractile agonist in smooth muscle preparations (Dimaio et al. 1995) so it was clear that the search for a PAR_1 selective antagonist was unlikely to be straightforward. As outlined in the paragraphs below, there are a number of factors, apart from the structure of the receptor activating peptides, that must be considered for the development of PAR antagonists.

One major question that must be answered is whether the platelet G protein-coupled thrombin receptor (PAR_1, or possibly also PAR_3) represents the *only* cell surface constituent activated by thrombin to cause platelet aggregation. Data obtained using PAR_1-blocking antibodies at low thrombin concentrations suggest that binding sites other than PAR_1/PAR_3 may also be involved in thrombin-regulated platelet function (summarized by JAMIESON 1997). The cellular activities of thrombin domains outside the catalytic site take on added significance in this regard (BAR-SHAVIT 1986a, 1986b; GLENN et al. 1988), since 'receptors' for these peptide sequences would be very distinct from the PARs and since such receptors may work synergistically with PAR_1 not only in cultured cells (HOLLENBERG et al. 1996) but also in human platelets (see KINLOUGH-RATHBONE et al. 1993b).

A second issue that must be considered for PAR antagonist development concerns the heterogeneity of the protease-activated receptor family, for which three members have been cloned (Sect. D.I). It is now clear that the pharmacophores on the PAR_2-activating peptide (e.g. Ser-Leu-Ile-Gly-Arg-Leu-NH_2) do not interact with either PAR_1 or PAR_3. The PAR_1 activating peptides, however, possess pharmacophores capable of activating PAR_2 (BLACKHART et al. 1996; HOLLENBERG et al. 1997). Thus, any search for PAR_1 selective antagonists (e.g. BERNATOWICZ et al. 1996) must also evaluate carefully the actions of such compounds in PAR_2 receptor systems. The development of ligand binding assays based on radiolabelled peptide receptor agonists or antagonists (AHN et al. 1997; BERNATOWICZ et al. 1996) will undoubtedly facilitate screening of a number of compound libraries, but each lead substance identified will still have to be evaluated in all three protease-activated receptor systems. In this regard, PAR_3 represents something of an enigma since it is the first of the PARs that does not appear to be amenable to activation by small peptides derived from the proteolytically revealed neoligand (e.g. TFRGA). This situation offers both an advantage and a disadvantage. The disadvantage is that PAR_3, which like PAR_1, may also regulate human platelet function, may not interact with its neoligand in the same way as PAR_1 and PAR_2, precluding the design of agonists/antagonists based on the TFRGA sequence. TFRGA would nonetheless be expected to activate PAR_1 and PAR_2 with low potency. The potential advantage of the insensitivity of PARs to activation by short peptides is that this property may make it easier to develop PAR_1 selective reagents (e.g. BLACKHART et al. 1996; HOLLENBERG et al. 1997).

A final issue to be faced in developing thrombin receptor antagonists (focused primarily on PAR_1), relates to the widespread distribution of PAR_1 in many tissues apart from the platelet. Consequently, any putative PAR_1 antagonist may not be platelet-selective and may affect many different tissues. This lack of selectivity could be an advantage, since the control of thrombin-regulated endothelial cell function and smooth muscle proliferation may be desirable at sites of inflammation or vascular damage (e.g. in the setting of atherosclerotic plaque formation or angioplasty). On the other hand, the differential cellular dynamics of the thrombin receptor pose a challenge in the design of tissue-selective PAR_1 inhibitors. For instance, platelet thrombin receptor activation leads to inactivation and shedding of the receptor, as opposed to the replenishment of cell surface receptors from intracellular stores such as occurs in endothelial cells (MOLINO et al. 1997; WOOLKALIS et al. 1995). Thus, certain cells with large intracellular stores of PAR_1 may be able to compensate rapidly for the action of a PAR_1 antagonist. Further, in the platelet, any PAR_1 antagonist would have to access cryptic receptors (about 40%) that might not be accessible until after platelet activation by other stimuli (MOLINO et al. 1997).

One common potential drawback to the design of PAR_1 antagonists relates to the relatively low affinity of the receptor-targeted peptides (micromolar), which must compete for the tethered neoligand. Despite this apparent drawback, it has been possible to synthesize peptides with very high receptor affinity with K_DS in the nanomolar range (BERNATOWICZ et al. 1996). Thus, one can be optimistic about the rational design of high affinity PAR_1 antagonists.

Taken together, the information summarized above can be used to outline the desired properties of a "hypothetical" PAR_1 antagonist, presuming that PAR_1 antagonists would have a major effect on controlling platelet function. One key for the design of such platelet-selective inhibitors (to be discussed below: Sect. H.II.1) relates to the unique absence in platelets (as opposed to nucleated cells) of the ability to resynthesize and replenish the complement of cell surface receptors. What one could aim for is a compound that would not only abrogate the tethered ligand activation mechanism, but might also, as a mild "partial agonist", be able to promote receptor shedding without activating other platelet processes such as aggregation. In that manner, the platelet would be disarmed selectively and irreversibly, but other PAR_1 bearing cells would be spared irreversible inhibition. Although challenging, both from a practical and theoretical point of view, the search for such "process-selective" antagonists represents an intriguing opportunity.

II. Signalling Pathways as Targets

A major limitation in targeting surface receptors as antiplatelet agents results from the redundancy of different receptors in mediating platelet activation. Most of these stimuli induce activation through a limited number of intracellular pathways, however, so the intracellular events may represent important targets for development of antiplatelet drugs that will be effective against several groups of stimuli. An important example is the family of adhesion molecules which are activated by clustering and signal through a route fundamentally different from that of G protein coupled receptors. This group of receptors may be activated pathologically through surface contact, perhaps at sites of atherosclerotic plaques, giving rise to activation. Inhibitors of the signalling events activated by the adhesion receptors have the potential to be powerful antithrombotic drugs without having the widespread bleeding problems of GPIIb-IIIa antagonists since G protein-mediated events will remain intact.

There will be major difficulties, however, in any attempt to target intracellular pathways that regulate only the platelet. The central role of signalling in nearly all cells within the body suggests that inhibitors will have widespread side effects unless they can be targeted against a process or pathway that is unique to the platelet.

Activation of platelets in the circulation may occur through clustering of proteins on the cell surface at sites of reduced blood flow, damaged atherosclerotic plaques or in the vicinity of prosthetic devices (valves, synthetic shunts). This interaction gives rise to kinase-dependent cascades that lead to activation, and thus one may highlight the kinases as possible targets. The lack of information on the mechanisms that give rise to activation for many of the platelet adhesion receptors, however, is a limitation in identifying appropriate targets. Platelet activation may also occur in vivo through exposure to high shear forces, osmotic stress and generation of free radicals. It is possible that such stimuli may prime platelets for activation through phosphorylation of $cPLA_2$ by sap kinases, pointing to this enzyme family as a potential target for therapeutic intervention.

A major challenge is proving the selectivity of the drug under study. This is a particular problem for inhibitors of kinases and phosphatases, the families of which are estimated to be made up of 1000 members or more. It is simply not practical to test each compound against all of these intracellular proteins.

1. Protein-Protein Interfaces

As outlined in Sects. B–F of this chapter, platelet signalling, triggered by a host of cell surface receptors, proceeds via a sequence of non-catalytic protein-protein interactions (e.g. receptor-G protein; G protein-effector; phosphotyrosine-SH2 domain) that ultimately result in the activation of a key signalling enzyme or enzymes (e.g. adenylyl cyclase, phospholipase C, tyrosine kinase). Remarkably, the protein domains responsible for the non-catalytic protein-protein interaction comprise relatively short peptide sequences, involving as few as four to six amino acids (e.g. phosphotyrosyl/SH2 domain interactions: PAWSON 1995; or receptor G-protein interfaces: BOURNE 1997). Many of these protein-protein interactions may turn out to be receptor- or pathway-specific. For example, the residues on PAR_1 that regulate G_i and G_q are probably distinct. These surfaces on PAR_1 may also, in principle, differ from other receptor sequences in terms of their interactions with G_i and G_q. This possibility raises the issue that peptidomimetics, designed to selectively abrogate PAR_1 G-protein interactions, might prove to be of use as PAR_1 antagonists. In this regard, the predominance of G_z in platelets might also be important. The same might be said for the development of agents that could selectively abrogate interactions between SH2 domains and signal-related phosphotyrosyl motifs that might predominate in the platelet.

2. Enzyme Targets

As pointed out above, many of the enzymes involved in signal amplification (including isoforms of adenylyl cyclase, PKC and PLC) are found in a wide variety of tissues apart from platelets. Thus, designing a targeted enzyme inhibitor to affect platelets selectively presents a considerable challenge. One superclass of enzymes that merits particular attention is represented by the multiple forms of non-receptor tyrosine kinases present in platelets. Over seven years ago, it was demonstrated that a relatively non-selective tyrosine kinase inhibitor of the tyrphostin family (Tyrphostin 47 or AG213) was able to inhibit both thrombin-induced aggregation and 5-HT release; the tyrphostin also blocked thrombin-mediated phosphorylation of the 43 kDa platelet kinase C substrate (presumably pleckstrin) (RENDU et al. 1990). Since, at the concentration of AG213 used in that study (25 μM), the tyrphostins are poor inhibitors of src family kinases (Hollenberg, unpublished observations), it would be interesting to identify the rate-limiting tyrosine kinase affected by AG213 in platelets. Tyrphostins in general are relatively nontoxic and such compounds already have been used for studies that show therapeutic promise in a number of settings (e.g. shock, leukaemias). Accordingly, a search for a platelet-targeted tyrosine kinase inhibitor of this family (summarized by LEVITZKI and GAZIT 1995) would appear to be warranted. In principle, it might prove possible to selectively inhibit tyrosine kinases (e.g. FAK-kinase versus syk) that may be involved in distinct platelet responses (e.g. secretion, shape change and aggregation). Inhibitors of other platelet signalling enzymes (SAPKA, PI 3-kinase γ) that may effect only a subset of responses could also prove to be valuable. Platelet selectivity would still represent a major challenge because of the ubiquity of signal pathway enzymes in a variety of tissues.

III. Development of New Drugs: Aspirin's Legacy

In view of the difficulty of identifying signal pathway targets that are unique for the platelet, in contrast with other tissues, it is possible to suggest an alternative strategy. This approach would take advantage of the inability of the platelet to replenish enzymes that are inactivated or otherwise degraded. For example, the reversible acetylation and inactivation of platelet cyclooxygenase (Coxl) by acetylsalicylic acid renders the platelet permanently disabled whereas in other tissues Coxl can regenerate. We propose that transition state inhibitors that covalently inactivate platelet signal pathway enzymes or permenantly block surface receptors would be suitable candidates as platelet-selective therapeutic agents.

Acknowledgements. Work in the laboratory of S.P.W. is supported by the British Heart Foundation and Wellcome Trust. S.P.W. is a Royal Society University Research Fellow. Work in the laboratory of M.D.H. is supported by the Medical Research Council of Canada, the Heart and Stroke Foundation of Alberta, and a National

Science and Engineering/National Research Council of Canada Partnership grant in conjunction with BioChem Therapeutic. We are grateful to Dr. David Severson for helpful discussions.

References

Abrams CS, Zhang J, Downes CP, Tang XW, Zhao W, Rittenhouse SE (1996) Phosphopleckstrin inhibits $G\beta\gamma$-activable platelet phosphatidylinositol-4,5-bisphosphate 3-kinase. J Biol Chem 271:25192–25197

Ahn HS, Foster C, Boykow G, Arik L, Smith, Torhan A, Hesk D et al (1997) Binding of a thrombin receptor tethered ligand analogue to human platelet thrombin receptor. Mol Pharmacol 51:350–356

Anderson GP, Anderson CL (1990) Signal transduction by the platelet Fc receptor. Blood 76:1165–1172

Anderson CL, Chacko GW, Osborne JM, Brandt JT (1995) The Fc receptor for immunoglobulin G ($Fc\gamma RII$) on human platelets. Semin Thromb Hemostas 21:1–9

Antiplatelet Trialists' Collaboration (1994a) Collaborative overview of randomised trials of antiplatelet therapy-I: Prevention of death, myocardial infarction, and stroke by prolonged antiplatelet therapy in various categories of patients. BMJ 308:81–106

Antiplatelet Trialists' Collaboration (1994b) Collaborative overview of randomised trials of antiplatelet therapy-II: Maintenance of vascular graft or arterial patency by antiplatelet therapy. BMJ 308:159–168.

Antiplatelet Trialists' Collaboration (1994c) Collaborative overview of randomised trials of antiplatelet therapy-III: Reduction in venous thrombosis and pulmonary embolism by antiplatelet prophylaxis among surgical and medical patients. BMJ 308:235–246

Asselin J, Gibbins JM, Achison M, Lee YH, Morton, LF, Farndale RW et al (1997) A collagen-like peptide stimulates tyrosine phosphorylation of syk and phospholipase $C\gamma 2$ in platelets independent of the integrin $\alpha_2\beta_1$. Blood 89:1235–1242

Atrial Fibrillation Investigators (1997) The efficacy of aspirin in patients with atrial fibrillation. Analysis of pooled data from 3 randomised trials. Arch Intern Med 157:1237–1240

Bar-Shavit R, Eldor A, Vlodavsky I (1989) Binding of thrombin to subendothelial extracellular matrix. Protection and expression of functional properties. J Clin Invest 84:1096–1104

Bar-Shavit R, Kahn AJ, Mann KG, Wilner GD (1986a) Identification of a thrombin sequence with growth factor activity on macrophages. Proc Natl Acad Sci USA 83:976–980

Bar-Shavit R, Hruska KA, Kahn AJ, Wilner GD (1986b) Hormone-like activity of human thrombin. Ann N Y Acad Sci 485:335–348

Bernatowicz MS, Klimas CE, Hartl KS, Peluso M, Allegretto NJ, Seiler SM (1996) Development of potent thrombin receptor antagonist peptides. J Med Chem 39:4879–4887

Blackhart BD, Emilsson K, Nguyen D, Teng W, Martelli AJ, Nystedt S et al (1996) Ligand cross-reactivity within the protease-activated receptor family. J Biol Chem 271:16466–16471

Blake RA, Asselin J, Walker T, Watson SP (1994a) $Fc\gamma$ receptor II stimulated formation of inositol phosphates in human platelets is blocked by tyrosine kinase inhibitors and associated with tyrosine phosphorylation of the receptor. FEBS Lett 342:15–18

Blake RA, Schieven GL, Watson SP (1994b) Collagen stimulates tyrosine phosphorylation of phospholipase C-$\gamma 2$ but not phospholipase C-$\gamma 1$ in human platelets. FEBS Letts 353:212–216

Boeynaems JM, Dumont JE (1977) The two-step model of ligand-receptor interaction. Mol Cell Endocrinol 7:33–47
Borsch-Haubold AG, Kramer RM, Watson SP (1995) Cytosolic phospholipase A_2 is phosphorylated in collagen-and thrombin-stimulated human platelets independent of protein kinase C and mitogen-activated protein kinase. J Biol Chem 270:25885–25892
Borsch-Haubold AG, Kramer RM, Watson SP (1996) Inhibition of mitogen-activated protein kinase does not impair primary activation of human platelets. Biochem J 318:207–212
Borsch-Haubold AG, Kramer RM, Watson SP (1997) Phosphorylation and activation of cytosolic phospholipase A_2 by 38-kDa mitogen-activated protein kinase in collagen-stimulated human platelets. Eur J Biochem 245:751–759
Bourne HR (1997) How receptors talk to trimeric G proteins. Curr Opin Cell Biol 9:134–142
Brass LF, Vassallo RR, Jr, Belmonte E, Ahuja M, Cichowski K, Hoxie JA (1992) Structure and function of the human platelet thrombin receptor. Studies using monoclonal antibodies against a defined domain within the receptor N terminus. J Biol Chem 267:13795–13798
Brass LF, Hoxie JA, Manning DR (1993) Signaling through G proteins and G protein-coupled receptors during platelet activation. Thromb Haemost 70:217–223
Buchanan MR, Brister SJ (1995) Individual variation in the effects of ASA on platelet function: implications for the use of ASA clinically. Can J Cardiol 11:221–227
Butt E, Walter U (1997) Platelet phosphodiesterases. Handb Exp Pharmacol 126: 219–231
CAPRIE Steering Committee (1996) A randomised, blinded, trial of clopidogrel versus aspirin in patients at risk of ischaemic events (CAPRIE). Lancet 348:1329–1339
CAST (Chinese Acute Stroke Trial) Collaborative Group (1997) CAST: randomised placebo-controlled trial of early aspirin use in 20,000 patients with acute ischaemic stroke. Lancet 349:1641–1649
Catella-Lawson F, Fitzgerald GA (1997) Therapeutic aspects of platelet pharmacology. Handb Exp Pharmacol 126:719–736
Chao BH, Kalkunte S, Maraganore JM, Stone SR (1992) Essential groups in synthetic agonist peptides for activation of the platelet thrombin receptor. Biochemistry 31:6175–6178
Chen JL, Herceg-Harjacek L, Groopman JE, Grabarek J (1995) Regulation of platelet activation in vitro by the c-Mpl ligand, thrombopoietin. Blood 86:4054–4062
Chow TW, Hellums JD, Moake JL, Kroll MH (1992) Shear stress-induced von Willebrand factor binding to platelet glycoprotein Ib initiates calcium influx associated with aggregation. Blood 80:113–120
Clemetson KJ (1997) Platelet GPIb-V-IX complex. Thromb Haemost 78:266–270
Connolly TM, Condra C, Feng DM, Cook N, Stranieri MT, Reilly CF et al (1994) Species variability in platelet and other cellular responsiveness to thrombin receptor-derived peptides. Thromb Haemost 72:627–633
Connolly AJ, Ishihara H, Kahn ML, Farese RV Jr, Coughlin SR (1996) Role of the thrombin receptor in development and evidence for a second receptor. Nature 381:516–519
Coughlin SR, Scarborough RM (1992a) Recombinant thrombin receptor and related pharmaceuticals. International Patent W0 92/14750
Coughlin SR, Vu TK, Hung DT, Wheaton VI (1992b) Characterization of a functional thrombin receptor. Issues and opportunities. J Clin Invest 89:351–355
Cuatrecasas P, Hollenberg MD (1976) Membrane receptors and hormone action. Adv Protein Chem 30:251–451
Dachary-Prigent J, Freyssinet I, Pasquet JM, Carron JC, Nurden AT (1993) Annexin V as a probe of aminophospholipid exposure and platelet membrane vesiculation: a flow cytometry study showing a role for free sulphydryl groups. Blood 81: 2554–2565

Daniel JL, Dangelmaier C, Smith JB (1994) Evidence for a role for tyrosine phosphorylation of phospholipase Cγ2 in collagen-induced platelet cytosolic calcium mobilization. Biochem J 302:617–622

De Haen C (1976) The non-stoichiometric floating receptor model for hormone-sensitive adenylate cyclase. J Theor Biol 58:383–400

De Marco L, Mazzucato M, Masotti A, Ruggeri ZM (1994) Localization and characterization of an α-thrombin-binding site on platelet glycoprotein Ibα. J Biol Chem 269:6478–6484

De Marco L, Girolami A, Zimmerman TS, Ruggeri ZM (1985) Interaction of purified type IIB von Willebrand factor with the platelet membrane glycoprotein Ib induces fibrinogen binding to the glycoprotein IIb/IIIa complex and initiates aggregation. Proc Natl Acad Sci USA 82:7424–7428

DiMaio J, Winocour P, Leblond L, Saifeddine M, Laniyonu A, Hollenberg MD (1995) Thrombin inhibitors and thrombin receptor agonists/antagonists. In: Giardina D, Piergentili A, and Pigini M (eds) Perspectives in Receptor Research Elsevier, Amsterdam, Vol. 24, pp 271–289

Dockrell ME, Webb DJ, Williams BC (1996) Activation of the endothelin B receptor causes a dose-dependent accumulation of cyclic GMP in human platelets. Blood Coagul Fibrinolysis 7:178–180

Eckhart W, Hutchinson MA, Hunter T (1979) An activity phosphorylating tyrosine in polyoma T antigen immunoprecipitates. Cell 18:925–933

Ferrell JE Jr, Martin GS (1989) Tyrosine-specific protein phosphorylation is regulated by glycoprotein IIb-IIIa in platelets. Proc Natl Acad Sci USA 86:2234–2238

Fields TA, Casey PJ (1995) Phosphorylation of Gz α by protein kinase C blocks interaction with the βγ complex. J Biol Chem 270:23119–23125

Fields TA, Casey PJ (1997) Signalling functions and biochemical properties of pertussis toxinresistant G proteins. Biochem J 321:561–571

Gachet C, Hechler B, Leon C, Vial C, Leray P, Ohlmann P et al (1997) Activation of ADP receptors and platelet function. Thromb Haemost 78:271–275

Gibbins J, Asselin J, Farndale R, Barnes M, Law CL, Watson SP (1996) Tyrosine phosphorylation of the Fc receptor γ-chain in collagen-stimulated platelets. J Biol Chem 271:18095–18099

Gibbins JM, Okuma M, Farndale R, Barnes M, Watson SP (1997) Glycoprotein VI is the collagen receptor in platelets which underlie tyrosine phosphorylation of the Fc recetor γ-chain. FEBS Lett 413:255–259

Gilman AG (1995) Nobel Lecture. G proteins and regulation of adenylyl cyclase. Biosci Rep 15:65–97

Glenn KC, Frost GH, Bergmann JS, Carney DH (1988) Synthetic peptides bind to high-affinity thrombin receptors and modulate thrombin mitogenesis. Pep Res 1:65–73

Grabarek J, Groopman JE, Lyles YR, Jiang S. Bennett L, Zsebo K et al (1994) Human kit ligand (stem cell factor) modulates platelet activation *in vitro.* J Biol Chem 269:21718–21724

Grand RJ, Turnell AS, Grabham PW (1996) Cellular consequences of thrombin-receptor activation. Biochem J 313:353–368

Grotemeyer KH, Scharafinski HW, Husstedt IW (1993) Two-year follow-up of aspirin responder and aspirin non responder. A pilot study including 180 post-stroke patients. Thromb Res 71:397–403

Guderman T, Nurnberg B (1995) Receptors and G proteins as primary components of transmembrane signal transduction. Part 1. G-protein-coupled receptors: Structure and function. J Mol Med 73:51–63

Halushka PV, Pawate S, Martin ML (1997) Thromboxane A_2 and other eicosanoids. Handb Exp Pharm 126:459–482

Harmon JT, Jamieson GA (1986a) Activation of platelets by a-thrombin is a receptor-mediated event. D-phenylalanyl-L-prolyl-L-arginine chloromethyl ketone–thrombin, but not N α-tosyl-L-lysine chloromethyl ketone-thrombin, binds to the high affinity thrombin receptor. J Biol Chem 261:15928–15933

Harmon JT, Jamieson GA (1986b) The glycocalicin portion of platelet glycoprotein Ib expresses both high and moderate affinity receptor sites for thrombin. A soluble radioreceptor assay for the interaction of thrombin with platelets. J Biol Chem 261:13224–13229

Hollenberg MD, Yang SG, Laniyonu AA, Moore GJ, Saifeddine M (1992) Action of thrombin receptor polypeptide in gastric smooth muscle: identification of a core pentapeptide retaining full thrombin-mimetic intrinsic activity. Mol Pharmacol 42:186–191

Hollenberg MD, Saifeddine M, Al-Ani B. Kawabata A (1997) Proteinase-activated receptor: structural requirements for activity, receptor cross-reactivity, and receptor selectivity of receptor-activating peptides. Can J Physiol Pharmacol 75:832–841

Hollenberg MD, Mokashi S, Leblond L, DiMaio J (1996) Synergistic actions of a thrombin-derived synthetic peptide and a thrombin receptor-activating peptide in stimulating fibroblast mitogenesis. J Cell Physiol 169:491–496

Horstrup K, Jablonka B,. Honig-Liedl P, Just M, Kochsiek K, Walter U (1994) Phosphorylation of focal adhesion vasodilator-stimulated phosphoprotein at Ser 157 in intact human platelets correlates with fibrinogen receptor inhibition. Eur J Biochem 225:21–27

Huang MM, Bolen JB, Barnwell JW, Shattil SJ, Brugge J (1991) Membrane glycoprotein IV (CD36) is physically associated with the Fyn, Lyn, and Yes protein-tyrosine kineses in human platelets. Proc Natl Acad Sci USA 88:7844–7848

Hui KY, Jakubowski JA, Wyss VL, Angleton EL (1992) Minimal sequence requirement of thrombin receptor agonist peptide. Biochem Biophys Res Commun 184:790–796

Humphries RG, Robertson MJ, Leff P (1995) A novel series of $P2_T$ purinoceptor antagonists: definition of the role of ADP in arterial thrombosis. Trends Pharmacol Sci 16:179–181

Hynes RO (1992) Integrins: versatility, modulation and signalling in cell adhesion. Cell 69:11–25

Ichinohe T, Takayama H, Ezumi Y, Yanagi S, Yamamura H, Okuma M (1995) Cyclic AMP-insensitive activation of c-Src and Syk protein-tyrosine kinases through platelet membrane glycoprotein VI. J Biol Chem 270:28029–28036

Ichinohe T, Takayama H, Ezumi Y, Arai M, Yamamoto N, Takahasi H et al (1997) Collagen-stimulated activation of Syk but not c-Src is severely compromised in human platelets lacking membrane glycoprotein VI. J Biol Chem 272:63–68

Ikeda Y, Handa M, Kawano K, Kamata T, Murata M, Araki Y et al (1991) The role of von Willebrand factor and fibrinogen in platelet aggregation under varying shear stress. J Clin Invest 87:1234–1240

International Stroke Trial Collaborative Group (1997) The International Stroke Trial (IST): a randomised trial of aspirin, subcutaneous heparin, both, or neither among 19,435 patients with acute ischaemic stroke. Lancet 349:1569–1581

Ishihara H, Connolly AJ, Zeng D, Kahn ML, Zheng YW, Timmons C et al (1997) Protease-activated receptor 3 is a second thrombin receptor in humans. Nature 386:502–506

ISIS-2 (Second International Study of Infarct Survival) Collaborative Group (1988) Randomised trial of intravenous streptokinase, oral aspirin, both, or neither among 17,187 cases of suspected acute myocardial infarction. Lancet 2:349–360

Jamieson GA (1997) Pathophysiology of platelet thrombin receptors. Thromb Haemost 78:242–246

Jap TS, Kwok CF, Wong MC, Chiang H (1988) The effects of in vitro and in vivo exposure to insulin upon prostaglandin E1 stimulation of platelet adenylate cyclase activity in healthy subjects. Diabetes Res 27:39–46

Kinlough-Rathbone RL, Rand ML, Packham MA (1993a) Rabbit and rat platelets do not respond to thrombin receptor peptides that activate human platelets. Blood 82:103–106

Kinlough-Rathbone RL, Perry DW, Guccione MA, Rand ML, Packham MA (1993b) Degranulation of human platelets by the thrombin receptor peptide

SFLLRN: comparison with degranulation by thrombin. Thromb Haemost 70:1019–1023
Koch CA, Anderson D, Moran MF, Ellis C, Pawson T (1991) SH2 and SH3 domains: elements that control interactions of cytoplasmic signaling proteins. Science 252:668–674
Koesling D, Nurnberg B (1997) Platelet G proteins and adenylyl and guanylyl cyclases. Handb Exp Pharm 126:181–218
Kozasa T, Gilman AG (1996) Protein kinase C phosphorylates G_{12} α and inhibits its interaction with G $\beta\gamma$. J Biol Chem 271:12562–12567
Lee SB, Rao AK, Lee KH, Yang X, Bae YS, Rhee SG (1996) Decreased expression of phospholipase C-beta 2 isozyme in human platelets with impaired function. Blood 88:1684–1691
Lemmon MA, Falasca M, Ferguson KM, Schlessinger J (1997) Regulatory recruitment of signalling molecules to the cell membrane by pleckstrin-homology domains. Trends Cell Biol. 7:237–242
Levitzki A, Bar-Sinai A (1991) The regulation of adenylyl cyclase by receptor-operated G proteins. Pharmacol Ther 50:271–283
Levitzki A, Marbach I, Bar-Sinai A (1993) The signal transduction between p–receptors and adenylyl cyclase. Life Sci 52:2093–2100
Levitzki A, Gazit A (1995) Tyrosine kinase inhibition: an approach to drug development. Science 267:1782–788
Lin LL, Wartmann M, Lin A, Knopf JL, Seth A, Davis RJ (1993) $cPLA_2$ is phosphorylated and activated by MAP kinase. Cell 72:269–278
Luttrell LM, Hawes BE, Touhara K, van Biesen T, Koch WJ, Lefkowitz RJ (1995) Effect of cellular expression of Pleckstrin homology domains on Gi-coupled receptor signaling. J Biol Chem 270:12984–12989
MacKenzie AB, Mahaut-Smith MP, Sage SO (1996) Activation of receptor-operated cation channels via $P2X_1$ not $P2_T$ purinoceptors in human platelets. J Biol Chem 271:2879–2881
Matsoukas JM, Panagiotopoulos D, Keramida M, Mavromoustakos T, Yamdagni R, Wu Q et al (1996) Synthesis and contractile activities of cyclic thrombin receptor-derived peptide analogues motif: Importance of Phe Arg relative conformation and the primary amino group for activity. J Med Chem 39:3585–3591
Molino M, Bainton DF, Hoxie JA, Coughlin SR, Brass LF (1997) Thrombin receptors on human platelets. Initial localization and subsequent redistribution during platelet activation. J Biol Chem 272:6011–6017
Moroi M, Jung SM (1997) Platelet receptors for collagen. Thromb Haemost 78:439–444
Moroi M, Jung SM, Okuma M, Shinmyozu K (1989) A patient with platelets deficient in glycoprotein VI that lack both collagen-induced aggregation and adhesion. J Clin Invest 84:1440–1445
Moroi M, Jung SM, Shinmyozu K, Tomiyama Y, Ordinas A, Diaz-Ricart M (1996) Analysis of platelet adhesion to a collagen–coated surface under flow conditions: the involvement of glycoprotein VI in the platelet adhesion. Blood 68:2081–2092
Morton LF, Hargreaves PG, Farndale RW, Young RD, Barnes MJ (1995) Integrin alpha 2 beta 1-independent activation of platelets by simple collagen-like peptides: collagen tertiary (triple-helical) and quaternary (polymeric) structures are sufficient alone for alpha 2 beta 1-independent platelet reactivity. Biochem J 306:337–344
Morton LF, Peachey AR, Barnes MJ (1989) Platelet-reactive sites in collagens type I and type III. Evidence for separate adhesion and aggregatory sites. Biochem J 258:157–163
Muramatsu I, Laniyonu A, Moore GJ, Hollenberg MD (1992) Vascular actions of thrombin receptor peptide. Can J Physiol Pharmacol 70:996–1003
Nakano T, Hanasaki K, Arita H (1989) Possible involvement of cytoskeleton in collagen-stimulated activation of phospholipases in human platelets. J Biol Chem 264:5400–5406

Natarajan S, Riexinger D, Peluso M, Seiler SM (1995) 'Tethered ligand' derived pentapeptide agonists of thrombin receptor: a study of side chain requirements for human platelet activation and GTPase stimulation. Int J Pep Protein Res 45:145–151

Nathan N, Denizot Y (1995) PAF and human cardiovascular disorders. J Lipid Mediat Cell Signal 11:103–104

Nystedt S, Emilsson K, Wahlestedt C, Sundelin J (1994) Molecular cloning of a potential proteinase activated receptor. Proc Natl Acad Sci USA 91:9208–9212

Nystedt S, Larsson AK, Aberg H, Sundelin J (1995) The mouse proteinase-activated receptor-2 cDNA and gene. Molecular cloning and functional expression. J Biol Chem 270:5950–5955

Oda A, Yokoyama K, Murata M, Tokuhira M, Nakamura K, Handa M et al (1995) Protein tyrosine phosporylation in human platelets during shear stress-induced platelet aggregation (SIPA) is regulated by glycoprotein (GP) Ib/Dk as well as GPIIb/IIIa and requires intact cytoskeleton and endogenous ADP. Thromb Haemost 74:736–742

Ono M, Okada H, Bolland S, Yanagi S, Kurosaki T, Ravetch JV (1997) Deletion of SHIP or SHP-1 reveals two distinct pathways for inhibitory signaling. Cell 90:293–301

Park D, Jhon DY, Lee CW, Lee KH, Rhee SG (1993) Activation of phospholipase C isozymes by G protein $\beta\gamma$ subunits. J Biol Chem 268:4573–4576

Pawson T (1995) Protein modules and signalling networks. Nature 373:573–580

Peterson DM, Stathopoulos NA, Giorgio TD, Hellums JD, Moake JL (1987) Shear-induced aggregation requires von Willebrand factor and platelet membrane glycoproteins Ib and IIb-IIIa. Blood 69:625–628

Peto R, Gray R, Collins R, Wheatley K, Hennekens C, Jamrozik K et al (1988) Randomised trial of prophylactic daily aspirin in British male doctors. Br Med J Clin Res Ed 296:313–316

Poole A, Gibbins JM, Turner M, van Vugt M, van de Winkel J, Saito T et al (1997) The Fc receptor γ-chain and the tyrosine kinase Syk are essential for activation of mouse platelets by collagen. EMBO J 16:2333–2341

Presek P, Martinson EA (1997) Platelet protein tyrosine kineses. Handb Exp Pharm 126:263–297

Rasmussen UB, Vouret-Craviari V, Jallat S, Schlesinger Y, Pages G, Pavirani A et al (1991) cDNA cloning and expression of a hamster or α-thrombin receptor coupled to Ca^{2+} mobilization. FEBS Lett 288:123–128

Razdan K, Hellums JD, Kroll MH (1993) Shear stress-induced von Willebrand factor binding to platelets causes the activation of tyrosine kinase(s). Biochem J 302:681–686

Rendu F, Eldor A, Grelac F, Levy-Toledano S, Levitzki A (1990) Tyrosine kinase blockers: new platelet activation inhibitors. Blood Coag Fibrinolysis 1:713–716

RISC Group (1990) Risk of myocardial infarction and death during treatment with low dose aspirin and intravenous heparin in men with unstable coronary artery disease. Lancet 336:827–830

Rodbell M (1995) Nobel Lecture. Signal transduction: evolution of an idea. Biosci Rep 15:117–133

Rodriguez-Linares B, Watson SP (1996) Thrombopoietin potentiates activation of human platelets in association with JAK2 and TYK2 phosphorylation. Biochem J 316:93–98

Ruggeri ZM (1997) Mechanisms initiating platelet thrombus formation. Thromb Haemost 78:611–616

Sabo T. Gurwitz D, Motola L, Brodt P, Barak R, Elhanaty E (1992) Structure-activity studies of the thrombin receptor activating peptide. Biochem Biophys Res Commun 188:604–610

Sagawa K, Swaim W, Zhang J, Unsworth E, Siraganian RP (1997) Aggregation of the high affinity IgE receptor results in the tyrosine phosphorylation of the surface adhesion protein PECAM-1 (CD31). J Biol Chem 272:13412–3418

Santoro SA, Walsh JJ, Staatz WD, Baranski KJ (1991) Distinct determinants on collagen support alpha 2 beta 1 integrin-mediated platelet adhesion and platelet activation. Cell Regul 2:905–913

Savage B,. Saldivar E, Ruggeri ZM (1996) Initiation of platelet adhesion by arrest onto fibrinogen or translocation on von Willebrand factor. Cell 84:289–297

Savarese TM, Fraser CM (1992) In vivo mutagenesis and the search for structure-function relationships among G protein-coupled receptors. Biochem J 283:1–19

Scarborough RM, Naughton MA, Jeng W, Hung DT, Rose J, Vu TK et al (1992) Tethered ligand agonist peptides. Structural requirements for thrombin receptor activation reveal mechanism of proteolytic unmasking of agonist function. J Biol Chem 267:13146–13149

Schlessinger J, Ullrich A (1992) Growth factor signaling by receptor tyrosine kineses. Neuron 9:383–391

Seiler SM, Peluso M, Michel IM, Goldenberg H, Fenton 2d JW, Riexinger D, Natarajan S (1995) Inhibition of thrombin and SFLLR-peptide stimulation of platelet aggregation, phospholipase A_2 and Na+/H+exchange by a thrombin receptor antagonist. Biochem Pharmacol 49:519–528

Severson DL, Hollenberg MD (1997) The plasma membrane as a transducer and amplifier. In Principles of Medical Biology Vol. 7B, Membranes and Cell Signalling. JAI Press, Greenwich Conn. USA, pp. 387–419

Shattil SJ, Gao J, Kashiwagi H (1997) Not just another pretty face: regulation of platelet function at the cytoplasmic face of integrin αIIbβ3. Thomb Haemost 78:220–225.

Simon MI, Strathmann MP, Gautam N (1991) Diversity of G proteins in signal transduction. Science 252:802–808

Simonet S, Bonhomme E, Laubie M, Thurieau C, Fauchere JL, Verbeuren TJ (1992) Venous and arterial endothelial cells respond differently to thrombin and its endogenous receptor agonist. Eur J Pharmacol 216:135–137

Sixma JJ, van Zanten GH, Huizinga KG, van der Plas RM, Verkley M, Wu YP et al (1997) Platelet adhesion to collagen: an update. Thromb Haemost 78:434–438

Sixma JJ, van Zanten GH, Saelman EU, Verkleij M, Lankhof H, Nieuwenhuis HK et al (1995) Platelet adhesion to collagen. Thromb Haemost 74:454–459

Steering Committee of the Physicians' Health Study Research Group (1989) Final report on the aspirin component of the ongoing Physicians' Health Study. N Engl J Med 321:129–135

Strosberg AD (1991) Structure/function relationship of proteins belonging to the family of receptors coupled to GTP-binding proteins. Eur J Biochem 196:1–10

Sugiyama T, Okuma M, Ushikubi F, Sensaki S, Kanaji K, Uchino H (1987) A novel platelet aggregating factor found in a patient with defective collagen-induced platelet aggregation and autoimmune thrombocytopenia. Blood 69:1712–1720

Sunahara RK, Dessauer CW, Gilman AG (1996) Complexity and diversity of mammalian adenylyl cyclases. Annu Rev Pharmacol Toxical 36:461–480

Thomason PA, James SR, Casey PJ, Downes CP (1994) A G-protein $\beta\gamma$-subunit-responsive phosphoinositide 3-kinase activity in human platelet cytosol. J Biol Chem 269:16525–16528

Toker A, Bachelot C, Chen CS, Falck JR, Hartwig JH, Cantley LC et al (1995) Phosphorylation of the platelet p47 phosphoprotein is mediated by the lipid products of phosphoinositide 3-kinase. J Biol Chem 270:29525–29531

Tuy FP, Henry J, Rosenfeld C, Kahn A (1983) High tyrosine kinase activity in normal nonproliferating cells. Nature 305:435–438

van Biesen T, Luttrell LM, Hawes BE, Lefkowitz RJ (1996) Mitogenic signalling via G protein-coupled receptors. Endocr Rev 17:698–714

Vassallo RR Jr, Kieber-Emmons T, Cichowski K, Brass LF (1992) Structure-function relationships in the activation of platelet thrombin receptors by receptor-derived peptides. J Biol Chem 267:6081–6085

Vassbotn FS, Havnen OK, Heldin CH, Holmsen H (1994) Negative feedback regulation of human platelets via autocrine activation of the platelet-derived growth factor α-receptor. J Biol Chem 269:13874–13879

Vu TK, Hung DT, Wheaton VI, Coughlin SR (1991a) Molecular cloning of a functional thrombin receptor reveals a novel proteolytic mechanism of receptor activation. Cell 64:1057–1068

Vu TK, Wheaton VI, Hung DT, Charo I, Coughlin SR (1991b) Domains specifying thrombin-receptor interaction. Nature 353:674–677

Walker TR, Watson SP (1993) Synergy between Ca^{2+} and protein kinase C is the major factor in determining the level of secretion from human platelets. Biochem J 289:277–282

Watson SP (1997) Protein phosphatases in platelet function. Handb Exp Pharm 126:297–325

Watson SP, Reep B, McConnell RT, Lapetina EG (1985) Collagen stimulates [3H] inositol trisphosphate formation in indomethacin-treated human platelets. Biochem J 226:831–837

Watson S, Arkinstall S (1994) The G protein-linked receptors factsbook. Academic Press, New York London Toronto Sydney San Francisco

Watson SP, Gibbins JM (in press) Collagen receptor signalling in platelets: extending the role of the ITAM. Immunol Today 19:260–264

Wedegaertner PB, Wilson PT, Bourne HR (1995) Lipid modifications of trimeric G proteins. J Biol Chem 270:503–506

Weiss A, Littman DR (1994) Signal transduction by lymphocyte antigen receptors. Cell 76:263–274.

Weiss HJ, Turitto VT, Baumgartner HR (1986) Platelet adhesion and thrombus formation on subendothelium in platelets deficient in glycoproteins IIb-IIIa, Ib, and storage granules. Blood 67:322–330

Woolkalis MJ, DeMelfi TM Jr, Blanchard N, Hoxie JA, Brass LF (1995) Regulation of thrombin receptors on human umbilical vein endothelial cells. J Biol Chem 270:9868–9875

Yanaga F, Poole A, Asselin J, Blake R, Schieven GL, Clark EA et al (1995) Syk interacts with tyrosine-phosphorylated proteins in human platelets activated by collagen and Cross-linking of the Fcγ-IIA receptor. Biochem J 311:471–478

Yatomi Y, Yamamura S, Ruan F, Igarashi Y (1997) Sphingosine 1-phosphate induces platelet activation through an extracellular action and shares a platelet surface receptor with lysophosphatidic acid. J Biol Chem 272:5291–5297

Yatomi Y, Ruan F, Hakomori S, Igarashi Y (1995) Sphingosine 1-phosphate: a platelet-activating sphingolipid released from agonist-stimulated human platelets. Blood 86:193–202

Zhang J, Zhang J, Shattil SJ, Cunningham MC, Rittenhouse SE (1996) Phosphoinositide 3-kinase gamma and p85/phosphoinositide 3-kinase in platelets. Relative activation by thrombin receptor or β-phorbol myristate acetate and roles in promoting the ligand-binding function of $\alpha_{IIb}\beta_3$ integrin. J Biol Chem 271:6265–6272

CHAPTER 9

Heparin and Other Indirect Antithrombin Agents

W.R. BELL and T.A. HENNEBRY

A. Introduction

I. Magnitude of the Problem of Intravascular Thrombosis and Thromboembolic Disease

Intravascular thrombotic disease is universal and despite improved therapies, the total burden of disease worldwide is increasing as the population ages and Western type of lifestyle proliferates. In the United States alone, over 150000 venous thromboses and 1.5 million myocardial infarctions occur per year. The majority of these patients require the administration of antithrombotic therapy. Heparin has the distinction of being the oldest antithrombotic and, along with a number of the agents, it is identified as an indirect antithrombotic agent (Table 1) because it inhibits thrombin via the potentiation of an endogenously occurring natural inhibitor.

B. Unfractionated Heparin: The Prototypical Indirect Antithrombin

I. History

Heparin is the oldest and most efficacious anticoagulant available. It is a mucopolysaccharide organic acid which was discovered by Jay McLean, a Johns Hopkins medical student, in 1915 when he was searching for a procoagulant in the liver and hearts of dogs (MCLEAN 1916, 1959). Heparin ranks as one of the most important therapeutic discoveries of all time. There is a wide range of diseases for which heparin is the most effective and often the sole therapeutic agent available. Many modern therapeutic modalities which depend on either instrumentation of the arterial tree or extracorporeal circulation such as cardiopulmonary bypass, hemodialysis and ultrafiltration are uniquely dependent on heparin. It is astonishing that when heparin was discovered, it was evaluated by two of the largest pharmaceutical manufacturers of the time and rejected as useless. It required a further quarter of a century before investigators in Canada and Sweden demonstrated its therapeutic benefit in patients with venous thrombosis and thromboembolic disease (JORPES 1939; MURRAY 1939).

Table 1. Indirect antithrombin agents

Agent	Source	Administration	Half-life	Mechanism of action	Antidote
Heparin	Animal mucosa	Intravenous; continuous intermittent subcutaneeous	70–120 minutes	Antithrombin III -major minor effect on heaparin cofactor II	Protamine sulphate
Pentosans	Wood chippings	intravenous, intramuscular oral (not acutely)	120–180 minutes	antithrombin III heparin cofactor II enhances fibrinolysis	Difficult to reverse if given intramuscularily
Dermatan sulfate	animal intestine and skin	intravenous intramuscular	45 minutes (IV) 300 minutes (IM)	Heparin cofactor II	Unclear; probably protamine sulfate
Sulodexide	porcine intestinal mucosa	intravenous intramuscular, oral	120–300 minutes	heparin cofactor II antithrombin III direct platelet inhibition	unclear
Danaparoid	animal mucosa	intravenous subcutaneous	90 minutes*	antithrombin III uneplained anti Xa effects	unclear
Pentasaccharide sequence	most animal tissue or chemical synthesis	intravenous	45–90 minutes	antithrombin III	experimental use only

*18 hours antifactor Xa effect

II. Source

There are two major sources of heparin available for commercial use today. Despite its name, heparin is obtained from either the lung or intestinal mucosa of animals as both tissues are particularly rich in glycosaminoglycans. The pulmonic tissue source is only obtained from cattle. Heparin from this source tends to have a higher mean molecular weight than that from other sources. Intestinal tissue from bovine, porcine, ovine and caprine animals is the other principal source of heparins including low molecular weight heparin. Since heparin is a biodiverse compound, considerable variability occurs from lot to lot. Following extraction and purification one is left with standard heparin or, the alternative designation, unfractionated heparin (UFH), which has a molecular weight range of 2750–57500 daltons (the majority of molecules being less than 20000 daltons) and a mean molecular weight of 15000 daltons (LAM et al. 1976; ANDERSSON et al. 1976; ENGLEBERT 1977). Lung and intestinal mucosa are the tissues in which heparin is found in the highest concentrations because these are the tissues with the highest concentration of mast cells. Heparin is endogenously synthesized and stored in the secretory granules of mast cells (LINDAHL et al. 1989). Heparin is normally found in human blood albeit at much lower concentrations than when it is employed therapeutically. Heparin quite likely plays a role as a natural antithrombotic agent and potentially as a natural inhibitor of atherosclerosis.

III. Structure

As heparin is a biodiverse compound, both with regard to its molecular weight and animal species of origin, it would be imprecise to say we know its exact structure but we now have a better knowledge and understanding than ever before. When heparin is first extracted from a tissue source it is a large macromolecular proteoglycan in which the disaccharide chains are covalently linked by a peptide matrix. It is subunits of this complex that are referred to as heparin, depolymerization of which yields low molecular weight heparin.

Heparin is a glycosaminoglycan (mucopolysaccharide) which implies it has a backbone consisting of sugar moieties which are highly sulfated either directly or indirectly via amine links. The disaccharide chain is composed of alternating uronic acid (α-L-iduronic acid or β-D-glucuronic acid) and α-D-glucosamine units joined by 1–4 glycosidic linkages. Each disaccharide unit is trisulfated with the result that heparin carries the highest negative charge of any biological substance known to man. Most of the glucosamine units are sulfated at the C2 and C6 positions and a proportion of the iduronic residues are sulfated at C2. Glucuronic acid, when evident, tends not to be sulfated (LINDAHL and AXELSSON 1971; PETITOU et al. 1987; KJELLEN et al. 1992). The structural criteria for heparin have been well characterized (KISS 1975; CHOAY 1989; HOYLAERTS et al. 1984).

Heparins possess no branching in the polysaccharide chain. A straight chain is required if one is to retain anticoagulant properties, as demonstrated for sulfonated polysaccharides with anticoagulant activity. Compounds with identical molecular weight lose most anticoagulant activity if branched and have practically none when globular (VON KAULLA and HUSEMAN 1946). Heparin has been separated based on its binding affinity for antithrombin III (HOOK et al. 1976; OLSON et al. 1981). One-third of commercial heparin has high affinity for antithrombin III. Structurally, the binding is dependent on the presence of a relatively specific pentasaccharide unit in the heparin molecule (Fig. 1) which presents a 3-*O*-sulfate group in a stereospecific fashion that binds to antithrombin III via a highly specific electrostatic bond (LINDAHL et al. 1980; VAN DELDEN et al. 1995). Whether this site is the active site in heparin for its many other biological functions, such as inactivation of factors VIII and IX, binding to heparin cofactor II, stimulation of tissue factor pathway inhibitor (TFPI) release, inactivation of histamine and activation of lipoprotein lipase, remains to be determined.

IV. Mechanism of Action

We know that the mechanism of action of heparin is complex and multifac torial; many of the details remain to be elucidated. It has been known for almost 60 years that the anticoagulant properties of heparin were largely dependent on the presence of an endogenous cofactor (BRINKHOUS et al. 1939). This cofactor has since been isolated and subsequently named antithrombin III (ROSENBERG and DAMUS 1973). The trend today is to call it *antithrombin*, but this may incorrectly imply there is only one antithrombin.

Antithrombin III is a glycoprotein of molecular weight 60000 daltons found normally in the plasma. It contains 432 amino acids which form a globular protein and has a protruding reactive center (STEIN et al. 1990). Its crystalline structure and nucleotide sequence are now known. The gene is found on chromosome 1 and extends over 14 kilobases (OLDS et al. 1993). Antithrombin functions as an inhibitor of serine proteases via a suicide mechanism. Its active site, which contains an arginine to serine peptide bond (position 394–395), acts as substrate for the major serine proteases such as thrombin. When the serine protease complexes with the antithrombin III it is unable to separate, thus forming a stable complex. This activity occurs in normal plasma in a time-dependent fashion.

Heparin chains which contain more than 18 (and ideally more than 21) saccharide residues possess the ability to bind both antithrombin III and thrombin simultaneously acting as an adsorption catalyst. This causes a local preponderance of substrate (antithrombin III) and enzyme (thrombin), thereby increasing the inactivation of the thrombin up to one thousand fold (OLSON and BJORK 1991, 1992; POMERANTZ and OWEN et al. 1978). This is known as the template model but probably only accounts for heparin's inhibition of thrombin via antithrombin III. Antithrombin III may also be

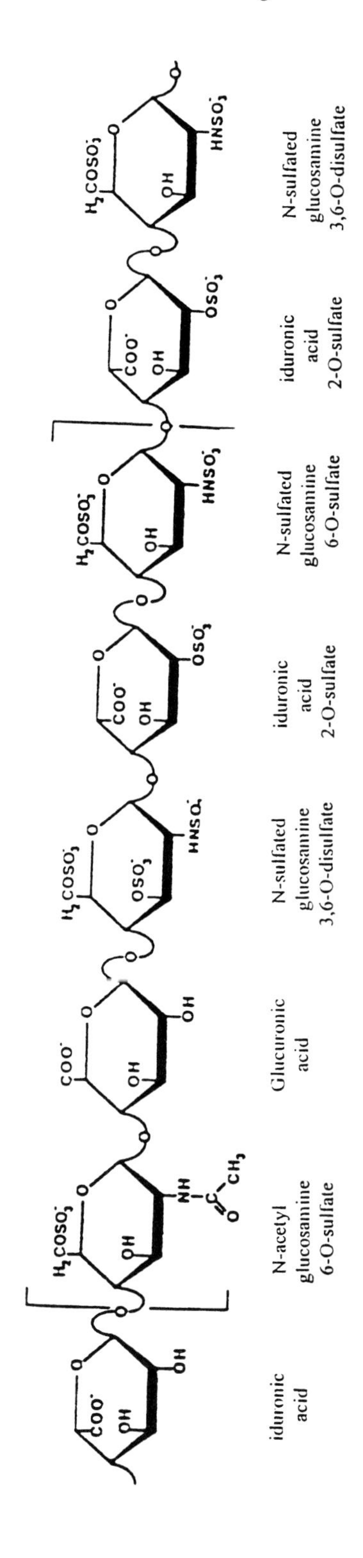

Fig. 1. The five residues within the brackets represent the pentasaccharide sequence

conformationally changed by heparin binding, which increases the availability or accessibility of its active site. The heparin-antithrombin III complex is up to 270 times more effective at inhibiting other serine proteases, especially factors IXa and Xa, than antithrombin III alone (Olson and Bjork 1992). It is probable that both mechanisms occur, the adsorption catalyst (template model) effect being best suited to longer chains of heparin as chains of less than 18 residues cannot form the heparin-antithrombin III-thrombin complex. The conformational change induced in antithrombin III by heparin may better explain the antithrombotic effect of lower molecular weight heparins which tend to inhibit factor Xa more than thrombin. On the other hand, inhibition of factor Xa is not the exclusive function of lower molecular weight heparins as it has been demonstrated that heparin chains containing more than 18 residues, the minimum for the template model of adsorption catalysis, can double the rate of factor Xa inhibition (Danielsson et al. 1986).

A second heparin cofactor, heparin cofactor II, has been identified as a 66000-dalton glycoprotein which shares some amino acid sequences with antithrombin (Tollefsen et al. 1982; Ragg 1986). Its reactive site contains a leucine-serine peptide group in contrast to antithrombin III's arginine-serine group (the latter being the preferred substrate for serine proteases) and it is believed that this difference accounts for heparin cofactor II having inhibitory effects on thrombin but not on factors IXa or Xa (Parker and Tollefsen 1985; Griffith et al. 1985). Heparin cofactor II requires heparin of chain length 26 saccharides or greater and has only one-tenth the affinity for heparin binding that antithrombin III exhibits. Heparin binds to heparin cofactor II in a nonspecific fashion and this induces a conformational change whereby the acidic N-terminus of heparin cofactor II is displaced, exposing a glycosaminoglycan binding site which can then bind thrombin (Van Deerlin and Tollefsen 1991; Rogers et al. 1992). At usual therapeutic doses the role of heparin cofactor II is minimal but it achieves greater significance, as explained later in the chapter, when we discuss dermatan sulfate and other heparinoids.

The unique and variable bioavailability of heparin when assayed by functional assays (when it can appear to have greater than 100% bioavailability) suggests that neither antithrombin III nor heparin cofactor II alone or together can explain all of the observed clinical effects of heparin. One postulated explanation for this greater-than-expected antithrombotic activity is that heparin stimulates the release of an endogenous serpin inhibitor. Human plasma has been shown on occasion to contain an inhibitor of tissue factor-induced coagulation. This inhibitor was originally called extrinsic pathway inhibitor (EPI) but is now known as tissue factor pathway inhibitor (TFPI) and is less commonly referred to as lipoprotein associated coagulation inhibitor (Broze et al. 1987; Rao and Rapaport 1987). TFPI is a 40000-dalton protein which directly inhibits factor Xa and indirectly inhibits the tissue factor–factor VIIa complex (Broze et al. 1987). It is important to note that this is the only mechanism by which heparin can inhibit factor VIIa. Intravenous heparin has been shown to increase blood levels of TFPI up to tenfold (WUN et al. 1988). It also increases the rate of TFPI binding to the factor VIIa–tissue factor complex, which further increases inhibition of the complex (WUN 1992). TFPI release from endothelial cells is not dependent on whether the heparin has high or low affinity for antithrombin III (JESKE et al. 1995). The precise contribution of TFPI to normal hemostatic balance or to thean-tithrombotic effect of heparin remains to be quantified (HOPPENSTEADT et al. 1995).

Heparin has also been shown to inhibit the coagulation system via inhibition of factor V and factor VIII activation, which is primarily a thrombin-induced phenomenon (OFOSU et al. 1987, 1989), which may explain some of the anticoagulant properties of heparin. Similarly it has been shown that heparin reduces generation of thrombin and has a thrombin scavenging effect which is not usually associated with low molecular weight heparin – hence the latter may not be as potent an antithrombotic agent (BEGUIN et al. 1988; ELGUE et al. 1994).

The influence of heparin on platelets is less well understood. The side effect of thrombocytopenia will be dealt with in a later section. It is known that heparin binds in a nonspecific manner to platelet factor 4 (BELL 1988) and that heparin can cause platelet dysfunction (KHURI et al. 1995; FERNANDEZ et al. 1986). It is probable that the major effect of heparin on platelet activation is through its antithrombin activity, thrombin being one of the more efficacious platelet aggregators available (EIKA 1971).

The more obtuse effects of heparin such as lipoprotein lipase activation and the prevention of atherosclerosis will be dealt with in the clinical uses section.

V. Pharmacokinetics

1. Administration

Aqueous heparin must be given parenterally to achieve an anticoagulant and antithrombotic effect because oral heparin is poorly absorbed from the gastrointestinal tract and to date there is no clinical indication approved for oral administration (LARSEN et al. 1986). Similarly, intramuscular administration of heparin is not advised as there is a real and significant risk of hemorrhage with formation of intramuscular hematomas and subsequent erratic absorption. For the same reason, intramuscular injections in general should be avoided in patients on heparin therapy. Studies in dogs have demonstrated achievement of a steady and sustained anticoagulant effect with intramuscular administration (PERRY and HORTON 1976), but it should not be considered safe nor effective in human patients.

a) Intravenous Route

Steady state prolongation of the clotting time is best achieved by intravenous administration using continuous infusion after an initial loading dose is given.

To achieve safe, therapeutic anticoagulation and an effective antithrombotic effect in the shortest time, a fairly standard regimen has evolved. Therapy is initiated with a loading dose of 5000 IU heparin given over 10 min in 10–20 ml diluent. Immediately after the loading dose, a continuous infusion is started which initially can be calculated on an empiric basis as 300–500 IU/kg lean body mass per 24 h. This dose can be adjusted subsequently based on the results of the whole blood clotting time or aPTT (activated partial thromboplastin time), which should be measured prior to initiation of therapy and should be repeated at 6-h intervals until a steady state is achieved in the desired therapeutic range. Thereafter, monitoring should be performed every 12 h for 2 days and, if stable, at daily intervals for the duration of therapy.

Heparin may also be given intravenously by intermittent infusion, but this method should only be employed when the facilities for safe and reliable continuous infusion are not available. Intermittent infusions are associated with an increased risk of hemorrhage, which may be due to higher daily doses being administered (LEVINE et al. 1989). The loading dose is given as for continuous infusion and subsequently at 4-h intervals, and an additional 50–100 IU heparin/kg lean body mass is administered over a 10-min period. During the first 24 h a whole blood clotting time (or aPPT) should be obtained prior to each dose and subsequent doses adjusted as required to achieve the desired therapeutic level. Once a steady state is reached, then the whole blood clotting time (or aPPT) should be determined before the second and fourth dose each day for the duration of the therapy.

b) Subcutaneous Route

There are two main objectives for the administration of subcutaneous heparin: therapeutic anticoagulation or prophylaxis for venous thrombosis. If one desires to use the subcutaneous route for therapeutic anticoagulation, it is important to remember that immediate anticoagulation will not be achieved unless an initial bolus of 5000 IU is given intravenously. At the time of the intravenous bolus heparin should also be given subcutaneously, the initial dose being half the calculated total daily dose (based on lean body mass as described in the continuous infusion section) and subsequent doses administered at 12-h intervals. The clotting time (or aPTT) should be checked 4 h after the bolus dose and then at 12-h intervals until a steady state is achieved in the desired therapeutic range. As a consequence of poor bioavailability, which ranges from 20% to 40% depending on the assay one uses (BARA et al. 1985; BRIANT et al. 1989), the initial subcutaneous dose needs to be sufficiently high if one is to achieve anticoagulation that is comparable to that which can be obtained with continuous intravenous infusion. Once this precaution is taken, however, both continuous intravenous infusion and subcutaneous administration are comparable in terms of safety and efficacy (PINI et al. 1990; BENTLEY et al. 1980; DOYLE et al. 1987).

Heparin may also be given subcutaneously at doses of 5000 IU every 8 h for prophylaxis against the formation of deep venous thrombosis (KAKKAR et al. 1975). In this case, therapeutic monitoring is not usually required but patients may become fully anticoagulated occasionally due to a cumulative effect (CLAGETT and SALZMAN 1974).

c) *Novel Methods of Administration*

More novel methods of administration include the application of a heparin coating to prosthetic heart valves, and indwelling arterial and venous catheters (especially those of the Swan-Ganz type) in which it has been shown to reduce the formation of fibrin sheaths. The fibrin sheath that forms on indwelling catheters can induce significant thrombus formation when the catheter is withdrawn. If this occurs in an artery, it could lead to distal occlusion of the vessel.

The utilization of inhaled heparin has recently attracted renewed interest. There are reports of its use as a treatment for exercise induced asthma (AHMED et al. 1993), for the prevention of hypoxia induced pulmonary hypertension in an animal model (SPENCE et al. 1993) and as an antithrombotic agent in diabetic women (EFIMOV et al. 1995; JACQUES 1979). Currently, buccal mucosal patches and rectal suppositories are being investigated but to date no significant reports of their efficacy have been documented in the literature.

2. Distribution

Parenterally administered heparin largely remains in the intravascular space but its volume of distribution more closely reflects the whole blood volume rather than the plasma volume to which it is mainly confined (ESTES 1975; KANDROTAS et al. 1989). This apparent anomaly is explained by the extent of heparin binding in both specific and nonspecific fashion to endothelial cells (BARZU et al. 1985), histidine rich glycoproteins and platelet factor 4 (LANE et al. 1986), to vitronectin (PREISSNER and MULLER-BERGHAUS 1987), fibronectin (DAWES and PAVUK 1991), lipoproteins and von Willebrand factor (SOBEL et al. 1991) and, of course, to antithrombin III and heparin cofactor II as discussed previously. A large proportion of this nonspecific binding is related to the huge negative charge conferred on heparin as a consequence of its polysulfation. This has the net effect of reducing heparin's bioavailability, occasionally in an unpredictable manner (HIRSH et al. 1976).

3. Clearance

The clearance of heparin is complex and depends on the route of administration, the dose and duration of therapy, the host, and the clinical disease process affecting that patient. At low doses given intravenously, heparin is eliminated initially by first order kinetics at a rate proportional to the dose. Half-life ranges from 60 min with a 100-IU/kg bolus to 150 min for a 400-IU/kg bolus (McAVOY 1979; DE SWART et al. 1982). There is an initial period of rapid elimination effected by a saturable mechanism that is attributed to heparin binding to various cells and receptors. Once bound to the endothelial cells and macrophages, heparin is internalized and denatured to smaller fragments by desulfation and deacetylation. At the more common therapeutic doses, this mechanism contributes significantly to the continued clearance of heparin (DAWES and PAPPER 1979; BJORNSSON et al. 1988). The nonsaturable mechanism appears to be due to renal excretion. At therapeutic doses, it plays a minor role but at high doses (such as during cardiopulmonary bypass surgery) it assumes greater significance and heparin may be detected in the urine (BONEU et al. 1989). As a general rule, it is reasonable to assume for routine clinical applications that the half-life of heparin is 70–120 min.

4. Effect of Physiological State

Heparin is not transported across the placenta nor secreted into breast milk and the frequency of bleeding in pregnancy is no different than in the nonpregnant population. However, all antithrombotics carry the risk of potentially undesirable sequelae for both mother and fetus; hence the need for use should be clearly established and attention paid to proper dose regimens and therapeutic monitoring (GINSBERG and HIRSH 1992).

Patients with established thrombosis need higher doses of heparin (SIMON et al. 1978), although it is not clearly established if this is due to greater clearance or altered pharmacokinetics (CIPOLLE et al. 1981; KANDROTAS et al. 1990a).

We are unaware of any studies performed in a prospective and controlled fashion examining the pharmacokinetics of heparin in the elderly, although studies in the pig have shown that piglets clear heparin more quickly than adult pigs. This correlates with observations in human neonates who also appear to clear heparin more rapidly (ANDREW et al. 1989; SIMON et al. 1978).

In patients with renal disease and reduced glomerular filtration rate, the half-life of heparin is prolonged, consistent with reduced clearance at higher doses (TEIEN et al. 1976). There is substantial interpatient variability but remarkably little intrapatient variability in heparin clearance in patients undergoing hemodialysis (KANDROTAS et al. 1990b).

The half-life of heparin is minimally prolonged in patients with liver disease, which is consistent with some recent evidence linking the liver with removal of heparin from the blood, in keeping with the role long established of the reticuloendothelial system in heparin metabolism (TEIEN 1977).

5. Drug Interactions

The most serious problem that occurs with heparin is bleeding. This is often associated with the concomitant use of other anticoagulants, antithrombotics, fibrinolytics or antiplatelet agents. The increased risk of bleeding is a direct result of inhibiting the coagulation cascade at multiple levels and, as such, is not a drug-drug interaction per se.

Despite the widespread use of heparin and the knowledge that it displaces other drugs from albumin and other plasma proteins, there have been very few reports of serious drug-drug interactions (COLBURN 1976; HODBY et al. 1972). The most commonly reported interaction is with nitroglycerin and, considering the common usage of both in acute coronary syndromes, the importance of determining the significance of any such interaction is considerable. Initial reports suggested that nitroglycerin reduced the anticoagulant activity of heparin (HABBAB and HAFT 1986; PIZULLI et al. 1988) through a qualitative antithrombin III defect (BECKER et al. 1990). A randomized trial published in 1990, however, concluded that nitroglycerin had no effect on heparin activity or clearance. Controversy still exists, though, because a more recent study revealed significant reductions in the aPTT ratio and heparin levels in the setting of unaltered antithrombin III or platelet factor 4 levels in patients who received nitroglycerin, suggesting that nitrates did indeed increase heparin clearance (BRACK et al. 1993). The current consensus is that the effect is minimal and should easily be corrected if one follows appropriate therapeutic monitoring of heparin dosing (KOH et al. 1995).

Other agents that have been reported to decrease the anticoagulant activity of heparin include antihistamines, digitalis, digitalis-like compounds and tetracycline. Similarly, reports have been compiled on many medications that reportedly enhance the anticoagulant nature of heparin: These include streptomycin, polymycins B and M, monomycin, gentamycin, erythromycin, phenylbutazone, indomethacin, sulfinpyrazone, clofibrate, EDTA, nicotine, penicillin, quinine and thyroxine (HANSTEN 1974). There is a dearth of literature on the significance of these interactions.

VI. Clinical Indications

Heparin remains the most universally employed parenteral anticoagulant and the treatment of choice (occasionally the sole treatment available) for certain disease processes. Its popularity and efficacy stem from its triple anticoagulant, antithrombotic and antiplatelet effects as discussed in Sect. B.IV. It has an immediate onset of action when given intravenously, which is essential in many of the disease states in which it is employed. We shall initially discuss its use as an anticoagulant and antithrombotic agent and then briefly describe certain other therapeutic uses in which the anticoagulant effect is not considered the primary mechanism of action.

Prior to the institution of any anticoagulant therapy one should hesitate before thoughtlessly instituting treatment and instead take the opportunity to perform appropriate studies to determine why pathological thrombus formation has occurred. Once anticoagulant therapy is initiated, the assessment of the coagulation, platelet and fibrinolytic systems by functional assays is not possible. Genetic testing is only available for a limited number of the known abnormalities that contribute to hypercoagulable states (BAKER and BICK 1994; NUCCI and BELL 1994).

1. Venous Thrombosis and Thromboembolic Disease

The successful use of heparin in the treatment of venous thrombosis and pulmonary embolism has been reported since 1939 (JORPES 1939). It remains a primary treatment for both of these disease processes and its efficacy in preventing embolism from occurring in the presence of existing deep venous thrombosis and in preventing recurrent thrombosis is well established (BARRITT and JORDAN 1960). Heparin prevents further thrombosis and allows endogenous fibrinolysis to occur, leading to clearance of the thrombus (POLLACK et al. 1973; COON and WILLIS 1973). In venous thrombosis, we suggest that the therapy should initially be intravenous and continued for at least 5 days until oral anticoagulation is established and in the therapeutic range (HULL et al. 1990; BRANDJES et al. 1992). There is some clinical evidence that in venous thrombosis without embolism, therapy after an initial intravenous loading dose can be given subcutaneously although the strength of this conclusion is limited due to the small size of the trials (DOYLE et al. 1987; PINI et al. 1990). In the treatment of pulmonary embolism, however, use of the subcutaneous route cannot be advised and continuous intravenous infusion is required with close attention to the clotting time (or aPTT) as the heparin requirement may need to be increased (SIMON 1978). One caveat is that in massive or submassive pulmonary embolism, alternative therapies such as fibrinolysis need to be considered (BELL et al. 1974; GOLDHABER et al. 1993).

Prevention of deep venous thrombosis must remain a major target of modern medical care. Heparin has been used effectively in this regard and many excellent studies in patients undergoing surgery have clearly shown the benefits of heparin for prophylaxis (KAKKAR et al. 1975; HALKIN et al. 1982; COLLINS et al. 1988). Unfortunately certain groups of patients, notably patients undergoing pelvic surgery or arthroplasty of the lower limb, continue to have an unacceptable frequency of deep venous thrombosis despite heparin prophylaxis (WILLE-JORGENSEN and OTT 1990). In hip arthroplasty there appears to be some benefit associated with the use of either adjusted dose heparin or low molecular weight heparin (DECHAVANNE et al. 1989). Despite much enthusiasm for low molecular weight heparin, one must remain cautious as different preparations are not interchangeable. In addition, some of the studies have used indirect methods for assessing the frequency of deep venous thrombosis. For example, one significant report has emerged which demonstrated much higher rates than expected of deep venous thrombosis in patients treated with low molecular weight heparin than would have been detected if venography had not been employed (BOUNAMEAUX et al. 1993). This reinforces the necessity to base clinical decision making only on clinical studies that employ the gold standards for diagnosis of the disease process involved. One can conclude that low dose subcutaneous heparin is effective prophylaxis for the prevention of deep venous thrombosis but certain high risk groups may need adjusted dose heparin or low molecular weight heparin.

2. Use of Heparin in Acute Coronary Syndromes

Heparin administered intravenously in anticoagulant doses is acknowledged as the single most efficacious and important therapy in the treatment of unstable angina, with a 50% reduction in cardiac events being possible (THEROUX et al. 1988, 1992). The benefits of the adding of aspirin in the acute setting are unproven but aspirin does lead to less rebound ischemia when the heparin is discontinued (FUSTER et al. 1993; THEROUX et al. 1992). Continuous intravenous infusion is preferable because intermittent infusion has been demonstrated to be relatively ineffective in unstable angina (NERI-SERNERI et al. 1990).

The role of heparin in acute myocardial infarction is less well defined. It has been suggested that all such patients should get subcutaneous heparin at doses similar to those used in the prophylaxis of deep venous thrombosis (TURPIE et al. 1989). The role of intravenously administered heparin is much less certain. Many studies that have examined the use of heparin in acute myocardial infarction have been confounded by the addition of 30 days of oral anticoagulation, leading to an increased risk of hemorrhage which may have offset any potential benefit (MRC TRIAL 1969; VA COOPERATIVE STUDY 1973). The combined data from ISIS 3 and GISSI 2 demonstrated 5 lives saved per 1000 treated in the patients who received high dose subcutaneous heparin versus aspirin alone in the first 7 days, but no difference in the mortality rate of either group was evident at 35 days. This is difficult to understand as few patients received heparin after the 7th day. Consequently any detrimental effect of the heparin should not have been evident after 7 days unless it were a pro-thrombotic rebound not totally prevented by aspirin (ISIS 3 1992; GISSI 2 1990). There is some evidence that higher dose heparin may significantly reduce the formation of mural thrombi in patients who have experienced a transmural myocardial infarction (TURPIE et al. 1989). The prolonged use of heparin administered subcutaneously postinfarction has been studied and a reduction in thrombotic events of almost 60% was achieved during a 1-month follow up, a promising result which awaits duplication and, if confirmed, could significantly alter postinfarction patient care (NERI-SERNERI et al. 1987).

The role of heparin as an adjunct to thrombolysis remains an ongoing area of considerable interest. Current thrombolytic regimes employed in acute myocardial infarction establish patency in 60%–80% of patients by 90 min. Only half achieve normal flow, however, and reocclusion occurs in 10% of patients (GUSTO ANGIOGRAPHIC INVESTIGATORS 1993). Thrombin plays a central role in reocclusion so it is expected that heparin should prevent or reduce the frequency of reocclusion. Heparin has been given post-thrombolysis with mixed results, although there tends to be more positive than negative studies (HSIA et al. 1990; NATIONAL HEART FOUNDATION 1989; GUSTO 1993). It is now routine clinical practice to use adjunctive heparin when TPA (tissue plasminogen activator) is used as the thrombolytic therapy, but no clear consensus has emerged when streptokinase is used (KAPLAN et al. 1987). Overall, the weight of evidence favors the use of heparin and it is required with some of the newer thrombolytics, especially those with a short half-life and those than do not deplete fibrinogen levels (TEBBE et al. 1995). The optimal duration of therapy with heparin post-thrombolysis is as yet unknown.

In myocardial infarction, heparin will continue to be used as the principal anticoagulant, especially given the recent tests of heparin versus hirudin in the GUSTO IIb and TIMI 9b trials, which demonstrated essentially equal efficacy and safety in this patient population.

3. Heparin for Trousseau Syndrome: A Unique Therapy

Trousseau syndrome is characterized by recurrent migratory venous and arterial thrombosis and nonbacterial thrombotic endocarditis in association with malignancy, the latter often occult. It is a devastating syndrome of unknown etiology which is uniquely sensitive to anticoagulation with heparin but, commonly, cannot be successfully treated with warfarin (SACK et al. 1977; BELL et al. 1985). It may require lifelong therapy with heparin, which should be administered subcutaneously in therapeutic rather than prophylactic doses (BELL 1986; BELL and PITNEY 1969).

Disseminated intravascular coagulation (DIC) may be part of a broad disease spectrum that includes Trousseau syndrome, and is characterized by bleeding at sites of minimal trauma, abnormal coagulation studies, potential antithrombin III deficiency and occasional pathological thrombus formation. It commonly occurs in the setting of critically ill patients and may be treated effectively with intravenous heparin, although the advice of a coagulation expert should be sought to guide therapy (SPERO et al. 1980; PRAGER et al. 1979). The use of heparin in patients with bleeding associated with acute promyelocytic leukemia is similarly an unresolved question, and current practice would be to use only low dose heparin in this setting (STONE and MAYER 1990; GOLDBERG et al. 1987; RODEGHIERO et al. 1990). The clinical studies are sparse and mainly retrospective; hence one must be guided by the individual clinical situation, and if the DIC process is confirmed heparin should be considered in full therapeutic doses.

There are some data supporting the use of heparin to prevent the development of hepatic veno-occlusive disease in patients being treated with heterologous bone marrow transplantation. One randomized prospective trial demonstrated its efficacy (ATTAL et al. 1992).

Heparin is also employed in large doses in situations in which aPTT is not used to guide therapy. These include circumstances in which heparin is utilized to maintain the thrombosis-free patency of extracorporeal circulation in cardiopulmonary bypass, dialysis and invasive therapeutic vascular procedures such as angioplasty. These uses present special problems with regard to therapeutic monitoring and reversal of anticoagulation and will be discussed in detail later.

4. Novel Uses of Heparin

Thre is a great deal of interest in using heparin for situations in which its anticoagulant effect is not central to its efficacy. Since heparin closely resembles heparan sulfate, an endogenous product found on almost every cell surface, it is postulated that it may play a role in cell adhesion, control of cell boundaries and possibly angiogenesis (FRITZE et al. 1985).

A major area of study has been the investigational use of heparin to prevent atherosclerosis and restenosis after angioplasty (where myointimal hyperplasia may be considered an accelerated form of "normal" atherosclerosis) (AUSTIN et al. 1995). It is known, for example, that in experimental models heparin reduces smooth muscle proliferation in injured arteries (CLOWES and KARNOWSKY 1977). However, in two recent clinical studies in which both standard heparin and low molecular weight heparin were used, no benefit was evident in preventing restenosis after transluminal angioplasty (CAIRNS et al. 1996; BRACK et al. 1995). This may be explained by the ability of heparin to displace basic fibroblast growth factor (bFGF) from damaged endothelium. The bFGF could then promote smooth muscle proliferation and possibly contribute to restenosis (MEDALION et al. 1997).

A new and potentially significant use of heparin may be as a bioactive semicatalytic coating on implanted coronary artery stents. It is now technically feasible to bind heparin to metallic surfaces by endpoint attachment. The bound heparin can prevent thrombosis on the start surface by inhibition of coagulation and by preventing platelet adhesion (KIM and JACOBS 1996). Efficacy has been demonstrated in a porcine model in which heparin coated stents were associated with no acute occlusions compared to over 30% acute occlusions in noncoated stents (HARDHAMMAR et al. 1996). The use of heparin-coated stents is being evaulated clinically in Europe where, in a pilot study, patients receiving heparin-coated stents without concurrent systemic heparin demonstrated no bleeding complications (SERRUYS et al. 1996). This may become universal therapy if it prevents the main problem associated with stenting of coronary arteries, acute thrombotic occlusion. It is estimated that by the year 2000 over one million stents will be utilized globally. The high success rate with heparin-coated stents is probably a function of heparin acting more to reduce the generation of thrombin rather than as a thrombin inhibitor.

Heparin has been demonstrated to activate lipoprotein lipase, which then cleaves lipoproteins, releasing complex fatty acids into plasma (COLBURN 1976; TULLIS 1976). This effect is independent of any anticoagulant activity (EHNHOLM et al. 1977). Heparin may also bind and inactivate histamine, complement, autoagglutinins and many allergic autoantibodies (ENGLEBERT 1977; AHMED et al. 1992; RAGAZZI and CHINELATTO 1995). Mention has already been made of the potential benefit of heparin in preventing exercise-induced asthma (AHMED et al. 1993). It is postulated that heparin blocks inositol triphosphate-mediated calcium release, thereby attenuating smooth muscle contraction (GHOSK et al. 1988). Similar benefits in animal models of hypoxic pulmonary hypertension have been reported (SPENCE et al. 1993).

VII. Therapeutic Monitoring

Heparin doses should always be specified in units, never milligrams or any other weight-based dimension. Depending on the source and manufacturer, 1 mg heparin may contain 100–180 USP units (anticoagulant units) (SAMAMA 1995). The therapeutic antithrombotic effect of heparin is best monitored by measurement of the whole blood clotting time, of which the Lee White method, utilizing venous blood in a plain glass tube, is the most reliable. For reasons of expediency, however, the PTT (partial thromboplastin time) or aPTT (activated partial thromboplastin time) is often substituted.

Prior to the initiation of therapy a control clotting time should be determined; this is particularly critical with the aPTT as there is considerable interpatient variability. If the Lee White whole blood clotting time is used, then the aim for therapeutic anticoagulation should be to achieve a clotting time of 25–35 min when samples are drawn at the correct intervals (as described in the dosing section). If heparin is used in doses sufficient to prolong the clotting time, a daily summary sheet should be used to record the infusion rate, brand of heparin, dose per 24 h, and the result of studies for occult blood in the urine and stool. At the present time, the Lee White whole blood clotting time has fallen from favor and has been replaced by the aPTT, which provides information mainly on the anticoagulant effects mediated via thrombin and activated factors IX and X. The aPTT is an excellent assay of *plasma* coagulation but it is not a whole blood, global assay of coagulation.

The aPTT is the time taken for plasma to clot after incubation with an activator such as kaolin and subsequent calcification (PROCTER and RAPAPORT 1961). The aPTT at baseline is characterized by considerable interpatient variability with moderate intrapatient variability. Consequently, the aPTT ratio is a safer measure of the level of anticoagulation than the absolute aPTT value in seconds (BJORNSSON and WOLFRAM 1982; WHITFIELD and LEVY 1980). The aPTT that is associated with therapeutic effect may vary by a factor of two (KITCHEN et al. 1996) so the aPTT should be standardized in relation to a level of heparin equivalent to 0.2–0.4 IU/ml (BRILL-EDWARDS et al. 1993; D'ANGELO et al. 1990). The usual recommended therapeutic aPTT ratio is 1.5–2.0 times baseline (CHIU et al. 1977). Evidence has accumulated for a relationship between therapeutic effects and the aPTT ratio in treatment of deep venous thrombosis (HULL et al. 1992), prevention of mural thrombi postmyocardial infarction (TURPIE et al. 1989) and the maintenance of coronary patency post-thrombolytic therapy for myocardial infarction (CAMILLERI et al. 1988; DE BONO et al. 1992).

The special assay most commonly used is factor Xa inhibition, which can be measured in two ways. One method involves the addition of excess bovine activated factor X. It is titrated carefully until a clot forms while noting the amount of exogenous Xa required to do so, which will be directly proportional to the degree of factor X inhibition induced by the heparin therapy (EGGLETON et al. 1979). A second and more easily automated method involves a chromogenic assay (TEIEN and BJORNSSON 1976). It has been recommended that a specific factor assay be used when the baseline aPTT is very abnormal. Current data do not allow us to establish whether more specific assays will enable more accurate or safer heparin dosing.

When heparin is utilized to prevent thrombosis in extracorporeal circuits or in patients undergoing percutaneous transluminal angioplasty, the aPTT is not a useful assay as ratios are not related in a linear fashion to the heparin dose. This problem can be overcome by using a heparin concentration derived from protamine titration, a chromogenic factor Xa assay or, more commonly, an activated whole blood clotting time (HATTERSLEY 1966). This is now semi-automated and correlates reasonably well with the Lee White whole blood clotting time and heparin blood levels.

The activated clotting time can be useful, although an activated clotting time-heparin dose curve should be established for each patient and one must remember that the absolute values are dependent on the activator, the strength of which may vary from batch to batch. This is a whole blood technique, reflecting the effects of all blood elements that play a role in thrombosis (SCHRIEVER et al. 1973).

The concept of heparin resistance means a failure to prolong in vitro coagulation studies with heparin doses normally associated with therapeutic effect. This phenomenon is largely due to heparin binding to plasma proteins and the physiological state of the patient. Another mechanism of resistance may be real or relative antithrombin III deficiency. Congenital antithrombin III deficiency can be ruled out by assays on blood drawn prior to the initiation of heparin in which the antithrombin III level is usually 40%–60% of normal (NIELSEN et al. 1987). Acquired antithrombin III deficiency may occur in renal or hepatic disease where the level of antithrombin III is approximately 60%–80% of normal (KNOT et al. 1984; KAUFFMANN et al. 1978) and even large doses of heparin may not prolong the aPTT ratio. Some newer etiologies for heparin resistance have been suggested that include heparin binding to the products of platelet degranulation (MANSON et al. 1996).

The platelet count must be monitored in patients receiving all types of heparin if one is to detect heparin-induced thrombocytopenia at an early and asymptomatic stage. Regular platelet counts are recommended for all hospitalized patients.

It is important to emphasize that heparin needs careful monitoring for optimal therapeutic effect. The aPTT is a reasonable if imperfect assay and one should consider the Lee White whole blood clotting time or chromogenic Xa assay in complex cases where imprecision may not be tolerable. Future monitoring will focus on bedside aPTT assays utilizing fingerstick capillary samples, which should allow for more rapid adjustment of doses. Such technology may also be used at home for extended therapeutic heparin anticoagulation (ANSELL et al. 1991).

We do not currently advocate the use of heparin nomograms despite the reported evidence that one achieves therapeutic aPTT levels more expediently, because the more commonly used nomograms were supratherapeutic for at least 24 h (HULL et al. 1992). The American College of Pathologists reported in 1994 that many nomograms had been prepared in institutions where a particularly sensitive thromboplastin had been used, making comparison between institutions very difficult. The method of dosing we described in Sect. B.V.I is adequate if clotting time/aPPT is assayed at 6-h intervals until a steady state is achieved. A very clear relationship between intracranial bleeding and supratherapeutic aPTT ratios was evident in the TIMI 9b and GUSTO IIa trials; therefore, we cannot recommend any nomogram if 50% of those treated would be exposed to supratherapeutic levels of anticoagulation and its attendant risks.

VIII. Toxicity (Table 2)

The most common adverse effect encountered with the therapeutic use of heparin is hemorrhage. The precise frequency is unknown but it ranges from 2% to 25% (most commonly 2%–4%), and it is a potential problem with all currently available heparin preparations including low molecular weight heparin (BORRIS and LASSEN 1995; ROSENDAAL et al. 1991). Fatal hemorrhagic episodes occur in about 7 per 1000 courses of treatment (LEVINE et al. 1989). The results of a meta-analysis suggest that bleeding is related to the total daily dose rather than the method of administration (MORABIA 1986), which may explain the increased risk of hemorrhage occasionally reported with intermittent intravenous infusion since a higher total daily dose is required to achieve a therapeutic level of anticoagulation (WILSON et al. 1981).

Table 2. Heparin; Side effects and Toxicity

Reaction	Purported mechanism
Hemorrhage	Dose related anticoagulant effect
Allergic reactions; Urticaria Asthma Rhinitis	Dose independant, possible mast cell degranulation, may be related to contaminants in heparin
Thrombocytopenia	Dose independent, antibodies either a platelet factor 4 or against NAP-2 or IL8
Osteoporosis	Dose dependent inhibition of osteoblasts
Arterial thrombosis	Platelet aggregation
Hyponatremia	Inhibition of aldosterone release
Resistance	Physiological excess of platelet factor 4 and acute phase reactants
Chest pain	Probable IgE immune mediated anaphylactoid type reaction

Concurrent use of antiplatelet therapy and thrombolytic agents also increase the risk of hemorrhage. Supratherapeutic aPTT ratios, as shown in recent multicenter trials in patients with acute coronary syndromes, increase the risk of hemorrhage (GUSTO IIa 1994).

The clinical condition of the patient also influences the likelihood of hemorrhage and in some studies was the major risk factor for hemorrhage (Nieuwenhuis et al. 1991). Renal failure, peptic ulcer disease, recent surgery, large body surface area and age greater than 60 years have all been linked to increased hemorrhagic risk (Jick et al. 1968; Kher and Samama 1984). Attention to close therapeutic monitoring and avoidance of large bolus doses or concomitant use of antiplatelet therapies or NSAIDs, unless strongly clinically indicated, should help to reduce the risk.

Thrombocytopenia associated with the administration of heparin was first recognized in animals (Copley and Robb 1941) and later in man (Quick et al. 1948). The true frequency is unclear and ranges from 2% to 30%, the latter more likely if heparin sourced from bovine pulmonic tissue is used (Bell et al. 1976; Bell and Royall 1980). In most patients it occurs between days 3 and 5, but it can occur as late as day 15 (Bell 1988; Chong 1995). In some patients it follows a severe course with marked thrombocytopenia and paradoxical arterial thrombosis (Weismann and Tobin 1958; Schuster et al. 1980). The clinical and laboratory studies are often consistent with a syndrome of disseminated intravascular coagulopathy (Klein and Bell 1974).

Heparin-induced thrombocytopenia is associated with the presence of antibodies to the heparin-platelet factor 4 complex in up to 90% of patients (Amiral et al. 1992; Visentin et al. 1994). It is proposed that the 10% of patients with heparin-induced thrombocytopenia, but no antibodies to the heparin-platelet factor 4 complex, have antibodies that react with either interleukin 8 or neutrophil activating peptide-2, both of which have some homology with platelet factor 4 and may be found at the site of vessel injury and platelet activation (Amiral et al. 1996). Low molecular weight heparin may be associated with a lower frequency of heparin-induced thrombocy-topenia but, since it exhibits cross reactivity in vitro with the antibodies mentioned above, it should not be used as a substitute for unfractionated heparin once thrombocytopenia has occurred (Greinacher et al. 1992). If immediate intravenous anticoagulation is needed in patients with a history of heparin-induced thrombocytopenia, then danaparoid (which has less than 10% cross reactivity) or epirudin should be considered (Magnani 1993). If less immediate anticoagulation is sufficient, then ancrod may be the agent of choice as it has no cross reactivity with heparin, although it is con-traindicated in disseminated intravascular coagulation (Cole and Bormanis 1988; Bell and Pitney 1969). The decline in the use of bovine lung as a source of commercial heparin has resulted in a significant decline in the frequency of heparin-induced thrombocytopenia, but it is important to remain aware of this potentially life-threatening complication.

Osteoporosis may develop in a small number of patients treated with heparin for more than 3 (but usually greater than 5) months. In the majority of reports, it occurs with therapeutic doses (usual dose >15000 IU/day) and not with prophylactic doses (GRFFITH et al. 1965). We are unaware of any proper randomized, controlled trials designed to examine this problem. The frequency, based on two small studies, appears to be less than 15% even at therapeutic doses (>15000 IU/day) with treatment durations of over 6 months (MONREAL et al. 1994; DAHLMAN et al. 1990). The mechanism by which heparin induces osteoporosis is unknown. Heparin may directly activate osteoclasts, decrease osteoblast activity, potentiate collagenase activity, and/or perhaps alter vitamin D metabolism. No effect on parathyroid metabolism is now believed to occur (GINSBERG et al. 1990; AARSKOG et al. 1980).

Hypoaldosteronism with persistent natriuresis and occasional hyperkalemia can also occur with heparin therapy (CEJKA et al. 1960; OSTER et al. 1995; WILSON and GOETZ 1964). It is most likely related to aldosterone suppression and may be partially due to a reduction in number and affinity of angiotensin II receptors in the zona glomerulosa of the nephron. The frequency is 5%–7%, but it is rarely a clinically evident problem unless there is concurrent renal impairment, diabetes mellitus or use of potassium sparing diuretics. The usual treatment is discontinuation of heparin, but when this cannot be done potassium-wasting diuretics may be utilized. Special caution is needed when electrolyte changes are encountered in the elderly.

Hypersensitivity reactions to heparin are rare and usually occur in those who have had prior exposure to heparin therapy. It may be related to methods of extraction or to preservatives in the commercial preparation (CURRY et al. 1973). One report cites an association with the IgG molecule that binds the heparin-platelet factor 4 complex but the importance of this possibility remains unproven (WARKENTIN et al. 1992). This report also noted that the same patients were prone to develop skin necrosis, which is a rare complication of heparin therapy characterized by subacute onset. It is postulated to be an immune-dependent response involving local angiitis with lesions that are typically painful (PLATELL and TAN 1986).

Elevation of liver transaminase enzymes has also been reported to occur with heparin therapy, albeit very infrequently (LAMBERT et al. 1986), and in one series was linked to thrombocytopenia (SHARATH et al. 1985). The mechanism is unknown but it does appear to resemble the condition of eclamptic patients who have the HELLP syndrome with thrombocytopenia, transaminitis and hypertension.

IX. Antidotes: Reversal of Anticoagulant Effect

The anticoagulant effect of heparin is relatively short (hours), so it is usually sufficient to stop administration and wait for its elimination. Occasionally, due to major bleeding or when large doses have been given inadvertently, it may be necessary to reverse the anticoagulant effect of heparin. Several antidotes are available including protamine sulfate, hexadimethrine bromide, polyprene, indocyanine green, clupeine, polylysine, lysozyme, toluidine blue, fushin and tryptophan. In clinical practice, an intravenous infusion of protamine sulfate (a heterogeneous mixture of polypeptides obtained from salmon and rainbow trout sperm) is most frequently used. Protamine carries a large positive charge due to a large proportion of basic amino acids, so it avidly binds the negatively charged heparin. It has an immediate effect but it may need to be given a second time if one is reversing the effect of a large subcutaneous depot of heparin. The recommended dose is 1 mg/100 IU heparin estimated to remain in the circulation. It should be administered slowly, intravenously and at a rate lower than 5 mg/min with no single treatment to exceed 50 mg. Its therapeutic effect lasts 2 h. A clotting time should be checked 15 min after the protamine is administered to decide if another dose is required. It should be noted that plasma products are not effective in reversing the anticoagulant action of heparin and if significant bleeding occurs, blood transfusion may be needed (WOLZT et al. 1995).

The major problem associated with the administration of protamine is anaphylaxis, which can occur in up to 1% of patients who have been exposed to protamine previously. This occurs most commonly when protamine has been added to insulin preparations to create a slowly dissociating form with a long half-life (SHARATH et al. 1985). An association between the occurrence of anaphylaxis and protamine specific antibodies has been reported in some patients (WEISS et al. 1989) and a prospective study has demonstrated that a single exposure to protamine can result in protamine specific IgE and IgG production (NYHAN et al. 1989, 1996). If a previous reaction has occurred, or if the presence of protamine specific IgE has been documented, then consideration should be given to the use of another antidote or perhaps an antifibrinolytic.

New antidotes proposed include platelet factor 4, which is found in platelet alpha granules and is now available in recombinant form. Platelet factor 4 binds to heparin, rendering it inactive, a mechanism recently demonstrated to be effective in a phase I clinical trial (DEHMER et al. 1995). It may also prove to have a role in patients with documented anti-protamine IgE, prior protamine induced complications, patients with fish allergy, or those who have received NPH insulin. Unfortunately, similar nonspecific binding of heparin to platelet factor 4 has also been suggested to occur at the epitope that induces antibody formation associated with heparin induced thrombocytopenia, microvascular injury and transaminitis.

C. Heparinoids and Related Anticoagulants

I. Pentosans (Sulfonated Xylans)

1. Source

Pentosan polysulfate is one of the most important heparinoid compounds available. It is a semisynthetic, highly sulfonated xylan obtained from beech and birch wood chippings (Sie et al. 1986; Ryde et al. 1981; Wagenvoord et al. 1988). In the natural world, xylans are the most important elements in the structure of plants and are believed to be very old in evolutionary terms. The xylan is modified by chemical reaction with chlorosulfonic acid, which leads to polysulfation of the organic acid xylan (Yin and Tangen 1976).

2. Mechanism of Action

The mechanism of action of pentosan polysulfate is not firmly established but is believed to be based on binding of the pentosan to either antithrombin III or heparin cofactor II. The complex so formed then inhibits activated factor X (Scully et al. 1986). There is no significant direct effect on thrombin as most pentosan polysulfate molecules are too small (molecular weights 2000–7000) to bind both antithrombin III and thrombin simultaneously (Vinazzer et al. 1980; Simmons et al. 1995). Uncertainty still exists, however, as it has also been demonstrated that pentosan can both inhibit thrombin and prolong the aPTT in plasma that is deficient in both antithrombin III and heparin cofactor II (Sie et al. 1986; Beguin and Hemker 1987). In addition, a modest ability to activate the endogenous fibrinolytic system has been demonstrated (Dunn et al. 1983; Oleson 1965; Vinazzer 1991). The contribution of both mechanisms to the antithrombotic effect is believed to be complementary.

Pentosan polysulfate is more efficacious than heparin as an activator of lipolysis. It is known that certain lipids and lipoproteins, especially lipoprotein(a), may serve as inhibitors of plasmin; hence the effect on clearing plasma lipids may enhance endogenous fibrinolysis (Von Schon and Sauer 1963). Pentosan polysulfate has been reported in both animal models and in human volunteers to increase the release of plasminogen activator, but the clinical significance of this observation is uncertain (Marsh et al. 1985).

3. Administration

When a therapeutic antithrombotic effect is required, pentosan polysulfate is usually given intravenously, although long term it may be given by intramuscular injection. Oral and rectal absorption has been demonstrated using a radiolabeled preparation. In Europe, it is commonly given as a once-daily intramuscular injection, facilitating treatment at home. Pentosan has also been administered orally but usually to produce therapeutic effects unrelated to thrombosis (see below).

4. Clinical Uses

Pentosan polysulfate has been used as an anticoagulant in Europe and South Africa for over 30 years. It has also been used as a local treatment for thrombophlebitis. Older studies demonstrated its efficacy in the prevention of postoperative deep venous thrombosis and in the treatment of cerebral apoplexy (JOFFEE 1975; BOKONJIE 1971). If these benefits are confirmed, then it may have a role in clinical practice especially if oral forms are proven to be effective in deep venous thrombosis. In France, it is considered to have efficacy equivalent to heparin in the prevention of deep venous thrombosis (SIE 1988).

Current interest in pentosan polysulfate centers on its use as a chemotherapeutic agent although no clinical responses were seen in a recent clinical trial (SWAIN et al. 1995). The FDA recently approved pentosan polysulfate as a treatment for interstitial cystitis, although its mechanism of action is not believed to be mediated through an antithrombotic effect.

Therapeutic monitoring of pentosan polysulfate is not clearly defined. The literature would suggest that 50 mg pentosan polysulfate has an anticoagulant effect equivalent to 5000 IU heparin. At high doses, the clotting time and the APPT are prolonged. Since its primary antithrombotic effect is mediated via inhibition of activated factor X, assays that measure factor X activity may be more accurate markers of the antithrombotic effect, yet no anti-factor Xa activity can be demonstrated if chromogenic factor Xa assays are utilized (FISCHER et al. 1982). No rational explanation for this paradox has been proposed.

5. Toxicity

The major side effects reported with pentosan polysulfate are thrombocytopenia, elevation of hepatic transaminases and, uncommonly, bleeding. Pentosan polysulfate has been associated with thrombocytopenia (FOLLEA et al. 1986) and intravascular thrombosis for many years (LECLERCQ et al. 1990). ELISA testing for the heparin-platelet factor 4 complex was positive in some of these cases, confirming that the antibody is not heparin-specific. This also suggests that the thrombocytopenia may be independent of any particular saccharide sequence and is more dependent on a high negative charge density (GREINACHER et al. 1992; GIRONELL et al. 1996; GOAD et al. 1994). The dose limiting effect in the clinical trial (SWAIN et al. 1995) was thrombocytopenia and transaminitis, both of which were reversible.

6. Clinical Relevance

To date, the sulfated xylans have failed to make a major impact in a therapeutic area which remains dominated by heparin. However, a paradigm shift has occurred in Europe with regard to the use of animal-derived products for human use. Reports in the Lancet in 1996 of a new type of Creutzfeldt-Jacob disease which could be a variant of bovine spongiform encephalopathy raise the possibility of transmission from cattle to humans if bovine derived heparin is used (WILL et al. 1996). Pentosan polysulfate does not have this potential liability. It is derived from wood so there is no possibility of transmission of animal disease. A second advantage of pentosan over heparin is that it has been shown to block prion protein from interacting with endogenous glycosaminoglycans, so it is reasonable to postulate it may prevent the accumulation of protease resistant, prion-induced amyloid. This has been demonstrated in scrapie (a prion disease found in sheep that represents the phylogenetic, prototypical prion disease) infected neuroblastoma cells (CAUGHEY and RAYMOND 1993). Other novel uses of pentosan polysulfate have been in the prevention of nephrolithiasis (FELLSTROM et al. 1994), inhibition of human immunodeficiency viral replication (BABA et al. 1988) and experimental treatment of osteoarthritis (GHOSH and HUTADILOK 1996). In terms of thrombosis, pentosan polysulfate may have a significant future if animal studies demonstrating significant thrombus resolution in venous thrombosis models treated with a combination of low dose heparin and pentosan are duplicated in man (MAFFRAND et al. 1991).

II. Dermatan Sulfate

1. Introduction

Dermatan sulfate is a glycosaminoglycan usually found in the extracellular matrix and on the cell surface of eukaryotes. It remains under active investigation for its potential use as an intravenous antithrombotic agent especially in situations where extracorporeal circulation is employed (TEIEN and BJORNSSON 1976).

2. Source

Dermatan sulfate that is used in clinical studies is obtained from bovine and porcine intestine or porcine skin. In its original form it is polymerized, with a molecular weight ranging from 15000 to 48000 daltons (mean 28000). Current interest is concentrated on a depolymerized low molecular weight dermatan sulfate which ranges in size from 1600 to 6000 daltons. The molecular structure of dermatan sulfate consists of repeating units of L-iduronic acid, sulfated at C2, linked 1 to 4 with *N*-acetyl-D-galactosamine, which is sulfated in the C4 and C6 positions (CONRAD 1989; MAIMONE and TOLLEFSEN 1990; LINHARDT et al. 1994).

3. Mechanism of Action

Dermatan sulfate acts as an indirect antithrombotic by potentiating heparin cofactor II, which then inhibits thrombin. As noted earlier, heparin cofactor II is a unique protease inhibitor in that its reactive site contains leucine in a peptide bond with serine at position 444–445, a location where one more typically expects arginine, which may explain why heparin cofactor II only inhibits thrombin and no other activated coagulation factor (GRIFFITH et al. 1985). Dermatan sulfate molecules that contain the triple repeat of the disaccharide iduronic acid-2-sulfate linked to *N*-acetyl galactosamine 4 sulfate bind to heparin cofactor II, causing a structural rearrangement which permits the cofactor to bind to the fibrinogen recognition site of thrombin up to one thousand times more effectively (MAIMONE and TOLLEFSEN 1990; RAGG et al. 1990; TOLLEFSEN et al. 1983). It has recently been demonstrated that the active site on dermatan sulfate is present in some depolymerized or low molecular weight dermatan sulfates and that this confers antithrombotic activity, especially in the presence of larger molecular weight heparinoids (FERRARI et al. 1994; MASCELLANI et al. 1996; THOMAS et al. 1990). Dermatan sulfate has no direct effect on platelets but indirectly it may inhibit platelet dependent prothrombinase formation indirectly and also thrombin induced platelet aggregation (HOPPENSTEADT et al. 1991; LIU et al. 1995).

4. Pharmacokinetics

The pharmacokinetics of dermatan sulfate have been studied in humans using a chromogenic assay based on the catalytic effect of dermatan sulfate to potentiate the heparin cofactor II–thrombin interaction (DUPOUY et al. 1988). The volume of distribution of dermatan sulfate in humans after intravenous injection is 1.8 times plasma volume, a dose-dependent consequence of its binding to plasma proteins. The intravenous half-life ranges from 35 to 45 min, increasing with escalating doses, which suggests a saturable clearance mechanism. When administered intramuscularly, the bioavailability is 25% and the half-life is about 6 h with wide variations. Subcutaneous administration results in a bioavailability of 12% with a half-life of approximately 7 h (DOL et al. 1989). There is marked variability in these ranges; hence the enthusiasm for the use of depolymerized low molecular weight dermatan sulfate. Low molecular weight dermatan sulfate is believed to interact in an identical fashion with heparin cofactor II and has similar effects when given intravenously. Its purported advantage is that, when given intramuscularly or subcutaneously, it has greater efficacy as indicated by prolonged heparin cofactor II-induced thrombin inhibition (BARBANTI et al. 1993).

It has recently been reported that Desmin 370, a low molecular weight dermatan sulfate manufactured by Alfa Wassermann SpA, has anti-factor Xa activity (DETTORI et al. 1994), which would not be easily explained by potentiation of heparin cofactor II since the latter is only active against thrombin. Some investigators have attributed the anti-Xa activity to the potential presence of 3 IU/mg low molecular weight heparin in Desmin 370 (BRIEGER and DAWES 1996). Although most studies on the pharmacokinetics of dermatan sulfate utilized the functional assay based on the effect of dermatan to alter the heparin cofactor II-thrombin interaction, it is unfortunate that the majority of clinical trials quote doses in milligrams and not units of activity.

5. Clinical Use

Dermatan sulfate remains under active clinical investigation as an antithrombotic agent. Its efficacy in the rat model of deep venous thrombosis is well established (MAGGI et al. 1987). One controlled clinical trial in patients with hip fractures randomized 206 patients to receive either intramuscular dermatan sulfate (Desmin 300 mg bid) or placebo and no bleeding episodes were noted. In the second phase of this study, venographically detected deep venous thrombosis occurred in 38% of patients receiving Desmin compared to a 64% frequency in the placebo group ($P < 0.01$) (AGNELLI et al. 1992). These results suggest a potential clinical role for dermatan sulfate, but it is too early to conclude that it is significantly safer than other heparinoids. The slow absorption from intramuscular or subcutaneous sites may indicate that in nonelective settings the initial dose should be given intravenously.

Current interest centers on the role of dermatan sulfate in maintaining the patency of extracorporeal circuits, and studies in humans have shown it to be effective in hemodialysis (RYAN et al. 1992; NURMOHAMED et al. 1993). The short half-life of dermatan sulfate makes it particularly suitable for intermittent hemodialysis or cardiopulmonary bypass. A study performed in pigs demonstrated that the drug was effective as an anticoagulant in cardiopulmonary bypass circuits (BRISTER et al. 1994).

There is one report of the use of dermatan sulfate in the treatment of the consumptive coagulopathy of acute leukemia in which it appeared to be more efficacious than heparin, but it should be acknowledged that this was a small study. The use of antithrombotics should not be considered routine in acute promyelocytic leukemia, but should be confined to those patients in whom the DIC process is confirmed (COFRANCESCO et al. 1992).

6. Clinical Relevance

Further large studies are needed to define the efficacy, safety and clinical situations where dermatan sulfate or low molecular weight dermatan sulfate are indicated. The ideal method of monitoring dermatan sulfate therapy has not been defined in humans although the clotting time or direct assays of heparin cofactor II-thrombin potentiation appear useful. At the higher doses used in cardiopulmonary bypass circuits, the activated clotting time may prove to be the most appropriate monitoring parameter.

III. Sulodexide

1. Introduction

Sulodexide is a biodiverse glycosaminoglycan obtained from porcine intestinal mucosa. Its potential importance as an antithrombotic agent is significant given its bioavailability when administered orally.

2. Source

The product is obtained from porcine mucosa and purified to yield two major components. Twenty percent is dermatan sulfate (described in the previous section) and the other 80% is a glycosaminoglycan chain composed of repeating iduronyl-glycosaminoglycan sulfate units (RADHAKRISHNAMURTHY et al. 1986; CALLAS et al. 1993).

3. Mechanism of Action

Sulodexide has many effects in vitro and in vivo. In vivo, sulodexide exerts its antithrombotic effect mainly through inhibition of activated factor X. This is believed to occur via the iduronyl-glycosaminoglycan component, which increases the catalytic activity of antithrombin III. No inhibition of thrombin occurs via this mechanism, but at higher doses the dermatan sulfate component of sulodexide potentiates heparin cofactor II, leading to inhibition of thrombin which may be significant (PALAZZINE and PROCIDA 1978; ANDRIUOLI et al. 1984). There is some experimental evidence that agents which accelerate thrombin inhibition by both antithrombin III and heparin cofactor II pathways may have a more beneficial ratio of antithrombotic to anticoagulant effect (BUCHANAN et al. 1994). Sulodexide inhibits platelet aggregation induced by thrombin and cathepsin G in a manner dependent on N-sulfation, suggesting it may be related to charge-dependent binding to platelet surface receptors (RAJTAR et al. 1993). In addition sulodexide induces profibrinolytic activity by decreasing the concentration of plasminogen activator inhibitor I (PAI-1) and reducing plasma and serum viscosity (AGRATI et al. 1992; CREPALDI et al. 1990). Administration of sulodexide leads to in vivo hypolipidemic activity with reduction of LDL levels (RADHAKRISHNAMURTHY et al. 1986). A reduction in the rate of smooth muscle cell proliferation has also been demonstrated (TIOZZO et al. 1989). The decrease in serum lipids is effected by release of lipoprotein lipase and more rapid clearance of LDL and VLDL by the liver.

4. Pharmacokinetics

Sulodexide can be administered either parenterally or orally. It may be given once daily by intramuscular injection or twice daily orally. As oral administration is associated with a lag before therapeutic efficacy is evident, it is usual to initiate therapy with 1–2 days of parenteral administration. Most studies have concentrated on the effects of long-term administration of sulodexide. Its chronic effects appear to be mediated by potentiation of the fibrinolytic system via reduction of PAI-1 and fibrinogen levels. Consequently, plasma half-life per se may not correlate closely with clinical effects (CREPALDI et al. 1990).

Sulodexide, unlike most other newer antithrombotics, has been studied in units of activity rather than milligrams. The unit used is lipoprotein-lipase releasing units (LRU) and the usual dose is 600 LRU/day intramuscular or 1000 LRU/day orally. The best means of therapeutic monitoring is unclear but one could suggest a factor X activity assay for parenteral therapy and either plasma viscosity or fibrinogen level for oral administration.

5. Clinical Indications

Sulodexide was originally used in patients with peripheral vascular disease in whom its viscosity reducing, lipid clearing, and antithrombotic actions were all considered beneficial. Its efficacy has been demonstrated in many European trials in which its unique oral absorption made it very suitable for long-term administration (BARTOLO et al. 1984; PISANO et al. 1986). One large multicenter study has been completed assessing the efficacy of sulodexide in preventing death, thromboembolic events and left ventricular mural thrombus formation in patients experiencing a myocardial infarction (CONDORELLI et al. 1994). The results demonstrated a significant reduction in total deaths in the group receiving sulodexide compared to placebo (86 fatalities out of 1769 sulodexide treated versus 134 fatalities in the 1788 who received placebo). A significant reduction was also seen in mural thrombus formation and recurrent cardiac events in those treated with sulodexide. This study awaits confirmation in larger multicenter trials and if confirmed may lead to a significant change in postinfarction care.

6. Toxicity

In the large study performed by CONDORELLI et al. (1994), no serious side effect was reported that resulted in death, hospitalization or permanent disability but one should note that it was not a double-blinded study. Of the 1769 patients treated, 2 developed hematomas at the site of intramuscular injection, 4 developed skin reactions and 12 had gastrointestinal upset, giving a total complications frequency of about 1%. We are not aware of the long-term effects of sulodexide on bone density, an endpoint which unfortunately was not assessed in the study.

7. Future

The role of sulodexide remains to be established, but if its potential benefits in the postinfarct patient are confirmed, then it may have an exciting role in clinical practice in the next century. We also await investigations on its antiatherogenic effects and potential efficacy in restenosis. But in terms of restenosis, however, overenthusiasm should be avoided, since both heparin and low molecular weight heparin failed to prevent restenosis despite proven inhibition of proliferative responses in vitro.

IV. Danaparoid (Organan 10172)

1. Introduction

Danaparoid sodium is a mixture of sulfated glycosaminoglycans derived from porcine mucosa after the removal of heparin (MEULEMAN 1992). It is unquestionably a heparinoid and not a low molecular weight heparin (NICHOLSON et al. 1994).

2. Source

Danaparoid is a mixture of 83% heparan sulfate, 12% dermatan sulfate and 5% chondroitin sulfate. Only 55% of the heparan sulfate has affinity for antithrombin III and only 4% of the heparan sulfate has the critical pentasacharride sequence (VAN DEDEM et al. 1993). The mean molecular weight of danaparoid is 5500 daltons.

3. Mechanism of Action

Danaparoid appears to be primarily a factor Xa inhibitor but the precise mechanism of action remains elusive. It is unknown what, if any, role the majority of the heparan sulfate that has low affinity for antithrombin III plays in the antithrombotic process (MEULEMAN 1992; ZAMMIT and DAWES 1994). The major mechanism of action of dermatan sulfate is known (see Sect. C.II) and it certainly contributes to the antithrombotic effect of danaparoid, but this does not explain why danaparoid has such potent anti-factor X activity, as heparin cofactor II potentiated by dermatan sulfate has no major inhibitory effect on activated factor X. In animal studies the ratio of factor Xa inhibition to thrombin inhibition is approximately 20 to 1. Danaparoid has little or no effect on platelet function.

4. Pharmacokinetics

Danaparoid may be given intravenously or subcutaneously with excellent bioavailability. Its volume of distribution and clearance have been calculated indirectly from assays of coagulation factor inhibition as no direct chemical assay is available to quantitate danaparoid directly. It has an antithrombin potentiation half-life of 90 min and an anti-factor Xa half-life of 18 h (NURMOHAMED et al. 1991). The major route of clearance is through renal excretion and the reticuloendothelial system is believed to contribute minimally (STIEKEMA et al. 1989). The unit of dosage is anti-factor X units and every 0.6 ml contains 750 anti-Xa units.

5. Clinical Uses

Danaparoid has been studied for both the treatment and prophylaxis of venous thromboembolic disease. In placebo-controlled trials, it was shown to be effective for prophyloxis of deep venous thrombosis in patients with thromboembolic stroke (TURPIE et al. 1987), hip fracture (BERGQVIST et al. 1992) and elective hip surgery (HOEK et al. 1992). It has also been compared to intravenous standard heparin in the treatment of venous thromboembolic disease, with danaparoid exhibiting an apparent reduction in recurrent events. The strength of the findings is limited because neither the pulmonary emboli nor deep venous thromboses were confirmed by angiography and control of the heparin dosing was uncertain (DE VALK et al. 1995). The recommended dose is 750 anti-Xa units SC twice daily.

6. Toxicity

Like all antithrombotic agents, bleeding is the most important side effect. Initial enthusiasm suggested it was a minor problem, but in the study by DE VALK et al. (1995) hemorrhage was as commonly associated with danaparoid as therapeutically dosed heparin. In surgical patients there are sporadic reports suggesting that bleeding may be a problem even at prophylactic doses, arguing against any proposed therapeutic safety factor. There are no reports of danaparoid inducing severe thrombocytopenia and it has been suggested as a substitute for heparin for cases of heparin-induced thrombocytopenia (MAGNANI 1993). As with any natural product, anaphylaxis can occur but often at a frequency too low to be detected in small studies.

7. Antidotes

Protamine is not very effective in reversing the effect of danaparoid and this together with its long half-life has raised some concern for its utility in treatment of thromboembolic disease since acute reversal may not be possible if and when it is needed.

8. Future

At the time of going to press the clinical role of danaparoid remains to be established. Based on its prolonged anti-factor Xa effect it is reasonable to presume its potential role will be focused on the prevention of deep venous thrombosis and, possibly, the treatment of severe heparin-induced thrombocytopenia. It is currently approved for use in the United States, specifically for prophylaxis of venous thrombosis in surgical patients.

V. Other Indirect Antithrombins

A great number of other agents that potentiate the activity of heparin cofactor II and antithrombin III have been described. These include fucoidan extracted from seaweed (MAURAY et al. 1995), pentasaccharide (the sequence of heparin that binds to the heparin binding site on antithrombin III), aprasulfate (a synthetic sulfated bis-lactobionic acid amide which binds heparin cofactor II) (JESKE and FAREED 1993), hemovasal (a heparin sulfate with weak antithrombotic activity but strongly profibrinolytic) (ZAWILSKA et al. 1995) and a sulfated *E. coli* K5 capsular polysaccharide which has a pentasaccharide sequence similar to heparin that enables it to bind to antithrombin III.

The unifying link is that all are sulfated glycosaminoglycans. Those that act via antithrombin III share a common pentasaccharide sequence and, now that this sequence is known, it may be used to define the role of antithrombin III in particular processes. It should be remembered that pentasaccharide alone is incapable of binding both antithrombin III and any coagulation factor simultaneously. Aprasulfate similarly is the smallest molecular structure that will potentiate heparin cofactor II and can be used experimentally to assess the contribution of heparin cofactor II to the antithrombotic process.

The physiology of glycosaminoglycans remains largely unknown. Space does not permit description of the above products or even a listing of the full range of compounds for which an indirect antithrombin effect has been claimed. Instead we have concentrated on those in clinical use or nearing approval for clinical applications.

D. Conclusion

It is probable that the future of indirect antithrombic agents will not center on the search for the elusive single agent that prevents all thrombosis while causing no hemorrhage. It might be better to consolidate the information on the vast array of agents currently available to ensure proper standardization of the products and of the assays used to monitor their efficacy. Indirect antithrombin agents, especially the prototype heparin, have been used well and effectively for a long time. Their liabilities, however, are also well known. Focusing on how to minimize the liabilities and improve application of current therapies may be the most important clinical objective for indirect antithrombotic agents in the near future. It is also critical that all patients for whom antithrombotic therapy is indicated receive it in a timely and correctly dosed manner. We may want to take a leaf from the studies on the use of fibrinolytic therapy in which efforts were concentrated on ensuring all eligible patients received therapy, rather than focusing on differences among the various therapies available.

The long term challenge for the pharmaceutical industry will be to employ the knowledge of structure and structure-specific binding in the design of nonsaccharide agents or other orally bioavailable agents that can catalyze and potentiate the effect of the natural antithrombins. This may be a difficult challenge as a certain chain length is required for inhibition of thrombin via antithrombin III. On the other hand, shorter chain molecules can be effective inhibitors of factor Xa, which may represent an even more appealing target than thrombin.

References

Aarskog D, Aksnes L, Lehmann V (1980) Low 1,25-dihydroxy vitamin D in heparin-induced osteopenia. Lancet 2:650–651

Agnelli G, Cosmi B, di Filippo P, Ranucci V, Veschi F, Longetti M (1992) A randomised, double-blind, placebo-controlled trial of dermatan sulfate for prevention of deep in thrombosis in hip fracture. Thromb Haemost 67:203–208

Agrati AM, Mauro M, Savasta C, Palmieri GC, Palazzini E (1992) A double-blind crossover placebo-controlled study of the profibrinolytic and antithrombotic effects of oral sulodexide. Adv Ther 9:147–155

Ahmed T, Abraham WM, D'Brot J (1992) Effects of inhaled heparin on immunologic and nonimmunologic bronchoconstrictor responses in sheep. Am Rev Respir Dis 145:566–570
Ahmed T, Garrigo J, Danta I (1993) Preventing bronchoconstriction in exercise-induced asthma with inhaled heparin. N Engl J Med 329:90–95
Amiral J, Bridey F, Dreyfus M, Vissoc AM, Fressinaud E, Wolf M (1992) Platelet factor 4 complexed to heparin is the target for antibodies generated in heparin-induced thrombocytopenia. Thromb Haemost 68:95–96
Amiral J, Marfaing-Koka A, Wolf M, Alessi MC, Tardy B, Boyer-Neumann C et al (1996) Presence of antibodies to interleukin-8 or neutrophil-activating peptide-2 in patients with heparin associated thrombocytopenia. Blood 88:410–416
Andersson LO, Barrowcliffe TW, Holmer E, Johnson EA, Sims GEC (1976) Anticoagulant properties of heparin fractionated by affinity chromatography on matrix-bound antithrombin III and by gel filtration. Thromb Res 9:575–583
Andrew M, Ofosu F, Schmidt B, Brooker L, Mirsh J, Buchanan MR (1989) Heparin clearance and ex vivo recovery in newborn piglets and adult pigs. Thromb Res 56:517–527
Andriuoli G, Mastacchi R, Barbanti M (1984) Antithrombotic activity of a glycosaminoglycan (sulodexide) in rats. Thromb Res 34:81–86
Ansell J, Tiarks C, Hirsh J, McGehee W, Adler D, Weibert R (1991) Measurement of the activated partial thromboplastin time from a capillary (finger stick) sample of whole blood. A new method for monitoring heparin therapy. Am J Clin Pathol 95:222–227
Attal M, Huguet F, Rubie H, Huynh A, Charlet JP, Payen JL et al (1992) Prevention of hepatic veno-occlusive disease after bone marrow transplantation by con tinuous infusion of low-dose heparin: a prospective, randomized trial. Blood 79:2834–2840
Austin GE, Ratliff NB, Hollman J, Tabei S, Phillips DT (1995) Intimal proliferation of smooth muscle cells as an explanation for recurrent coronary artery stenosis after percutaneous transluminal coronary angioplasty. J Am Coll Cardiol 6: 369–375
Baba M, Nakajima M, Schols D, Pauwels R, Balzarimi J, De Clercq E (1988) Pentosan oligosulphate, a sulphated polysaccharide, is a potent and selective anti-HIV agent in vitro. Antiviral Res 9:335–343
Baker WF Jr, Bick RL (1994) Deep vein thrombosis. Diagnosis and management. Med Clin North Am 78:685–712
Bara L, Billaud E, Kher A, Samama M (1985) Increased anti-Xa bioavailability for a low molecular weight heparin (PK 10169) compared with unfractionated heparin. Semin Thromb Hemost 11:316–317
Barbanti M, Calanni F, Babbini M, Bergonzini G, Parma B, Marchi E et al (1993) Antithrombotic activity of Desmin 370. Comparison with a high molecular weight dermatan sulfate. Thromb Res 71:417–422
Barritt DW, Jordan SC (1960) Anticoagulant drugs in the treatment of pulmonary embolism. Lancet 1:1309–1312
Bartolo M, Antignani PL, Eleuteri P (1984) Experiences with sulodexide in the arterial peripheral diseases. Curr Ther Res 36:979–988
Barzu T, Molho P, Tobelem G, Petitou M, Caen J (1985) Binding and endocytosis of heparin by human endothelial cells in culture. Biochim Biophys Acta 845:196–203
Becker RC, Corrao JM, Bovill EG, Gore JM, Baker SP, Miller ML et al (1990) Intravenous nitroglycerin-induced heparin resistance: a qualitative antithrombin III abnormality. Am Heart J 119:1254–1261
Beguin S, Hemker HC (1987) The mode of action of Pentosan polysulphate in plasma. Thromb Haemost 58:126
Beguin S, Lindhout T, Hemker HC (1988) The mode of action of heparin in plasma. Thromb Haemost 60:457–462
Bell WR (1986) Trousseau's syndrome. Grand Rounds Johns Hopkins Hospital. Vol XII Prog., 3 update 1996

Bell WR (1988) Heparin-associated thrombocytopenia and thrombosis. J Lab Clin Med 111:600–605
Bell WR, Pitney WR (1969) Management of priapism by therapeutic defibrination. N Engl J Med 280:649–650
Bell WR, Royall RM (1980) Heparin associated thrombocytopenia: a comparison of three heparin preparations. N Engl J Med 303:902–907
Bell WR, Simon TL, Stengle JM, Sherry S (1974) The urokinase – streptokinase pulmonary embolism trial (phase II) results. Circulation 50:1070–1071
Bell WR, Tomasulo PA, Alving BM, Duff TP (1976) Thrombocytopenia occurring during the administration of heparin. A prospective study in 52 patients. Ann Intern Med 85:155–160
Bell WR, Starksen NF, Tong S, Porterfield JK (1985) Trousseau's syndrome. Devastating coagulopathy in the absence of heparin. Am J Med 79:423–430
Bentley PG, Kakkar VV, Scully MF, MacGregor Jr, Webb P, Chan P et al (1980) An objective study of alternative methods of heparin administration. Thromb Res 18:177–187
Bergqvist D, Kettunen K, Fredin H, Fauno P, Suomalainen O, Soimakallio S et al (1992) Thromboprophylaxis in patients with hip fractures: a prospective, randomized, comparative study between Org10172 and Dextran 70. Surgery 109:617–622
Bjornsson TD, Wolfram KM (1982) Intersubject variability in the anticoagulant response to heparin in vitro. Eur J Clin Pharmacol 21:491–497
Bjornsson TD, Schneider DE, Hecht AR (1988) Effects of N-deacetylation and N-desulfation of heparin on its anticoagulant activity and in vivo disposition. J Pharmacol Exp Ther 245:804–808
Bokonjie R (1971) Haemostatics. Arztl Forsch 23:2033
Boneu B, Dol F, Caranobe C, Sie P, Houin G (1989) Pharmacokinetics of heparin and related polysaccharides. Ann NY Acad Sci 556:282–291
Borris L, Lassen MR (1995) A comparative review of the adverse effect profiles of heparins and heparinoids. Drug Saf 12:26–31
Bounameaux H, Huber O, Khabiri E, Schneider PA, Didier D, Rohner A (1993) Unexpectedly high rate of phlebographic deep venous thrombosis following elective general abdominal surgery among patients given prophylaxis with low-molecular weight heparin. Arch Surg 128:326–328
Brack MJ, More RS, Hubner PJ, Gershlick AH (1993) The effect of low dose nitroglycerin on plasma heparin concentrations and activated partial thromboplastin times. Blood Coagul Fibrinolysis 4:183–186
Brack MJ, Ray S, Chauhan A, Fox J, Hubner PJ, Schofield P et al (1995) The Subcutaneous Heparin and Angioplasty Restenosis Prevention (SHARP) trial. Results of a multicenter randomized trial investigating the effects of high dose unfractionated heparin on angiographic restenosis and clinical outcome. J Am Coll Cardiol 26:947–954
Brandjes DP, Heijboer H, Buller HR, de Rijk M, Jagt H, ten Cate JW (1992) Acenocoumarol and heparin compared with acenocoumarol alone in the initial treatment of proximal-vein thrombosis. N Engl J Med 327:145–148
Briant L, Caranobe C, Saivin S, Sie P, Bayrou B, Houin G et al (1989) Unfractionated heparin and CY216: pharmacokinetics and bioavailabilities of the antifactor Xa and IIa effects after intravenous and subcutaneous injection in the rabbit. Thromb Haemost 61:348–353
Brieger D, Dawes J (1996) Low molecular weight heparin is responsible for the anti-Xa activity of Desmin 370. Thromb Haemost 75:286–291
Brill-Edwards P, Ginsberg JS, Johnston M, Hirsh J (1993) Establishing a therapeutic range for heparin therapy. Ann Intern Med 119:104–109
Brinkhous KM, Smith HP, Warner ED, Seegers WH (1939) The inhibition of blood clotting: an unidentified substance which acts in conjunction with heparin to prevent the convert ion of prothrombin into thrombin. Am J Physiol 125:683

Brister SJ, Ofosu FA, Heigenhauser GJ, Gianese F, Buchanan MR (1994) Is heparin the ideal anticoagulant for cardiopulmonary bypass? Dermatan sulfate may be an alternate choice. Thromb Haemost 71:468–473
Broze GJ Jr, Warren LA, Girard JJ, Miletich JP (1987) Isolation of the lipoprotein associated coagulation inhibitor produced by HepG2 (human hepatoma) cells using bovine factor Xa affinity chromatography. Thromb Res 48:253–259
Broze GJ Jr, Warren LA, Novotny WF, Higuchi DA, Girard JJ, Miletich JP (1988) The lipoprotein-associated coagulation inhibitor that inhibits the factor VII-tissue factor complex also inhibits factor Xa: insight into its possible mechanism of action. Blood 71:335–343
Buchanan MR, Liao P, Smith LJ, Ofosu FA (1994) Prevention of thrombus formation and growth by antithrombin III and heparin cofactor II-dependent thrombin inhibitors: importance of heparin cofactor II. Thromb Res 74:463–475
Cairns JA, Gill J, Morton B, Roberts R, Gent M, Hirsh J et al (1996) Fish oils and low-molecular-weight heparin for the reduction of restenosis after percutaneous transluminal coronary angioplasty. The EMPAR Study. Circulation 94:1553–1560
Callas DD, Hoppensteadt DA, Jeske W, Igbal O, Bacher P, Ahsan A et al (1993) Comparative pharmacologic profile of a glycosaminoglycan mixture, Sulodexide, and a chemically modified heparin derivative, Suleparoide. Semin Thromb Hemost 19 Suppl 1:49–57
Camilleri JF, Bonnet JL, Bouvier JL, Levy G, Dijiane P, Bory M et al (1988) Thrombolyse intraveineuse dans l'infarctus du myocarde. Influence de la qualite de l'antiocoagulation sur le taux de recidives precoces d'angor ou d'infarctus (French). Arch Mal Coeur Vaiss 81:1037–1041
Caughey B, Raymond GJ (1993) Sulfated polyanion inhibition of scrapie-associated PrP accumulation in cultured cells. J Virol 67:643–650
Cejka V, De Vries LA, Smorenbery-Schoorl ME, Van Daatselaar JJ, Borst JGG, Majoor CLH et al (1960) Effect of heparinoid and spirolactone on the renal excretion of sodium and aldosterone. Lancet 1:317–319
Chiu HM, Hirsh J, Yung WL, Regoeczi E, Gent M (1977) Relationship between the anticoagulant and antithrombotic effects of heparin in experimental venous thrombosis. Blood 49:171–184
Choay J (1989) Chemically synthesized heparin-derived oligosaccharides. Ann NY Acad Sci 556:61–74
Chong BH (1995) Heparin-induced thrombocytopenia. Br J Haematol 89:431–439
Cipolle RJ, Seifert RD, Neilan BA, Zaske DE, Haus E (1981) Heparin kinetics: variables related to disposition and dosage. Clin Pharmacol Ther 29:387–393
Clagett GP, Salzman EW (1974) Prevention of venous thromboembolism in surgical patients. N Engl J Med 290:3–96
Clowes AW, Karnowsky MJ (1977) Suppression by heparin of smooth muscle cell proliferation in injured arteries. Nature 265:625–626
Cofrancesco E, Boschetti C, Leonardi P, Cortellaro M (1992) Dermatan sulfate in acute leukemia. Lancet 339:1177–1178
Colburn WA (1976) Pharmacologic implications of heparin interactions with other drugs. Drug Metab Rev 5:281–293
Cole CW, Bormanis J (1988) Ancrod: a practical alternative to heparin. J Vasc Surg 8:59–63
College of American Pathologists (1994) CAP Surveys, set CG@-A. College of American Pathologists, Northfield, Illinois, pp 5–9
Collins R, Scrimgeour A, Yusuf S, Peto R (1988) Reduction in fatal pulmonary embolism and venous thrombosis by perioperative administration of subcutaneous heparin overview of results of randomized trials in general, orthopedics and urologic surgery. N Engl J Med 318:1162–1173
Condorelli M, Chiarello M, Dagianti A, Penco M, Dalla Volta S, Pengo V et al (1994) IPO-V2: a prospective, multicenter, randomized, comparative clinical investigation of the effects of sulodexide in preventing cardiovascular accidents in the first year after acute myocardial infarction. J Am Coll Cardiol 23:27–34

Conrad HE (1989) Structure of heparan sulfate and dermatan sulfate. Ann NY Acad Sci 556:18–28
Coon WW, Willis PW 3d (1973) Recurrence of venous thromboembolism. Surgery 73:823
Copley AL, Robb TP (1941) The effect of heparin on the platelet count in dogs and mice. Am J Physiol 133:248
Crepaldi G, Fellin R, Calabro A, Rossi A, Ventura A, Mannarino E et al (1990) Double-blind multicenter trial on a new medium molecular weight glycosaminoglycan. Current therapeutic effects and perspectives for clinical use. Atherosclerosis 81:233–243
Curry N, Bardana EJ, Pirofsky B (1973) Heparin sensitivity. Report of a case. Arch Intern Med 132:744–745
Dahlman T, Lindvall N, Hellgren M (1990) Osteopenia in pregnancy during long-term heparin treatment: a radiological study post partum. Br J Obstet Gynaecol 97: 221–228
D'Angelo A, Seveso MP, D'Angelo SV, Gilardoni F, Dettori AG, Bonini P (1990) Effect of clot-detection methods and reagents on activated partial thromboplastin time (aPTT). Implications in heparin monitoring by aPTT. Am J Clin Pathol 94:297–306
Danielsson A, Raub E, Lindahl U, Bjork I (1986) Role of ternary complexes, in which heparin binds both antithrombin and proteinase, in the acceleration of the reactions between antithrombin and thrombin or factor Xa. J Biol Chem 261: 15467–15473
Dawes J, Papper DS (1979) Catabolism of low-dose heparin in man. Thromb Res 14:845–860
Dawes J, Pavuk N (1991) Sequestration of therapeutic glycosaminoglycans by plasma fibronectin. Thromb Haemost 65:929
De Bono DP, Simoons ML, Tijssen J, Arnold AE, Betriu A, Burgersdijk C et al (1992) Effect of early intravenous heparin on coronary patency, infarct size, and bleeding complications after alteplase thrombolysis: results of a randomized double blind European Cooperative Study Group trial. Br Heart J 67:122–128
de Swart CA, Nijmeyer B, Roelofs JM, Sixma JJ (1982) Kinetics of intravenously administered heparin in normal humans. Blood 60:1251–1258
de Valk HW, Banga JD, Wester JW, Brouwer CB, van Hessen MW, Meuwissen OJ et al (1995) Comparing subcutaneous danaparoid with intravenous unfractionated heparin for the treatment of venous thromboembolism. A randomized controlled trial. Ann Intern Med 123:1–9
Dechavanne M, Ville D, Berruyer M, Trepo F, Dalery F, Clermont N et al (1989) Randomized trial of a low-molecular-weight heparin (Kabi 2165) versus adjusted-dose subcutaneous standard heparin in the prophylaxis of deep-vein thrombosis after elective hip surgery. Haemostasis 11:5–12
Dehmer GJ, Fisher M, Tate DA, Teo S, Bonnem EM (1995) Reversal of heparin anticoagulation by recombinant platelet factor 4 in humans. Circulation 91: 2188–2194
Dettori AG, Galli G, Manotti C, Palazzini E (1994) Pharmacological activity of a low molecular weight dermatan sulfate (Desmin) in healthy volunteers. Semin Thromb Hemost 20:259–265
Dol F, Houin G, Rostin M, Montastruc JL, Dupouy D, Gianese F et al (1989) Pharmacodynamics and pharmacokinetics of dermatan sulfate in humans. Blood 74:1577–1582
Doyle DJ, Turpie AG, Hirsh J, Best C, Kinch D, Levine MN et al (1987) Adjusted subcutaneous heparin or continuous intravenous heparin in patients with acute deep venous thrombosis. A randomized trial. Ann Intern Med 107: 441–445
Dunn F, Soria J, Soria C, Thomaidis A, Tobelem G, Caen JP (1983) Fibrinogen binding on human platelets. Influence of different heparins and pentosane polysulfate. Thromb Res 29:141–148

Dupouy D, Sie P, Dol F, Boneu B (1988) A simple method to measure dermatan sulfate at sub-microgram concentrations in plasma. Thromb Haemost 60: 236–239

Efimov VS, Gradusov KA, Podolian VA, Gritsin VL, Koptsov VP (1995) The diagnostic value of tests for recalcification and blood tolerance for heparin when their performance is automated (Russian). Klin Lab Diagn 6:84–86

Eggleton CA, Barrowcliffe TW, Merton RE (1979) In vitro and in vivo studies of the anti-Xa activity of heparin. Thromb Haemost 42:1446–1451

Ehnholm C, Heaf DJ, Kaijser L, Kinnunen PK (1977) Heparin-induced release of lipase activity in the human forearm: an immunological study. Atherosclerosis 27:35–39

Eika C (1971) Inhibition of thrombin induced aggregation of human platelets by heparin. Scand J Haematol 8:216–222

Elgue G, Sanchez J, Egberg N, Olsson P (1994) Coagulation inhibition capacities of low-molecular mass and unfractionated heparin, as determined by thrombin generation. Thromb Res 75:539–549

Englebert H (1977) Probable physiologic functions of heparin. Fed Proc 36:70–72

Estes JW (1975) The fate of heparin in the bod. Curr Ther Res Clin Exp 18:45–57

Fellstrom B, Backman U, Danielson B,Wikstrom B (1994) Treatment of renal calcium stone disease with the synthetic glycosaminoglycan pentosan polysulphate. World J Urol 12:52–54

Fernandez F, Nguyen P, van Ryn J, Ofosu FA, Hirsh J, Buchanan MR (1986) Hemorrhagic doses of heparin and other glycosaminoglycans induce a platelet defect. Thromb Res 43:491–495

Ferrari GP, Marchesini D, Maggi AP (1994) Preliminary chemical, biochemical, and pharmacological characterization of a low molecular weight dermatan sulfate. Carbohydr Res 255:125–132

Fischer AM, Barrowcliffe TW, Thomas DP (1982) A comparison of pentosan polysulphate (SP54) and heparin. I: Mechanism of action on blood coagulation. Thromb Haemost 47:104–108

Follea G, Hamandjian I, Trzeciak MC, Nedey C, Streichenberger R, Dechavanne M (1986) Pentosan polysulphate associated thrombocytopenia. Thromb Res 42: 413–418

Fritze LMS, Reilly CF, Rosenberg RD (1985) An antiproliferative heparan sulfate species produced by postconfluent smooth muscle cells, J Cell Biol 100: 1041–1049

Fuster V, Dyken ML, Vokonas PS, Hennekens C (1993) Aspirin as a therapeutic agent in cardiovascular disease. Special Writing Group. Circulation 87:659–675

Ghosh P, Hutadilok N (1996) Interactions of pentosan polysulfate with cartilage matrix proteins and synovial fibroblasts derived from patients with osteoarthritis. Osteoarthritis Cartilage 4:43–53

Ghosk TK, Eis PS, Mullaney JM (1988) Competitive, reversible and potent antagonism of inositol 1,4,5 triphosphate activated calcium release by heparin. J Biol Chem 263:11075–11079

Ginsberg JS, Hirsh J (1992) Use of antithrombotic agents during pregnancy. Chest 1024. Suppl 385S–390S

Ginsberg JS, Kowalchuk G, Hirsh J, Brill-Edwards P, Burrows R, Coates G (1990) Heparin effect on bone density. Thromb Haemost 64:286–289

Gironell A, Altes A, Arboix A, Fontcuberta J, Munoz C, Marti-Vilalta JL (1996) Pentosan polysulfate-induced thrombocytopenia: a case diagnosed with an ELISA test used for heparin-induced thrombocytopenia. Ann Hematol 73:51–52

GISSI 2 (1990) 01SS1–2: a factorial randomised trial of alteplase versus streptokinase and heparin versus no heparin among 12,490 patients with acute myocardial infarction. Gruppo Italiano per lo Studio della Sopravvivenza nell'Infarto Miocardico. Lancet 336:65–71

Goad KE, Horne MK 3d, Gralnick HR (1994) Pentosan-induced thrombocytopenia: support for an immune complex mechanism. Br J Haematol 88:803–808

Goldberg MA, Ginsburg D, Mayer RJ, Stone RM, Maguire M, Rosenthal DS et al (1987) Is heparin administration necessary during induction chemotherapy for patients with acute promyelocytic leukemia? Blood 69:187–191
Goldhaber SZ, Haire WD, Feldstein ML, Miller M, Toltzis R, Smith JL et al (1993) Alteplase versus heparin in acute pulmonary embolism: randomised trial assessing right-ventricular function and pulmonary perfusion. Lancet 341:507–511
Greinacher A, Michels I, Mueller-Eckhardt C (1992) Heparin-associated thrombocytopenia: the antibody is not heparin specific. Thromb Haemost 67: 545–549
Griffith CC, Nichols G, Asher JD, Flanagan B (1965) Heparin osteoporosis. JAMA 193:91–94
Griffith MJ, Noyes CM, Tyndall JA, Church FC (1985) Structural evidence for leucine at the reactive site of heparin cofactor II. Biochemistry 24:6777–6782
GUSTO Angiographic Investigators (1993) The effects of tissue plasminogen activator, streptokinase, or both on coronary-artery patency, ventricular function, and survival after acute myocardial infarction. N Engl J Med 329:1615–1622
GUSTO IIa Investigators (1994) A randomized trial of intravenous heparin versus recombinant hirudin for acute coronary syndromes. The Global Use of Strategies to Open Occluded Coronary Arteries (GUSTO) IIa Investigators. Circulation 90:1631–1637
GUSTO IIb (1996) A comparison of recombinant hirudin with heparin for the treatment of acute coronary syndromes. The Global Use of Strategies to Open Occluded Coronary Arteries (GUSTO) IIb investigators. N Engl J Med 335: 775–782
Habbab MA, Haft JI (1986) Heparin resistance induced by intravenous nitroglycerin. Circulation Suppl II:74:321
Halkin H, Goldberg J, Modan M, Modan B (1982) Reduction of mortality in general medical in-patients by low-dose heparin prophylaxis. Ann Intern Med 96:561–565
Hansten PD (1974) In: Morsellin PL et al. (eds) Drug interconnections, 2nd edn. Lea & Febiger, Philadelphia, pp 230–231
Hardhammar PA, van Beusekom HM, Emanuelsson HU, Hofmash H, Albertsson PA, Verdouw PD et al (1996) Reduction in thrombotic events with heparin-coated Palmaz-Schatz stents in normal porcine coronary arteries. Circulation 93: 423–430
Hattersley PG (1966) Activated coagulation time of whole blood. JAMA 196: 436–440
Hirsh J, van Aken WG, Gallus AS, Dollery CT, Cade JF, Yung WL (1976) Heparin kinetics in venous thrombosis and pulmonary embolism. Circulation 53:691–695
Hodby ED, Hirsh J, Adeniyi-Jones C (1972) The influence of drugs upon the anticoagulant activity of heparin. Can Med Assoc J 106:562–564
Hoek JA, Nurmohamed MT, Hamelynck KJ, Marti RK, Knipscheer HC, ten Cate H et al (1992) Prevention of deep vein thrombosis following total hip replacement by a low molecular weight heparinoid. Thromb Haemost 67:28–32
Hook M, Bjork I, Hopwood J, Lindahl U (1976) Anticoagulant activity of heparin: separation of high-activity and low-activity heparin species by affinity chromatography on immobilized antithrombin. FEBS Lett 66:90–93
Hoppensteadt D, Walenga JM, Fareed J (1991) Effect of dermatan sulfate and heparan sulfate on platelet activity compared to heparin. Semin Thromb Hemost 17 Suppl 1:60–64
Hoppensteadt DA, Jeske W, Fareed J, Bermes EW Jr et al (1995) The role of tissue factor pathway inhibitor in the mediation of the antithrombotic actions of heparin and low-molecular weight-heparin. Blood Coagul Fibrinolysis 6:S57–S64
Hoylaerts M, Owen WG, Collen D (1984) Involvement of heparin chain length in the heparin-catalyzed inhibition of thrombin by antithrombin III. J Biol Chem 259:5670–5677
Hsia J, Hamilton WB, Kleiman N, Roberts R, Chaitman BR, Ross AM (1990) A comparison between heparin and low-dose aspirin as adjunctive therapy with

tissue plasminogen activator for acute myocardial infarction. Heparin-Aspirin Reperfusion Trial (HART) Investigators. N Engl J Med 323:1433–1437
Hull RD, Raskob GE, Rosenbloom D, Panju AA, Brill-Edwards P, Ginsberg JS et al (1990) Heparin for 5 days as compared with 10 days in the initial treatment of proximal venous thrombosis. N Engl J Med 322:1260–1264
Hull RD, Raskob GE, Rosenbloom D, Lemaire J, Pineo GF, Baylis B et al (1992) Optimal therapeutic level of heparin therapy in patients with venous thrombosis. Arch Intern Med 152:1589–1595
ISIS 3 (1992) ISIS-3: a randomized comparison of streptokinase versus tissue plasminogen activator versus anistreptase and of aspirin plus heparin versus aspirin alone among 41299 cases of suspected acute myocardial infarction. ISIS-3 (Third International Study of Infarct Survival) collaborative Group. Lancet 339:753–770
Jacques LB (1979) Heparin: an old drug with a new paradigm. Science 206:528–533
Jeske W, Fareed J (1993) Antithrombin III- and heparin cofactor II-mediated anticoagulant and antiprotease actions of heparin and its synthetic analogues. Semin Thromb Hemost 19:S241–S247
Jeske W, Lormeau JC, Callas D, Igbal O, Hoppensteadt D, Fareed J (1995) Antithrombin III affinity dependence on the anticoagulant, antiprotease, and tissue factor pathway inhibitor actions of heparins. Semin Thromb Hemost 21: 193–200
Jick H, Slone D, Borda IT, Shapiro S (1968) Efficacy and toxicity of heparin in relation to age and sex. N Engl J Med 279:284–286
Joffee SN (1975) The prevention of post operative deep venous thrombosis. South Afr J Surg 49:153–156
Jorpes E (1939) Heparin, its chemistry, physiology and application in medicine. Oxford University Press, London
Kakkar VV, Corrigan TP, Fossard DP, Schraibman IG, Evans CM, Rosenberg IL et al (1975) Prevention of fatal postoperative pulmonary embolism by low doses of heparin. An international multicentre trial. Lancet 2:45–51
Kandrotas RJ, Gal P, Douglas JB, Deterding J (1989) Heparin pharmacokinetics during hemodialysis. Ther Drug Monit 11:674–679
Kandrotas RJ, Gal P, Douglas JB, Deterding J (1990a) Pharmacokinetics and pharmacodynamics of heparin during hemodialysis: interpatient and intrapatient variability. Pharmacotherapy 10:349–355
Kandrotas RJ, Gal P, Groce JB, Douglas JB (1990b) Altered heparin pharmacodynamics in pulmonary embolism. Pharmacotherapy 10:241
Kaplan K, Davison R, Parker M, Mayberry B, Feiereisel P, Salinger M (1987) Role of heparin after intravenous thrombolytic therapy for acute myocardial infarction. Am J Cardiol 59:241–244
Kauffmann RH, Veltkamp JJ, van Tilburg NH, Van Es LA (1978) Acquired antithrombin III deficiency and thrombosis in the nephrotic syndrome. Am J Med 65:607–613
Kher A, Samama MM (1984) Maniement des anticoagulants chez le sujet age (French). Gaz Med 91:57–61
Khuri SF, Valeri CR, Loscalzo J, Weinstein MJ, Birjiniuk V, Healey NA et al (1995) Heparin causes platelet dysfunction and induces fibrinolysis before cardiopulmonary bypass. Ann Thorac Surg 50:1008–1014
Kim SW, Jacobs (1996) Design of nonthrombogenic polymer surfaces for blood-contacting medical devices. Blood Purif 14:357–372
Kiss J (1975) Chemistry of heparin. A short review on recent chemical trends. Thromb Diath Haemorrh 33:20–25
Kitchen S, Jennings I, Woods TA, Preston JE (1996) Wide variability in the sensitivity of aPTT reagents for monitoring of heparin dosage. J Clin Pathol 49:10–14
Kjellen L, Pettersson I, Unger S et al (1992) In: Lane DA, Björk I, Lindahl U (eds) Heparin and related polysaccharides, vol 313. Plenum, New York, pp 107–120
Klein HG, Bell WR (1974) Disseminated intravascular coagulation during heparin therapy. Ann Intern Med 80:477–481

Knot E, Ten-Cate JW, Drijfhout HR, Kahle LH, Tytgat GN (1984) Antithrombin III metabolism in patients with liver disease. J Clin Pathol 37:523–530
Koh KK, Park GS, Song JH, Moon TH, In HH, Kim JJ et al (1995) Interaction of intravenous heparin and organic nitrates in acute ischemic syndromes. Am J Cardiol 76:706–709
Lam LH, Silbert JE, Rosenberg RD (1976) The separation of active and inactive forms of heparin. Biochem Biophys Res Commun 69:570–577
Lambert M, Laterne PF, Leroy C, Lavenne E, Coche E, Moriau M (1986) Modifications of liver enzymes during heparin therapy. Acta Clin Belg 41:307–310
Lane DA, Pejler G, Flynn AM, Thompson EA, Lindahl U (1986) Neutralization of heparin-related saccharides by histidine-rich glycoprotein and platelet factor 4. J Biol Chem 261:3980–3986
Larsen AK, Lund DP, Langer R, Folkman J (1986) Oral heparin results in the appearance of heparin fragments in the plasma of rats. Proc Natl Acad Sci USA 83: 2964–2968
Leclercq C, de Place C, Rioux C, Mabo P, Paillard F, Daubert JC (1990) Infarctus du myocarde et thrombose biventriculaire massive au cours d'une thrombopenie induite par le pentosane polysulfate et l'heparine (French). Arch Mal Coeur Vaiss 83:117–120
Levine MN, Raskob G, Hirsh J (1989) Hemorrhagic complications of long-term anticoagulant therapy. Chest 95:26S–36S
Lindahl U, Axelsson O (1971) Identification of iduronic acid as the major sulfated uronic acid of heparin. J Biol Chem 246:82–74
Lindahl U, Backstrom G, Thurnberg L, Leder IG (1980) Evidence for a 3-*O*-sulfated D-glucosamine residue in the antithrombin-binding sequence of heparin. Proc Natl Acad Sci USA 77:6551–6555
Lindahl U, Kusche M, Lidholt K, Oscarsson LG (1989) Biosynthesis of heparin and heparan sulfate. Ann NY Acad Sci 556:36–50
Linhardt RJ, Desai UR, Liu J, Pervin A, Hoppensteadt D, Fareed J (1994) Low molecular weight dermatan sulfate as an antithrombotic agent. Structure-activity relationship studies. Biochem Pharmacol 47:1241–1252
Liu L, Freedman J, Hornstein A, Ofosu FA (1995) Catalytically equivalent doses of heparin and dermatan sulfate inhibit platelet dependent prothrombinase production to the same extent in plasma. (Abstract) #3507 Blood 88:880
Maffrand JP, Herbert JM, Bernat A, Defreyn G, Delebassee D, Savi P et al (1991) Experimental and clinical pharmacology of pentosan polysulfate. Semin Thromb Hemost 17 Suppl 2:186–198
Maggi A, Abbadini N, Pagella PG, Borowska A, Pangrazzi J, Donati MB (1987) Antithrombotic properties of dermatan sulphate in a rat venous thrombosis model. Haemostasis 17:329–335
Magnani HN (1993) Heparin-induced thrombocytopenia (HIT): an overview of 230 patients treated with Orgaran (Org 10172). Thromb Haemost 70:554–561
Maimone MM, Tollefsen DM (1990) Structure of a dermatan sulfate hexasaccharide that binds to heparin cofactor II with high affinity. J Biol Chem 265:18263–18271
Manson L, Young E, Hirsh J (1996) Altered recovery but not clearance contributes to the variation in anticoagulant response to heparin in vivo. (Abstract) #3511 Blood 88:881
Marsh NA, Peyser PM, Creighton LJ, Mahmoud M, Gaffney PJ (1985) The effect of pentosan polysulphate (SP54) on the fibrinolytic enzyme system – a human volunteer and experimental animal study. Thromb Haemost 54:833–837
Mascellani G, Livenni L, Parma B, Bergonzini G, Bianchini P (1996) Active site for heparin cofactor II in low molecular mass dermatan sulfate. Contribution to the antithrombotic activity of fractions with high affinity for heparin cofactor II. Thromb Res 84:21–32
Mauray S, Sternberg C, Theveniaux J, Milllet J, Sinquin C, Tapon-Bretaudiere J et al (1995) Venous antithrombotic and anticoagulant activities of a fucoidan fraction. Thromb Haemost 74:1280–1285

McAvoy TJ (1979) Pharmacokinetic modeling of heparin and its clinical implications, J Pharmacokinet Biopharm 7:331–354
McLean J (1916) The thromboplastic action of cephalin. Am J Physiol 41:250–267
McLean J (1959) The discovery of heparin. Circulation 19:75–78
Medalion B, Merin G, Aingorn H, Miao H, Nagler A, Elami A et al (1997) Endogenous basic fibroblast growth factor displaced by heparin from the lumenal surface of human blood vessels is preferentially sequestered by injured regions of the vessel wall. Circulation 95:1853–1862
Meuleman DG (1992) Orgaran (Org 10172): its pharmacological profile in experimental models. Haemostasis 22:58–65
Monreal M, Lafoz E, Olive A, del Rio L, Vedia C (1994) Comparison of subcutaneous unfractionated heparin with a low molecular weight heparin (Fragmin) in patients with venous thromboembolism and contraindications to Coumarin. Thromb Haemost 71:7–11
Morabia A (1986) Heparin doses and major bleedings. Lancet I:1278–1279
MRC Trial (1969) Assessment of short-anticoagulant administration after cardiac infarction. Report of the Working Party on Anticoagulant Therapy in Coronary Thrombosis to the Medical Research Council. Br Med J 1:335–342
Murray DWG (1939) Heparin in thrombosis and embolism. Br J Surg 27:567–598
National Heart Foundation of Australia Coronary Thrombolysis Study Group (1989) A randomized comparison of oral aspirin/dipyridamole versus intravenous heparin after rtPA for acute myocardial infarction. Circulation 80 Suppl III, p 14
Neri-Serneri GG, Rovelli F, Gensini GF, Pirelli S, Carnovali M, Fortini A (1987) Effectiveness of low-dose heparin in prevention of myocardial reinfarction. Lancet 1:937–942
Neri-Serneri GG, Gensini SF, Poggesi L, Trotta F, Modesti PA, Boddi M et al (1990) Effect of heparin, aspirin, or alteplase in reduction of myocardial ischaemia in refractory unstable angina. Lancet 335:615–618
Nicholson CD, Meuleman DG, Magnani HN, Egberts JF, Leibowitz DA, Spinler SA et al (1994) Danaparoid is not a low-molecular-weight heparin. Am J Hosp Pharm 51:2049–2050
Nielsen LE, Bell WR, Borkon AM, Neill CA (1987) Extensive thrombus formation with heparin resistance during extracorporeal circulation. A new presentation of familial antithrombin III deficiency. Arch Intern Med 147:149–152
Nieuwenhuis K, Albada J, Banga JD, Sixma JJ (1991) Identification of risk factors for bleeding during treatment of acute venous thromboembolism with heparin or low molecular weight heparin. Blood 78:2337–2343
Nucci MR, Bell WR (1994) Acquired hypercoagulable states. In: Loscalzo J, Schafer AI (eds) Thrombosis and hemorrhage. Blackwell Scientific Publishers, Boston, pp 835–860
Nurmohamed MT, Fareed J, Hoppensteadt D, Walenga JM, ten Cate JW (1991) Pharmacological and clinical studies with Lomaparan, a low molecular weight glycosaminoglycan. Semin Thromb Hemost 17 Suppl 2: 205–213
Nurmohamed MT, Knipscheer HC, Stevens P, Krediet RT, Roggekamp MC, Berckmans RJ (1993) Clinical experience with a new antithrombotic (dermatan sulfate) in chronic hemodialysis patients. Clin Nephrol 39:166–171
Nyhan DP, Weiss M, Hirshman CA (1989) Immunological aspects of anesthetic agents. Anesth Rev Vol XVI Suppl:10–18
Nyhan DP, Shampaine EL, Hirshman CA, Hamilton RG, Frank SM, Baumgartner WA et al (1996) Single doses of intravenous protamine result in the formation of protamine-specific IgE and IgG antibodies. J Allergy Clin Immunol 97:991–997
Ofosu FA, Sie P, Modi GJ, Fernandez F, Buchanan MR, Blajchman MA et al (1987) The inhibition of thrombin dependent positive-feedback reactions is critical to the expression of the anticoagulant effect of heparin. Biochem J 243:579–588
Ofosu FA, Hirsh J, Esmon CT, Modi GJ, Smith LM, Anvari N et al (1989) Unfractionated heparin inhibits thrombin-catalysed amplification reactions of

coagulation more efficiently than those catalysed by factor Xa. Biochem J 257: 143–150

Olds RJ, Lane DA, Chowdhury V, de Stefano V, Leone G, Thein SL (1993) Complete nucleotide sequence of the antithrombin gene: evidence for homologous recombination causing thrombophilia. Biochemistry 32:4216–4224

Oleson ES (1965) Activation of the blood fibrinolytic system. Munksgaard, Copenhagen, pp 111–117

Olson ST, Bjork I (1991) Predominant contribution of surface approximation to the mechanism of heparin acceleration of the antithrombin-thrombin reaction. Elucidation from salt concentration effects. J Biol Chem 266:6353–6364

Olson ST, Bjork I (1992) Role of protein conformational changes, surface approximation and protein cofactors in heparin-accelerated antithrombin-proteinase reactions. Adv Exp Med Biol 313:155–165

Olson ST, Srinivasan KR, Bjork I, Shore JD (1981) Binding of high affinity heparin to antithrombin III. Stopped flow kinetic studies of the binding interaction. J Biol Chem 256:11073–11079

Oster JR, Singer I, Fishman LM (1995) Heparin-induced aldosterone suppression and hyperkalemia. Am J Med 98:575–586

Palazzine E, Procida C (1978) Effect of some mucopolysaccharides on activated factor X. Biochem Pharmacol 27:608–610

Parker KA, Tollefsen DM (1985) The protease specificity of heparin cofactor II. Inhibition of thrombin generated during coagulation. J Biol Chem 260:3501–3505

Perry MO, Horton J (1976) Kinetics of heparin administration. Arch Surg 111:403–409

Petitou M, Duchaussoy I, Lederman I, Choay J, Jacquinet JC, Sinay P et al (1987) Synthesis of heparin fragments: a methyl alpha-pentaoside with high affinity for antithrombin III. Carbohydr Res 167:67–75

Pini M, Pattachini C, Quintavalla R, Poli T, Megha A, Tagliaferri A et al (1990) Subcutaneous versus intravenous heparin in the treatment of deep venous thrombosis – a randomized clinical trial. Thromb Haemost 64:222–226

Pisano L, Moronesi F, Falco F, Stipa E, Fabbiani N, Dolfi R et al (1986) The use of sulodexide in the treatment of peripheral vasculopathy accompanying metabolic diseases. Controlled study in hyperlipidemic and diabetic subjects. Thromb Res 41:23–31

Pizzulli L, Nitsch J, Luderitz B (1988) Hemmung der Heparinwirkung durch Glyceroltrinitrat. Dtsch Med Wochenschr 133:1837–1840

Platell CF, Tan EG (1986) Hypersensitivity reactions to heparin: delayed onset thrombocytopenia and necrotizing skin lesions. Aust N Z J Surg 56:621–623

Pollack EW, Sparks FC, Barker WF (1973) Pulmonary embolism: an appraisal of therapy in 516 cases. Arch Surg 107:492

Pomerantz MW, Owen WG (1978) A catalytic role for heparin. Evidence for a ternary complex of heparin cofactor thrombin and heparin. Biochim Biophys Acta 535: 66–77

Prager RL, Dunn EL, Kirsh MM, Penner JA (1979) Endotoxin-induced intravascular coagulation (DIC) and its therapy. Adv Shock Res 2:277–287

Preissner KT, Muller-Berghaus G (1987) Neutralization and binding of heparin by Sprotein/vitronectin in the inhibition of factor Xa by antithrombin III. Involvement of an inducible heparin-binding domain of S protein/vitronectin. J Biol Chem 262:12247–12253

Proctor RR, Rapaport SI (1961) The partial thromboplastin time with kaolin. Am J Clin Pathol 36:212–219

Quick AJ, Shanberge JN, Stefanini M (1948) The effect of heparin on platelets in vivo. J Lab Clin Med 33:1424–1430

Radhakrishnamurthy B, Sharma C, Bhandaru RR, Berenson GS, Stanzani L, Mastacchi R (1986) Studies of chemical and biologic properties of a fraction of sulodexide, a heparin-like glycosaminoglycan. Atherosclerosis 60:141–149

Ragazzi E, Chinelatto A (1995) Heparin: pharmacological potentials from atherosclerosis to asthma. Gen Pharmacol 26:697–701

Ragg H (1986) A new member of the plasma protease inhibitor gene family. Nucleic Acids Res 14:1073–1088
Ragg H, Ulshofer T, Gerewitz J (1990) Glycosaminoglycan-mediated leuserpin-2/thrombin interaction. Structure-function relationships. J Biol Chem 265: 22386–22391
Rajtar G, Marchi E, de Gaetano G, Cerletti C (1993) Effects of glycosaminoglycans on platelet and leucocyte function: role of N-sulfation. Biochem Pharmacol 46: 958–960
Rao LV, Rapaport SI (1987) Studies of a mechanism inhibiting the initiation of the extrinsic pathway of coagulation. Blood 69:645–651
Rodeghiero F, Avvisati G, Castaman G, Barbui T, Mandelli F (1990) Early deaths and antihemorrhagic treatments in acute promyelocytic leukemia. A GINEMA retrospective study in 268 consecutive patients. Blood 75:2112–6505
Rogers SJ, Pratt CW, Whinna HC, Church FC (1992) Role of thrombin exosites in inhibition by heparin cofactor II. J Biol Chem 267:3613–3617
Rosenberg RD, Damus PS (1973) The purification and mechanism of action of human antithrombin-heparin cofactor. J Biol Chem 248:6490–6505
Rosendaal FR, Nurmohamed MT, Bueller HR, Dekker E, Vande brouke JP, Brieet E (1991) Low molecular weight heparin in the prophylaxis of venous thrombosis: a meta-analysis. Thromb Haemost 65:927
Ryan KE, Lane DA, Flynn A, Ireland M, Boisclair M, Shepperd J et al (1992) Antithrombotic properties of dermatan sulfate (MF701) in hemodialysis for chronic renal failure. Thromb Haemost 68:563–569
Ryde M, Eriksson H, Tange O (1981) Studies on the different mechanisms by which heparin and polyunsulphated xylan (PZ68) inhibit blood coagulation in man. Thromb Res 23:435–445
Sack GH Jr, Levin J, Bell WR (1977) Trousseau's syndrome and other manifestations of chronic disseminated coagulopathy in patients with neoplasms: clinical, pathophysiologic, and therapeutic features. Medicine Baltimore 56: 1–37
Samama MM (1995) Laboratory monitoring of unfractionated heparin treatment. Clin Lab Med 15:109–117
Schriever HG, Epstein SE, Mintz MD (1973) Statistical correlation and heparin sensitivity of activated partial thromboplastin time, whole blood coagulation time, and an automated coagulation time. Am J Clin Pathol 60:323–329
Schuster EH, Achuff SC, Bell WR, Bulkley BH (1980) Multiple coronary thromboses in previously normal coronary arteries: a rare cause of acute myocardial infarction. Am Heart J 99:506–509
Scully MF, Ellis V, Kakkar VV (1986) Pentosan polysulphate: activation of heparin cofactor II or antithrombin III according to molecular weight fractionation. Thromb Res 41:489–499
Serruys PW, Emanuelsson H, Van der Giessen W, Lunn AC, Kiemeney F, Macaya C et al (1996) Heparin-coated Palmaz-Schatz stents in human coronary arteries. Early outcome of the Benestent-II Pilot Study. Circulation 93:412–422
Sharath MD, Metzger WJ, Richerson HB, Scupham RK, Meng RL, Ginsberg BH et al (1985) Protamine-induced fatal anaphylaxis. Prevalence of antiprotamine immunoglobulin E antibody. J Thorac Cardiovasc Surg 90:86–90
Sie P (1988) Les Heparinoides (French). Rev Prat 38:952–956
Sie P, Albarede JL, Robert M, Bouloux C, Lansen J, Chigot C et al (1986) Tolerance and biological activity of pentosan polysulfate after intramuscular or subcutaneous administration for ten days in humans volunteers. Thromb Haemost 55:86–89
Simmons RB, Newton GR, Doctor VM (1995) Effect of sulfated xylcans during the interaction of (^{125}I)-thrombin with antithrombin III or heparin cofactor II of human plasma. Eur J Drug Metab Pharmacokinet 20:73–77
Simon TL (1978) The rationale for continuous heparin infusion. Drug Ther 3:17
Simon TL, Hyers TM, Gaston JP, Harker LA (1978) Heparin pharmacokinetics: increased requirements in pulmonary embolism. Br J Haematol 39:111–120

Sobel M, McNeill PM, Carlson PL, Kermode JC, Adelman B, Conroy R et al (1991) Heparin inhibition of von Willebrand factor-dependent platelet function in vitro and in vivo. J Clin Invest 87:1787–1793
Spence CR, Thompson BT, Janssens SP, Steigman DM, Hales CA (1993) Effect of aerosol heparin on the development of hypoxic pulmonary hypertension in the guinea pig. Am Rev Respir Dis 148:241–244
Spero JA, Lewis JH, Hasiba U (1980) Disseminated intravascular coagulation. Findings in 346 patients. Thromb Haemost 43:28–33
Stein PE, Leslie AGW, Finch JT, Turnell WG, McLaughlin PJ, Carrell RW (1990) Crystal structure of ovalbumin as a model for the reactive centre of serpins. Nature 347:99–102
Stiekema JC, Wijnand HP, Van Dinther TG, Moelker HC, Dawes J, Vinchenzo A et al (1989) Safety and pharmacokinetics of the low molecular weight heparinoid Org 10172 administered to healthy elderly volunteers. Br J Clin Pharmacol 27: 39–48
Stone RM, Mayer RJ (1990) The unique aspects of acute promyelocytic leukemia. J Clin Oncol 8:1913–1921
Swain SM, Parker B, Wellstein A, Lippmann ME, Steakley C, DeLap R (1995) Phase I trial of pentosan polysulfate. Invest New Drugs 13:55–62
Tebbe U, Windeler J, Boesl I, Hoffmann H, Wojcik J, Ashmawy M et al (1995) Thrombolysis with recombinant unglycosylated single chain urokinase-type plasminogen activator in acute myocardial infarction: influence of heparin on early patency rate (LIMITS study). Liquemin in Myocardial Infarction During Thrombolysis with Saruplase. J Am Coll Cardiol 26:365–373
Teien AN (1977) Heparin elimination in patients with liver cirrhosis. Thromb Haemost 38:701–706
Teien AN, Bjornsson J (1976) Heparin elimination in uraemic patients on haemodialysis. Scand J Haematol 17:29–35
Teien AN, Abildgaard U, Hook M (1976a) The anticoagulant effect of heparan sulfate and dermatan sulfate. Thromb Res 8:859–867
Teien AN, Lie M, Abildgaard U (1976b) Assay of heparin in plasma using a chromogenic substrate for activated factor X. Thromb Res 8:413–416
Theroux P, Oumet H, McCans J, Latour JG, Joly P, Levy G et al (1988) Aspirin, heparin, or both to treat acute unstable angina. N Engl J Med 319:1105–1111
Theroux P, Waters D, Lam J, Juneau M, McCans J (1992) Reactivation of unstable angina after the discontinuation of heparin. N Engl J Med 327:141–145
Thomas DP, Gray E, Merton RE (1990) Potentiation of the antithrombotic action of dermatan sulphate by small amounts of heparin. Thromb Haemost 64: 290–293
Tiozzo R, Cingi MR, Pietrangelo A, Albertazzi L, Calandra S, Milani MR (1989) Effect of heparin-like compounds on the in vitro proliferation and protein synthesis of various cell types. Arzneimittelforschung 39:15–20
Tollefsen DM, Majerus DW, Blank MK (1982) Heparin cofactor II. Purification and properties of a heparin-dependent inhibitor of thrombin in human plasma. J Biol Chem 257:2162–2169
Tollefsen DM, Pestka CA, Monafo WJ (1983) Activation of heparin cofactor II by dermatan sulfate. J Biol Chem 258:6713–6716
Tullis TL (1976) Clot. Charles C Thomas, Springfield, Illinois, p 495
Turpie AG, Levine MN, Hirsh J, Carter J, Jay RM, Powers PJ et al (1987) A double-blind randomised trial of Org 10172 low-molecular-weight heparinoid in the prevention of deep-vein thrombosis thrombotic stroke. Lancet 332:523–526
Turpie AG, Robinson JG, Doyle DJ, Mulji AS, Miskel GJ, Sealey BJ et al (1989) Comparison of high dose with low-dose subcutaneous heparin to prevent left ventricular mural thrombosis in patients with acute transmural anterior myocardial infarction. N Engl J Med 320:352–357
VA Cooperative Study (1973) Anticoagulants in acute myocardial infarction: results of a cooperative clinical trial. JAMA 225:724–729

Van Dedem G, de Leeuw den Bouter H (1993) The nature of the glycosaminoglycan in Organan (Org 10172). Thromb Haemost 69:652

Van Deerlin VM, Tollefsen DM (1991) The N-terminal acidic domain of heparin cofactor II mediates the inhibition of alpha-thrombin in the presence of glycosaminoglycans. J Biol Chem 266:20223–20231

Van Delden CJ, Engbers GHM, Feijen J (1995) Interaction of antithrombin III with surface-immobilized albumin-heparin conjugates. J Biomed Mater Res 29: 1317–1329

Vinazzer H (1991) Effect of pentosan polysulfate on fibrinolysis: basic tests and clinical application. Semin Hem Thromb Hem 17:375–378

Vinazzer H, Haas S, Stemberger A (1980) Influence on the clotting mechanism of sodium pentosan polysulfate (SP54) in comparison to commercial beef lung sodium heparin. Thromb Res 20:57–68

Visentin GP, Ford SE, Scott JP, Aster RH (1994) Antibodies from patients with heparin-induced thrombocytopenia/thrombosis are specific for platelet factor 4 complexed with heparin or bound to endothelial cells. J Clin Invest 93:81–88

Von Kaulla KN, Huseman NE (1946) Über Beziehungen der Molekülgröße und -gestalt zur physiologischen Wirkung. Experienta 2:222–224

Von Schon H, Sauer M (1963) Die lipolytische Aktivität einiger antiatherosklerotischer Medikamente. Arzneimittelforsch 13:718–722

Wagenvoord R, Hendrix H, Soria C, Hemker HC (1988) Localization of the inhibitory site(s) of pentosan polysulphate in blood coagulation. Thromb Haemost 60: 220–225

Warkentin TE, Soutar RL, Panju A, Ginsberg JS (1992) Acute systemic reactions to intravenous bolus heparin therapy: characterization and relationship to heparin-induced thrombocytopenia. Blood 80:160a

Weismann RI, Tobin RW (1958) Arterial embolisation occurring during systemic heparin therapy. Arch Surg 76:219

Weiss ME, Nyhan D, Peng Z, Horrow JC, Lowenstein E, Hirshman C (1989) Association of protamine IgE and IgG antibodies with life-threatening reactions to intravenous protamine. N Engl J Med 320:886–892

Whitfield LR, Levy G (1980) Relationship between concentration and anticoagulant effect of heparin in plasma of normal subjects: magnitude and predictability of interindividual differences. Clin Pharmacol Ther 28:509–516

Will RG, Ironside JW, Zeidler M, Cousens SN, Estibeiro K, Alperovitch A et al (1996) A new variant of Creutzfeldt-Jakob disease in the UK. Lancet 347: 921–925

Wille-Jorgensen P, Ott P (1990) Predicting failure of low-dose prophylactic heparin in general surgical procedures. Surg Gynecol Obstet 171:126–130

Wilson ID, Goetz FC (1964) Selective hypoaldosteronism after prolonged heparin administration. Am J Med 36:635–640

Wilson JE 3d, Bynum LJ, Parkey RW (1981) Heparin therapy in venous thromboembolism. Am J Med 70:808–816

Wolzt M, Weltermann A, Nieszpaur-Los M, Schneider B, Fassolt A, Lechner K et al (1995) Studies on the neutralizing effects of protamine on unfractionated and low molecular weight heparin (Fragmin) at the site of activation of the coagulation system in man. Thromb Haemost 73:439–443

Wun TC (1992) Lipoprotein-associated coagulation inhibitor (LACI) is a cofactor for heparin: synergistic anticoagulation action between LACI and sulfated polysaccharides. Blood 79:430–438

Wun TC, Kretzmer KK, Girard TJ, Miletich JP, Broze GJ Jr (1988) Cloning and characterization of a cDNA coding for the lipoprotein-associated coagulation inhibitor shows that it consists of three tandem Kunitz-type inhibitory domains. J Biol Chem 263:6001–6004

Yin ET, Tangen O (1976) Heparin, heparinoids and blood coagulation. In: Kakkar VV, Thomas DP (eds) Heparin, chemistry and clinical usage. Academic Press, London, pp 121–124

Zammit A, Dawes J (1994) Low-affinity material does not contribute to the antithrombotic activity of Orgaran (Org 10172) in human plasma. Thromb Haemost 71:759–767

Zawilska K, Elikowski W, Turowiecka Z, Zozulinska M, Grzywacz A, Przybyl L et al (1995) On the action of a heparan-like glycosaminoglycan (Hemovasal) on the mechanism of haemostasis and fibrinolysis. Thromb Res 78:211–216

CHAPTER 10

Low Molecular Weight Heparin

G.F. PINEO and R.D. HULL

A. Introduction

Unfractionated heparin is used widely for the prevention of venous thromboembolism in medical patients or in patients undergoing various surgical procedures. Furthermore, unfractionated heparin, given by a continuous intravenous infusion with laboratory monitoring using the activated partial thromboplastin time (APTT), with warfarin starting on day 1 or day 2 and continued for 3 months, has been the standard treatment of established venous thromboembolism. Heparin is used in a number of other clinical settings and constitutes one of the most frequently used agents in hospital medicine. Over the past 15 years, various low molecular weight heparins have been evaluated against a number of different controls, including unfractionated heparin for many of these clinical problems. In a number of countries, the low molecular weight heparins have replaced unfractionated heparin for both the prevention and treatment of venous thromboembolism. The low molecular weight heparins have also been used in clinical trials for the prevention and treatment of arterial thrombosis. In this chapter, we review the problems related to the use of unfractionated heparin, compare the low molecular weight heparins with unfractionated heparin, and discuss their role in the prevention and treatment of venous thromboembolism, as well as in arterial thrombosis.

B. Discovery and Development of Low Molecular Weight Heparins

I. Properties of Unfractionated Heparin

Unfractionated heparin from either porcine or bovine sources has been available for clinical use for several decades. Although heparin has been studied extensively, much remains uncertain about its mode of action, particularly the non-anticoagulant properties, and some of the complications have only recently been better understood (HIRSH et al. 1995).

The anticoagulant activity of unfractionated heparin depends upon a unique pentasaccharide which binds to antithrombin III (ATIII) and potentiates the inhibition of thrombin and activated factor X (X_a) by ATIII (LANE 1989; LINDAHL et al. 1984, 1979). About one-third of all heparin molecules contain the unique pentasaccharide sequence regardless of whether they are low or high in molecular weight fractions. It is the pentasaccharide sequence which confers the molecular high affinity for ATIII (Lindahl et al. 1979; Rosenberg and Lam 1979). The remaining two-thirds of heparin has minimal anticoagulant activity at the therapeutic concentrations that are used clinically. For the inhibition of thrombin, heparin must form a bridge between thrombin and ATIII, but for the inhibition of factor X_a this bridging is not necessary (Lindahl et al. 1984; Rosenberg and Lam 1979). It has been shown that molecules of heparin with fewer than 18 saccharide units are unable to bind thrombin and ATIII simultaneously and as a result cannot catalyze thrombin inhibition (Casu et al. 1981). Heparin fragments with smaller numbers of saccharide units are capable of catalyzing the inhibition of factor X_a by ATIII providing the high-affinity pentasaccharide sequence is present. Unfractionated heparin is unable to inhibit thrombin bound to fibrin whereas the specific antithrombin agents do so (Weitz et al. 1990). Heparin does not inhibit factor X_a bound to platelets (Salzman et al. 1980).

Heparin also catalyzes the inactivation of thrombin by another plasma cofactor (cofactor II), which acts independently of ATIII (Tollefesen et al. 1982). Heparin has a number of other effects. Those related to the anticoagulant effects of heparin include the release of tissue factor pathway inhibitor (Hoppensteadt et al. 1995), binding to numerous plasma and platelet proteins, endothelial cells and leukocytes (Lane 1989; Barzu et al. 1984), suppression of platelet function (Salzman et al. 1980) and increase in vascular permeability (Blajchman et al. 1989).

The anticoagulant response to a standard dose of heparin varies widely between patients. Heparin is poorly absorbed from the subcutaneous site especially at lower doses (Bara et al. 1985). The plasma clearance of heparin depends on a dose-related renal clearance and a non-dose-related saturable cellular mechanism (Bjornsson et al. 1982). The binding of heparin to plasma proteins, endothelial cells and platelets contributes to the unpredictable response (Hoppensteadt et al. 1995; Barzu et al. 1984). Some patients develop relative heparin resistance and require a large dose of heparin to achieve a response in the APTT (Levine et al. 1994). Recent studies have documented a rebound thrombin generation when heparin is abruptly stopped (Theroux et al. 1992; Granger et al. 1995). It is therefore necessary to frequently monitor the anticoagulant response of heparin using either the APTT or heparin levels and to titrate the dose to the individual patient.

Unless a prescriptive heparin nomogram is used, many patients receive inadequate heparin in the initial 24–48 h of treatment (Fennerty et al. 1985; Wheeler et al. 1988; Cruickshank et al. 1991). This inadequate therapy has been shown to increase the incidence of venous thromboembolism during follow-up (Hull et al. 1986; Raschke et al. 1993; Brandjes et al. 1992). Treatment is further complicated by the fact that there is a diurnal variation in the APTT response in patients on a constant infusion of intravenous heparin (Hull et al. 1992b). A peak response is seen at 3 a.m. and a reduction of heparin infusion in response to the high APTT could result in subtherapeutic treatment later in the day (HULL et al. 1992). There is a wide variation in the sensitivity of various thromboplastins used in performing the APTT, and even with the same thromboplastin, different coagulometers may yield different results (BRILL-EDWARDS et al. 1993). It is necessary for each laboratory to define a therapeutic range with respect to APTT in terms of heparin blood levels (therapeutic range 0.2–0.4 units/ml heparin) (BRILL-EDWARDS et al. 1993).

The use of heparin is associated with a number of complications, the most serious of which is bleeding. Bleeding is primarily related to underlying clinical risk factors but is also increased in females and individuals over the age of 65 years (CAMPBELL et al. 1996). The relationship to heparin dosage and APTT levels is less clear-cut. A second significant complication of heparin is heparin-induced thrombocytopenia (WARKENTIN and KELTON 1996; BOSHKOV et al. 1993; KELTON et al. 1984; AREPALLY et al. 1995). When accompanied by venous or arterial thrombosis, heparin-induced thrombocytopenia carries a significant morbidity and mortality rate (WARKENTIN and KELTON 1996; BOSHKOV et al. 1993; KELTON et al. 1984). Awareness of this complication and constant vigilance will permit the early diagnosis of heparin-induced thrombocytopenia and permit the institution of appropriate therapy (WARKENTIN and KELTON 1996; BOSHKOV et al. 1993; KELTON et al. 1984). Agents thought to be effective in case series include the defibrinogenating agent ancrod (Arvin) (DEMERS et al. 1991), the heparinoid Danaparoid (MAGNANI 1993; WARKENTIN 1996) and specific thrombin inhibitors such as hirudin (GREINACHER et al. 1996) or argatroban (MATSUO et al. 1992).

Other complications of heparin include: osteoporosis most frequently documented in pregnancy, elevated liver enzymes, hypoaldosteronism, hypersensitivity and allergic skin reactions and heparin induced skin necrosis (HIRSH et al. 1995).

The hope that the low molecular weight heparins may overcome some of the problems related to heparin therapy and decrease some of the complications has stimulated a large number of trials comparing low molecular weight heparins with unfractionated heparin.

II. Antithrombotic Properties of Low Molecular Weight Heparin

Over the past 15 years a number of low molecular weight heparin fractions of unfractionated heparin have become available for commercial use (HIRSH and LEVINE 1992). The low molecular weight heparins are manufactured from unfractionated heparin (usually of porcine origin) by controlled depolymerization using either chemical (nitrous oxide, alkaline hydrolysis or peroxidative cleavage) or enzymatic (heparinase) techniques (FAREED et al. 1988a). The low molecular weight fractions have a molecular weight between 4000 and 6000, with 60% of the polysaccharide chains having a molecular weight between 2000 and 8000. The various low molecular weight heparins

Table 1. Some commercial low molecular weight heparins and some of their properties

Trade name	International nonproprietary name (INN)	Method of production	Mean molecular weight	Anti-$X_{a/IIa}$ ratio
Logiparin Innohep	Tinzaparin	HD	5866	1.9:1
Fragmin	Dalteparin	NAP	5819	2.1:1
Lovenox	Enoxaparin	AH	4371	2.7:1
Fraxiparin	Nadroparin	NAP	4855	3.2:1
Reviparin	Clivarin	NAP	4653	3.6–6.1[a]
Normoflo	Ardeparin	PC	6000	2.0:1

HD, heparinase digestion; NAP, nitrous acid depolymerization; AH, alkaline hydrolysis; PC, peroxidative cleavage.
Manufacturers: Logiparin, Novo Nordisk; Innohep, Leo Laboratories; Fragmin, Pharmacia-Upjohn; Lovenox, Rhone Poulenc Rorer; Fraxiparin, Sonofi; Reviparin, Knoll AG, Normoflo, Wyeth-Ayerst.
[a] Range provided by Knoll AG.

differ in terms of mean molecular weight, glycosoaminoglycan content and anticoagulant activity in terms of anti X_a and anti II_a activity (FAREED et al. 1988a,b; BARA and SAMAMA 1988; BRIANT et al. 1989; ANDERSSON et al. 1979; BARROWCLIFFE et al. 1988; HOLMER et al. 1986; CARTER et al. 1982). The various fractions have different pharmacologic profiles in terms of bioavailability, plasma clearance, and release of tissue factor pathway inhibitor and in experimental models they have different antithrombotic and hemorrhagic properties (FAREED et al. 1988a,b; BARA and SAMAMA 1988; BRIANT et al. 1989; ANDERSSON et al. 1979; BARROWCLIFFE et al. 1988; HOLMER et al. 1986; CARTER et al. 1982). The low molecular weight heparins which have been most thoroughly studied are shown in Table 1; the method of production, molecular weight and anti X_a to anti II_a ratio are shown as well. Because the low molecular weight heparins are different compounds with distinct pharmacologic properties (FAREED et al. 1988a,b; BARA and SAMAMA 1988; BRIANT et al. 1989; ANDERSSON et al. 1979; BARROWCLIFFE et al. 1988) and because different regimens have been used in clinical trials, it may be inappropriate to use meta-analyses for comparing the effects of low molecular weight heparin with placebo, unfractionated heparin, dextran or warfarin.

Despite the various differences between the low molecular weight heparins, the clinical outcomes in clinical trials are very similar particularly in prophylactic studies using lower doses. In the higher doses used in treatment of thrombotic disorders it is quite possible that differences in outcomes will become apparent.

There has been a hope that the low molecular weight heparins will have fewer serious complications such as bleeding (CARTER et al. 1982; CADE et al. 1984; ANDRIUOLI et al. 1985; LENSING et al. 1995), osteoporosis (MONREAL et al.

1989b, 1990; MATZSCH et al. 1990; SHAUGHNESSY et al. 1995) and heparin-induced thrombocytopenia (WARKENTIN et al. 1995) when compared with unfractionated heparin. Evidence is accumulating that these complications are indeed less serious and less frequent with the use of low molecular weight heparin. Low molecular weight heparin has not been approved for the prevention or treatment of venous thromboembolism in pregnancy. These drugs do not cross the placenta (FORESTIER et al. 1984; OMRI et al. 1989) and small case series suggest that they are both effective and safe (MELISSARI et al. 1992; WAHLBERG and KHER 1994; HUNT et al. 1997). However, at the present time the standard treatment for venous thromboembolism in pregnancy is twice daily adjusted doses of subcutaneous unfractionated heparin (GINSBERG and HIRSH 1995). The low molecular weight heparins all cross-react with unfractionated heparin and they can therefore not be used as alternative therapy in patients who develop heparin-induced thrombocytopenia. The heparinoid Danaparoid possesses a 10%–20% cross-reactivity with heparin and it can be safely used in patients who have no cross-reactivity (MAGNANI 1993).

III. Advantages of Low Molecular Weight Heparin over Unfractionated Heparin

The low molecular weight heparins differ from unfractionated heparin in numerous ways as demonstrated in Table 2. Of particular importance are the following: increased bioavailability (ANDERSSON et al. 1979; FAREED et al. 1988) (>90% after subcutaneous injection), prolonged half-life (FAREED et al. 1988; ANDERSSON et al. 1979), predictable clearance, enabling once or twice daily injection (BONEU et al. 1988), and predictable antithrombotic response based on body weight, permitting treatment without laboratory monitoring (MATZSCH et al. 1990). Patients or other caregivers in the home can be readily taught to inject low molecular weight heparin, particularly from pre-filled syringes, and numerous clinical trials involving self-injection for 3 or 4 weeks,

Table 2. Comparison between low molecular weight heparin and unfractionated heparin. (Modified with permission from HIRSH and LEVINE 1992)

	Low molecular weight heparin	Unfractionated heparin
Mean molecular weight	4000–6500	12000–15000
Saccharide units	13–22	40–50
Anti X_a : anti II_a activity	2.1 : 1–4 : 1	1 : 1
Bioavailability at lower doses	High	Low
Dose-dependent clearance	–	+
Inhibited by platelet factor 4	+++	+
Inhibits platelet bound X_a	+	–
Inhibits platelet function	++	++++
Increases vascular permeability	–	+

or even longer, have documented that patient compliance is very high. Other possible advantages are their ability to bind to platelet-bound factor X_a (BONEU et al. 1985), resistance to inhibition by platelet factor 4 (SALZMAN et al. 1980; BARZU et al. 1984) and their decreased effect on platelet function (SALZMAN et al. 1980) and vascular permeability (BLAJCHMAN et al. 1989) (possibly accounting for less hemorrhagic effects at comparable anti-thrombotic doses) (CARTER et al. 1982; CADE et al. 1984; ANDRIUOLI et al. 1985; LENSING et al. 1995).

C. Use of Low Molecular Weight Heparins

I. Prevention of Venous Thromboembolism

There are two approaches to the prevention of fatal pulmonary embolism: (a) secondary prevention involves the early detection and treatment of subclinical venous thrombosis by screening postoperative patients with objective tests that are sensitive for venous thrombosis; and (b) primary prophylaxis is carried out using either drugs or physical methods that are effective for preventing deep-vein thrombosis. The latter approach, primary prophylaxis, is preferred in most clinical circumstances. Furthermore, prevention of deep-venous thrombosis and pulmonary embolism is more cost-effective than treatment of the complications when they occur (SALZMAN and DAVIES 1980; HULL et al. 1982; OSTER et al. 1987; BERGQVIST et al. 1990; HAUCH et al. 1991). Secondary prevention by case-finding studies should never replace primary prophylaxis. It should be reserved for patients in whom primary prophylaxis is either contraindicated or relatively ineffective.

The ideal thromboprophylactic agent would prevent all deep venous thromboses (DVTs), be free from side effects, be simple to apply or administer, need no laboratory monitoring and be cost-effective. The prophylactic measures most commonly used are low-dose or adjusted dose unfractionated heparin, low molecular weight heparin (LMWH), oral anticoagulants [International Normalized Ratio (INR) of 2 to 3], and intermittent pneumatic leg compression. Low molecular weight heparin fulfils many of the properties of an ideal prophylactic agent, in that it can be given by a once or twice daily subcutaneous injection, without laboratory monitoring. It is effective and safe, and it is cost-effective (HULL et al. 1997a).

It has become standard practice to commence prophylaxis, for example, with low-dose heparin, prior to anesthesia in patients undergoing thoraco-abdominal surgery, and in Europe prophylaxis is started the night before surgery in patients undergoing total hip or total knee replacement surgery. In North America, because of the concern related to postoperative bleeding, prophylaxis for patients having total knee or total hip replacement has been started postoperatively (KEARON and HIRSH 1995). This difference in the patterns of practice may account for the differences in the rates of postoperative venous thrombosis in Europe and North America (HULL et al. 1996). Clinical trials are currently underway to compare the efficacy and safety of preoperative prophylaxis with postoperative commencement of prophylaxis within the same trial.

The duration of prophylaxis required after high-risk procedures such as total joint replacement is currently under intensive study. There is evidence that thrombin can be generated after the discontinuation of prophylaxis 1 week after total hip replacement and these laboratory abnormalities can be demonstrated up to 35 days postoperatively (DAHL et al. 1995). Studies from Europe have demonstrated the presence of venous thrombosis at days 28–35 (BERGQVIST et al. 1996; PLANES et al. 1996; DAHL et al. 1997; LASSEN and BORRIS 1995). In three of these studies, negative venograms were required at discharge before patients could continue on either low molecular weight heparin or placebo (PLANES et al. 1996; DAHL et al. 1997; LASSEN and BORRIS 1995). Thrombosis rates were lower in patients who continued on low molecular weight heparin as compared to those on placebo. Similar studies are underway in North America. These studies support the need for extended prophylaxis after such procedures.

II. Orthopedic Surgery

A number of randomized clinical trials have been performed comparing low molecular weight heparin with placebo, intravenous dextran, low doses or adjusted dose unfractionated heparin or warfarin for the prevention of venous thrombosis following total hip replacement (TURPIE et al. 1986; TØRHOLM et al. 1991; LASSEN et al. 1991; THE DANISH ENOXAPARIN STUDY GROUP 1991; LEVINE et al. 1991; ERIKSSON et al. 1991; PLANES et al. 1991; COLWELL et al. 1994; HULL et al. 1993; HAMULYAK et al. 1995) (Table 3). The drugs under investigation and their dosage schedules vary from one clinical trial to another, making comparison across trials difficult. Furthermore it has been shown that even within the same clinical trial there can be considerable intercenter variability (HULL et al. 1993). Low molecular weight heparin is usually started the night before surgery in the European trials, in contrast to North American trials, where it is started 12–24 h postoperatively. Total bleeding rates vary quite widely across trials as well, making comparisons difficult.

Two recent decision analyses compared the cost-effectiveness of enoxaparin with warfarin in patients undergoing hip replacement (O'BRIEN et al. 1994; MENZIN et al. 1995). The cost-effectiveness compared well with other medical interventions, although enoxaparin was more expensive than warfarin. A recent economic evaluation of low molecular weight heparin vs. warfarin prophylaxis after total hip or knee replacement identified that low molecular weight heparin was cost-effective (HULL et al. 1997).

Although the number of patients undergoing total knee replacement now equals those undergoing total hip replacement, there have been fewer trials in this patient population. Clinical trials in North America and Europe comparing low molecular weight heparins started either preoperatively or postopera-

Table 3. Randomized trials of low molecular weight heparin prophylaxis for deep vein thrombosis following hip replacement surgery: total deep vein thrombosis and bleeding

Reference	Treatment	No. of patients	Total deep vein thrombosis (%)	Total bleeding (%)
TURPIE et al. (1955)	Enoxaparin	40	10.0	4.0
	Placebo	40	60.6	4.0
TORHOLM et al. (1991)	Dalteparin	58	16.0	NA
	Placebo	54	35.0	NA
LASSEN et al. (1991)	Tinzaparin	93	31.0	9.5
	Placebo	97	45.0	12.6
Danish Enoxaparin Study Group	Enoxaparin	108	6.5	13.9
	Dextran 70	111	21.6	23.4
LEVINE et al.	Enoxaparin	258	19.4	5.1
	Unfractionated heparin	263	23.2	9.3
LEYVRAZ et al.	Nadroparin (Fraxiparin)	198	12.6	0.5
	Unfractionated heparin	199	16.0	1.5
ERIKSSON et al. (1991)	Dalteparin (Fragmin)	67	30.2	1.5
	Unfractionated heparin	68	42.4	7.4
PLANES et al.	Enoxaparin	120	12.5	2.4
	Heparin	108	25.0	1.8
COLWELL et al. (1994)	Enoxaparin	136	21.0	10.0
	Enoxaparin	136	6.0	12.0
	Heparin	142	1.5	12.0
HULL et al.	Tinzaparin (Logiparin)	332	21.0	4.1
	Warfarin	340	23.0	3.8
HAMULYAK et al. (1995)	Nadroparin	195	13.8	2.4[a]
	Warfarin	196	13.8	5.2

NA, not available.
[a] Clinically important plus minor bleeding for combined hip and knee replacement patients.

tively have shown lower rates of total and proximal deep vein thrombosis when compared with warfarin (HULL et al. 1993; HAMULYAK et al. 1995; LECLERC et al. 1992, 1995; HEIT et al. 1997) (Table 4). Rates of major bleeding are somewhat lower with warfarin (HULL et al. 1993; LECLERC et al. 1995).

D. Trauma

A recent study showed that low molecular weight heparin is superior to low-dose unfractionated heparin in the prevention of venous thromboembolism in patients suffering multiple trauma (GEERTS et al. 1996).

Table 4. Randomized control trials of low molecular weight heparin prophylaxis for deep vein thrombosis bleeding following total knee replacement: total deep vein thrombosis and bleeding

Reference	Treatment	No. of patients	Total deep vein thrombosis (%)	Total bleeding (%)
LECLERC et al.	Enoxaparin	41	20.0	6.1
	Placebo	54	65.0	6.2
HULL et al.	Tinzaparin	317	45.0	4.4
	Warfarin	324	54.0	2.4
LECLERC et al.	Enoxaparin	206	37.0	33.0
	Warfarin	211	52.0	30.2
HEIT et al. (1997)	Ardeparin	232	27.0[a]	7.9
	Warfarin	222	38.0	4.4
HAMULYAK et al. (1995)	Nadroparin	65	24.6	2.4[b]
	Warfarin	61	37.7	5.2

[a] Venogram on operated leg only.
[b] Clinical important and minor bleeding for combined hip and knee replacement patients.

For the prevention of venous thromboembolism following hip fracture, low molecular weight heparin was superior to low-dose heparin (MONREAL et al. 1989), the heparinoid Danaparoid was more effective than dextran (BERGQVIST et al. 1991) or aspirin (POWERS et al. 1993) and dermatan sulfate was superior to placebo (AGNELLI et al. 1992).

For patients suffering acute spinal cord injury with paralysis, low molecular weight heparin was superior to low-dose heparin (GREEN et al. 1990), adjusted dose heparin or intermittent pneumatic compression (GREEN 1992). In patients undergoing elective neurosurgical procedures, low molecular weight heparin started postoperatively, along with graduated compression stockings, was superior to graduated compression stockings alone (NURMOHAMED et al. 1996).

E. General Surgery

A number of low molecular weight heparin fractions have been evaluated by randomized clinical trials in moderate-risk general surgical patients (KAKKAR et al. 1993; KAKKAR and MURRAY 1985; BERGQVIST et al. 1988, 1995; SAMAMA et al. 1988; THE EUROPEAN FRAXIPARIN STUDY GROUP 1988; CAEN 1988; LEIZOROVICZ et al. 1991; NURMOHAMED et al. 1995). The low molecular weight heparins that have been most extensively evaluated include Dalteparin, Nadroparin, Enoxaparin, and Tinzaparin. In randomized clinical trials comparing low molecular weight heparin with unfractionated heparin, the low molecular weight heparins given once or twice daily have been shown to be as effective or more effective in preventing thrombosis (KAKKAR et al. 1993;

Kakkar and Murray 1985; Bergqvist et al. 1988, 1995; Samama et al. 1988; The European Fraxiparin Study Group 1988; Caen 1988; Leizorovicz et al. 1991; Nurmohamed et al. 1995). In most of the trials, similar low frequencies of bleeding for low molecular weight heparin and low-dose unfractionated heparin were documented, although the incidence of bleeding was lower in the low molecular weight heparin group as evidenced by a reduction in the incidence of wound hematoma, severe bleeding, and the number of patients requiring reoperation for bleeding (Kakkar et al. 1993).

A meta-analysis showed low molecular weight heparin to be more effective than unfractionated heparin in the prevention of venous thrombosis, but the risk of bleeding was slightly higher (Nurmohamed et al. 1992). It should be noted, however, that the findings of meta-analyses evaluating the low molecular weight heparins should be interpreted with caution, because all the low molecular weight heparins differ.

I. Medical Patients

For the prevention of deep vein thrombosis following acute thrombotic stroke, the heparinoid Danaparoid (Organon) was superior to low-dose heparin (Turpie et al. 1987), low molecular weight heparin was superior to placebo in one study (Prins et al. 1987), whereas there was no difference between low molecular weight heparin and placebo in another study (Sandset et al. 1990). In medical inpatients, low molecular weight heparin is as effective as low-dose heparin in preventing venous thromboembolism (Harenberg et al. 1996; Bergmann and Neuhart 1996).

II. Low Molecular Weight Heparinoid

The low molecular weight heparinoid Danaparoid (Organon) has been evaluated in patients undergoing surgery for cancer (Gallus et al. 1993), hip fractures (Bergqvist et al. 1991; Gerhart et al. 1991), and total hip replacement (Hoek et al. 1989). The thrombosis rates were similar with Danaparoid and unfractionated heparin in patients undergoing cancer surgery (Gallus et al. 1993). In patients undergoing surgery for hip fracture, the deep-vein thrombosis rates were significantly lower compared with intravenous dextran (Bergqvist et al. 1991) and with low-intensity warfarin (Gerhart et al. 1991). More blood transfusions were required in the dextran group. Compared with placebo, the rates of deep-vein thrombosis following total hip replacement were significantly lower (Hoek et al. 1989).

F. Treatment of Venous Thromboembolism

In the treatment of established venous thromboembolism, low molecular weight heparin given by subcutaneous injection has a number of advantages over continuous intravenous unfractionated heparin; it can be given by once or twice daily subcutaneous injection and the antithrombotic response to low molecular weight heparin is highly correlated with body weight, permitting administration of a fixed dose without laboratory monitoring.

In a number of early clinical trials (some of which were dose finding), low molecular weight heparin given by subcutaneous or intravenous injection was compared with continuous intravenous unfractionated heparin, with repeat venography at day 7–10 being the primary end point (BRATT et al. 1985; HOLM et al. 1986; ALBADA et al. 1989; BRATT et al. 1990; HARENBERG et al. 1990; SIEGBAHN et al. 1989). These studies demonstrated that low molecular weight heparin was at least as effective as unfractionated heparin in preventing extension or increasing resolution of thrombi on repeat venography (GRATT et al. 1985; HOLM et al. 1986; ALBADA et al. 1989; BRATT et al. 1990; HARENBERG et al. 1990; SIEGBAHN et al. 1989). More recently the more relevant clinical end points of recurrent venous thromboembolism or death during follow-up have been used as end points (HULL et al. 1992; PRANDONI et al. 1992; LOPACIUK et al. 1992; SIMONNEAU et al. 1993; LINDMARKER et al. 1994). These studies are not all comparable because different regimens of low molecular weight heparins were used, not all studies ensured that adequate intravenous heparin therapy was given or properly monitored and some studies entered patients with distal as well as proximal deep vein thrombosis. Only one study was double blinded although others used blinded assessment of outcome measures for both efficacy and safety. Low molecular weight heparin was given for 6–10 days with warfarin therapy starting either on day 2 (HULL et al. 1992) or on day 7–10 (PRANDONI et al. 1992; LOPACIUK et al. 1992; SIMONNEAU et al. 1993; LINDMARKER et al. 1994). Warfarin was continued for 3 months, with the target INR range being 2.0–3.0. The outcomes in terms of recurrent venous thromboembolism, major bleeding and mortality for five clinical trials using clinical end points are summarized in Table 5. When the results of two clinical trials were pooled there was a striking decrease in mortality in the patients receiving low molecular weight heparin particularly for patients with cancer (GREEN et al. 1992). Most of the abrupt deaths could not be attributed to thromboembolic events, suggesting that the benefits of low molecular weight heparin may not be entirely related to thrombotic events.

A cost-effectiveness analysis indicated that low molecular weight heparin was cost-effective when compared with continuous intravenous heparin under the study protocol conditions because monitoring was not necessary and there were fewer complications requiring rehospitalization and treatment (HULL et al. 1997). These findings were verified by a sensitivity analysis. It was estimated that 37% of patients on low molecular weight heparin could have been discharged on day 2, which would have further increased the cost-effectiveness of the low molecular weight heparin (HULL et al. 1997).

As the clinical trials comparing low molecular weight heparin with intravenous heparin for the treatment of proximal venous thrombosis were designed quite differently, certain questions remain unanswered. Most of the

Table 5. Randomized trials of low molecular weight heparin vs. unfractionated heparin for the in-hospital treatment of proximal deep vein thrombosis: results of long-term follow-up

Reference	Treatment	Recurrent venous thromboembolism [no. (%)]	Major bleeding [no. (%)]	Mortality [no. (%)]
HULL et al.	Tinzaparin	6/213 (2.8)	1/213 (0.5)[a]	10/213 (4.7)[a]
	Heparin	15/219 (6.8)	11/219 (5.0)	21/219(9.6)
PRANDONI et al. (1992)	Fraxiparine	6/85 (7.1)	1/85 (1.2)	6/85 (7.1)
	Heparin	12/85 (14.1)	3/85 (3.8)	12/85 (14.1)
LOPACIUK et al. (1992)[b]	Fraxiparine	0/74 (0)	0/74	0/74
	Heparin	3/72 (4.2)	1/72 (1.4)	1/72 (1.4)
SIMONNEAU et al. (1993)	Enoxaparin	0/67	0/67	3/67 (4.5)
	Heparin	0/67	0/67	2/67 (3.0)
LINDMARKER et al. (1994)[c]	Dalteparin	5/101 (5.0)	1/101	2/101 (2.0)
	Heparin	3/103 (2.9)	0/103	3/103 (2.9)

[a] $p < 0.05$ vs. heparin.
[b] In 19.5% there was calf vein deep vein thrombosis + involvement of the popliteal.
[c] In 42.6% there was distal deep vein thrombosis only.

studies have used twice-daily dosing, but the outcomes were not better than those seen with once-daily dosing. Furthermore, when comparable doses of the same agents have been given in twice-daily as opposed to once-daily injections within the same trial, the outcomes have been similar. Finally, as with unfractionated heparin, it may be beneficial to deliver a large dose of low molecular weight heparin in a once-daily fashion, in order to adequately suppress thrombin generation in patients with thromboembolic disease. Further studies should clarify these questions regarding dosage and scheduling, but at the present time it will be necessary to consider the low molecular weight heparins as distinct agents, rather than as a class.

G. Out-of-Hospital Treatment of Venous Thromboembolism with Low Molecular Weight Heparin

Two recently reported studies indicate that in selected patients low molecular weight heparin treatment can be administered safely out of hospital (LEVIN et al. 1996; KOOPMAN et al. 1996). Patients who met the entry criteria were randomized to receive twice daily low molecular weight heparin either entirely out of hospital or with early discharge or continuous intravenous heparin in hospital. Warfarin was started on day 1 or 2 and continued for 3 months. Both studies showed equivalence with respect to the incidence of recurrent venous thromboembolism, major bleeding and mortality rates (LEVINE et al. 1996; KOOPMAN et al. 1996). In these studies 31% and 33% of patients with proximal venous thrombosis were eligible for entry.

A third study published in abstract form gave similar results (The Columbus Investigators 1996). In that study, 30% of the patients presented with pulmonary embolism, and the efficacy and safety of low molecular weight heparin was equal to that of unfractionated heparin. The increased use of out-of-hospital low molecular weight heparin for the initial treatment of venous thromboembolism will clearly improve bed utilization and decrease health care costs, but there will always be a population of patients who will require in-hospital treatment.

H. Role of Low Molecular Weight Heparin in the Prevention and Treatment of Arterial Thrombosis

I. Unstable Angina

The short-term and long-term use of low molecular weight heparin for the prevention of vascular events in patients with unstable angina is currently under active study. In a small, open study in patients with unstable angina, low molecular weight heparin plus aspirin was compared with low-dose heparin plus aspirin or aspirin alone for a period of 5–7 days or sooner if an end point occurred (Gurfinkel et al. 1995). For the prevention of recurrent angina, myocardial infarction and revascularization procedures, low molecular weight heparin plus aspirin was superior to the two other groups.

In three large multicenter studies, low molecular weight heparin has been compared with either placebo or intravenous unfractionated heparin for the prevention of recurrent angina, myocardial infarction, urgent revascularization or death in patients with unstable angina or non-Q-wave myocardial infarction [Fragmin during Instability in coronary Artery Disease (FRISC) Study Group 1996; Turpie et al. 1995; Cohen et al. 1996]. In the FRISC study (Fragmin during instability in coronary artery disease) LMWH (Dalteparin) was compared with an identical subcutaneous placebo twice daily for the initial 5–8 days following a diagnosis of unstable angina or non-Q-wave myocardial infarction (in-hospital phase) (Fragmin during Instability in coronary Artery Disease (FRISC) Study Group 1996). Subsequently, Dalteparin once a day or a corresponding placebo was continued for 35–45 days (out-of-hospital phase). All patients took aspirin and conventional antiangina drugs throughout the study. During the in-hospital phase, the incidence of death and new myocardial infarction was significantly lower in the low molecular weight heparin group, as was the need for intravenous heparin and the need for revascularization. During the 35- to 40-day long-term treatment (out-of-hospital phase), the incidence of death and new myocardial infarction, and the need for revascularization or heparin infusion, remained lower with the low molecular weight heparin group, although the difference was not significant. The addition of low molecular weight heparin to aspirin did not increase the incidence of major bleeding.

In the FRIC study (Fragmin in unstable coronary heart disease) twice daily subcutaneous low molecular weight heparin (Dalteparin) was compared with intravenous heparin by continuous infusion for a period of 6 days, followed by long-term subcutaneous low molecular weight heparin once a day, with an identical placebo for 40 days (TURPIE et al. 1995). As with the FRISC study, the FRIC study was double-blind. During the in-hospital phase (6-day outcomes), the risk of death, myocardial infarction and recurrent angina, and the need for revascularization were similar. Also during the prolonged treatment phase (day 6–45) there was no difference in the composite end points of recurrent angina, myocardial infarction, death or revascularization. The incidence of major bleeding was similar in the two groups.

The initial report from the ESSENCE study (Efficacy and Safety of Subcutaneous Enoxaparin in non Q Wave Coronary Events) was recently announced (COHEN et al. 1996). In this study, low molecular weight heparin (Enoxaparin) given by twice-daily subcutaneous injection was compared with intravenous heparin by continuous infusion in patients with unstable angina and non-Q-wave myocardial infarction for an average of 3.5 days. All patients received aspirin. At 14 days, the incidence of recurrent angina, myocardial infarction and death were significantly lower in the low molecular weight heparin group than in the intravenous heparin group. Again at 30 days, there was still a significant difference in the composite end points, favoring the low molecular weight heparin group. There was no increase in the risk of major bleeding with the low molecular weight heparin.

II. Thrombotic Stroke

In a recent study, low molecular weight heparin in two different doses was compared with an identical placebo in patients with acute ischemic stroke (KAY et al. 1995). The main outcome measures were poor outcome (death or dependency in daily activities) at 6 months, and secondary outcomes included death, hemorrhagic transformation of the infarct, other complications at 10 days, and poor outcome at 3 months. At the 6-month follow-up, patients who received the high-dose low molecular weight heparin for 10 days had significantly fewer poor outcomes when compared with the placebo group. The difference for the low-dose low molecular weight heparin compared with placebo did not reach statistical significance. There was no significant difference in the secondary outcomes. The hope that the use of low molecular weight heparin heparin can decrease the long-term morbidity associated with ischemic stroke has stimulated the development of further such studies, which are currently underway.

III. Peripheral Vascular Disease

Low molecular weight heparin was compared with aspirin and dipyridamole for the maintenance of graft patency in patients undergoing femoropopliteal bypass grafting (EDMONDSON et al. 1994). Stratified survival analysis at 12 months showed that low molecular weight heparin was superior to aspirin and dipyridamole for patients requiring limb salvage surgery, whereas there was no significant benefit for those patients having surgery for claudication. No major bleeding events occurred in either group.

IV. Hemodialysis

A number of low molecular weight heparins and the heparinoid Danaparoid have been used in a number of open studies in patients undergoing hemodialysis, but there have been no formal comparisons with standard heparin (NURMOHAMED et al. 1994). Studies suggest that low molecular weight heparin is as effective and safe as standard heparin, and has the advantage that a single predialysis bolus injection can be given without the need for laboratory monitoring. At this stage, it is not known whether the low molecular weight heparins will be safer than unfractionated heparin in patients at high risk of bleeding. The heparinoid Danaparoid has been successfully used in patients with known heparin-induced thrombocytopenia (NURMOHAMED et al. 1994).

V. Other Vascular Problems

Low molecular weight heparins have been used in an attempt to prevent restensois following primary angioplasty. Low molecular weight heparin used for 28 days (FAXON et al. 1994) or 6 weeks (CAIRNS et al. 1996) has failed to demonstrate a beneficial effect. In patients undergoing coronary stenting without ultrasound guidance, anticoagulation with warfarin was replaced with aspirin and Ticoplidine treatment, and low molecular weight heparin was progressively reduced in four consecutive stages from 1 month treatment to none (KARRILLON et al. 1996). Whereas aspirin and Ticopidine treatment seemed to be comparable to warfarin with respect to subacute closure and bleeding, there was no evidence that low molecular weight heparin altered the rates of either of these events.

I. Current Recommendations for the Use of Low Molecular Weight Heparin

I. Prevention of Venous Thromboembolism

In assessing the literature relating to the prevention of venous thromboembolism, the rules of evidence as defined by COOK et al. (1995) have been used. They are summarized as follows:

Level I Randomized trials with low false-positive (a) and low false-negative (β) errors

Level II Randomized trials with high false-positive (a) and high false-negative (β) errors
Level III Nonrandomized concurrent cohort studies
Level IV Nonrandomized historical cohort studies
Level V Case series

Based on level I evidence, low molecular weight heparin can be recommended in the following situations:

1. General surgery – thoracoabdominal, gynecologic, urologic, and cancer surgery
2. Orthopedic surgery – hip fracture, total hip replacement, and total knee replacement
3. Multiple trauma
4. Following neurosurgery, following acute spinal cord injury with paralysis
5. Medical patients – hospitalized patients (particularly elderly and debilitated), following acute ischemic stroke

II. Treatment of Venous Thromboembolism

Low molecular weight heparin can replace intravenous unfractionated heparin for the treatment of distal and proximal deep vein thrombosis in hospital. For patients at low risk of bleeding and who do not have other serious medical problems, low molecular weight heparin can now be used in the out-of-hospital setting. Low molecular weight heparin can replace intravenous unfractionated heparin in patients with unstable angina or non-Q-wave myocardial infarction.

J. Summary and Conclusions

The low molecular weight heparins have made a major impact on the management of thrombotic disease. The ease of administration by subcutaneous injection without laboratory monitoring, and the large body of evidence that the low molecular weight heparins are both effective and safe, have led to their widespread use wherever they are registered. A large number of ongoing clinical trials will further expand the indications for the low molecular weight heparins. If low molecular weight heparin proves to be safer that unfractionated heparin with respect to the complications of heparin-induced thrombocytopenia and osteoporosis, their use will be further expanded. Recent studies indicate that low molecular weight heparin is cost-effective when compared with unfractionated heparin for the treatment of venous thromboembolism, and when compared with warfarin for the prevention of venous thrombosis following total joint replacement. Market forces will tend to lower the costs of these agents, making them even more cost effective. It is likely that the low molecular weight heparins will replace unfractionated heparin in all of the indications where heparin has been used. For many workers in the field, the sooner this happens the better it will be for patient care.

References

Agnelli G, Cosmi B, di Filippo P, Ranucci V, Veschi F, Longetti M et al (1992) A randomized double-blind, placebo-controlled trial of dermatan sulphate for prevention of deep vein thrombosis in hip fracture. Thromb Haemost 67:203–208

Albada J, Nieuwenhuis HK, Sixma JJ (1989) Treatment of acute venous thromboembolism with low molecular weight heparin (Fragmin): results of a double-blind randomized study. Circulation 80:935–940

Andersson LO, Barrowcliffe TW, Holmer E, Johnson EA, Soderstrom G (1979) Molecular weight dependency of the heparin potentiated inhibition of thrombin and activated factor X. Effect of heparin neutralization in plasma. Thromb Res 115:531–541

Andriuoli G, Mastacchi R, Barbanti M, Sarret M (1985) Comparison of the antithrombotic and haemorrhagic effects of heparin and a new low molecular weight heparin in rats. Haemostasis 15:324–330

Arepally G, Reynolds C, Tomaski A, Amiral J, Jawad A, Poncz M et al (1995) Comparison of PF4/heparin ELISA assay with the $^{(14)}$C-serotonin release assay in the diagnosis of heparin-induced thrombocytopenia. Am J Clin Pathol 104:648–654

Bara L, Samama M (1988) Pharmacokinetics of low molecular weight heparins. Acta Chir Scand [Suppl] 543:65–72

Bara L, Billaud E, Gramond G, Kher A, Samama M (1985) Comparative pharmacokinetics of low molecular weight heparin (PK 10169) and unfractionated heparin after intravenous and subcutaneous administration. Thromb Res 39:631–636

Barrowcliffe TW, Curtis AD, Johnson EA, Thomas DP (1988) An international standard for low molecular weight heparin. Thromb Haemost 60:1–7

Barzu T, Molho P, Tobelem G, Petitou M, Caen JP (1984) Binding of heparin and low molecular weight heparin fragments to human vascular endothelial cells in culture. Nouv Rev Fr Haematol 26:243–247

Bergmann JF, Neuhart E (1996) A multicenter randomized double-blind study of Enoxaparin compared with unfractionated heparin in the prevention of venous thromboembolic disease in elderly in-patients bedridden for an acute medical illness. Thromb Haemost 7:529–534

Bergqvist D, Matzsch T, Burmark US, Frisell J, Guilbaud O, Hallbook T et al (1988) Low molecular weight heparin given the evening before surgery compared with conventional low-dose heparin in prevention of thrombosis. Br J Surg 75:888–891

Bergqvist D, Matzsch T, Jendteg S, Lindgren B, Persson U (1990) The cost-effectiveness of prevention of post-operative thromboembolism. Acta Chir Scand [Suppl] 556:36–41

Bergqvist D, Kettunen K, Fredin H, Fauno P, Suomalainen O, Soimakallio S et al (1991) Thromboprophylaxis in patients with hip fractures: a prospective, randomized, comparative study between ORG 10172 and dextran 70. Surgery 109:617–622

Bergqvist D, Burmark US, Flordal PA, Frisell J, Hallbook T et al (1995) Low molecular weight heparin started before surgery as prophylaxis against deep vein thrombosis: 2500 versus 500 XaI units in 2070 patients. Br J Surg 82:496–501

Bergqvist D, Benoni G, Bjorgell O, Fredin H, Medlundh U, Nicolas S et al (1996) Low-molecular-weight heparin (Enoxaparin) as prophylaxis against venous thromboembolism after total hip replacement. N Engl J Med 335:696–700

Bjornsson TD, Wolfram KM, Kitchell BB (1982) Heparin kinetics determined by three assay methods. Clin Pharmacol Ther 31:104–113

Blajchman MA, Young E, Ofosu FA (1989) Effects of unfractionated heparin, dermatan sulfate and low molecular weight heparin on vessel wall permeability in rabbits. Ann NY Acad Sci 556:245–254

Boneu B, Buchanan MR, Cade JF, VanRyn J, Fernandez FF, Ofosu FA et al (1985) Effects of heparin, its low molecular weight fractions and other glycosaminoglycans on thrombus growth in vivo. Thromb Res 40:81–89

Boneu B, Caranobe C, Cadroy Y, Dol F, Gabaig AM, Dupouy D et al (1988) Pharmacokinetic studies of standard unfractionated heparin, and low molecular weight heparins in the rabbit. Semin Thromb Hemost 14:18–27

Boshkov LK, Warkentin TE, Hayward CP, Andrew M, Kelton JF (1993) Heparin-induced thrombocytopenia and thrombosis: clinical and laboratory studies. Br J Haematol 84:322–328

Brandjes DP, Heijboer H, Buller HR, de Rijk M, Jagt H, ten Cate JW (1992) Acenocoumarol and heparin compared with acenocoumarol alone in the initial treatment of proximal-vein thrombosis. N Engl J Med 327:1485–1489

Bratt G, Tornebohme, Grangvist S, Aberg W, Lockner D (1985) A comparison between low molecular weight heparin (KABI 2165) and standard heparin in the intravenous treatment of deep venous thrombosis. Thromb Haemost 54:813–817

Bratt G, Aberg W, Johansson M, Torneboh ME, Grangvist S, Lockner D (1990) Two daily subcutaneous injections of fragmin as compared with intravenous standard heparin in the treatment of deep venous thrombosis (DVT). Thromb Haemost 64:506–510

Briant L, Caranobe C, Saivin S, Sie P, Bayrou B, Houin G et al (1989) Unfractionated heparin and CY216: pharmacokinetics and bioavailabilities of the anti Factor X_a and II_a Effects after intravenous and subcutaneous injection in the rabbit. Thromb Haemost 61:348–353

Brill-Edwards P, Ginsberg JS, Johnston M, Hirsh J et al (1993) Establishing a therapeutic range for heparin therapy. Ann Intern Med 119:104–109

Cade JF, Buchanan MR, Boneu B, Ockelford P, Carter CJ, Cerskus AL et al (1984) A comparison of the antithrombotic and haemorrhagic effects of low molecular weight heparin fractions: the influence of the method of preparation. Thromb Res 35:613–625

Caen JP (1988) A randomized double-blind study between a low molecular weight heparin Kabi 2165 and standard heparin in the prevention of deep vein thrombosis in general surgery. A French multicenter trial. Thromb Haemost 59:216–220

Cairns JA, Gill J, Morton B, Roberts R, Gent M, Hirsh J et al (1996) Fish oils and low-molecular-weight heparin for the reduction of restenosis after percutaneous transluminal coronary angioplasty. The EMPAR Study. Circulation 94:1553–1560

Campbell NR, Hull RD, Brant R, Hogan DB, Pineo GF, Raskob GE (1996) Aging and heparin-related bleeding. Arch Intern Med 156:857–860

Carter CJ, Kelton JG, Hirsh J, Cerskus A, Santos AV, Gent M (1982) The relationship between the hemorrhagic and antithrombotic properties of low molecular weight heparin in rabbits. Blood 59:1239–1245

Casu B, Oreste P, Torri G, Zoppetti G, Choay J, Lormeau JC et al (1981) The structure of heparin oligosaccharide fragments with high anti-(factor X_a) activity containing the minimal antithrombin III-binding sequence. Biochem J 197:599–609

Cohen M, Demers C, Gurfinkel E, Fromell G, Langer A, Turpie AGG, ESSENCE Group (1996) Primary end point analysis from the ESSENCE trial: enoxaparin vs. unfractionated heparin in unstable angina and non-Q wave infarction. Circulation 94:I-554

Colwell CW Jr, Spiro TE, Trowbridge AA, Morris BA, Kwaan HC, Blaha JD et al (1994) Use of enoxaparin, a low-molecular-weight heparin, and unfractionated heparin for the prevention of deep venous thrombosis after elective hip replacement. A clinical trial comparing efficacy and safety. Enoxaparin Clinical Trial Group. J Bone Joint Surg Am 76:3–14

Cook DJ, Guyatt GM, Laupacis A, Sackett DL, Goldberg RJ (1995) Clinical recommendations using levels of evidence for antithrombotic agents. Chest 108:227S–230S

Cruickshank MK, Levine MN, Hirsh J, Roberts R, Siguenza M (1991) A standard heparin nomogram for the management of heparin therapy. Arch Intern Med 151:333–337

Dahl OE, Aspelin T, Arnesen H, Seljeflot I, Kierulf P, Ruyter R et al (1995) Increased activation of coagulation and formation of late deep venous thrombosis following discontinuation of thromboprophylaxis after hip replacement surgery. Thromb Res 80:299–306

Dahl OE, Andreassen G, Aspelin T, Muller C, Mathiesen P, Nyhus S et al (1997) Prolonged thromboprophylaxis following hip replacement surgery – results of a double-blind prospective, randomized, placebo-controlled study with dalteparin (fragmin). Thromb Haemost 77:26–31

Demers C, Ginsberg JS, Brill-Edwards P, Panju A, Warkentin TE, Anderson DR (1991) Rapid anticoagulation using ancrod for heparin-induced thrombocytopenia. Blood 78:2194–2197

Edmondson RA, Cohen AT, Das SK, Wagner MB, Kakkar VV (1994) Low molecular weight heparin vs. aspirin and dipyridamole after femoropopliteal bypass grafting. Lancet 344:914–918

Eriksson BI, Kalebo P, Anthymyr BA, Wadenvik H, Tengborn L, Risberg B (1991) Prevention of deep-vein thrombosis and pulmonary embolism after total hip replacement. Comparison of low-molecular-weight heparin and unfractionated heparin. J Bone Joint Surg Am 73:484–493

Fareed J, Walenga JM, Hoppensteadt D, Huan X, Racanelli A (1988a) Comparative study on the in vitro and in vivo activities of seven low-molecular-weight heparins. Haemostasis 18 [Suppl 3]:3–15

Fareed J, Walenga JM, Racanelli A, Hoppensteadt D, Huan X, Messmore HL et al (1988b) Validity of the newly established low-molecular-weight heparin standard in cross-referencing low-molecular-weight heparins. Haemostasis 18 [Suppl 3]:33–47

Faxon DP, Spiro TE, Minors, Cote G, Douglas J, Gottlieb R et al (1994) Low molecular weight heparin in prevention of restenosis after angioplasty. Results of Enoxparin Restenosis (ERA) Trial. Circulation 90:908–914

Fennerty AG, Thomas P, Backhouse G, Bentley P, Campbell IA, Routledge PA (1985) Audit of control of heparin treatment. Br Med J Clin Res Ed 290:27–28

Forestier F, Daffos F, Capella-Pavlovsky M (1984) Low molecular weight heparin (PK 10169) does not cross the placenta during the second trimester of pregnancy: study by direct fetal blood sampling under ultrasound. Thromb Res 34:557–560

Fragmin during Instability in Coronary Artery Disease (FRISC) Study Group (1996) Low-molecular-weight heparin during instability in coronary artery disease. Lancet 347:561–568

Gallus A, Cade J, Ockelford P, Hepburn S, Maas M, Magnani H et al (1993) Orgaran (Org 10172) or heparin for preventing venous thromboembosis after elective surgery for malignant disease? A double-blind, randomised, multicentre comparison. ANZ-Organon Investigators' Group. Thromb Haemost 70:562–567

Geerts WH, Jay RM, Code KI, Chen E, Szalai JP, Saibil EA et al (1996) A comparison of low-dose heparin with low-molecular-weight heparin as prophylaxis against venous thromboembolism after major trauma. N Engl J Med 335:701–707

Gerhart TN, Yeh HS, Robertson LK, Lee MA, Smith M, Salzman EW (1991) Low-molecular-weight heparinoid compared with warfarin for prophylaxis of deep-vein thrombosis in patients who are operated on for fracture of the hip. A prospective, randomized trial. J Bone Joint Surg Am 73:494–502

Ginsberg JS, Hirsh J (1995) Use of antithrombotic agents during pregnancy. Chest 108:305S–311S

Granger CB, Miller JM, Bovill EG, Gruber A, Tracy RP, Krucoff MW, et al (1995) Rebound increase in thrombin generation and activity after cessation of

intravenous heparin in patients with acute coronary syndromes. Circulation 91:1929–1935
Green D (1992) Prophylaxis of thromboembolism in spinal cord-injured patients. Chest 102:649S–651S
Green D, Lee MY, Lim AC, Chmiel JS, Vetter M, Pang T et al (1990) Prevention of thromboembolism after spinal cord injury using low-molecular-weight heparin. Ann Intern Med 113:571–574
Green D, Hull RD, Brant R, Pineo GF (1992) Lower mortality in cancer patients treated with low-molecular-weight versus standard heparin. Lancet 339:1476
Greinacher A, Volpel H, Potzsch B (1996) Recombinant hirudin in the treatment of patients with heparin-induced thrombocytopenia (HIT). Blood 88:281a
Gurfinkel EP, Manos EJ, Mejail RI, Cerda MA, Duronto EA, Garcia CN et al (1995) Low molecular weight heparin vs. regular heparin or aspirin in the treatment of unstable angina and silent ischemia. J Am Coll Cardiol 26:313–318
Hamulyak K, Lensing AW, Van der Meer J, Smid WM, van Ooy A, Hoek JA (1995) Subcutaneous low-molecular-weight heparin or oral anticoagulants for the prevention of deep vein thrombosis in elective hip and knee replacement? Fraxiparine Oral Anticoagulant Study Group. Thromb Haemost 74:1428–1431
Harenberg J, Huck K, Bratsch H, Stehle G, Dempfle CE, Mall K et al (1990) Therapeutic application of subcutaneous low-molecular-weight heparin in acute venous thrombosis. Haemostasis 20 [Suppl 1]:205–219
Harenberg J, Roebruck P, Heene DL (1996) Subcutaneous low-molecular-weight heparin versus standard heparin and the prevention of thromboembolism in medical inpatients. The Heparin Study in Internal Medicine Group. Haemostasis 26:127–139
Hauch O, Khattar SC, Jorgensen LN (1991) Cost-benefit analysis of prophylaxis against deep vein thrombosis in surgery. Semin Thromb Hemost 17 [Suppl 3]:280–283
Heit JA, Berkowitz SD, Bona R, Cabanas V, Corson JD, Elliott CG et al (1997) Efficacy and safety of low molecular weight heparin (ardeparin sodium) compared to warfarin for the prevention of venous thromboembolism after total knee replacement surgery: a double-blind, dose-ranging study. Ardeparin Arthroplasty Study Group. Thromb Haemost 77:32–38
Hirsh J, Levine MN (1992) Low molecular weight heparin. Blood 79:1–17
Hirsh J, Raschke R, Warkentin TE, Dalen JE, Deykin D, Poller L (1995) Heparin: mechanism of action, pharmacokinetics, dosing considerations, monitoring, efficacy, and safety. Chest 108:4:258S–275S
Hoek J, Nurmohamed MT, ten Cate JW, Buller HR (1989) Prevention of deep vein thrombosis following total hip replacement by a low molecular weight heparinoid (ORG 10172). Thromb Haemost [Suppl] 62:520
Holm HA, Ly B, Handeland GF, Abildgaard U, Arnesen KE, Gott Schalk P et al (1986) Subcutaneous heparin treatment of deep venous thrombosis: a comparison of unfractionated and low molecular weight heparin. Haemostasis 16 [Suppl 2]:30–37
Holmer E, Soderberg K, Bergqvist D, Lindahl U (1986) Heparin and its low molecular weight derivatives: anticoagulant and antithrombotic properties. Haemostasis 16 [Suppl 2]:1–7
Hoppensteadt DA, Walenga JM, Fasanella A, Jeske W, Fareed J (1995) TFPI antigen levels in normal human volunteers after intravenous and subcutaneous administration of unfractionated heparin and a low molecular weight heparin. Thromb Res 77:175–185
Hull RD, Hirsh J, Sackett DL, Stoddart GL (1982) Cost-effectiveness of primary and secondary prevention of fatal pulmonary embolism in high-risk surgical patients. Can Med Assoc J 127:990–995
Hull RD, Raskob GE, Hirsh J, Jay RM, Leclerc JR, Geerts WH et al (1986) Continuous intravenous heparin compared with intermittent subcutaneous heparin in the initial treatment of proximal-vein thrombosis. N Engl J Med 315:1109–1114

Hull RD, Raskob GE, Pineo GF, Green D, Trowbridge AA, Elliott CG et al (1992a) Subcutaneous low-molecular-weight heparin compared with continuous intravenous heparin in the treatment of proximal-vein thrombosis. N Engl J Med 326:975–982

Hull RD, Raskob GE, Rosenbloom D, Lemaire J, Pineo GF, Baylis B, et al (1992b) Optimal therapeutic level of heparin therapy in patients with venous thrombosis. Arch Intern Med 152:1589–1595

Hull RD, Raskob G, Pineo G, Rossenbloom D, Evans W, Mallory T et al (1993) A comparison of subcutaneous low-molecular-weight heparin with warfarin sodium for prophylaxis against deep-vein thrombosis after hip or knee implantation. N Engl J Med 329:1370–1376

Hull RD, Pineo GF, Valentine KA, Stagg V, Brant RF (1996) Preoperative low-molecular-weight heparin (LMWH) is more effective than postoperative LMWH in the prevention of deep vein thrombosis (DVT) in patients undergoing total hip replacement (THR): a pooled analysis. Blood 88S:168a

Hull RD, Raskob GE, Pineo GF, Feldstein W, Rosenbloom D, Gafni A et al (1997a) Subcutaneous low-molecular-weight heparin vs warfarin for prophylaxis of deep vein thrombosis after hip or knee implantation. An economic perspective. Arch Intern Med 157:298–303

Hull RD, Raskob GE, Rosenbloom D, Pineo GF, Lerner RG, Gafni A et al (1997b) Treatment of proximal vein thrombosis with subcutaneous low-molecular-weight heparin vs intravenous heparin. An economic perspective. Arch Intern Med 157:289–294

Hunt BJ, Doughty HA, Majumdar G, Copplestone A, Kerslakes S, Buchanan N et al (1997) Thromboprophylaxis with low molecular weight heparin (Fragmin) in high risk pregnancies. Thromb Haemost 77:39–43

Kakkar VV, Murray WJ (1985) Efficacy and safety of low-molecular-weight heparin (CY216) in preventing postoperative venous thromboembolism: a co-operative study. Br J Surg 72:786–791

Kakkar VV, Cohen AT, Edmonson RA, Phillips MJ, Cooper DJ, Das SK et al (1993) Low molecular weight versus standard heparin for prevention of venous thromboembolism after major abdominal surgery. The Thromboprophylaxis Collaborative Group. Lancet 341:259–265

Karrillon GJ, Morice MC, Benveniste E, Bunouf P, Aubry P, Cattan S et al (1996) Intracoronary stent implantation without ultrasound guidance and with replacement of conventional anticoagulation by antiplatelet therapy. 30-day clinical outcome of the French Multicenter Registry. Circulation 94:1519–1527

Kay R, Wong KS, Yu YL, Chan YW, Tsoi TM, Ahuja AT et al (1995) Low-molecular-weight heparin for the treatment of acute ischemic stroke. N Engl J Med 333:1588–1593

Kearon C, Hirsh J (1995) Starting prophylaxis for venous thromboembolism postoperatively. Arch Intern Med 155:366–372

Kelton JG, Sheridan D, Brain H, Powers PJ, Turpie AG, Carter CJ (1984) Clinical usefulness of testing for a heparin-dependent platelet-aggregating factor in patients with suspected heparin-associated thrombocytopenia. J Lab Clin Med 103:606–612

Koopman MM, Prandoni P, Piovella F, Ockelford PA, Brandjes DP, Vander Meer J et al (1996) Treatment of venous thrombosis with intravenous unfractionated heparin administered in the hospital as compared with subcutaneous low-molecular-weight heparin administered at home. The Tasman Study Group. N Engl J Med 334:682–687

Lane DA (1989) Heparin binding and neutralizing protein. In: Lane DA, Lindahl U (eds) Heparin, chemical, and biological properties, clinical applications, vol 189. Arnold, London, pp 363–391

Lassen MR, Borris LC (1995) Prolonged thromboprophylaxis with low molecular weight heparin (Fragmin) after elective total hip arthroplasty – a placebo controlled study. Thromb Haemost 73:1104

Lassen MR, Borris LC, Christiansen HM, Boll KL, Eiskjaer SP et al (1991) Prevention of thromboembolism in 190 hip arthroplasties. Comparison of LMW heparin and placebo. Acta Orthop Scand 62:33–38
Leclerc JR, Geerts WH, Desjardins L, Jobin F, Laroche F, Delorme F et al (1992) Prevention of deep vein thrombosis after major knee surgery – a randomized, double-blind trial comparing a low molecular weight heparin fragment (enoxaparin) to placebo. Thromb Haemost 67:417–423
Leclerc JR, Geerts WH, Desjardins L, l'Esperance B, Cruickshank M, Demers C et al (1995) Prevention of venous thromboembolism (VTE) after knee arthroplasty – a randomized, double-blind trial comparing a low molecular weight heparin fragment (Enoxparin) to Warfarin. Thromb Haemost 73:1103
Leizorovicz A, Picolet H, Peyrieux JC, Boissel JP (1991) Prevention of perioperative deep vein thrombosis in general surgery: a multicentre double-blind study comparing two doses of Logiparin and standard heparin. H. B. P. M. Research Group. Br J Surg 78:412–416
Lensing AW, Prins MH, Davidson BL, Hirsh J (1995) Treatment of deep venous thrombosis with low-molecular-weight heparins. A meta-analysis. Arch Intern Med 155:601–607
Levine MN, Hirsh J, Gent M, Turpie AG, Leclerc J, Powers PJ et al (1991) Prevention of deep-vein thrombosis after elective hip surgery. A randomized trial comparing low molecular weight heparin with standard unfractionated heparin. Ann Intern Med 114:545–551
Levine MN, Hirsh J, Gent M, Turpie AG, Cruickshank M, Weitz J et al (1994) A randomized trial comparing activated thromboplastin time with heparin assay in patients with acute venous thromboembolism requiring large daily doses of heparin. Arch Intern Med 154:49–56
Levine M, Gent M, Hirsh J, Leclerc J, Anderson D, Weitz J et al (1996) A comparison of low-molecular-weight heparin administered primarily at home with unfractionated heparin administered in the hospital for proximal deep-vein thrombosis. N Engl J Med 334:677–681
Lindahl U, Backstrom G, Hook M, Thunberg L, Fransson LA, Linker A (1979) Structure of the antithrombin-binding site in heparin. Proc Natl Acad Sci USA 76:3198–3202
Lindahl U, Thunberg L, Backstrom G, Riesenfeld J, Nordling K, Bjork I (1984) Extension and structural variability of the antithrombin-binding sequence in heparin. J Biol Chem 259:12368–12376
Lindmarker P, Holmstrom M, Grangvist S, Johnsson H, Lockner D (1994) Comparison of once-daily subcutaneous Fragmin with continuous intravenous unfractionated heparin in the treatment of deep vein thrombosis. Thromb Haemost 72:186–190
Lopaciuk S, Meissner AJ, Filipecki S, Zawilska K, Sowier J, Ciesielski L et al (1992) Subcutaneous low-molecular-weight-heparin versus subcutaneous unfractionated heparin in the treatment of deep vein thrombosis: a Polish multicentre trial. Thromb Haemost 68:14–18
Magnani HN (1993) Heparin-induced thrombocytopenia (HIT): an overview of 230 patients treated with Orgaran (Org 10172). Thromb Haemost 70:554–561
Matsuo T, Kario K, Chikahira Y, Nakao K, Yamada T (1992) Treatment of heparin-induced thrombocytopenia by use of argatroban, a synthetic thrombin inhibitor. Br J Haematol 82:627–629
Matzsch T, Bergqvist D, Hedner U, Nilsson B, Ostergaard P et al (1990) Effects of low molecular weight heparin and unfragmented heparin on induction of osteoporosis in rats. Thromb Haemost 63:505–509
Melissari E, Parker CJ, Wilson NV, Monte G, Kanthou C, Pemberton KD et al (1992) Use of low molecular weight heparin in pregnancy. Thromb Haemost 68:652–656
Menzin J, Colditz GA, Regan MM, Richner RE, Oster G (1995) Cost-effectiveness of enoxaparin vs low-dose warfarin in the prevention of deep-vein thrombosis after total hip replacement surgery. Arch Intern Med 155:757–764

Monreal M, Lafoz E, Navarro A, Granero X, Caja V, Caceres E et al (1989a) A prospective double-blind trial of a low molecular weight heparin once daily compared with conventional low-dose heparin three times daily to prevent pulmonary embolism and venous thrombosis in patients with hip fracture. J Trauma 29:873–875

Monreal M, Lafoz E, Salvador R, Roncales J, Navarro A (1989b) Adverse effects of three different forms of heparin therapy: thrombocytopenia, increased transaminases, and hyperkalemia. Eur J Clin Pharmacol 37:415–418

Monreal M, Vinas L, Monreal L, Lavin S, Lafoz E, Angles AM (1990) Heparin-related osteoporosis in rats. A comparative study between unfractionated heparin and a low-molecular-weight heparin. Haemostasis 20:204–207

Nurmohamed MT, Rosendaal FR, Buller HR, Dekker E, Hommes DW, Vandenbroucke JP (1992) Low-molecular-weight heparin versus standard heparin in general and orthopaedic surgery: a meta-analysis. Lancet 340:152–156

Nurmohamed MT, Buller HR, ten Cate JW (1994) Low-molecular-weight heparins in extracorporeal circuits. In: Bounameaux H (ed) Low-molecular-weight heparins in prophylaxis and therapy of thromboembolic disease. Dekker, New York, p 247 (Fundamental and clinical cardiology, vol 19)

Nurmohamed MT, Verhaeghe R, Haas S, Iriarte JA, Bogel G, Van Rij AM et al (1995) A comparative trial of a low molecular weight heparin (enoxaparin) versus standard heparin for the prophylaxis of postoperative deep vein thrombosis in general surgery. Am J Surg 169:567–571

Nurmohamed MT, van Riel AM, Henkens CM, Koopman MM, Que GT, d'Azemar P et al (1996) Low molecular weight heparin and compression stockings in the prevention of venous thromboembolism in neurosurgery. Thromb Haemost 75:233–238

O'Brien BJ, Anderson DR, Goeree R (1994) Cost-effectiveness of enoxaparin versus warfarin prophylaxis against deep-vein thrombosis after total hip replacement. Can Med Assoc J 150:1083–1090

Omri A, Delaloye JF, Andersen H, Bachmann F (1989) Low molecular weight heparin Novo (LHN-1) does not cross the placenta during the second trimester of pregnancy. Thromb Haemost 61:55–56

Oster G, Tuden RL, Colditz GA (1987) A cost-effectiveness analysis of prophylaxis against deep-vein thrombosis in major orthopedic surgery. JAMA 257:203–208

Planes A, Vochelle N, Fagola M, Feret J, Bellaud M (1991) Prevention of deep-vein thrombosis after total hip replacement. The effect of low-molecular-weight heparin with spinal and general anaesthesia. J Bone Joint Surg Br 73:418–422

Planes A, Vochelle N, Darmon JY, Fagola M, Bellaud M, Huet Y (1996) Risk of deep-venous thrombosis after hospital discharge in patients having undergone total hip replacement: double-blind randomised comparison of enoxaparin versus placebo. Lancet 348:224–228

Powers P, Gent M, Klimek M, Levin M, Geerts W, Neemeh J (1993) A randomized, double-blind trial comparing a low molecular weight heparinoid, Orgaran, with aspirin in the prevention of venous thromboembolism following surgery for hip fracture. Thromb Haemost 69:1116

Prandoni P, Lensing AW, Buller HH, Carta M, Cogo A, Vigo M et al (1992) Comparison of subcutaneous low-molecular-weight heparin with intravenous standard heparin in proximal deep-vein thrombosis. Lancet 339:441–445

Prins MH, den Ottolander GJH, Gelsema R, van Woerkom TCM, Sing AK, Meller I et al (1987) Deep venous thrombosis prophylaxis with a LMW heparin (Kabi 2165) in stroke patients. Thromb Haemost 58:117

Raschke RA, Reilly BM, Guidry JR, Fontana JR, Srinivas S (1993) The weight-based heparin dosing nomogram compared with a "standard care" nomogram. A randomized controlled trial. Ann Intern Med 119:874–881

Rosenberg RD, Lam L (1979) Correlation between structure and function of heparin. Proc Natl Acad Sci USA 76:1218–1222

Salzman EW, Davies GC (1980) Prophylaxis of venous thromboembolism: analysis of cost effectiveness. Ann Surg 191:207–218
Salzman EW, Rosenberg RD, Smith MH, Lindon JN, Favreau L (1980) Effect of heparin and heparin fractions on platelet aggregation. J Clin Invest 65:64–73
Samama M, Bernard P, Bonnardot JP, Combe-Tamzali S, Lanson Y (1988) Low-molecular-weight heparin compared with unfractionated heparin in prevention of postoperative thrombosis. Br J Surg 75:128–131
Sandset PM, Dahl T, Stiris M, Rostad B, Scheel B, Abildgaard U (1990) A double-blind and randomized placebo-controlled trial of low molecular weight heparin once daily to prevent deep-vein thrombosis in acute ischemic stroke. Semin Thromb Hemost 16 [Suppl]:25–33
Shaughnessy SG, Young E, Deschamps P, Hirsh J (1995) The effects of low molecular weight and standard heparin on calcium loss from fetal rat calvaria. Blood 86:1368–1373
Siegbahn A, Y-Hassan S, Boberg J, Bylund H, Neerstrand HS, Ostergaard P et al (1989) Subcutaneous treatment of deep venous thrombosis with low molecular weight heparin. A dose finding study with LMWH-Novo. Thromb Res 55:767–778
Simonneau G, Charbonnier B, Decousus H, Planchon B, Ninet J, Sie P et al (1993) Subcutaneous low-molecular-weight heparin compared with continuous intravenous unfractionated heparin in the treatment of proximal deep vein thrombosis. Arch Intern Med 153:1541–1546
The Columbus Investigators (1996) Low molecular weight heparin is an effective and safe treatment for deep-vein thrombosis and pulmonary embolism. Blood 88:626a
The Danish Enoxaparin Study Group (1991) Low-molecular weight heparin (enoxaparin) vs Dextran 70. The prevention of postoperative deep vein thrombosis after total hip replacement. Arch Intern Med 151:1621–1624
The European Fraxiparin Study (EFS) Group (1988) Comparison of a low molecular weight heparin and unfractionated heparin for the prevention of deep vein thrombosis in patients undergoing abdominal surgery. Br J Surg 75:1058–1063
Theroux P, Waters D, Lam J, Juneau M, McCans J (1992) Reactivation of unstable angina after the discontinuation of heparin. N Engl J Med 327:141–145
Tollefsen DM, Majerus DW, Blank MK (1982) Heparin cofactor II. Purification and properties of a heparin-dependent inhibitor of thrombin in human plasma. J Biol Chem 257:2162–2169
T¨rholm C, Broeng L, Jorgensen PS, Bjerregaard P, Josephsen L, Jorgensen PK et al (1991) Thromboprophylaxis by low-molecular-weight heparin in elective hip surgery. A placebo controlled study. J Bone Joint Surg Br 73:434–438
Turpie AG, Levin MN, Hirsh J, Carter CJ, Jay RM, Powers PJ et al (1986) A randomized controlled trial of low-molecular-weight heparin (enoxaparin) to prevent deep-vein thrombosis in patients undergoing elective hip surgery. N Engl J Med 315:925–929
Turpie AG, Levine MN, Hirsh J, Carter CJ, Jay RM, Powers PJ et al (1987) Double-blind randomised trial of ORG 10172 low-molecular-weight heparinoid in prevention of deep-vein thrombosis in thrombotic stroke. Lancet 1:523–526
Turpie AGG (1995) Low molecular weight heparin (Dalteparin) in unstable coronary artery disease (FRIC). Can J Cardiol 11 [Suppl] E:110E
Wahlberg TB, Kher A (1994) Low molecular weight heparin as thromboprophylaxis in pregnancy. A retrospective analysis from 14 European clinics. Haemostasis 24:55–56
Warkentin TE (1996) Danaparoid (Orgaran) for the treatment of heparin-induced thrombocytopenia (HIT) and thrombosis: effects on in vivo thrombin and cross-linked fibrin generation, and evaluation of the clinical significance of in vitro cross-reactivity (XR) of danaparoid for HIT-IgG. Blood 88:626a
Warkentin TE, Kelton JG (1996) A 14-year study of heparin-induced thrombocytopenia. Am J Med 101:502–507

Warkentin TE, Levin MN, Hirsh J, Horsewood P, Roberts RS, Gent M et al (1995) Heparin-induced thrombocytopenia in patients treated with low-molecular-weight heparin or unfractionated heparin. N Engl J Med 332:1330–1335

Weitz JI, Hudoba M, Massel D, Maraganore J, Hirsh J (1990) Clot-bound thrombin is protected from inhibition by heparin-antithrombin III but is susceptible to inactivation by antithrombin III-independent inhibitors. J Clin Invest 86:385–391

Wheeler AP, Jaquiss RD, Newman JH (1988) Physician practices in the treatment of pulmonary embolism and deep venous thrombosis. Arch Intern Med 148:1321–1325

CHAPTER 11

Parenteral Direct Antithrombins

M.A. LAUER and A.M. LINCOFF

A. Introduction

The development of thrombosis within a diseased artery or vein is responsible for acute coronary syndromes, deep venous thrombosis, pulmonary embolism, cerebral vascular accidents, and limb and bowel ischemia. Thrombin plays a central role in the process of thrombosis. Heparin, which is currently the mainstay of antithrombotic therapy for the acute phase of all of these clinical syndromes, is only partially effective. The limitations of heparin therapy may be linked to several features of its mode of action, which may be overcome by the use of direct thrombin inhibitors, a new class of agents that specifically and potently antagonize the actions of thrombin.

B. Thrombin: Structure and Function

Thrombin is a glycosylated, trypsin-like serine protease, generated by the cleavage of prothrombin by the prothrombinase complex consisting of factors Xa, Va, calcium, and membrane phospholipids. It has recently become clear that thrombin has multiple actions which regulate thrombosis, hemostasis and the vessel wall's response to injury. Aside from the classically described role of catalyzing the conversion of fibrinogen to fibrin, thrombin further activates factors XIII to cross-link fibrin and stabilize the clot and further activates the extrinsic pathway by inducing thromboplastin (tissue factor) synthesis in endothelial cells (GALDAL et al. 1985).

There are complex reactions between thrombin and the endothelium to locally regulate the hemostatic mechanism. Where the endothelium is damaged, it causes vasoconstriction through liberation of endothelin (SCHINI et al. 1989), and promotes coagulation by activating factors V and VIII and promoting the secretion of plasminogen activator inhibitor (PEARSON 1994). Where the endothelium is intact, thrombin promotes vasodilatation through release of prostacyclin and nitric oxide (PEARSON 1993), and exerts an anticoagulant effect by combining with thrombomodulin to activate protein C, which joins with protein S to inhibit factors Va and VIIIa and promotes the secretion of tissue type plasminogen activator (t-PA) (ESMON et al. 1982; JURD et al. 1996).

Thrombin is the most potent stimulus to platelet recruitment and aggregation (CHAO et al. 1974; PHILLIPS 1974). Through a variety of mechanisms (including stimulating the endothelial cells to release platelet-derived growth factor, platelet-activating factor, and P-selectin), thrombin stimulates the migration and proliferation of smooth muscle cells and the chemotaxis, adhesion, and activation of neutrophil, lymphocytes, and monocytes (THEROUX and LIDON 1994).

The thrombin receptor present on platelets and other cells was recently cloned and its mechanism of activation characterized (VU et al. 1991). Thrombin binds its receptor via an extracellular binding site near its amino terminal and cleaves the receptor at a specific site (arginine-41) within this extended substrate binding site. This cleavage releases an inactive fragment of the receptor's amino terminus and unmasks a new amino terminus which, acting as a tethered ligand, activates the receptor. The molecule of thrombin remains free to activate many receptors, contributing to its self-amplification effect. This mechanism accounts for thrombin's activation of platelets as well as its assorted actions on nonplatelet cells via the same or very similar receptors (HUNG et al. 1992).

The location of this catalytic binding site in a deep narrow canyon on the thrombin molecule is believed to account for its high specificity for its substrates, by restricting access to other macromolecules through steric hindrance (Fig. 1). Part of this extended binding site is a positively charged area carboxyl to the catalytic cleavage site known as the anion-binding

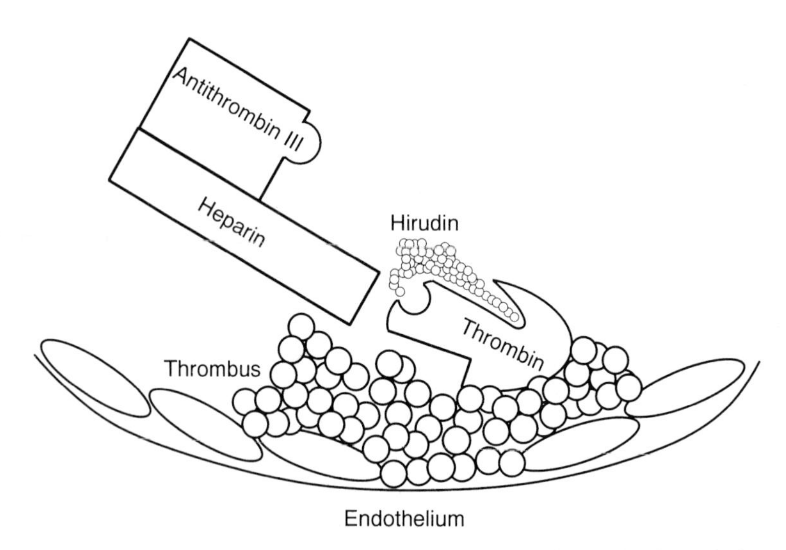

Fig. 1. Hirudin is able to penetrate thrombus and inactivate clot-bound thrombin. The heparin-antithrombin III complex, however, is unable to inactivate clot-bound thrombus, most likely due to steric hindrance of the heparin binding site on thrombin when the enzyme is bound to fibrin. During fibrinolysis, a continuos source of newly exposed clot bound thrombin is generated as layers of thrombus are removed. Modified from CANNON and BRAUNWALD with permission from the American College of Cardiology (Journal of the American College of Cardiology, 1995, 25, 30S–37S)

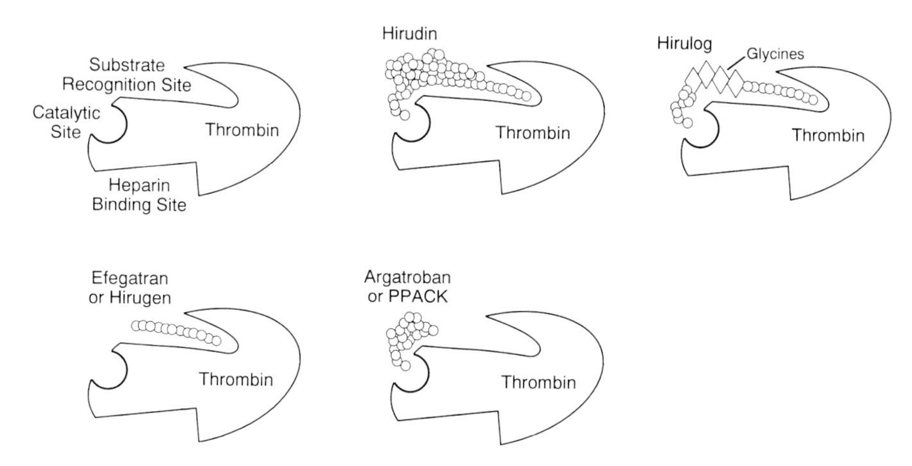

Fig. 2. Hirudin binds to thrombin at two sites within an extended binding region. This binding region also binds the thrombin receptor on platelets and other cells. The carboxyl terminal of hirudin attaches to the substrate recognition site within this region, while the amino terminal of hirudin attaches to the adjacent catalytic cleavage site. The second generation direct thrombin inhibitors interact with one or both of these sites

exosite (Fig. 2), which binds the thrombin receptor as well as fibrinogen, thrombomodulin, and most of the direct thrombin inhibitors, as will be discussed further. Thrombin has separate binding sites for fibrin and for heparin (STUBBS and BODE 1993).

C. Limitations of Current Antithrombotic Therapy

Heparin is a glycosaminoglycan commonly extracted from bovine lung or porcine gut which is composed of a mixture of polysaccharides of molecular weight varying between 300 and 40000 daltons. Heparin acts as an anticoagulant by complexing with antithrombin III and increasing its affinity to bind and inactivate thrombin and factors Xa, XII, XI, and IX as well as tissue factor-VIIa complex (HIRSH 1991).

One limitation of heparin is that its antithrombotic action is completely dependent on antithrombin III as a cofactor. This dependence on antithrombin III is underscored by the fact that heparin therapy is completely ineffective in patients with hereditary antithrombin III deficiency. Acquired states of decreased antithrombin III, such as may occur with prolonged heparin therapy, can also decrease the antithrombotic activity of heparin. At higher doses heparin may also inactivate thrombin by potentiating a second cofactor known as heparin cofactor II (LINDAHL and HOOK 1978).

Another limitation of heparin is the inability of the heparin-antithrombin III complex to inactivate clot-bound thrombin (WEITZ et al. 1990), most likely due to steric hindrance of the heparin-binding site on thrombin when the enzyme is bound to fibrin (Fig. 1). Thus, preformed thrombus acts as a protected source of thrombogenesis, which may be particularly important in the setting of thrombolytic therapy, where a continuous source of clot-bound thrombin is exposed as layers of thrombus are removed.

Heparin therapy is also limited by its variable anticoagulant activity, which requires careful laboratory monitoring. Some of this variability is inherent to the use of unfractionated mixtures which are variably bound to plasma proteins such as vitronectin and fibronectin (WEITZ et al. 1990). The more homogeneous low molecular weight heparin formulations recently introduced bind less to plasma proteins, resulting in somewhat more consistent antithrombotic action, but are still hampered by variable amounts of inactivation by circulating inhibitors such as platelet factor 4 and heparinase, both of which are released by activated platelets (LOSCALZO et al. 1985).

Finally heparin has a complex direct interaction with platelets which varies from a mild decrease in platelet counts and prolongation of bleeding time to the idiosyncratic immunologically mediated heparin-induced thrombocytopenia. This complication, which develops in 5% of patients treated with porcine heparin and more often with bovine heparin, can lead to profound thrombocytopenia and, in a small percentage of patients, a syndrome of venous and arterial thrombus related to a heparin-dependent platelet-aggregating factor (KELTON et al. 1984).

D. Direct Thrombin Inhibitors

I. Hirudin

The prototypic direct thrombin inhibitor is hirudin, first isolated from the saliva of the leech *Hirudo medicinalis* by MARKWARDT in 1950 (MARKWARDT 1991). The molecular structure was characterized and cloned by HARVEY et al. in 1986. It is a 65 amino acid protein with three disulfide bridges and a molecular weight of about 7000 daltons. Multiple isoforms of natural hirudin exist, but all have similar anticoagulant activities despite small differences in amino acid sequence (SCHARF et al. 1989). Recombinant DNA technology has made possible the production of several proprietary desulfatohirudins which are identical to natural hirudin except for a missing sulfate group on tyrosine 63 (TALBOT et al. 1989), all of which inhibit with similar potency the major actions of thrombin, including its autocatalytic reaction, the generation of fibrin, and the activation of platelets (LONGSTAFF et al. 1993).

Hirudin is the most potent and specific known inhibitor of thrombin, selectively and rapidly binding thrombin in a slowly reversible 1:1 stereochemical bond at two sites (Fig. 2). The amino-terminal core domain of hirudin (residues 1–52) binds to the catalytic site of thrombin, and the carboxyl-terminal tail of hirudin (residues 53–65) binds to the substrate recognition site of thrombin, which also recognizes fibrinogen and the platelet surface (THEROUX and LIDON 1994).

Hirudin can be administered intravenously, intramuscularly, or subcutaneously. It is excreted mainly by the kidneys and has a half-life of approximately 40 min when given intravenously (BICHLER et al. 1993).

Hirudin overcomes many of the functional limitations of heparin. The inhibition of thrombin by hirudin is direct, without a cofactor, and is not influenced by protein binding or inhibition by local or systemic inhibitors. Hirudin is able to penetrate thrombus and inactivate clot-bound thrombin. Because of the specificity of the carboxyl-terminus binding to thrombin, hirudin does not inhibit other enzymes in the coagulation or fibrinolytic pathways, such as factors Xa, IX, kallikrein, activated protein C, or plasminogen activators (CANNON and BRAUNWALD 1995). Finally, as opposed to heparin, hirudin has no direct effects on platelets, avoiding the complication of heparin-induced thrombocytopenia and possibly reducing bleeding complications.

II. Other Direct Thrombin Inhibitors

Both the anion-binding site and the catalytic sites of thrombin and the domains on hirudin which bind these two sites have been isolated, leading to the synthesis of hirudin-like peptides with biologic properties similar to that of hirudin (Fig. 2). One of these second-generation agents is Hirulog, a 20 amino acid peptide that contains the two terminal domains joined by a linker domain of four glycines (MARAGANORE 1993). The half-life of Hirulog is approximately 35 min, and only 20% is excreted in the urine, suggesting that it undergoes more extensive metabolism than hirudin (FOX et al. 1993). Other second-generation agents include argatroban, which reversibly blocks only the catalytic site, and PPACK, which irreversibly blocks this site. Hirugen is a 12 amino acid peptide which blocks only the anion-binding site of thrombin, resulting in much weaker antithrombotic activity, and has not progressed to clinical testing (LEFKOVITS and TOPOL 1994). Efegatran has been developed more recently and is a reversible inhibitor of the anion-binding site which also may be capable of blocking the generation of the factor Xa (CALLAS et al. 1995; SMITH et al. 1996).

E. Potential Roles for Direct Thrombin Inhibition

The direct thrombin inhibitors have been evaluated in most clinical situations for which heparin is presently utilized. The indications most extensively studied to date are the cardiovascular settings of acute coronary syndromes and adjunctive therapy to percutaneous revascularization. Other areas of recent investigation include deep venous thrombosis, heparin-induced thrombocytopenia, and as an adjunct to hemodialysis.

I. Acute Myocardial Infarction, Adjunct to Thrombolysis

Intravenous thrombolytic therapy improves ventricular function and reduces mortality during acute myocardial infarction. However, failure to achieve early patency and reocclusion after successful thrombolysis occur in many patients despite the use of heparin and aspirin with various thrombolytic regimens. The Global Use of Streptokinase and T-PA to Open Occluded Artery (GUSTO I) investigators showed that adding intravenous heparin to streptokinase in the treatment of acute myocardial infarction offered no advantage to subcutaneous heparin (Gusto Investigators 1993). This lack of additional benefit of intravenous heparin may be related to the protected nature of the clot-bound thrombin. Therefore the use of direct thrombin inhibitors seemed a potential means to overcome the limitations of this therapy and to improve the effectiveness of thrombolytic therapy.

1. Preclinical Studies

In a rabbit model of the femoral artery thrombosis followed by thrombolysis with streptokinase, hirudin was shown to reduce the incidences of both primary thrombus formation and reocclusion more effectively than heparin (Kaiser et al. 1990). A rat model of aortic thrombosis treated with t-PA found hirudin, Hirulog, and PPACK all to be superior to heparin and control with respect to patency rate, time to patency, and reocclusion rate (Klement et al. 1992). In canine models of coronary thrombosis, hirudin was found to be more effective than heparin as an adjunct to thrombolysis by either streptokinase (Rigel et al. 1993) or t-PA (Haskel et al. 1991) in the acceleration of lysis and the prevention of reocclusion.

2. Phase II Clinical Trials

In an early clinical study of direct thrombin inhibition, argatroban was administered for 72h to 22 patients with acute myocardial infarction immediately after successful reperfusion therapy. There was no reocculusion of the infarct-related artery observed at 1 month with argatroban, compared with a 15% reocclusion rate in 74 similar patients treated with heparin (Tabata et al. 1992).

The feasibility and safety of hirudin in conjunction with accelerated t-PA was tested in the Hirudin for the Improvement of Thrombolysis (HIT-I) trial, in which 40 patients received a relatively low dose of hirudin (0.07-mg/kg bolus followed by an infusion of 0.05 mg/kg/h over 48h), with a patency and safety profile comparable to that previously observed with heparin (Zeymer et al. 1995). The Thrombolysis in Myocardial Infarction (TIMI) 5 trial also examined the use of hirudin in conjunction with front-loaded t-PA in 246 patients with acute myocardial infarction. Patients received either intravenous heparin or hirudin at one of four ascending doses for 5 days. There was no significant difference between hirudin- and heparin-treated patients with respect to patency at 90 min, but at 18–36 h, 97.8% of hirudin-treated patients had a patent infarct-related artery compared with only 89.2% of heparin-treated patients ($p = 0.01$). Death or reinfarction during the hospital period was also significantly reduced in the hirudin-treated patients. Interestingly, a dose response was not observed, as all four doses of hirudin led to similar findings in the angiographic and clinical end points (CANNON et al. 1994).

The use of hirudin as adjunct to streptokinase was examined in the TIMI 6 trial, which randomized 196 patients with acute myocardial infarction to a 5-day infusion of either heparin or one of three doses of hirudin. At hospital discharge the incidence of death, reinfarction, congestive heart failure, or cardiogenic shock was lower in patients receiving the higher doses of hirudin (9.7% and 11.4%) than those receiving either the lowest dose of hirudin (21.6%) or heparin (17.6%) (LEE 1995).

Two small angiographic studies evaluated the use of Hirulog as an adjunct to streptokinase. LIDON et al. (1994) found that 77% of the 30 patients treated with Hirulog had TIMI 2 or 3 flow at 90 min, compared with only 47% of 15 patients treated with heparin ($p < 0.05$). THEROUX et al. (1995) randomized 68 patients to heparin, low-dose Hirulog for 24 h, or high-dose Hirulog for 12 h. TIMI grade 3 flow at 90 min was observed in 85% of patients receiving high-dose hirudin, 61% of patients receiving low-dose hirudin, and 31% of patients receiving heparin ($p = 0.008$).

3. Phase III Clinical Trials

The promising results of phase II studies led to the development of three similarly designed large-scale phase III trials of hirudin in acute myocardial infarction (Table 1): the Global Use of Strategies to Open Occluded Arteries (GUSTO) II, TIMI-9, and HIT-III trials. GUSTO II randomized a broad spectrum of patients with acute ischemic syndromes, with or without ST-segment elevation, while TIMI 9 and HIT III focused only on acute infarction patients with ST-segment elevation or new left bundle branch block who were eligible for thrombolytic therapy. All three of these trials were halted by their safety monitoring committees in their early phases due to higher than expected rates of hemorrhagic events, including intracranial hemorrhage. In GUSTO IIa and TIMI 9a, increased bleeding rates were seen in both the heparin and hirudin groups, and the trials were restarted as GUSTO IIb and TIMI 9b utilizing lower dosages of hirudin and heparin. Bleeding in HIT III was increased only in the hirudin arm, and this trial was not continued.

In GUSTO IIa the incidence of intracerebral hemorrhagic patients receiving thrombolytic therapy was 1.8% (GUSTO IIa INVESTIGATORS 1994). In TIMI 9a, intracranial hemorrhage occurred in 1.7% of patients treated with hirudin and 1.9% of those treated with heparin (ANTMAN 1994). In the HIT study, intracranial bleeding was observed in 3.4% of the 148 patients in the hirudin group, compared with none of the 154 heparin patients (NEUHAUS et al. 1994). These rates of intracranial hemorrhage in the thrombolytic patients were

Table 1. Phase III trials of hirudin in acute myocardial infarction

	GUSTO II	TIMI 9	HIT
EKG inclusion criteria	ST elevation ST depression T-wave inversion	ST elevation New LBBB	ST elevation New LBBB
Symptom duration	<12h	<12h	<6h
Thrombolytic agent	t-PA or SK	t-PA or SK	t-PA
Planned enrollment	12000 patients	3000 patients	7000 patients
Hirudin dose prior to suspension	0.6-mg/kg bolus 0.2 mg/kg/h	0.6-mg/kg bolus 0.2 mg/kg/h	0.4-mg/kg bolus 0.15 mg/kg/h
Heparin dose prior to suspension	5000-IU bolus 1000–1300 IU/h (weight adjusted)	5000-IU bolus 1000–1300 IU/h (weight adjusted)	70-IU/kg bolus 15 IU/kg/h
APTT target	60–90 s	60–90 s	2–3.5×baseline
Duration of therapy	72–102 h	96 h	48–72 h
Enrollment prior to suspension	2564 patients	757 patients	302 patients
Hirudin dose after interim analysis	0.1-mg/kg bolus 0.1-mg/kg/h	0.1-mg/kg bolus 0.1 mg/kg/h	NA
Heparin dose after interim analysis	5000-IU bolus 1000-IU/h	5000-IU bolus 1000/IU/h	NA
APTT target after interim analysis	55–85 s	55–85 s	NA
Enrollment after interim analysis	12142 total 4131 ST↑ 8011 non-ST↑	3002	NA

GUSTO = Global Use of Strategies to Open Occluded Arteries; TIMI = Thrombolysis in Myocardial Infarction; EKG = electrocardiogram; LBBB = left bundle branch block; t-PA = tissue-type plasminogen activator; SK = streptokinase; IU = international units; APTT = activated partial thromboplastin time; ST↑ = ST elevation; non-ST↑ = ST depression and T wave inversion

significantly higher than among similar patients receiving thrombolytic therapy and intravenous heparin in the GUSTO I trial (30892 patients with a rate of 0.7%, 95% CI 0.6%–0.8%), which utilized a dose of heparin approximately 20% lower than that used in GUSTO IIa and TIMI 9b. These findings led to a reduction of the doses of hirudin and heparin in both GUSTO IIb and TIMI 9b. As both trials correlated a higher activated partial thromboplastin time (APTT) with hemorrhagic complications, the target APTT range for both hirudin and heparin was also lowered in GUSTO IIb and TIMI 9b (Table 1).

With the reduced doses of heparin and hirudin, TIMI 9b enrolled 3002 patients. Patients were treated with aspirin and either accelerated-dose t-PA (64%) or streptokinase (36%) at the treating physician discretion and randomized to either heparin (n = 1491) or hirudin (n = 1511) infused for 96h. There was no significant difference in the primary combined end point of death, recurrent myocardial infarction, or the development of severe congestive heart failure or cardiogenic shock by 30 days, occurring in 11.9% of patients in the heparin group and 12.9% of patients in the hirudin group.

There was a 18% relative reduction in recurrent myocardial infarction at 30 days with hirudin which did not reach statistical significance (3.6% vs. 4.4%, odds ratio (OR) 0.81, CI 0.56–1.18). Patients randomized to hirudin were significantly more likely than those receiving heparin to have an APTT measurement in the target range at any time interval during the infusion (ANTMAN 1996).

The GUSTO IIb study enrolled 4131 patients with ST-segment elevation (74% receiving thrombolytic therapy, consisting of t-PA in 70% and streptokinase in 30%). Study drug was administered for a mean of 75 ± 29 h. There was a significant reduction in the incidence of death or myocardial infarction within the first 48 h. There was a small reduction in the primary combined end point of death or myocardial infarction at 30 days (Fig. 3): 9.9% of patients in the hirudin group compared with 11.3% in the heparin group ($p = 0.13$). Thus, hirudin provided a small advantage over heparin, principally related to early events and a reduction in the risk of nonfatal myocardial infarction.

When the patients in GUSTO IIb with ST elevation were combined with the 8011 patients without ST elevation, the reduction in the primary end point was of borderline significance (8.9% vs. 9.8%, $p = 0.058$) (Fig. 3). The predominant effect of hirudin was on myocardial infarction and reinfarction, which occurred in 5.4% of all hirudin-treated patients and in 6.3% of heparin-treated patients ($p = 0.04$). After 48 h of therapy, the risk of death or myocardial infarction was significantly lower in all the patients assigned to hirudin (2.3%) as compared to heparin (3.1%, $p = 0.001$). There was no increase in major bleeding, although in the total group there was a significant increase in moderate bleeding requiring transfusion among patients receiving hirudin (GUSTO IIb INVESTIGATORS 1996).

Although neither GUSTO II nor TIMI 9 were designed to have sufficient power to detect differences in noncombined end points or within subgroups, a prospective meta-analysis was planned which would provide the statistical power to calculate reliable estimates of treatment effects on mortality and reinfarction for all patients and within specified subgroups. The populations of thrombolytic-eligible patients were similar in the two studies. Among the 6054 patients from these two studies who received thrombolytics, there was no significant difference in mortality for heparin versus hirudin at either 24 h or 30 days. There was, however, a 19% relative reduction in reinfarction at 30 days, which was of borderline significance (4.7% vs. 5.8%, $p = 0.055$) (SIMES et al. 1996). It may be concluded from these large-scale studies that a modest reduction in early cardiac events following acute myocardial infarction is offered by hirudin in comparison to heparin, but this effect is not sustained at 30 days.

4. Further Acute Myocardial Infarction Studies

A subset analysis of the GUSTO IIb patients who received streptokinase showed a 34% relative reduction in the primary end points of death or reinfarction at 30 days in the patients receiving hirudin as compared to those

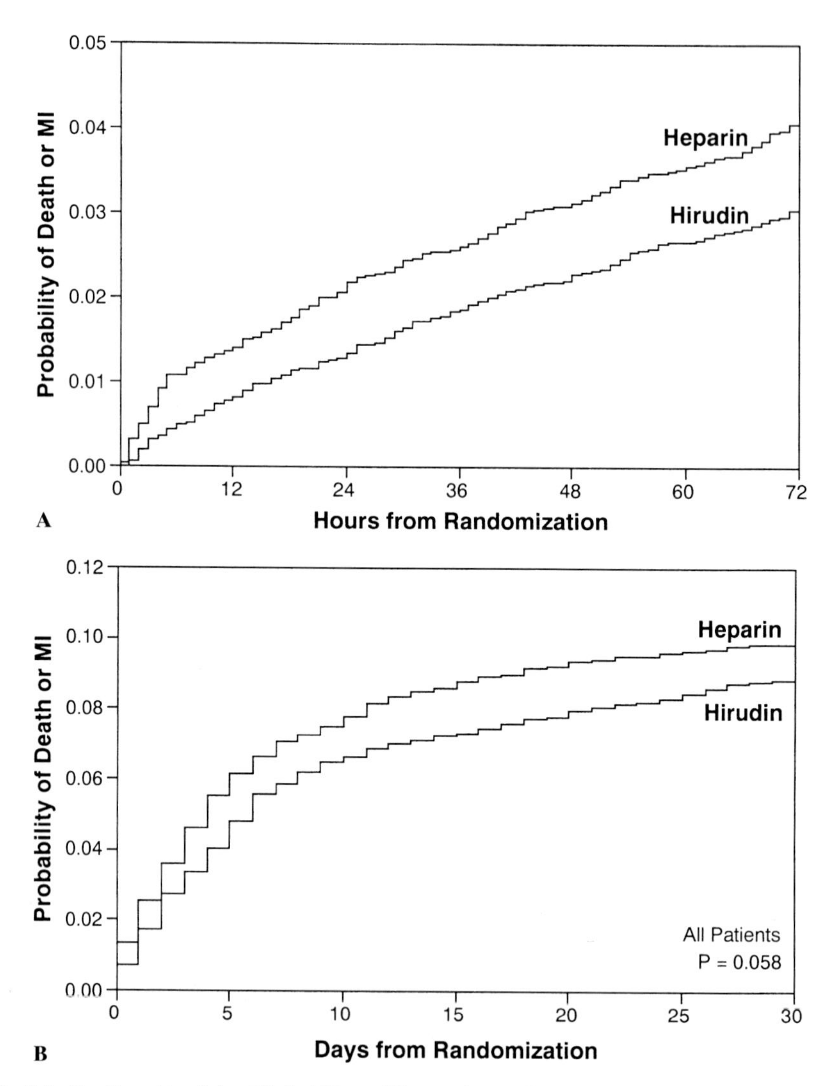

Fig. 3A–D. Results of the Global Use of Strategies to Open Occluded Arteries IIb trial comparing hirudin to heparin in patients with acute coronary syndromes. Kaplan-Meier estimate of the probability of death or myocardial infarction or reinfarction during the first 72 hours in all patients **A**, and during the first 30 days in all patients **B**, those with ST-segment elevation **C**, and those without ST-segment elevation **D**. Reprinted with permission from GUSTO IIb INVESTIGATORS. Copyright 1996 Massachusetts Medical Society. All rights reserved

receiving heparin (9.6% vs. 14.7%, $p = 0.01$), mostly attributable to a reduction in nonfatal myocardial infarctions. Although the choice of thrombolytic agent was not randomized, patient demographics were balanced in the two antithrombin therapy groups within the streptokinase cohort. There were, however, differences in the baseline features between the streptokinase and

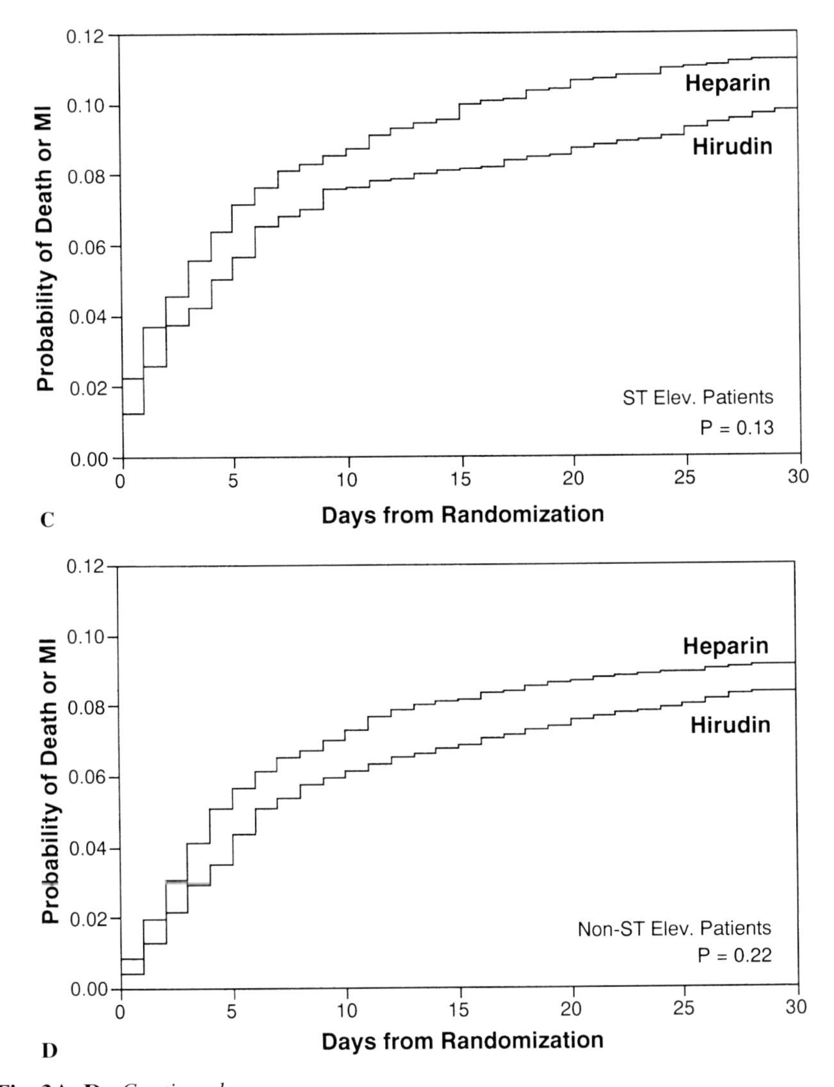

Fig. 3A–D. *Continued*

t-PA patients (METZ et al. 1996). It should be noted that the statistical significance of this observation was lost when the thrombolytic-treated patients from GUSTO IIb were combined with those from TIMI IIb (9.3% vs. 11.0%, OR 0.83, 95% CI 0.62–1.11, $p = 0.72$) (Simes, personal communication). The use of hirudin compared with heparin in conjunction with streptokinase will be studied further in the HIT IV trial, which has completed enrollment of 12000 patients. The HERO II study which is currently underway will compare Hirulog with heparin in 17000 patients receiving streptokinase for acute myocardial infarction.

The Argatroban in Myocardial Infarction (AIM) trial was a randomized, double-blind, phase II angiographic study of argatroban versus placebo in 1800 patients treated with streptokinase. The frequency of TIMI III flow at 90 min or of clinical events was not significantly different between patients receiving argatroban as compared to placebo.

II. Unstable Angina and Myocardial Infarction Without ST Elevation

Although unstable angina is a precursor to and part of the clinical continuum of acute myocardial infarction, proven therapies for myocardial infarctions such as thrombolytic agents and β-adrenergic receptor blockers have not shown to be efficacious in unstable angina. Therefore, it is important to separate these two indications when evaluating new therapies.

1. Phase II Clinical Trials

The feasibility of direct thrombin inhibition in unstable angina was first tested in a pilot study of 20 patients with unstable angina treated with intravenous Hirulog for 5 days, resulting in a steady prolongation of the activated partial thromboplastin time without any hemorrhagic or other adverse effects (SHARMA et al. 1993). In a dose escalation study of argatroban in 43 patients with unstable angina, a 4-h infusion of the drug resulted in a significant dose-related increase in activated partial thromboplastin time without bleeding time prolongation or spontaneous bleeding. Myocardial ischemia did not occur during therapy, but 21% of patients experienced an episode of unstable angina within hours of argatroban discontinuation. This early recurrent angina was correlated with higher argatroban dose and a greater prolongation of APTT, adding to the growing body of evidence for a rebound effect (GOLD et al. 1993).

A study of 55 patients with unstable angina receiving intravenous Hirulog showed a dose-related increase in activated partial thromboplastin time which was correlated with a reduction in in vivo thrombin activity and clinical events (LIDON et al. 1993). The TIMI 7 trial was a randomized, double-blind but uncontrolled dosing study of Hirulog given to patients with unstable angina. Patients received a constant infusion of Hirulog for 72 h at one of four doses. There were no differences between treatment groups in the primary efficacy end point of death, myocardial infarction, rapid clinical deterioration, or recurrent ischemia by 72 h. However, there was a significant reduction in the secondary end point of death or nonfatal myocardial infarction through hospital discharge in patients treated with the three higher doses of Hirulog (FUCHS and CANNON 1995).

TOPOL et al. (1994) reported a randomized study of direct thrombin inhibition with hirudin in unstable angina or non-ST-elevation infarction. Patients presenting with rest pain, electrocardiographic changes and baseline angiogram showing a significant culprit lesion with associated thrombus were randomized to one of two doses of heparin or one of four doses of hirudin. At 72- to 120-h repeat angiography, the 116 patients treated with hirudin tended to show more improvement than the 50 patients receiving heparin in regards to all primary angiographic variables.

2. Phase III Clinical Trials

Of the 12142 patients enrolled in GUSTO IIb, 8011 suffered from acute ischemia without ST-segment elevation at presentation (unstable angina or non-Q-wave myocardial infarction). There were small nonsignificant reductions in myocardial infarction (5.6% vs. 6.4%, $p = 0.152$) and the combination of death and myocardial infarction (8.3% vs. 9.1%, $p = 0.22$) at 30 days among patients receiving hirudin at bolus dose of 0.1 mg/kg followed by a 0.1-mg/kg/h infusion as compared with heparin (Fig. 3). In this subset of patients, there was a significant increase in the rate of transfusion with hirudin as compared with heparin (10.2% vs. 8.4%, $p = 0.01$), and increased rates of moderate and severe bleeding and intracranial hemorrhage, all with borderline significance, that were not evident in the group with ST elevation. It can be concluded from this trial that, at the dose studied, direct thrombin inhibition with hirudin offers no advantage over heparin therapy in the treatment of unstable angina. The OASIS trial, which has a planned enrollment of 12000 patients with unstable angina, will compare hirudin at a higher dose than that used in GUSTO IIb (0.4-mg/kg bolus followed by a 0.15-mg/kg/h infusion) with standard heparin therapy.

III. Adjunct to Percutaneous Revascularization

Heparin has traditionally been administered during percutaneous transluminal coronary angioplasty to avoid thrombotic complications during and after the procedure. Despite treatment with aspirin and relatively high doses of heparin with intraprocedural monitoring of the activated clotting time, the abrupt closure rate remains at 6%–8% due to intracoronary thrombosis, dissection, or both. An inverse relationship between the degree of anticoagulation and frequency of acute thrombosis has been shown (Narins et al. 1996), but administering high doses of heparin has been associated with an increased rate of bleeding complications and transfusions. The mechanism of acute thrombosis has not been firmly elucidated, but likely involves plaque compression, fracture and vessel wall injury inducing a prothrombotic environment by the exposure of tissue factor, extracellular matrix, subintimal collagen, and clot-bound thrombin. The thrombogenicity of the intracoronary equipment has also been proposed to contribute to acute thrombosis. Whatever the mechanism, the use of direct thrombin inhibitors as an adjunct to percutaneous revascularization offers several theoretical advantages over heparin. These agents would not only provide potent systemic anticoagulation, but also would inactivate clot-bound thrombin in the newly exposed lesion and inhibit the recruitment and aggregation of platelets.

1. Preclinical Studies

In porcine models of carotid angioplasty (HERAS et al. 1989) and coronary stenting (BUCHWALD et al. 1993), hirudin was shown to decrease the aggregation of platelets and deposition of fibrin compared with heparin. Hirudin was shown to be superior to heparin in inhibiting thrombus in rat (DOUTREMEPUICH et al. 1989) and rabbit (AGNELLI et al. 1990) models of thrombosis.

In a rabbit angioplasty model, SAREMBOCK et al. (1991) showed that animals receiving hirudin at the time of angioplasty had significantly less restenosis by angiography and by quantitative histopathology than rabbits receiving heparin, suggesting that aside from reducing acute thrombotic events, direct thrombin inhibition may also provide long-term benefits with respect to restenosis.

2. Phase II Clinical Trials

The use of direct thrombin inhibition in the setting of coronary intervention was tested in a multicenter dose escalation study of Hirulog in 291 patients undergoing elective coronary angioplasty (TOPOL et al. 1993). The abrupt closure rate, the primary end point of the study, was 3.9% for patients receiving the higher doses of Hirulog compared with 11.3% in the lower dose groups ($p = 0.05$), while bleeding complications were not increased. Interestingly, only 12 patients in the study had an ACT greater than 300 s, now the accepted standard for anticoagulation during angioplasty in the absence of glycoprotein IIb/IIa inhibition.

In a randomized, double-blind trial, VAN DEN BOS et al. (1993) compared the effects of a 24-h infusion of hirudin relative to heparin among 113 patients with stable angina undergoing coronary angioplasty. Myocardial infarction or emergency coronary bypass surgery occurred in 1.4% of the hirudin-treated patients compared with 10.3% of the heparin-treated patients (relative risk, 7.6; 95% CI, 0.9, 65.6). There was also a reduction with hirudin in ischemia as measured by continuous ST-segment monitoring. The promising results of these studies led to the two phase III clinical trials of direct thrombin inhibition as an adjunct to coronary intervention.

3. Phase III Clinical Trials

The Hirulog Angioplasty Study was a double-blind, randomized trial of Hirulog versus heparin in 4098 patients undergoing angioplasty for unstable or postinfarction angina. In the total study group, Hirulog did not significantly reduce the incidence of the primary composite end point of in-hospital death, myocardial infarction, abrupt vessel closure, or rapid clinical deterioration of cardiac origin (11.4% vs. 12.2% for heparin), but did result in a lower incidence of bleeding (3.8% vs. 9.8%, $p < 0.001$). In a prospectively stratified subgroup of 704 patients with postinfarction angina, however, Hirulog therapy was associated with a lower incidence of the primary end point (9.1% vs. 14.2%, $p = 0.04$) and a lower incidence of bleeding (3.0% vs. 11.1%, $p < 0.001$). Despite the acute benefits of Hirulog in this subgroup, at 6 months the cumulative rates of death, myocardial infarction, and repeated revascularization were similar in the two treatment arms (20.5% vs. 25.1%, $p = 0.17$) (BITTL et al. 1995).

The HELVETICA (Hirudin in a European Trial versus Heparin in the Prevention of Restenosis after PTCA) trial evaluated the impact of hirudin compared with heparin on the incidence of restenosis and ischemic complications among 1141 patients undergoing angioplasty for unstable angina. Patients were randomized to receive: (1) intravenous heparin for 24 h followed by subcutaneous placebo twice daily for 3 days, (2) intravenous hirudin for 24 h followed by subcutaneous placebo twice daily for 3 days, or (3) intravenous hirudin for 24 h and subcutaneous hirudin twice daily for 3 days. There were no differences among treatment groups in the primary end point of event-free survival at 7 months (67.3% with heparin, 63.5% with intravenous hirudin, and 68.0% with both intravenous and subcutaneous hirudin, $p = 0.61$) or in the angiographic restenosis rates at 6 months. Administration of hirudin was associated with a significant reduction in cardiac events within 4 days, occurring in 11.0%, 7.9%, and 5.6% of patients in the respective groups, representing a combined relative risk with hirudin of 0.61 (95% confidence interval, 0.41–0.90; $p = 0.023$) (SERRUYS et al. 1995). The observed loss of this early benefit over time in both of these trials again raises the question of a rebound phenomenon.

IV. Deep Venous Thrombosis

The use of direct thrombin inhibitors for the treatment and prevention of deep venous thrombosis has only been explored recently, but the results to date have been very promising. In a jugular vein thrombus model in rabbits utilizing a combination of endothelial damage and flow reduction, MATTHIASSON et al. (1995) showed that hirudin was equally or more effective than low molecular weight heparin in reducing thrombus weight and the frequency of occlusive thrombosis. In a pilot study of hirudin administered subcutaneously to ten patients with recent deep vein thrombosis at a dose of 0.75 mg/kg twice daily for 5 days, clinical evolution was uneventful other than a probable recurrence of pulmonary embolism in one patient (SCHIELE et al. 1994).

For the prevention of venous thrombosis after orthopedic surgery, phase II, dose-escalating studies found that subcutaneous regimens of either Hirulog (GINSBERG et al. 1994) or hirudin (ERIKSSON et al. 1994) were both safe and potentially efficacious. In a double-blind study of three doses of twice daily subcutaneous hirudin in comparison with unfractionated heparin in 1119 patients undergoing elective hip surgery, the frequency of proximal thrombosis was 19.6% in the heparin group as compared to 8.5% ($p < 0.001$), 3.1% ($p < 0.001$), and 2.4% ($p < 0.001$) in the 10-mg, 15-mg, and 20-mg hirudin groups, respectively (ERIKSSON et al. 1996). These extremely encouraging findings suggest that further studies in this area are warranted.

V. Heparin-Induced Thrombocytopenia

Heparin-induced thrombocytopenia is a rare but severe complication of heparin therapy that can result in venous or arterial thromboembolic events. In an early report by MATSUO et al. (1992), a patient with heparin-induced thrombocytopenia and thrombosis was successfully treated with argatroban. SCHIELE et al. (1995) described the treatment of six patients suffering from heparin-induced thrombocytopenia with hirudin therapy until oral anticoagulation with acenocoumarol was achieved. After initiation of hirudin, clinical evolution was uneventful in all patients, without recurrence of thromboembolism, limb ischemia, or hemorrhagic complication. Platelet counts rose from a median nadir value of 60000/dl (range 15000–90000/dl) to above 100000/dl in every patient within 3–6 days. LEWIS et al. (1996) reported their experience using argatroban in 11 patients with heparin-induced thrombocytopenia who required coronary intervention. Eight of the 11 patients had either a history of or an active syndrome of thrombocytopenia with thrombosis. All coronary lesions were treated successfully and no patient suffered acute closure or periprocedural complication. A multicenter trial is currently underway using argatroban in patients undergoing coronary intervention who have a history of heparin-induced thrombocytopenia. This application may be one of the most promising uses of direct thrombin inhibitors, in that these agents represent the only currently available means of providing acute antithrombin therapy for these patients.

F. Rebound Phenomenon

The existence of a profound rebound prothrombotic phenomenon after the discontinuation of direct thrombin inhibitor therapy was suggested by GOLD et al. (1993), who found that patients with unstable angina receiving higher doses of argatroban experienced more recurrent angina in the hours following discontinuation of therapy. This observation was supported by THEROUX et al. (1995), who noted that patients with acute myocardial infarction who received high-dose Hirulog experienced more clinical events following discontinuation than patients receiving either low-dose Hirulog or placebo.

The study by GOLD et al. (1993) offers some insight into the mechanism of this observed effect. Recurrent angina was correlated significantly with a higher argatroban dose and with greater prolongation of APTT. Pretreatment plasma concentrations of thrombin-antithrombin III complex (a measure of in vivo thrombin formation) and fibrinopeptide A (a measure of thrombin activity) were both elevated two to three times above normal values, consistent with an activation of coagulation during the acute ischemic syndromes. During argatroban infusion, thrombin-antithrombin III complex levels remained unchanged, whereas a significant 2.3-fold decrease in fibrinopeptide A concentrations was observed. By contrast, 2h after infusion, thrombin-antithrombin III complex concentrations increased 3.9-fold over baseline measurements together with return of fibrinopeptide A levels to baseline values. Thus, argatroban therapy was associated with inhibition of thrombin activity but not its formation, and cessation of therapy was associated with rebound thrombin generation and recurrence of unstable angina.

The presence of a significant rebound effect is further supported by the findings in the phase III trials of acute ischemic syndromes and coronary interventions, which found early reductions in adverse events, but loss of treatment effect over time. It was noted a preponderance of the reinfarctions in GUSTO IIb occurred after discontinuation of antithrombin therapy. Although there was no abrupt clinical rebound in thrombosis after discontinuing hirudin, there was a clustering of reinfarction upon stopping heparin therapy, with greatest risk during the first 8h (C. Granger, Duke University, personal communication). In the 2nd and 3rd days after stopping antithrombin therapy, however, there were more reinfarctions in the hirudin group, such that by the end of the 3rd day following therapy discontinuation, one-third of all infarctions in both treatment groups had occurred. This observation is consistent with the findings from GUSTO I, in which there was a clustering of reinfarction in the first 10h after discontinuation of intravenous heparin (GRANGER et al. 1996). Thus although there does seem to be a rebound increase in ischemic events following termination of even the slowly reversible direct thrombin inhibitor hirudin, such rebound seems to be a more delayed and gradual process than with either heparin or the highly reversible direct thrombin inhibitor, argatroban.

Further insight into the mechanism of the observed clinical rebound is offered by a hemostatic substudy of 368 patients from GUSTO IIb, which showed that, during infusion, there was a greater suppression of thrombin activity by hirudin than with heparin as measured by a decrease in fibrinopeptide A. After therapy termination, there was a rebound in thrombin generation (as measured by prothrombin fragment 1.2) and in thrombin activity (as measured by fibrinopeptide A), both of which rose to peaks in the heparin group within the first 24h, but seemed to be slowly rising in the hirudin group by the end of the 24-h study period (KOTTKE-MARCHANT et al. 1996). This laboratory finding correlates well with the timing of clinical events in the two treatment groups. Interestingly, there was a dramatic increase in levels of the endogenous anticoagulant, activated protein C, following hirudin discontinuation which was not observed following heparin therapy. These findings suggest that accumulation of prothrombotic factors during antithrombin therapy leads to a relative hypercoagulable state following drug withdrawal. However, the slowly reversible binding of hirudin and counteractive increases in endogenous anticoagulants may tend to dampen this effect in comparison to heparin.

G. Summary and Future Directions

Direct thrombin inhibitors offer several theoretical advantages over heparin as antithrombotic agents. Initial phase II studies appeared promising for the use of these agents in the treatment of acute myocardial infarction, unstable angina, and as an adjunct to percutaneous coronary revascularization, with reductions in surrogate end points and a more stable level of anticoagulation. However, phase III trials for all three indications have been disappointing, showing modest reductions in early cardiac events which have been lost over time.

The relatively modest improvement over heparin afforded by direct thrombin inhibitors in the large phase III trials have dampened enthusiasm for the use of these agents and called into question the pivotal role of thrombin in acute arterial disease and restenosis following coronary intervention. In contrast, recent trials have shown platelet glycoprotein IIb/IIIa inhibitors to be highly effective in both of these settings, suggesting that potent inhibition of platelet activity may be of critical importance in interrupting arterial thrombus formation. The combination of a direct thrombin inhibitor and a glycoprotein IIb/IIa inhibitor has shown promise in an experimental model (NICOLINI et al. 1994), and clinical trials are planned which will explore this combination therapy.

An alternative explanation for the lack of a durable benefit of direct thrombin inhibition over heparin therapy is the presence of rebound phenomenon due to buildup of thrombin precursors or the prothrombotic effect of thrombin inhibition involving thrombomodulin. Combined therapy with a direct thrombin inhibitor and one of the tissue factor pathway inhibitors discussed in Chaps. 14 and 15, this volume, or prolonged treatment with an oral direct thrombin inhibitor as discussed in Chap. 13 offer possible solutions to this theoretical limitation.

The use of direct thrombin inhibitors for deep venous thrombosis or in patients with heparin-induced thrombocytopenia are relatively unexplored areas which deserve further study based on the promising results of preliminary studies.

References

Agnelli G, Pascucci C, Cosmi B, Nenci GG (1990) The comparative effects of recombinant hirudin (CGP 39393) and standard heparin on thrombus growth in rabbits. Thrombosis Haemost 63:204–207

Antman EM (1994) Hirudin in acute myocardial infarction. Safety report from the Thrombolysis and Thrombin Inhibition in Myocardial Infarction (TIMI) 9A Trial. Circulation 90:1624–1630

Antman EM (1996) Hirudin in acute myocardial infarction. Thrombolysis and Thrombin Inhibition in Myocardial Infarction (TIMI) 9B trial. Circulation 94: 911–921

Bichler J, Baynes JW, Thorpe SR (1993) Catabolism of hirudin and thrombin-hirudin complexes in the rat. Biochemistry 296:771–776

Bittl JA, Strony J, Brinker JA, Ahmed WH, Meckel CR, Chaitman BR, Maraganore J et al (1995) Treatment with bivalirudin (Hirulog) as compared with heparin during coronary angioplasty for unstable or postinfarction angina. Hirulog Angioplasty Study Investigators. N Engl J Med 333:764–769

Buchwald AB, Sandrock D, Unterberg C, Ebbecke M, Nebendahl K, Luders S et al (1993) Platelet and fibrin deposition on coronary stents in minipigs: effect of hirudin versus heparin. J Am Coll Cardiol 21:249–254

Callas DD, Hoppensteadt D, Fareed J (1995) Comparative studies on the anticoagulant and protease generation inhibitory actions of newly developed site-directed thrombin inhibitory drugs. Efegatran, argatroban, hirulog, and hirudin. Semin Thromb Hemost 21:177–183

Cannon CP, Braunwald E (1995) Hirudin: initial results in acute myocardial infarction, unstable angina and angioplasty. J Am Coll Cardiol 25 [Suppl 7]:30S–37S

Cannon CP, McCabe CH, Henry TD, Schweiger MJ, Gibson RS, Mueller HS et al (1994) A pilot trial of recombinant desulfatohirudin compared with heparin in conjunction with tissue-type plasminogen activator and aspirin for acute myocardial infarction: results of the Thrombolysis in Myocardial Infarction (TIMI) 5 trial. J Am Coll Cardiol 23:993–1003

Chao FC, Tullis JL, Kenney DM, Conneely GS, Doyle JR (1974) Concentration effects of platelets, fibrinogen and thrombin on platelet aggregation and fibrin clotting. Thromb Diath Haemorrh 32:216–231

Doutremepuich C, Deharo E, Guyot M, Lalanne MC, Walenga J, Fareed J (1989) Antithrombotic activity of recombinant hirudin in the rat: a comparative study with heparin. Thromb Res 54:435–445

Eriksson BI, Kalebo P, Ekman S, Lindbratt S, Kerry R, Close P (1994) Direct thrombin inhibition with Rec-hirudin CGP 39393 as prophylaxis of thromboembolic complications after total hip replacement. Thromb Haemost 72:227–231

Eriksson BI, Ekman S, Kalebo P, Zachrisson B, Bach D, Close P (1996) Prevention of deep-vein thrombosis after total hip replacement: direct thrombin inhibition with recombinant hirudin, CGP 39393 Lancet 347:635–639

Esmon NL, Owen WG, Esmon CT (1982) Isolation of a membrane-bound cofactor for thrombin-catalyzed activation of protein C. J Biol Chem 257:859–864

Fox I, Dawson A, Loynds P, Eisner J, Findlen K, Levin E et al (1993) Anticoagulant activity of Hirulog, a direct thrombin inhibitor, in humans. Thromb Haemost 69:157–163

Fuchs J, Cannon CP (1995) Hirulog in the treatment of unstable angina. Results of the Thrombin Inhibition in Myocardial Ischemia (TIMI) 7 trial. Circulation 92:727–733

Galdal KS, Lyberg T, Evensen SA, Nilsen E, Prydz H (1985) Thrombin induces thromboplastin synthesis in cultured vascular endothelial cells. Thromb Haemost 54:373–376

Ginsberg JS, Nurmohamed MT, Gent M, MacKinnon B, Sicurella J, Rill-Edwards P et al (1994) Use of hirulog in the prevention of venous thrombosis after major hip or knee surgery. Circulation 90:2385–2389

Global Use of Strategies to Open Occluded Coronary Arteries (GUSTO) IIa Investigators (1994) Randomized trial of intravenous heparin versus recombinant hirudin for acute coronary syndromes. Circulation 90:1631–1637

Global Use of Strategies to Open Occluded Coronary Arteries (GUSTO) IIb Investigators (1996) A comparison of recombinant hirudin with heparin for the treatment of acute coronary syndromes. N Engl J Med 335:775–782

Gold HK, Torres FW, Garabedian HD, Werner W, Jang IK, Khan A et al (1993) Evidence for a rebound coagulation phenomenon after cessation of a 4-hour

infusion of a specific thrombin inhibitor in patients with unstable angina pectoris. J Am Coll Cardiol 21:1039–1047
Granger CB, Hirsch J, Califf RM, Col J, White HD, Betriu A et al (1996) Activated partial thromboplastin time and outcome after thrombolytic therapy for acute myocardial infarction: results from the GUSTO-I trial. Circulation 93:870–878
GUSTO Investigators (1993) An international randomized trial comparing four thrombolytic strategies for acute myocardial infarction. N Engl J Med 329:673–682
Harvey RP, Degryse E, Stefani L, Schamber F, Cazenave JP, Courtney M et al (1986) Cloning and expression of a cDNA coding for the anticoagulant hirudin from the bloodsucking leech, Hirudo medicinalis. Proc Nat Acad Sci USA 83:1084–1088
Haskel EJ, Prager NA, Sobel BE, Abendschein DR (1991) Relative efficacy of antithrombin compared with antiplatelet agents in accelerating coronary thrombolysis and preventing early reocclusion. Circulation 83:1048–1056
Heras M, Chesebro JH, Penny WJ, Bailey KR, Badimon L, Fuster V (1989) Effects of thrombin inhibition on the development of acute platelet-thrombus deposition during angioplasty in pigs. Heparin versus recombinant hirudin, a specific thrombin inhibitor. Circulation 79:657–665
Hirsh J (1991) Heparin. N Engl J Med 324:1565–1574
Hung DT, Vu TH, Nelken NA, Coughlin SR (1992) Thrombin-induced events in non-platelet cells are mediated by the unique proteolytic mechanism established for the cloned platelet thrombin receptor. J Cell Biol 116:827–832
Jurd KM, Stephens CJ, Black MM, Hunt BJ (1996) Endothelial cell activation in cutaneous vasculitis. Clin Exp Dermatol 21:28–32
Kaiser B, Simon A, Markwardt F (1990) Antithrombotic effects of recombinant hirudin in experimental angioplasty and intravascular thrombolysis. Thromb Haemost 63:44–47
Kelton JG, Sheridan D, Brain H, Powers PJ, Turpie AG, Carter CJ (1984) Clinical usefulness of testing for a heparin-dependent platelet-aggregating factor in patients with suspected heparin-associated thrombocytopenia. J Lab Clin Med 103:606–612
Klement P, Borm A, Hirsh J, Maraganore J, Wilson G, Weitz J (1992) The effect of thrombin inhibitors on tissue plasminogen activator induced thrombolysis in a rat model. Thromb Haemost 68:64–68
Kottke-Marchant K, Zoldhelyi P, Zaramo C, Brooks L, Cianciolo C, Janssens S. (1996) The effect of desirudin vs. heparin on hemostatic parameters in acute coronary syndromes: the GUSTO IIb hemostasis substudy. Circulation 94:I-742
Lee LV (1995) Initial experience with hirudin and streptokinase in acute myocardial infarction: results of the Thrombolysis in Myocardial Infarction (TIMI) 6 trial. Am J Cardiol 75:7–13
Lefkovits J, Topol EJ (1994) Direct thrombin inhibitors in cardiovascular medicine. Circulation 90:1522–1536
Lewis BE, Ferguson JJ, Grassman ED, Fareed J, Walenga J, Joffrion JL et al (1996) Successful coronary interventions performed with argatroban anticoagulation in patients with heparin-induced thrombocytopenia and thrombosis syndrome. J Invasive Cardiol 8:410–417
Lidon RM, Theroux P, Juneau M, Adelman B, Maraganore J (1993) Initial experience with a direct antithrombin, hirulog, in unstable angina. Anticoagulant, antithrombotic, and clinical effects. Circulation 88:1495–1501
Lidon RM, Theroux P, Lesperance J, Adelman B, Bonan R, Duval D et al (1994) A pilot, early angiographic patency study using a direct thrombin inhibitor as adjunctive therapy to streptokinase in acute myocardial infarction. Circulation 89:1567–1572
Lindahl U, Hook M (1978) Glycosaminoglycans and their binding to biological macromolecules. Annu Rev Biochem 47:385–417

Longstaff C, Wong MY, Gaffney PJ (1993) An international collaborative study to investigate standardisation of hirudin potency. Thromb Haemost 69:430–435
Loscalzo J, Melnick B, Handin RI (1985) The interaction of platelet factor four and glycosaminoglycans. Arch Biochem Biophys 240:446–455
Maraganore JM (1993) Pre-clinical and clinical studies on Hirulog: a potent and specific direct thrombin inhibitor. Adv Exp Med Biol 340:227–236
Markwardt F (1991) Past, present and future of hirudin. Haemostasis 21 [Suppl 1]:11–26
Matsuo T, Kario K, Chikahira Y, Nakao K, Yamada T (1992) Treatment of heparin-induced thrombocytopenia by use of argatroban, a synthetic thrombin inhibitor. Br J Haematol 82:627–629
Matthiasson SE, Lindblad B, Bergqvist D (1995) Prevention of experimental venous thrombosis in rabbits with different low molecular weight heparins, dermatan sulphate and hirudin. Haemostasis 25:124–132
Metz BK, Granger CB, White HD, Simes J, Topol EJ (1996) Streptokinase and hirudin reduces death and reinfarction in acute myocardial infarction compared with streptokinase and heparin: results from GUSTO-IIb. Circulation 94:I-430
Narins CR, Hillegass WB Jr, Nelson CL, Tcheng JE, Harrington RA, Phillips HR et al (1996) Relation between activated clotting time during angioplasty and abrupt closure. Circulation 93:667–671
Neuhaus KL, von Essen R, Tebbe U, Jessel A, Heinrichs H, Maurer W et al (1994) Safety observations from the pilot phase of the randomized r-Hirudin for Improvement of Thrombolysis (HIT-III) study. A study of the Arbeitsgemeinschaft Leitender Kardiologischer Krankenhausärzte (ALKK). Circulation 90:1638–1642
Nicolini FA, Lee P, Rios G, Kottke-Marchant K, Topol EJ (1994) Combination of platelet fibrinogen receptor antagonist and direct thrombin inhibitor at low doses markedly improves thrombolysis. Circulation 89:1802–1809
Pearson JD (1993) The control of production and release of haemostatic factors in the endothelial cell. Baillieres Clin Haematol 6:629–651
Pearson JD (1994) Endothelial cell function and thrombosis. Baillieres Clin Haematol 7:441–452
Phillips DR (1974) Thrombin interaction with human platelets. Potentiation of thrombin-induced aggregation and release by inactivated thrombin. Thromb Diath Haemorrh 32:207–215
Rigel DF, Olson RW, Lappe RW (1993) Comparison of hirudin and heparin as adjuncts to streptokinase thrombolysis in a canine model of coronary thrombosis. Circ Res 72:1091–1102
Sarembock IJ, Gertz SD, Gimple LW, Owen RM, Powers ER, Roberts WC (1991) Effectiveness of recombinant desulphatohirudin in reducing restenosis after balloon angioplasty of atherosclerotic femoral arteries in rabbits. Circulation 84:232–243
Scharf M, Engels J, Tripier D (1989) Primary structures of new 'iso-hirudins'. FEBS Lett 255:105–110
Schiele F, Vuillemenot A, Kramarz P, Kieffer Y, Soria J, Soria C et al (1994) A pilot study of subcutaneous recombinant hirudin (HBW 023) in the treatment of deep vein thrombosis. Thromb Haemost 71:558–562
Schiele F, Vuillemenot A, Kramarz P, Kieffer Y, Anguenot T, Bernard Y et al (1995) Use of recombinant hirudin as antithrombotic treatment in patients with heparin-induced thrombocytopenia. Am J Hematol 50:20–25
Schini VB, Hendrickson H, Heublein DM, Burnett JC Jr, Vanhoutte PM (1989) Thrombin enhances the release of endothelin from cultured porcine aortic endothelial cells. Eur J Pharmacol 165:333–334
Serruys PW, Herrman JP, Simon R, Rutsch W, Bode C, Laarman GJ et al (1995) A comparison of hirudin with heparin in the prevention of restenosis after coronary angioplasty. Helvetica Investigators. N Engl J Med 333:757–763

Sharma GV, Lapsley D, Vita JA, Sharma S, Coccio E, Adelman B et al (1993) Usefulness and tolerability of hirulog, a direct thrombin-inhibitor, in unstable angina pectoris. Am J Cardiol 72:1357–1360

Simes RJ, Granger CB, Antmann EM, Califf RM, Braunwald E, Topol EJ (1996) Impact of hirudin versus heparin on mortality and (re)infarction in patients with acute coronary syndromes: a prospective meta-analysis of the GUSTO-IIb and TIMI 9b trials. Circulation 94:I-430

Smith GF, Shuman RT, Craft TJ, Gifford DS, Kurz KD, Jones ND et al (1996) A family of arginal thrombin inhibitors related to efegatran. Semin Thromb Hemost 22:173–183

Stubbs MT, Bode W (1993) A player of many parts: the spotlight falls on thrombin's structure. Thromb Res 69:1–58

Tabata H, Mizuno K, Miyamoto A, Etsuda H, Isojima K, Satomura K et al (1992) The effect of a new thrombin inhibitor (argatroban) in the prevention of reocclusion after reperfusion therapy in patients with acute myocardial infarction. Circulation 86:I-260

Talbot MD, Ambler J, Butler KD, Findlay VS, Mitchell KA, Peters RF et al (1989) Recombinant desulphatohirudin (CGP 39393) anticoagulant and antithrombotic properties in vivo. Thromb Haemost 61:77–80

Theroux P, Lidon R-M (1994) Anticougulants and their use in acute ischemic syndromes. In: Topol EJ (ed) Textbook of interventional cardiology, vol 1. Saunders, Philadelphia, pp 23–45

Theroux P, Perez-Villa F, Waters D, Lesperance J, Shabani F, Bonan R (1995) Randomized double-blind comparison of two doses of Hirulog with heparin as adjunctive therapy to streptokinase to promote early patency of the infarct-related artery in acute myocardial infarction. Circulation 91:2132–2139

Topol EJ, Bonan R, Jewitt D, Sigwart U, Kakkar VV, Rothman M et al (1993) Use of a direct antithrombin, Hirulog, in place of heparin during coronary angioplasty. Circulation 87:1622–1629

Topol EJ, Fuster V, Harrington RA, Califf RM, Kleiman NS, Kereiakes DJ et al (1994) Recombinant hirudin for unstable angina pectoris. A multicenter, randomized angiographic trial. Circulation 89:1557–1566

van den Bos AA, Deckers JW, Heyndrickx GR, Laarman GJ, Suryapranata H, Zijlstra F et al (1993) Safety and efficacy of recombinant hirudin (CGP 39 393) versus heparin in patients with stable angina undergoing coronary angioplasty. Circulation 88:2058–2066

Vu TK, Wheaton VI, Hung DT, Charo I, Coughlin SR (1991) Domains specifying thrombin-receptor interaction. Nature 353:674–677

Weitz JI, Hudoba M, Massel D, Maraganore J, Hirsh J (1990) Clot-bound thrombin is protected from inhibition by heparin-antithrombin III but is susceptible to inactivation by antithrombin III-independent inhibitors. J Clin Invest 86:385–391

Zeymer U, von Essen R, Tebbe U, Michels HR, Jessel A, Vogt A et al (1995) Recombinant hirudin and front-loaded alteplase in acute myocardial infarction: final results of a pilot study. HIT-I (Hirudin for the Improvement of Thrombolysis). Eur Heart J 16 Suppl D:22–27

CHAPTER 12

Anticoagulant Therapy with Warfarin for Thrombotic Disorders

L.A. HARKER

A. Introduction

Important advances have been made during the past 20 years in the clinical use of oral anticoagulants. It is now established that oral anticoagulation provides effective antithrombotic therapy for the prevention and management of venous thromboembolism, the prevention of systemic embolism in patients with atrial fibrillation, mechanical prosthetic heart valves or acute myocardial infarction, and the secondary prevention of myocardial infarction. Evidence-based recommendations provide optimal efficacy and safety by maintaining a narrow range of anticoagulation using laboratory-monitored dosing (HIRSH et al. 1994a; DALEN and HIRSH 1995).

B. Pathogenesis

Thrombogenic processes responding to therapy that modestly reduces functional levels of substrate coagulation serine proteases (prothrombin and factors VII, IX and X) involve static venous flow conditions combined with the local generation of thrombin (HIRSH et al. 1994a). Factors contributing to venous stasis include bed rest, congestive heart failure, immobilization, paralysis, anesthesia, pregnancy and external compression (HIRSH et al. 1994a). The local generation of thrombin involves: (a) exposure of tissue factor (trauma, surgery, malignancy, infections, endothelial denudation during orthopedic surgery or intravascular devices, etc.); (HIRSH et al. 1994b); (b) genetic deficiencies in important regulatory plasma proteins (congenital deficiencies of plasma antithrombin, protein C, protein S, and factor V resistance to activated protein C; and (c) acquired prothrombotic blood factors, including hyperhomocysteinemia, and antiphospholipid antibodies (GINSBERG et al. 1995; DEN HEIJER et al. 1995; DAHLBACK 1995; HEIJBOER et al. 1990; HIRSH et al. 1994a,b). Laboratory testing should be considered in patients under 50 years of age, recurrent unexplained venous thrombosis, or family history (one or more first- or second-degree relatives) of venous thromboembolism.

C. Pharmacology

I. Mechanism of Action

Oral anticoagulants are 4-hydroxycoumarin compounds that inhibit the hepatic synthesis of four coagulation serine proteases, prothrombin, and factors VII, IX and X, and two coagulation inhibitor proteins, protein C and protein S. Coumarins interfere with the cyclic interconversion of vitamin K and vitamin K epoxide. Vitamin K is an essential cofactor for the post-translational carboxylation of glutamate residues to γ-carboxyglutamates (Glas). By inhibiting vitamin K epoxide reductase, vitamin K antagonists, such as warfarin, limit the γ-carboxylation of these proteins, thereby preventing the essential calcium-dependent conformational changes required for their complexing with cofactors on phospholipid surfaces. When the average number of Gla residues in each of these molecules falls below the critical 5–6 from the usual 10–13, calcium-dependent complex formation with critical cofactors on phospholipid surfaces is abolished (Fig. 1) (MALHOTRA et al. 1985; HIRSH et al. 1994a; DALEN and HIRSH 1995).

Warfarin is a racemic mixture comprising approximately equal amounts of *R* and *S* optical isomers. It is rapidly and extensively absorbed from the gastrointestinal tract, reaching maximal blood levels in approximately 90min, with a half-life of 36–42h. *R*-Warfarin is metabolized primarily by reduction of the acetenyl sidechain into warfarin alcohols excreted in the urine, and *S*-warfarin is metabolized by oxidation to 7-hydroxy-*S*-warfarin and is eliminated in the bile. Variations in metabolism occur in different individuals and in the same individual due to multiple factors, including drug interactions, intercurrent illness, variable dietary vitamin K intake, changes in receptor affinity, vitamin K availability, and pharmacokinetics, leading to major practical problems in monitoring. Diets rich in green vegetables and nutritional fluid supplements rich in vitamin K reduce the anticoagulant response to warfarin, whereas vitamin K absorption is impaired in fat malabsorption states. Increased synthesis of vitamin-K-dependent coagulation factors may cause mild resistance to warfarin during pregnancy or hepatic congestion secondary to cardiac failure. The many factors affecting the anticoagulant effects of warfarin are listed in Table 1. Unexpected fluctuations in dose response may also be due to changes in diet, inaccuracy in prothrombin time testing, undisclosed drug use, poor patient compliance, surreptitious self-medication, or intermittent alcohol consumption (HIRSH et al. 1994a; DALEN and HIRSH 1995).

II. Assessment of Clinical Efficacy

The clotting test generally used to assess the effects of warfarin is the one-stage prothrombin time (PT), because it detects decreased activities for three of the four vitamin-K-dependent coagulation factors, prothrombin, factor VII and factor X. The test is performed by adding calcium and thromboplastin (containing phospholipid and tissue factor) to citrated plasma. Unfortunately, the

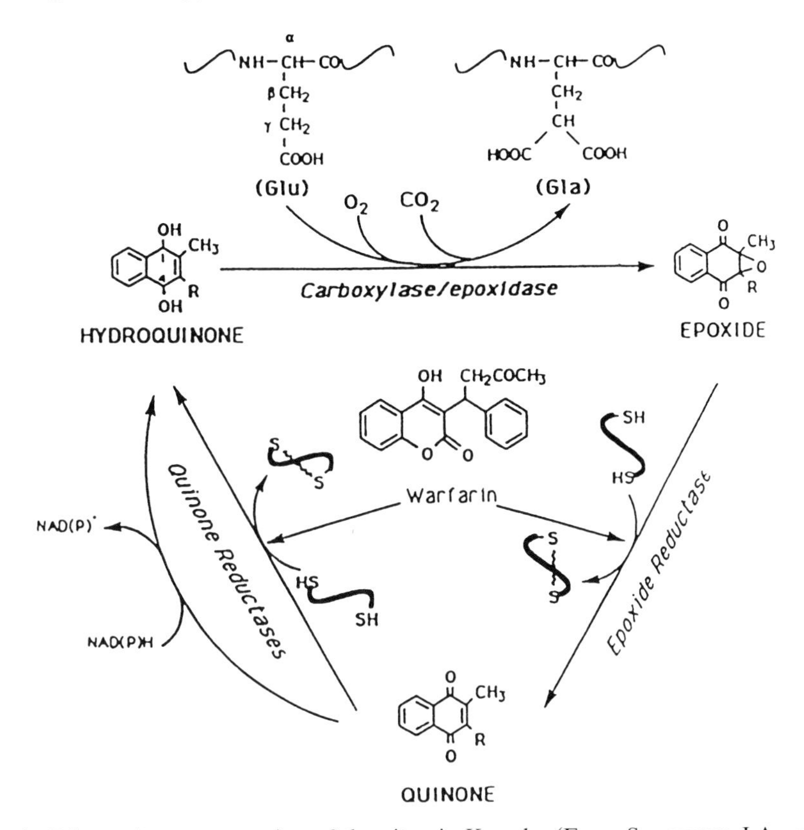

Fig. 1. Schematic representation of the vitamin K cycle. (From Sadowski, J.A., et al. [1991] Warfarin and the metabolism function of vitamin K. *Recent Advances in Blood Coagulation*, No. 5 [ed. by L. Poller], pp. 93–118. Churchill Livingstone, Edinburgh, with permission.)

various thromboplastins used in this test vary in their responsiveness to reductions in vitamin-K-dependent coagulation factors. For example, human brain thromboplastin is substantially more sensitive to warfarin-induced effects than commercial thromboplastins prepared from animal sources. To resolve differences in thromboplastin sensitivities, the World Health Organization has introduced International Normalized Ratios (INRs). The INR normalizes the PT results to what would have been obtained if the sensitive international reference thromboplastin had been used in performing the PT, and is dependent on obtaining the international sensitivity index (ISI) for each thromboplastin (Fig. 2) (Poller 1988; British Society for Haematology 1990; Hirsh et al. 1994a,b; Dalen and Hirsh 1995).

There is evidence that warfarin exerts its antithrombotic effects by specifically decreasing functional prothrombin, the substrate for thrombin. This conclusion is based on the importance of bound thrombin in mediating thrombus formation, improved predictive usefulness of prothrombin antigen

Table 1. Drug and food interactions with warfarin by level of supporting evidence and direction of interaction[a]. (Reprinted with permission from Hirsh et al. 1995)

Level of evidence	Potentiation	Inhibition	No effect
I. Randomized trials or meta-analyses in which the lower limit of the CI for the treatment effect exceeds the minimal clinically important benefit	Alcohol (if concomitant liver disease), amiodarone (anabolic steroids, cimetidine[b], clofibrate, cotrimoxazole, erythromycin, fluconazole, isoniazid [600 mg daily] metronidazole), miconazole, omeprazole, *phenylbutazone*, piroxicam, propafenone, propranolol[b], *sulfinpyrazone (biphasic with later inhibition)*	Barbiturates, carbamazepine, chlordiazepoxide, cholestyramine, *griseofulvin*, nafcillin, rifampin, sucralfate, high vitamin K content foods/enteral feeds, large amounts of avocado	Alcohol, antacids, atenolol, bumetadine, enoxacin, famotidine, fluoxetine, ketorolac, metoprolol, naproxen, nizatidine, psyllium, ranitidine[c]
II. Randomized trials or meta-analyses in which the CI for the treatment effect overlaps the minimal clinically important benefit	Acetaminophen, chloral hydrate, ciprofloxacin, dextropropoxyphene, disulfiram, itraconazole, quinidine, phenytoin (biphasic with later inhibition), tamoxifen, tetracycline, flu vaccine	Dicloxacillin	Ibuprofen, ketoconazole
III: Nonrandomized concurrent cohort comparisons between contemporaneous patients who did and did not receive antithrombotic agents	Acetylsalicylic acid, disopyramide, fluorouracil, ifosfhamide, ketoprofen, lovastatin, metozalone, moricizine, nalidixic acid, norfloxacin, ofloxacin, propoxyphene, sulindac, tolmetin, topical salicylates	Azathioprine, cyclosporine, etretinate, trazodone	
IV: Nonrandomized historic cohort comparisons between current patients who received antithrombotic agents and former patients (from the same institution or from the literature) who did not	Cefamandole, cefazolin, gemfibrozil, heparin, indomethacin, sulfisoxazole		Diltiazem, tobacco, vancomycin

[a] Drugs in italics are those that have supporting level I evidence from both patients and volunteers.
[b] In a small number of volunteer subjects, an inhibitory drug interaction occurred.
[c] I and II evidence of potentiation in patients

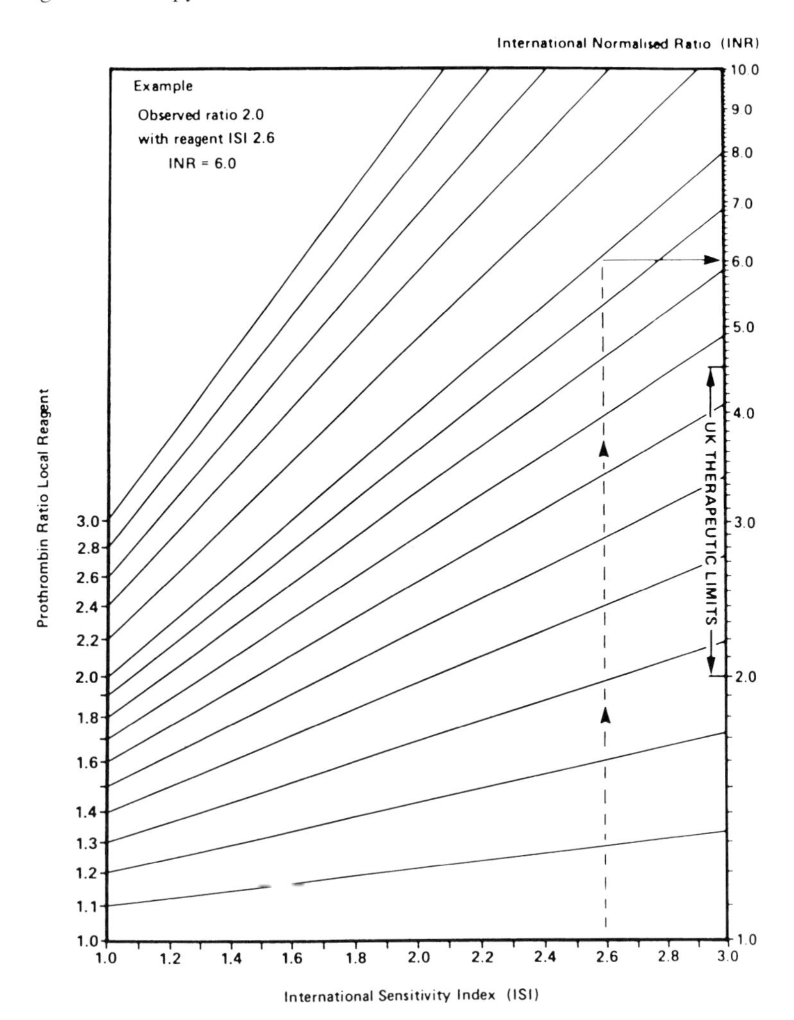

Fig. 2. Relationship between the PT ratio and the INR over a range of ISI values. At an ISI value of 1.0, the PT ratio is identical to the INR. As the ISI value of the thromboplastin increases, the INR for a given PT ratio also increases. (From POLLER, L. and HIRSH, J. [1992] Optimal therapeutic ranges for oral anticoagulation. *Thrombosis in Cardiovascular Disorders* [ed. by V. FUSTER, and M. VERSTRAETE], p. 166. W. B. SAUNDERS, Philadelphia, with permission.)

measurements over PT results, and the need to overlap heparin therapy with warfarin therapy for several days (until the prothrombin levels fall to inhibitor levels following warfarin dosing) (FURIE et al. 1990; WEITZ et al. 1990).

III. Optimal Therapeutic Regimens

The current recommendations for achieving optimal antithrombotic efficacy and hemostatic safety have been largely established on the basis of controlled clinical trials, and are summarized in Table 2. Overall, effective antithrombotic therapy is achieved by INR values between 2.0 and 3.0, except for mechanical heart valves, for which the recommended INR is 2.5–3.5. These INR values are associated with minimal risks of abnormal bleeding (Research Committee of the British Thoracic Society 1992; Poller 1988; British Society for Haematology 1990; Hirsh et al. 1994a,b; Dalen and Hirsh 1995).

Following warfarin administration, observable anticoagulant effects are delayed until newly synthesized dysfunctional vitamin-K-dependent coagulation factors replace the normal factors as they are cleared from the circulation over several days. Accordingly, heparin should be administered initially and overlapped with the beginning of warfarin therapy for at least 4 days. Warfarin should be started with an estimated maintenance dose of about 5 mg/day. This approach usually results in patients reaching an INR of 2.0 in approximately 4–5 days, at which time heparin therapy can be discontinued while the full effects of warfarin on prothrombin levels are produced during the subsequent day or two. PT monitoring is usually performed daily until the therapeutic range has been achieved and maintained for at least 2 consecutive days, then twice or three times weekly for 1–2 weeks. Thereafter, monitoring can be

Table 2. Therapeutic recommendations in international normalized ratios by three representative groups. (Reprinted with permission from Poller and Hirsh 1992)

	British Society for Haematology	Leuven Conference	American College of Chest Physicians/NHL BI Consensus
Prophylaxis of postoperative deep vein thrombosis (general surgery)	2.0–2.5	1.5–2.5	2.0–3.0
Prophylaxis of postoperative deep vein thrombosis in hip surgery and fractures	2.0–3.0	2.0–3.0	2.0–3.0
Myocardial infarction prevention of venous thromboembolism	2.0–3.0	2.0–3.0	2.0–3.0
Treatment of venous thrombosis	2.0–3.0	2.0–4.0	2.0–3.0
Treatment of pulmonary embolism	2.0–3.0	2.0–4.0	2.0–3.0
Transient ischemic attacks	2.0–3.0		
Tissue heart valves	2.0–3.0	?	2.0–3.0
Atrial fibrillation	2.0–3.0	?	2.0–3.0
Valvular heart disease	2.0–3.0	?	2.0–3.0
Recurrent deep vein thrombosis and pulmonary embolism	3.0–4.5	2.0–4.0	2.0–3.0
Arterial disease including myocardial infarction	3.0–4.5	3.0–4.5	2.0–3.0
Mechanical prosthetic valves	3.0–4.5	3.0–4.5	3.0–4.5
Recurrent systemic embolism	3.0–4.5	3.0–4.0	3.0–4.5

less often, perhaps every 1–2 months, depending on the stability of PT results (DALEN and HIRSH 1995).

IV. Benefits of Monitoring

Efficacy and safety of warfarin therapy are critically dependent on maintaining the INR in the therapeutic range, as emphasized by the benefits of good anticoagulant control in patients with atrial fibrillation. Control is improved by having the patient: (a) attend an anticoagulant clinic, similar to the ones conducted in many European countries; (b) participate in a computerized system for dosing and patient records; and (c) perhaps use one of the self-monitoring home kits, although this possibility has yet to be tested objectively (POLLER et al. 1993; CORTELAZZO et al. 1993; DALEN and HIRSH 1995).

D. Prevention and Management of Venous Thromboembolism

I. Prevention of Venous Thrombosis Following Orthopedic Surgery

Oral anticoagulants are effective in preventing venous thrombosis when targeting INRs of 2.0–3.0 after hip surgery, and major gynecologic surgery, beginning the first postoperative day (FRANCIS et al. 1983; POWERS et al. 1989; DALEN and HIRSH 1995; GINSBERG 1996). However, treatment with low molecular weight heparins, commencing 12h after surgery, has been shown to be more effective than warfarin (HULL et al. 1993). Despite the appeal of safety and ease of administration, it has now been firmly established that very low fixed doses of warfarin (1mg/day) are ineffective in preventing postoperative venous thrombosis in patients having hip surgery (FORDYCE et al. 1991; DALE et al. 1991).

II. Prevention of Stroke and Venous Thromboembolism in Acute Myocardial Infarction

Three early studies demonstrated that INRs of 2.0–3.0 reduced the frequency of stroke and venous thromboembolism in patients with acute myocardial infarction (MEDICAL RESEARCH COUNCIL GROUP 1969; DRAPKIN and MERSKEY 1972; VETERANS ADMINISTRATION COOPERATIVE STUDY 1973). Subsequent analysis of pooled data from seven randomized trials confirmed these results, and estimated the decrease in combined end points of mortality and nonfatal reinfarction to be approximately 20% (CAIRNS et al. 1992). Three contemporary studies targeting INRs of 3.0–4.5 also support the use of oral anticoagulation in acute myocardial infarction (ASPECT RESEARCH GROUP 1994; SMITH et al. 1990; SIXTY-PLUS REINFARCTION STUDY GROUP 1980). Combined low-dose aspirin and high-intensity warfarin produced clinically impressive reductions in cardiovascular mortality and stroke, although the addition of aspirin increased minor bleeding (TURPIE et al. 1993).

III. Treatment of Deep Venous Thrombosis

Therapy with oral anticoagulants is indicated for at least 3 months in patients with proximal vein thrombosis (HULL et al. 1979, 1982a) and for up to 3 months for patients with symptomatic calf vein thrombosis (LAGERSTEDT et al. 1985). Targeting an INR of 2.0–3.0 is as effective as a more intense INR of 3.0–4.5, but is associated with significantly fewer bleeding complications (HULL et al. 1982b).

E. Antithrombotic Therapy for Atrial Fibrillation

Five similarly designed studies have been reported: (a) the Stroke Prevention in Atrial Fibrillation trial (SPAF) (THE STROKE PREVENTION IN ATRIAL FIBRILLATION INVESTIGATORS 1991); (b) the Boston Area Anticoagulation Trial for Atrial Fibrillation (BAATAF) (THE BOSTON AREA ANTICOAGULATION TRIAL FOR ATRIAL FIBRILLATION INVESTIGATORS 1990); (c) the Stroke Prevention in Atrial Fibrillation trial (SPINAF) (EZEKOWITZ et al. 1991); (d) the Copenhagen Atrial Fibrillation, Aspirin, Anticoagulation study (AFASAK) (PETERSEN et al. 1989); and (e) the Canadian Atrial Fibrillation Anticoagulation study (CAFA) (CONNOLLY et al. 1991). The SPAF and AFASAK trials also included randomization to aspirin therapy. All trials reached consistent conclusions with a pooled rate of risk reduction averaging 68% on an intention to treat analysis. In patients remaining on warfarin therapy, stroke reduction was greater than 80%. The rates of hemorrhagic complications were low; minor hemorrhage was increased by 3%/year in the warfarin group. The two studies with aspirin treatment arms showed a small benefit of aspirin. Two additional studies have recently been completed: (f) the European Atrial Fibrillation Trial (EAFT) (EUROPEAN ATRIAL FIBRILLATION TRIAL STUDY GROUP 1993) and (g) the second phase of the Stroke Prevention in Atrial Fibrillation trial (SPAF II) (STROKE PREVENTION IN ATRIAL FIBRILLATION INVESTIGATORS 1994). In the SPAF II trial, warfarin was more effective than aspirin for preventing ischemic stroke, although the benefit was almost offset by an increased rate of intracranial hemorrhage in the warfarin-treated patients, especially patients older than 75 years, i.e., 1.8%/year; perhaps related to the higher intensity of anticoagulation, INR exceeded 3.0 at the time of hemorrhage. The EAFT study reported comparable risk reduction in the warfarin group, and no significant benefit in the aspirin-treated patients.

F. Antithrombotic Therapy for Prosthetic Heart Valves

Warfarin anticoagulation is clearly beneficial in patients with prosthetic heart valves as shown by studies comparing warfarin with two-different aspirin-containing antiplatelet therapy only groups (risk reductions of 60% and 79% for warfarin vs. aspirin) (Mok et al. 1985) although the frequency of bleeding complications was highest in the warfarin-treated patients. Patients receiving substitute tissue heart valves demonstrate that an INR of 2.25 provides antithrombotic protection and less hemorrhagic risk than more intense anticoagulation (INRs 3.0–4.5) (Turpie et al. 1988).

In patients with mechanical heart valves, INRs of 3.0–4.5 prevented thromboembolic events equally well as high-intensity anticoagulation (INRs 7.4–11) (Altman et al. 1991; Saour et al. 1990). However, there was significantly more bleeding observed with the more intense anticoagulant therapy. The addition of aspirin (100 mg/day) to warfarin (INR 3.0–4.5) markedly improved antithrombotic efficacy compared with warfarin only, producing a clinically significant reduction in mortality, cardiovascular mortality and stroke (Turpie et al. 1993). However, the benefit was associated with a significant increase in minor bleeding and a nonsignificant trend for an increase in major bleeding, including intracranial bleeding. While it is generally assumed on the basis of clinical experience and retrospective survey (Fuster et al. 1982; Loeliger et al. 1985) that mechanical valves require more intense anticoagulation (INR 2.5–3.5), this impression has not been validated by randomized trials.

G. Other Indications for Oral Anticoagulant Therapy

Although properly designed randomized placebo-controlled trials have not been carried out in patients with valvular heart disease, long-term oral anticoagulation is recommended for patients with native valvular heart disease with or without atrial fibrillation, or patients with at least one documented episode of systemic embolism, by analogy with atrial fibrillation patients. It is reasonable to target an INR of 2.0–3.0 until additional data are available. Anticoagulants are not indicated in patients with nonembolic cerebrovascular disease (Sherman et al. 1995). There is also suggestive evidence that oral anticoagulation (INR 2.0–3.0) is effective in preventing systemic embolism in patients with atrial fibrillation who undergo direct current cardioversion.

H. Complications of Warfarin Therapy

Bleeding, the principle complication of oral anticoagulant therapy, is directly dependent on the intensity of anticoagulant dosing (Levine et al. 1995; Landefeld and Goldman 1989; Landefeld et al. 1989; Turpie et al. 1988; Hirsh et al. 1994a,b; Dalen and Hirsh 1995) by the patient's underlying clinical disorder (Hylek and Singer 1994; Levine et al. 1995), and by the concomitant use of aspirin, which both impairs platelet function and produces gastric erosions, and in very high doses impairs synthesis of vitamin-K-dependent coagulation factors (Chesebro et al. 1983; Dale et al. 1980). The risk of clinically important bleeding is reduced by lowering the therapeutic

INRs from 3.0–4.5 to 2.0–3.0 (ALTMAN et al. 1991; SAOUR et al. 1990; TURPIE et al. 1988; HULL et al. 1982b) While the difference in daily dosing averaged only 1 mg/day, the reduction in bleeding was substantial.

The increased risk of bleeding reported in older patients is often associated with a history of gastrointestinal bleeding, and the presence of serious comorbid conditions, such as renal insufficiency or anemia (LANDEFELD and GOLDMAN 1989; LANDEFELD et al. 1989) Bleeding occurring with INRs less than 3.0 is often associated with an underlying cause, such as occult gastrointestinal source or renal lesion (SAOUR et al. 1990). The elderly may show exaggerated responses to warfarin, perhaps because the rate at which warfarin is cleared declines with age. However, even after controlling for the intensity of anticoagulation, the elderly are more prone to bleeding (LEVINE et al. 1995; ALTMAN et al. 1991; SAOUR et al. 1990). It follows that lower doses of warfarin are indicated when initiating anticoagulant therapy in older patients.

Skin necrosis is the most important nonhemorrhagic complication of warfarin therapy (WEINBERG et al. 1983; VERHAGEN 1954). This infrequent complication commonly develops between day 3 and 8 of therapy, and is caused by thrombotic occlusion of the venules and capillaries within the subcutaneous fat. This condition has been associated with deficiencies of protein C or protein S (SAMAMA et al. 1984; ZAUBER and STARK 1986; BROEKMANS et al. 1983), although it also occurs in the absence of these deficiencies. Microvascular thrombosis has been attributed to propensity for thrombosis produced by the early disparate drop in protein C and protein S before the levels of prothrombin fall after initiating warfarin therapy (GRIMAUDO et al. 1989; SAMAMA et al. 1984; ZAUBER and STARK 1986). If this hypothesis is true, early, adequate and overlapping heparin therapy during the start of oral anticoagulation should prevent this complication. Unfortunately, this postulate has not been formally tested. If chronic anticoagulation is strongly indicated in patients with a history of warfarin skin necrosis, it is reasonable to begin warfarin therapy under heparin coverage, and with small doses, i.e., 2 mg/day, and increase warfarin gradually over several weeks (SAMAMA et al. 1984; ZAUBER and STARK 1986).

During the first trimester of pregnancy, warfarin is contraindicated because it causes embryopathy and CNS abnormalities (HALL et al. 1980). Since warfarin may also produce fetal bleeding, warfarin should be avoided throughout pregnancy if possible (GINSBERG and HIRSH 1995). However, if the patient has a high risk of thromboembolism and heparin is contraindicated or cannot be given at full antithrombotic doses, or when a temporary loss of antithrombotic coverage would be life-threatening, it may be justified to use warfarin during pregnancy, particularly during the second trimester. Nevertheless, heparin is preferred when anticoagulation is indicated during pregnancy. Fortunately, warfarin does not produce anticoagulant effects in infants breast-fed by mothers receiving warfarin (MCKENNA et al. 1983).

References

Altman R, Rouvier J, Gurfinkel E, D'Ortencio O, Manzanel R, de La Fuente L et al (1991) Comparison of two levels of anticoagulant therapy in patients with substitute heart valves. J Thorac Cardiovasc Surg 101:427–431

Anticoagulants in the Secondary Prevention of Events in Coronary Thrombosis Research Group (ASPECT) (1994) Effect of long-term oral anticoagulant treatment on mortality and cardiovascular morbidity after myocardial infarction. Lancet 343:499–503

British Society for Haematology (1990) Guidelines on oral anticoagulation, 2nd edn. British Committee for Standards in Haematology. Haemostasis and Thrombosis Task Force. J Clin Pathol 43:177–183

Broekmans AW, Bertina RM, Loeliger EA, Hofmann V, Kingeman HG (1983) Protein C and the development of skin necrosis during anticoagulant therapy. Thromb Haemost 49:251

Cairns JA, Hirsh J, Lewis HD Jr, Resnekov L, Theroux P (1992) Antithrombotic agents in coronary artery disease. Chest 102:456S–481S

Chesebro JH, Fuster V, Elveback LR, McGoonn DC, Pluth JR, Puga FJ et al (1983) Trial of combined warfarin plus dipyridamole on aspirin therapy in prosthetic heart valve replacement: danger of aspirin compared with dipyridamole. Am J Cardiol 51:1537–1541

Connolly SJ, Laupacis A, Gent M, Roberts RS, Cairns JA, Joyner, C (1991) Canadian Atrial Fibrillation Anticoagulation (CAFA) study. J Am Coll Cardiol 18:349–355

Cortelazzo S, Finazzi G, Viero P, Galli M, Remuzzi A, Parenzan L et al (1993) Thrombotic and hemorrhagic complications in patients with mechanical heart valve prosthesis attending an anticoagulation clinic. Thromb Haemost 69:316–320

Dahlback B (1995) Inherited thrombophilia: resistance to activated protein C as a pathogenic factor of venous thromboembolism. Blood 85:607–614

Dale C, Gallus AS, Wycherley A, Langlois S, Howie D (1991) Prevention of venous thrombosis with minidose warfarin after joint replacement. BMJ 303:224

Dale J, Myhre E, Loew D (1980) Bleeding during acetylsalicylic acid and anticoagulant therapy in patients with reduced platelet reactivity after aortic valve replacement. Am Heart J 99:746–752

Dalen JE, Hirsh J (1995) Fouth ACCP Consensus Conference on Antithrombotic Therapy. Chest 108:225S–522S

den Heijer M, Blom HJ, Gerrits WBJ, Rosendaal FR, Haak HL, Wijermans PW et al (1995) Is hyperhomocysteinaemia a risk factor for recurrent venous thrombosis? Lancet 345:882–885

Drapkin A, Merskey C (1972) Anticoagulant therapy after acute myocardial infarction. Relation of therapeutic benefit to a patient's age, sex, and severity of infarction. JAMA 222:541–548

European Atrial Fibrillation Trial Study Group (1993) Secondary prevention in non-rheumatic atrial fibrillation after transient ischaemic attack or minor stroke. Lancet 342:1255–1262

Ezekowitz MD, Bridgers SL, James KE (1991) Interim analysis of VA Co-operative Study, Stroke Prevention in Non-rheumatic Atrial Fibrillation (SPINAF). Circulation 84:II-450

Fordyce MJF, Baker AS, Staddon GE (1991) Efficacy of fixed minidose warfarin prophylaxis in total hip replacement. BMJ 303:219–220

Francis CW, Marder VJ, Evarts CM, Yaukoolbodi S (1983) Two-step warfarin therapy. Prevention of postoperative venous thrombosis without excessive bleeding. JAMA 249:374–378

Furie B, Diuguid CF, Jacobs M, Diguid DL, Furie BC (1990) Randomized prospective trial comparing the native prothrombin antigen with the prothrombin time for monitoring anticoagulant therapy. Blood 75:344–349

Fuster V, Pumphrey CW, McGoon MD, Chesebro JH, Pluth JR, McGoon DC (1982) Systemic thromboembolism in mitral and aortic Starr-Edwards prosthesis: a 10–19 year follow-up. Circulation 66:I-157–I-161

Ginsberg JS (1996) Management of venous thromboembolism. N Engl J Med 335:1816–1828

Ginsberg JS, Hirsh J (1995) Use of antithrombotic agents during pregnancy. Chest 108:305S–311S

Ginsberg JS, Wells PS, Brill-Edwards P, Donovan D, Moffatt K, Johnston M et al (1995) Antiphospholipid antibodies and venous thromboembolism. Blood 86:3685–3691

Grimaudo V, Gueissaz F, Hauert J, Sarraj A, Kruithaf EKO, Bachmann F (1989) Necrosis of skin induced by coumarin in a patient deficient in protein S. BMJ 298:233–234

Hall JG, Pauli RM, Wilson KM (1980) Maternal and fetal sequelae of anticoagulation during pregnancy. Am J Med 68:122–140

Heijboer H, Brandjes DP, Buller, HR, Sturk A, ten-Cate JW (1990) Deficiencies of coagulation-inhibiting and fibrinolytic proteins in outpatients with deep-vein thrombosis. N Engl J Med 323:1512–1516

Hirsh J, Ginsberg JS, Marder VJ (1994a) Anticoagulant therapy with coumarin agents. In: Colman RW, Hirsh J, Marder VJ, Salzman EW (eds) Hemostasis and thrombosis: basic principles and clinical practice, 3rd edn. Lippincott, Philadelphia, pp 1567–1583

Hirsh J, Prins MH, Samama M (1994b) Therapeutic agents and their practical use in thrombotic disorders. Approach to the thrombophilic patient for hemostasis and thrombosis: basic principles and clinical practice. In: Colman RW, Hirsh J, Marder VJ, Salzman EW (eds) Hemostasis and thrombosis: basic principles and clinical practice, 3rd edn. Lippincott, Philadelphia, pp 1543–1561

Hirsh J, Dalen JE, Deykin D, Poller L, Bussey H (1995) Oral anticoagulants. Mechanism of action, clinical effectiveness, and optimal therapeutic range. Chest 108:231S–246S

Hull R, Delmore T, Genton E, Hirsh J, Gent M, Sackett D et al (1979) Warfarin sodium versus low dose heparin in the long-term treatment of venous thrombosis. N Engl J Med 301:855–858

Hull R, Delmore T, Carter C, Hirsh J, Genton E, Gent M et al (1982a) Adjusted subcutaneous heparin versus warfarin sodium in the long-term treatment of venous thrombosis. N Engl J Med 306:189–194

Hull R, Hirsh J, Jay R, Carter C, England C, Gent M et al (1982b) Different intensities of oral anticoagulant therapy in the treatment of proximal-vein thrombosis. N Engl J Med 307:1676–1681

Hull R, Raskob G, Pineo G, Rosenbloom D, Evans W, Mallory T et al (1993) A comparison of subcutaneous low-molecular-weight heparin with warfarin sodium for prophylaxis against deep-vein thrombosis after hip or knee implantation. N Engl J Med 329:1370–1376

Hylek EM, Singer DE (1994) Risk factors for intracranial hemorrhage in outpatients taking warfarin. Ann Intern Med 120:897–902

Lagerstedt CI, Olsson CG, Fagher BO, Oqvist BW, Albrechtsson U (1985) Need for long-term anticoagulant treatment in symptomatic calf-vein thrombosis. Lancet 2:515–518

Landefeld CS, Goldman L (1989) Major bleeding in outpatients treated with warfarin: incidence and prediction by factors known at the start of outpatient therapy. Am J Med 87:144–152

Landefeld CS, Rosenblatt MW, Goldman L (1989) Bleeding in outpatients treated with warfarin: relation to the prothrombin time and important remediable lesions. Am J Med 87:153–159

Levine MN, Raskob G, Landefeld, S, Hirsh J (1995) Hemorrhagic complications of long-term anticoagulant therapy. Chest 108:291–301

Loeliger EA, Poller L, Samama M, Thomson JM, Van den Besselaar AM, Vermylen J et al (1985) Questions and answers on prothrombin time standardisation in oral anticoagulant control. Thromb Haemost 54:515–517

Malhotra OP, Nesheim ME, Mann KG (1985) The kinetics of activation of normal and gamma-carboxyglutamic acid-deficient prothrombins. J Biol Chem 260:279–287

McKenna R, Cole ER, Vasan U (1983) Is warfarin sodium contraindicated in the lactating mother? J Pediatr 103:325–327

Medical Research Council Group (1969) Assessment of short-term anticoagulant administration after cardiac infarction. Report of the Working Party on Anti-coagulant Therapy in Coronary Thrombosis. BMJ 1:335–342

Mok CK, Boey J, Wang R, Chan TK, Cheung KL, Lee PK et al (1985) Warfarin versus dipyridamole-aspirin and pentoxifylline-aspirin for the prevention of prosthetic heart valve thromboembolism: a prospective randomized clinical trial. Circulation 72:1059–1063

Petersen P, Godtfredsen J, Andersen B, Boysen G, Andersen ED (1989) Placebo-controlled, randomised trial of warfarin and aspirin for prevention of thromboembolic complications in chronic atrial fibrillation. The Copenhagen AFASAK study. Lancet i:175–179

Poller L (1988) A simple nomogram for the derivation of international normalised ratios for the standardisation of prothrombin times. Thromb Haemost 60:18–20

Poller L, Hirsh J (1992) Optimal therapeutic ranges for oral anticoagulation. In: Fuster V, Verstraete M (eds) Thrombosis in cardiovascular disorders. Saunders, Philadelphia

Poller L, Wright D, Rowlands M (1993) Prospective comparative study of computer programs used for management of warfarin. J Clin Pathol 46:299–303

Powers PJ, Gent M, Jay RM, Julian DH, Turpie AGG, Levine M (1989) A randomized trial of less intense postoperative warfarin or aspirin therapy in the prevention of venous thromboembolism after surgery for fractured hip. Arch Intern Med 149:771–774

Research Committee of the British Thoracic Society (1992) Optimum duration of anticoagulation for deep-vein thrombosis and pulmonary embolism. Lancet 340:873–876

Sadowski JA et al (1991) Warfarin and the metabolism function of vitamin K. In: Poller L (ed) Recent advances in blood coagulation, no 5.Churchill Livingstone, Edinburgh, pp 93–118

Samama M, Horellou MH, Soria J, Conard J, Nicolas G (1984) Successful progressive anticoagulation in a severe protein C deficiency and previous skin necrosis at the initiation of oral anticoagulation treatment. Thromb Haemost 51:132–133

Saour JN, Sieck JO, Mamo LA, Gallus AS (1990) Trial of different intensities of anticoagulation in patients with prosthetic heart valves. N Engl J Med 322:428–432

Sherman DG, Dyken ML Jr, Gent M, Harrison JG, Hart RG, Hohr JP (1995) Antithrombotic therapy for cerebrovascular disorders. An Update. Chest 108: 444–456

Sixty-Plus Reinfarction Study Group (1980) A double-blind trial to assess long-term oral anticoagulant therapy in elderly patients after myocardial infarction. Lancet ii:989–994

Smith P, Arnesen H, Holme I (1990) The effect of warfarin on mortality and reinfarction after myocardial infarction. N Engl J Med 323:147–151

Stroke Prevention in Atrial Fibrillation Investigators (1994) Warfarin versus aspirin for prevention of thromboembolism in atrial fibrillation: Stroke Prevention in Atrial Fibrillation II Study. Lancet 343:687–691

The Boston Area Anticoagulation Trial for Atrial Fibrillation Investigators (1990) The effect of low-dose warfarin on the risk of stroke in patients with nonrheumatic atrial fibrillation. N Engl J Med 323:1505–1511

The Stroke Prevention in Atrial Fibrillation Investigators (1991) The stroke prevention in atrial fibrillation trial: final results. Circulation 84:527–539

Turpie AGG, Gunstensen J, Hirsh J, Nelson H, Gent M (1988) Randomised comparison of two intensities of oral anticoagulant therapy after tissue heart valve replacement. Lancet i:1242–1245

Turpie AGG, Gent M, Laupacis A, Latour Y, Gunstensen J, Basile F (1993) A comparison of aspirin with placebo in patients treated with warfarin after heart-valve replacement. N Engl J Med 329:524–529

Verhagen H (1954) Local haemorrhage and necrosis of the skin and underlying tissues during anti-coagulant therapy with dicumarol or dicumacyl. Acta Med Scand 148:455–467

Veterans Administration Cooperative Study (1973) Anticoagulants in acute myocardial infarction. Results of a cooperative clinical trial. JAMA 225:724–729

Weinberg AC, Lieskovsky G, McGehee WG, Skinner DG (1983) Warfarin necrosis of the skin and subcutaneous tissue of the male external genitalia. J Urol 130:352–354

Weitz JI, Hudoba M, Massel D, Maraganore J, Hirsh J (1990) Clot-bound thrombin is protected from inhibition by heparin-antithrombin III but is susceptible to inactivation by antithrombin III-independent inhibitors. J Clin Invest 86:385–391

Zauber NP, Stark MW (1986) Successful warfarin anticoagulation despite protein C deficiency and a history of warfarin necrosis. Ann Intern Med 104:659–660

CHAPTER 13

Oral Thrombin Inhibitors: Challenges and Progress

S.D. KIMBALL

A. Introduction

The effective modulation of coagulation would significantly improve the treatment of cardiovascular problems secondary to imbalances in the equilibrium between hemostasis and fibrinolysis. These problems include venous and arterial thrombosis, atrial fibrillation, stroke, restenosis, and recurrent myocardial infarction. The central position of thrombin in hemostasis and thrombosis makes it a compelling target for drug discovery. The discovery and development of oral thrombin inhibitors therefore presents a notable medical opportunity for improving the treatment of these cardiovascular disorders and currently represents a major challenge for pharmaceutical researchers.

Significant advances have been made in the discovery of potent and selective thrombin inhibitors. The large number of small molecule direct thrombin inhibitors that have been prepared and studied have been discussed in recent reviews (KIMBALL 1995a,b; HAUPTMANN and MARKWARDT 1992; DAS and KIMBALL 1995; CLAESON 1994; MAFFRAND 1992; TAPPARELLI et al. 1993; KAISER and HAUPTMANN 1992; LYLE 1994). However, there is no compound currently in clinical use or clinical development that clearly satisfies all of the criteria posed by Sixma for an ideal antithrombotic drug which "... should inhibit thrombosis without affecting hemostasis ... should have a long half life ... should be absorbed after oral administration ... should be safe and ... should have a wide therapeutic range" (SIXMA and DE GROOT 1992). Present limitations on the oral bioavailability, plasma half-life, safety and efficacy of thrombin active site inhibitors appear to be related to the presence of a highly basic guanidino or amidino group, or to chemically reactive electrophilic functional groups in most small molecule inhibitors. The criteria defined by Sixma highlight obstacles that must be overcome in developing clinically relevant and useful oral anticoagulants from the current collection of potent small molecule inhibitors which have served so well as research tools.

I. Role of Thrombin in Hemostasis and Thrombosis

The pivotal role of thrombin in hemostasis and thrombosis has long been appreciated from both a scientific and practical point of view (JACKSON and NEMERSON 1980; FENTON 1986; HARKER et al. 1995). Thrombin is the central enzyme that controls the balance between hemostasis and fibrinolysis; defects in the various pathways leading to the generation of thrombin are associated with bleeding diatheses or thrombotic complications. In addition to its pivotal role in hemostasis and thrombosis, thrombin may be involved in atherosclerosis, inflammation and neurodegenerative diseases (TAPPARELLI et al. 1993b).

The blood coagulation pathway consists of an intricate labyrinth of concatenated enzymatic reactions involving sequential activation of proenzymes and cofactors. The rate and localization of hemostasis is tightly controlled by the interplay of procoagulant and anticoagulant processes that are activated. The controlling events in blood coagulation are heterogeneous catalytic processes that require the assembly of the appropriate enzyme, zymogen (substrate) and cofactor reactants on phospholipid surfaces. The intricacy of the process is such that only the most general overview can be given here; the interested reader is referred to other chapters in this text, to the articles and reviews cited, and also to excellent and thorough texts that have been written (COLMAN et al. 1987; BEUTLER et al. 1995).

The traditional view of coagulation is a "cascade" of reactions that can amplify a small input and rapidly effect the generation of large amounts of localized, activated thrombin (DAVIE et al. 1991). In this view of blood coagulation, a controlled sequence of cleavages of inactive proenzymes to the active proteases occurs via the so-called intrinsic or extrinsic pathways. Both the intrinsic and extrinsic enzyme cascades converge in the generation of factor Xa, which, as part of the prothrombinase complex, activates thrombin from its zymogen, prothrombin. In Fig. 1 are illustrated the critical membrane-mediated complexes in each pathway that lead to the activation of the vitamin K dependent factor X. The extrinsic pathway "tenase" complex is formed between tissue factor (constitutively expressed on cells not normally exposed to the blood) via tissue damage or cell activation, the activated serine protease factor VIIa, and calcium on an activated anionic phospholipid surface. The membrane bound calcium-VIIa-tissue factor complex is catalytically competent to activate the serine protease factor Xa from its zymogen, factor X. The intensity of the procoagulant response is dependent upon the amounts of tissue factor and anionic phospholipid membrane that are present (LAWSON et al. 1994; JONES and MANN 1995).

The prothrombinase complex can also be generated via the intrinsic or "contact" pathway, so named because by this pathway coagulation requires only proteins that are present in the blood in conjunction with an anionic surface. The intrinsic pathway is initiated through a process involving activation of serine protease factors XII and XI (not shown) which lead to the activation of factor IX. Factor IXa and its cofactor VIIIa in association with calcium on an anionic phospholipid surface provides a complex that is capable of catalyzing the conversion of factor X to factor Xa. Although the clotting factors in both the extrinsic and intrinsic pathway are essential for normal hemostasis, the relative contribution of the two pathways in vivo is not clear.

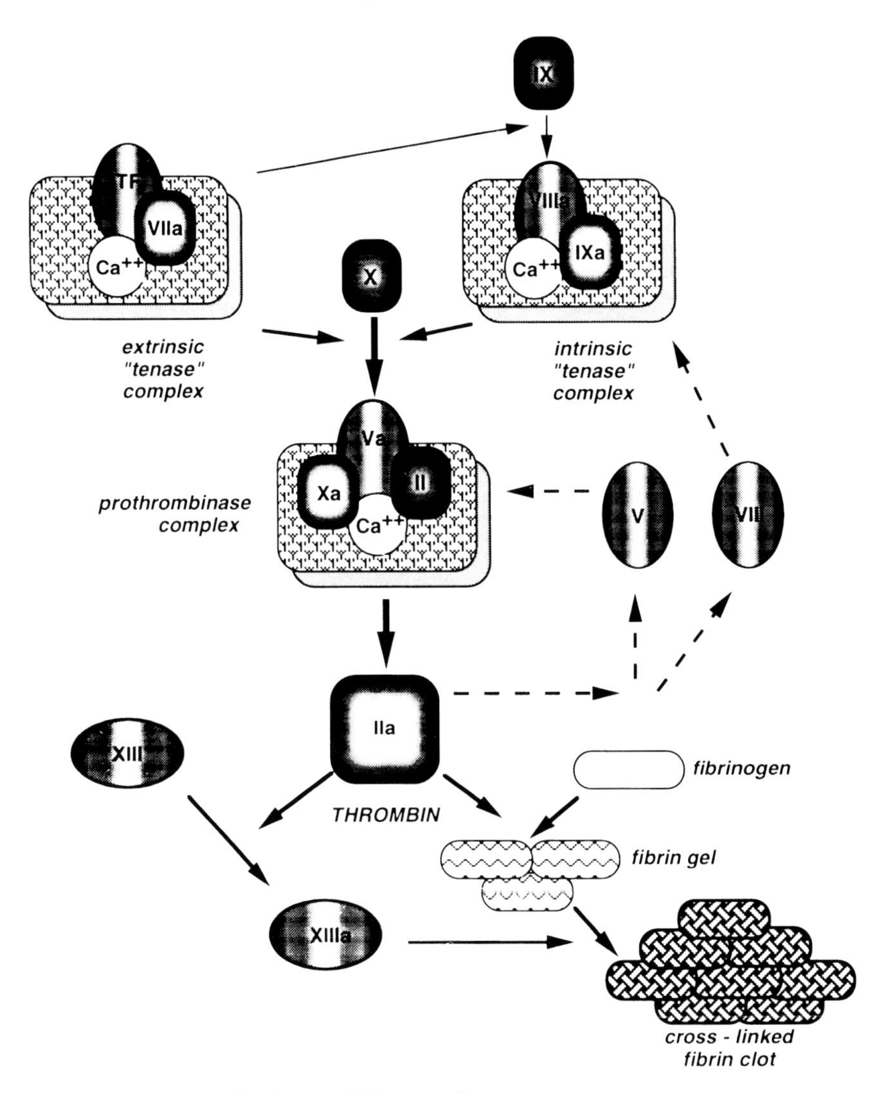

Fig. 1. Thrombin Activation and Hemostasis

Activation of factor X leads to the assembly of the prothrombinase complex between factor Xa, factor Va, and calcium on an activated phospholipid surface. This complex catalyzes the activation of thrombin (factor IIa) from its precursor, prothrombin (factor II). It has been calculated that activation of only 0.2% of the factor X in the blood provides sufficient enzymatic activity that one second's worth of thrombin generation by the resulting prothrombinase complex will clot blood in 10 s. In the absence of activating cofactors, catalytically active factor Xa would take 3.5 days to generate the same amount of thrombin; this corresponds to a 300 000-fold rate increase of factor Xa activity in the presence of Va, Ca^{2+}, and phospholipid (SCULLY 1992). Thrombin directly amplifies its own production through a positive feedback loop via proteolytic activation of the blood coagulation factors V and VIII (dashed arrows in Fig. 1). The net result of these pathways is that a relatively small signal quickly gives rise to an explosive burst of catalytically active thrombin.

Upon activation, thrombin proteolytically cleaves fibrinopeptides A and B from the soluble plasma protein fibrinogen to give the insoluble polymeric protein fibrin. The linear fibrin monomers self-aggregate through association of the D and E subunits to form a fibrin gel which is stabilized by internal hydrogen bonding. At the same time, thrombin activates the transglutaminase factor XIIIa from its zymogen (MOSESSON et al. 1989; SCULLY 1992; LORAND and RADEK 1992). Factor XIIIa, in the presence of calcium, rapidly cross-links the fibrin strands of the fibrin monomers, forming amide bonds between the γ-glutaminyl and ε-aminolysyl groups of specific residues in the α- and β-chains of fibrin (MCDONAGH 1987). The resulting network of insoluble, cross-linked fibrin polymer adds mechanical stability and resistance to lysis to the forming blood clot. Thrombin itself binds to fibrin (SIEBENLIST et al. 1990) and may also be trapped within the clot; both free and clot-bound thrombin can convert fibrinogen to fibrin, thus allowing stabilization and localized propagation of the thrombus at a site of injury. In addition, factor XIIIa forms amide bonds between Glu-2 of α_2 plasmin inhibitor (α_2PI) and the α-chains of fibrin (TAMAKI and AOKI 1982). The serpin, α_2 plasmin inhibitor, is the principal physiologic inhibitor of plasmin activity, and inactivates plasmin by forming a covalent complex at the active site of the enzyme (WIMAN and COLLEN 1979; HARPEL 1987). The localization of α_2PI in the clot may play an important role in stabilizing clot formation by minimizing the effects of the fibrinolytic enzyme plasmin, which otherwise degrades the fibrin matrix.

II. Medical Need for Anticoagulant and Antithrombotic Drugs

A great need exists for both acute and chronic anticoagulation in the treatment of cardiovascular disease (WEITZ 1994). Indirect thrombin inhibitors such as heparin (WEITZ 1994; HIRSH 1993) and warfarin (Coumadin) (HIRSH 1991) are important therapeutic agents (FIORE and DEYKIN 1995; LENGFELDER 1995; LITIN and GASTINEAU 1995); however, the use of these anticoagulants is limited by their shortcomings. The direct-acting thrombin inhibitors hirudin (WALENGA et al. 1991; MARKWARDT 1957) and Hirulog (FENTON et al. 1993; FENTON 1992) are being studied for use as parenteral agents in acute care settings; a widespread interest in their application underscores the broader need for improved anticoagulant and antithrombotic therapy. The development of safer, more predictable, and more effective anticoagulants is eagerly anticipated by the medical profession (BEAUFILS 1995; AGNOLA and GUIOMARD 1995).

1. Acute Anticoagulation

Anticoagulants are particularly important in the treatment of venous thrombosis and prevention of pulmonary embolism (HYERS et al. 1992), medical problems which are frequent complications of hospitalization and major orthopedic surgery. Venous thrombosis may develop when stasis in the deep veins of the legs occurs at times of increased coagulability of the blood, leading to the local generation of thrombin and the growth of a small platelet-fibrin nidus into an occlusive thrombus (THOMAS 1994). The increasing use of hip and knee replacement surgery (CUSHNER and FRIEDMAN 1988) in an aging population magnifies the importance of preventing venous thrombosis and pulmonary embolism. Deep-vein thrombosis (DVT) is difficult to detect and treat (KOOPMAN et al. 1995; WEINMANN and SALZMAN 1994). The incidence of DVT in untreated patients is estimated to be 40%–50% for total hip replacement, 50% for fractured hip, and 72% for total knee replacement; the incidence of fatal pulmonary embolism is ca. 1%–5% for the entire group (MERLI 1993; LOWE et al. 1992). A large proportion of hospitalized patients who are at high risk for venous thromboembolism (VTE) do not receive prophylaxis. Reluctance to use VTE prophylaxis in surgical patients may be due to fear of perioperative bleeding when anticoagulants are given (KEARON and HIRSH 1995).

Heparin is the first agent to be administered in situations requiring acute anticoagulation. In the case of venous thrombosis or pulmonary embolism, a 7- to 10-day course of heparin is usually followed by prolonged administration of an oral anticoagulant, e.g., warfarin. The parenteral agent heparin is generally coadministered with warfarin for a few days prior to cessation of heparin therapy (HULL et al. 1979).

Heparin is a naturally occurring glycosaminoglycan anticoagulant of average molecular weight ca. 15000 that is particularly useful in the treatment and prevention of venous thromboembolism. The antithrombin effects of heparin are mediated by a specific pentasaccharide chain on the molecule that binds to antithrombin III (AT III) and to the anion-binding exosite II of thrombin, increasing the inactivation of thrombin more than 1000-fold. The heparin-AT III complex is an effective inhibitor of both thrombin and factor Xa. While the inhibition of thrombin by heparin-AT III requires a ternary complex in which the heparin must have 18 or more saccharide units, the inhibition of factor Xa shows less dependence on the size of the heparin polysaccharide chains (HIRSH 1993).

Heparins that have been depolymerized to contain molecules of molecular weight <6500 are termed low molelcular weight heparins (LMWHs) (EICHINGER et al. 1994). These smaller molecules have a longer half-life, greater bioavailability, and more predictable anticoagulant effects than unfractionated heparin (BENDETOWICZ et al. 1994; BONEAU 1994; WALLENTIN 1996). There is evidence in general surgery patients that LMWH, compared with standard heparin, generates a clinically relevant decrease in bleeding complications (KAKKAR et al. 1997). LMWHs may be preferable for orthopedic surgery patients, in view of a larger absolute risk of venous thrombosis (NURMOHAMED et al. 1992; FAUNO et al. 1994). In general, LMWHs are at least as effective as heparin, and may be more conveniently and cost effectively administered compared with unfractionated heparin (BERGMANN and NEUHART 1996; LINDMARKER and HOLMSTROM 1996).

There are several disadvantages to the use of heparin, aside from the fact that heparin is a parenteral agent requiring i.v. or s.c. administration. The anticoagulant dose-response curve for heparin is not linear, and ex vivo coagulation parameters, e.g., APTT, must be followed to monitor the degree of anticoagulation. There are reports of a "rebound" reactivation of unstable angina subsequent to discontinuation of heparin therapy (THEROUX et al. 1992; GRANGER et al. 1995), and heparin has been associated with thrombocytopenia, requiring monitoring of platelet counts. Finally, heparin is ineffective at inhibiting clot-bound thrombin (WEITZ et al. 1990).

2. Chronic Anticoagulation with Warfarin

Chronic dosing of an anticoagulant is important in the treatment of thrombotic complications resulting from atrial fibrillation (PETERSEN et al. 1989; WHEELDON 1995), the prevention of myocardial infarction, and in the longer-term management of venous thrombosis. Atrial fibrillation affects an estimated 2%–4% of the population older than 60 years, and is associated with an incidence of ca. 4%/year of stroke. Chronic anticoagulation may also be important in the management of atherosclerosis and restenosis (FALK and FERNANDEZ-ORTIZ 1995). The indirect-acting anticoagulant warfarin (Coumadin) is currently the most efficacious oral anticoagulant available (HIRSH 1991). Warfarin has been shown to be effective in the secondary prevention of myocardial infarction (SMITH et al. 1990; JAFRI et al. 1992), and was recently approved for this indication by the FDA. Warfarin is also approved in the treatment of venous thromboembolism and stroke (POLLER et al. 1987; FRANCIS et al. 1983; HULL et al. 1979), and it is an agent of choice for prevention of stroke secondary to atrial fibrillation (MCBRIDE et al. 1994; DISCH et al. 1994). Recent studies with warfarin have demonstrated that prophylactic long-term warfarin therapy may reduce the recurrence of venous thrombosis (SCHULMAN et al. 1997; DIUGUID 1997). Currently, warfarin is in clinical studies in conjunction with low-dose aspirin, which will help to further define the limits of this drug in primary prevention of the sequelae of ischemic heart disease (MEADE and MILLER 1995).

While warfarin is an effective oral anticoagulant, it is also subject to important limitations. Warfarin and related coumarins act by blocking the vitamin K dependent γ-carboxylation of glutamic acids to form γ-carboxyglutamic acid (Gla) residues, which are essential in the function of a number of enzymes in the coagulation cascade. These include prothrombin, factor VII, factor IX, factor X and the anticoagulant proteins C and S. Since the mechanism of action is the inhibition of de novo synthesis of all vitamin K dependent coagulation factors, warfarin requires prolonged dosing to establish a stable level of anticoagulation. Because of this mechanism, the anticoagulant effects of warfarin are not quickly reversible. The blanket inhibition of the synthesis of vitamin K dependent, Gla-containing proteins may be responsible for the rare, but serious skin necrosis that is found in ca. 0.1% of patients taking warfarin. This syndrome has been thought to reflect the rapid disappearance of the anticoagulant protein C due to its shorter plasma half-life relative to the procoagulant clotting factors, with a resulting increase in the potential for thrombosis (COLE et al. 1988; BERNARD et al. 1992); hence the use of heparin early during the course of warfarin therapy. Warfarin is also highly bound to plasma albumin, and drug interactions are common (FREEDMAN 1992; SERLIN and BRECKENRIDGE 1983; SHETTY et al. 1989). The narrow therapeutic window of warfarin requires that its anticoagulant effect be subject to frequent monitoring in order to avoid the possibility of excessive bleeding.

In spite of these drawbacks, warfarin is an inexpensive and effective drug which, if used appropriately, addresses a significant medical need. With experience in the clinical application of warfarin, effective doses have decreased (POLLER et al. 1987), increasing the safety margin, opening new applications for warfarin therapy and tending to simplify the requirements for monitoring (POLLER and SAMAMA 1994; BERN 1992; PINEDE et al. 1994). For orally bioavailable direct thrombin inhibitors to be competitive with warfarin, they will have to unambiguously improve upon its limitations: a narrow therapeutic index, poor side-effect profile, and requirement for patient monitoring.

B. Potential Advantages of Direct, Small Molecule Inhibitors

The high level of interest in orally bioavailable small molecule thrombin inhibitors derives from the fact that an agent which selectively and reversibly inhibits thrombin could possess a distinct advantage over warfarin with respect to side effects and monitoring (TOPOL 1995; PHILIPPIDES and LOSCALZO 1996). The selectivity of a thrombin inhibitor compared with warfarin should allow it to be used with relative safety in settings of both arterial and venous thrombosis, as has been demonstrated in animal models (SCHUMACHER et al. 1993a,b, 1994, 1996; HARKER 1994; KAISER and HAUPTMANN 1992; TAPPARELLI et al. 1993). Compared with warfarin, the reversibility of thrombin active site inhibitors theoretically allows for superior therapeutic control of the anticoagulant and antithrombotic effects with minimal effects on bleeding. In contrast, following cessation of warfarin therapy, one has to wait for the resynthesis of vitamin K dependent coagulation factors by the liver to restore the hemostatic balance.

Thrombin active site inhibition will also block the thrombomodulin-dependent activation of protein C by thrombin on endothelial cells (ESMON

1993). Experimentally, this potential *procoagulant* effect of a thrombin active site inhibitor appears to be more than compensated for by the acute anticoagulant effects on fibrinogen cleavage and platelet thrombin receptor activation. Recently, it has been discovered that a mutation in the factor V gene leads to resistance to activated protein C, which is associated with an increased risk of venous thrombosis (KALAFATIS et al. 1995; BERTINA et al. 1994; DAHLBÄCK (1995). In this thrombosis-prone population, a component of the anticoagulant protein C system is effectively disabled, and administration of a thrombin inhibitor could not lead to undesirable procoagulant effects based on the inhibition of protein C activation.

A distinct advantage of small molecule direct thrombin inhibitors is their potentially important ability to inhibit clot-bound thrombin as well as fluid-phase thrombin. This capability of small molecule inhibitors has been clearly demonstrated in vitro (WEITZ et al. 1990; BERRY et al. 1994a), where small molecules such as argatroban (1) are significantly better than hirudin or heparin at blocking the enzymatic activity of thrombin which has been incorporated into fibrin clots. While there is some disagreement over the clinical significance of clot-bound thrombin (NASKI and SHAFER 1993), the ability of small molecule direct thrombin inhibitors to block both fluid-phase and clot-bound thrombin should ensure equal or greater anticoagulant and antithrombotic efficacy than heparin (WEITZ 1994, 1995).

C. Pharmacological and Pharmacokinetic Issues

I. Safety

Safety will always be a key issue in the use of anticoagulants, especially when chronic use outside the hospital setting is involved. With any anticoagulant drug, the maintenance of highly predictable peak/trough ratios of drug to constrain the anticoagulant effects within therapeutic bounds is critically important. To achieve tight peak/trough ratios, pharmaceutical researchers will have to optimize the delivery of orally bioavailable thrombin inhibitors. In order to attain the goal of once or twice-daily dosing, a new thrombin inhibitor with high oral bioavailability must also possess the appropriate pharmacokinetic and pharmacodynamic half-life in vivo. An oral and long-acting thrombin inhibitor, with an improved side effect profile relative to warfarin, may substantially decrease monitoring requirements; this would decrease both the direct and indirect costs of current anticoagulation with warfarin.

II. Selectivity and Fibrinolytic Compromise

The importance of enzyme specificity for the efficacy in vitro and in vivo of thrombin inhibitors has recently been demonstrated (CALLAS et al. 1994; CALLAS and FAREED 1995). In an animal model of venous thrombosis the electrophilic thrombin inhibitors DuP 714 **(6)** and efegatran **(4)** are capable of

Table 1. Enzyme selectivity.[a]

	Thrombin IC_{50} (μM)	Trypsin IC_{50} (μM)	t-PA IC_{50} (μM)	Plasmin IC_{50} (μM)	Urokinase IC_{50} (μM)	Streptokinase IC_{50} (μM)
Argatroban	0.008	7.9	370	400	>500	>500
activity relative to thrombin	*1*	*1000*	*46000*	*50000*	*>60000*	*>60000*
Efegatran	0.016	0.032	1.7	0.70	2.3	1.3
activity relative to thrombin	*1*	*2*	*100*	*40*	*140*	*80*

[a] Data from MARYANOFF et al. 1993; 90: 8052.

inhibiting streptokinase or urokinase-induced clot lysis. This "fibrinolytic compromise," resulting from inhibition of thrombolytic enzymes, demonstrates a potential danger – that a nonselective thrombin active site inhibitor might actually *promote* hemostasis if fibrinolytic enzymes are inhibited in addition to thrombin. Such a situation could arise in the acute intravenous coadministration of a poorly selective thrombin inhibitor combined with thrombolytic agents. In this model of fibrinolytic compromise, the highly selective thrombin inhibitor hirudin did not hinder clot lysis. As shown in Table 1, reversible thrombin inhibitors such as argatroban **(1)** (Fig. 2) are significantly more selective than slow, tight-binding electrophilic inhibitors such as efegatran **(4)** (Fig. 3) and might be anticipated to be safer in a setting of acute coadministration.

III. Pharmacodynamics: Efficacy and Kinetics of Inhibition

A number of inhibitors related to the tripeptide D-Phe-Pro-Arg incorporate chemically reactive electrophilic groups at the C-terminus. This leads to covalent modification of the active site of thrombin, and potent inhibition by a slow, tight-binding mechanism. However, this reliance on chemically reactive functionality brings with it several concerns. First is the expectation that chemically reactive groups might cause nonspecific side effects or lead to chronic toxicity that would only be revealed later in the drug development process. In addition, one might anticipate that a highly reactive inhibitor would have insufficient selectivity for thrombin compared to other serine proteases, since the thermodynamics of binding are driven by chemical reactivity. This may explain why the phenomenon of "fibrinolytic compromise" has only been described for electrophilic compounds such as efegatran **(4)** (Fig. 3).

There are also experimental data to indicate that the slow, tight-binding kinetics of electrophilic thrombin inhibitors may lead to a gross

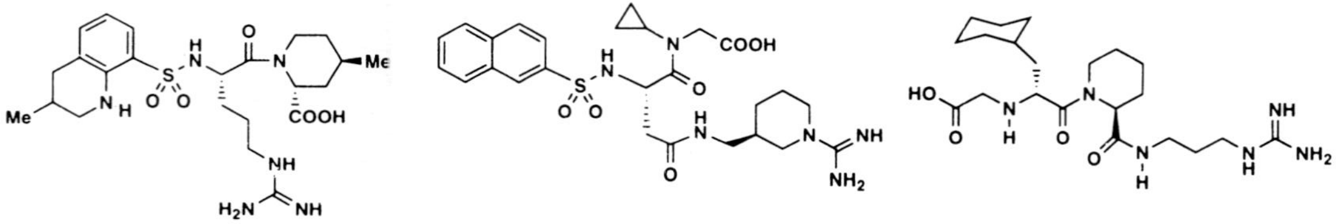

2 **Argatroban (1)** **Napsagatran (2)** **Inogatran (3)**

Fig. 2. Reversible Thrombin Inhibitors

3 **Efegatran (4)** **CVS-1123 (5)** **DuP714 (6)**

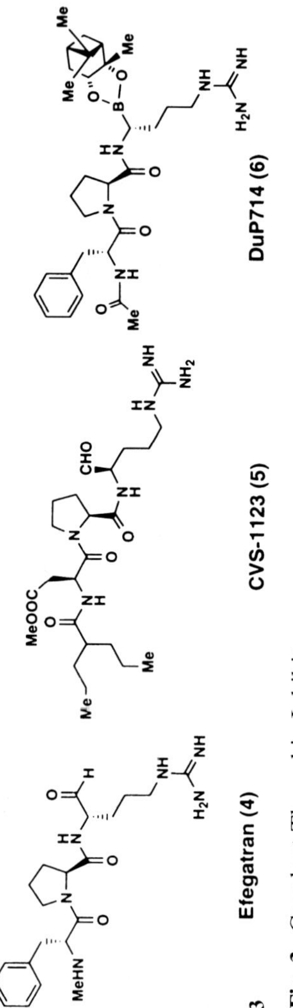

Fig. 3. Covalent Thrombin Inhibitors

overestimation of their true efficacy. Thrombin activation, and the resulting hemostatic or thrombotic events, are the result of a "burst" or pulse generation of thrombin that is highly localized in time and space (JACKSON 1987; NASKI and SHAFER 1991; LAWSON et al. 1994; JONES and MANN 1995). In order to be pharmacologically efficacious, direct thrombin inhibitors will of necessity have to inhibit quickly and completely the high local concentrations of activated thrombin that are generated at, for example, the sites of acute plaque rupture or thrombus formation.

The thrombin inhibitor efegatran **(4)** (Fig. 3) has the slow, tight-binding kinetics typical of electrophilic thrombin inhibitors as a class. This compound has been shown to be ineffective at blocking platelet activation by thrombin, as measured by intracellular calcium release (BAKONYI et al. 1994). When efegatran and thrombin were sequentially added to a rabbit platelet preparation in vitro, there was no effect on intracellular calcium release. However, a 5-min preincubation of the thrombin with efegatran abolished the calcium response. Based upon their kinetic analysis, the authors concluded that thrombin triggers the activation of platelets much more rapidly than the enzyme can be inhibited by efegatran in this assay.

This phenomenon has been described in more detail in studies evaluating the potent electrophilic boronic ester thrombin inhibitors related to DuP714 **(6)** (TAPPARELLI et al. 1993a). The authors postulated that despite the ultimate time-dependent inhibitory potency of slow, tight-binding boronates, they are not efficacious in vivo because they do not inhibit thrombin rapidly enough. Based upon these results, TAPPARELLI et al. concluded that platelet activation can only be inhibited under conditions in which the half-life of thrombin activity would be less than 10 s, and proposed comparing the association rate constants of potential thrombin inhibitors as a better predictor of efficacy in vitro and in vivo.

This analysis has been extended in a recent discussion of the importance of rapid inhibition of thrombin by active site inhibitors (STONE and TAPPARELLI 1995), and has led to proposals for screening in vitro to differentiate classical competitive inhibitors from slow, tight-binding inhibitors of thrombin (RUPIN et al. 1995). Previously, the potency of slow, tight-binding inhibitors has been their singular advantage and the source of their attractiveness to pharmaceutical researchers. If current prejudices are upheld by more research, the slow, tight-binding mechanism that leads to the potent inhibition of thrombin in vitro by electrophilic inhibitors may prove to have been little more than a distraction from the road to discovering clinically useful drugs.

IV. The Rebound Phenomenon

In a phase I study in patients with unstable angina, GOLD et al. (1993) described an apparent rebound coagulation phenomenon following cessation of argatroban treatment. Argatroban infusion in patients presenting with unstable angina led to dose-dependent increases in APTT, and also to a decrease in the amount of fibrinopeptide A, indicative of reduced thrombin activation. Two hours after cessation of the infusion, the thrombin-antithrombin III (TAT) levels increased beyond baseline, indicating a rebound activation of thrombin despite the peristence of supranormal APTT values and residual levels of argatroban in the blood (ca. 0.3 *M*). Since the study was not designed to address the possibility of rebound coagulation, it could not be ruled out that this result is artifactual (WILLERSON and CASSCELLS 1993), or related to the previously described rebound coagulation phenomenon described on cessation of heparin treatment (THEROUX et al. 1992). Heparin rebound has also been described in the setting of coronary angioplasty (SMITH et al. 1996). However, further studies with these patients indicated that there was increased activation of prothrombin *during* argatroban infusion, which persisted after discontinuation of the drug (GARABEDIAN et al. 1993). A more recent study describing the persistent generation of thrombin throughout the course of treatment with hirudin (ZOLDHELYI et al. 1994) contrasts with the argatroban study in documenting the expected decrease in TAT complex during the infusion period. Despite the presence of anticoagulant and antithrombotic concentrations of hirudin in the blood, there was continued generation of thrombin from prothrombin, as measured by the presence of both the thrombin-hirudin complex and prothrombin fragment 1.2. This inability of thrombin inhibitors to suppress the generation of thrombin may ultimately put a limit on their utility (LEFKOVITS and TOPOL 1994; CRENSHAW et al. 1995). It is possible that inhibition of thrombin generation by intervention at points earlier in the coagulation cascade (e.g., factor Xa) may then prove to be more desirable.

These studies raise important issues that are relevant in the clinical use of thrombin active site inhibitors. The short half-life of argatroban in vivo (CLARKE et al. 1991; HAUPTMANN and MARKWARDT 1992) leads to the rapid decline of blood concentrations of this thrombin inhibitor once drug infusion is stopped. Since argatroban is a reversible inhibitor of thrombin, on cessation of dosing, dissociation of the argatroban-thrombin complex and clearance of argatroban from the blood might be expected to free clinically significant amounts of thrombin, leading to the observed rebound increase in TAT levels and recurrence of unstable angina. Another possibility is that the rebound activation of the coagulation system may be related to selective inhibition of the anticoagulant effects of thrombin on thrombomodulin/protein C at low concentrations of the thrombin inhibitor. The rebound phenomenon described for argatroban and its mechanism have not yet been reported for electrophilic inhibitors of thrombin's active site, nor have animal models of the rebound phenomenon been developed.

V. Oral Bioavailability and Pharmacokinetics

The potential for side effects during treatment with anticoagulants is significant. To achieve the therapeutic ratio necessary for chronic dosing of anticoagulant agents in an elderly population (e.g., in the treatment of patients with atrial fibrillation) it is important to minimize any source of inter- and intraindividual variability, so that predictable and reproducible blood levels of the antithrombin agent can be maintained. Thus, the absorption, bioavailability, metabolism, and pharmacokinetics of orally delivered thrombin inhibitors will have to be evaluated critically early in the discovery and development process.

In general, the thrombin active site inhibitors that have been described to date are plagued by poor bioavailability and short plasma half-lives. In order to minimize variability in oral bioavailability, clinically useful thrombin inhibitors must be well-absorbed and stable to metabolism. Based upon the potential for rebound coagulation described above, it is reasonable to consider that a relatively long plasma half-life may be important to the clinical success of reversible thrombin inhibitors. The safety and efficacy of a thrombin inhibitor should be optimized if the pharmacokinetic half-life in the plasma is sufficiently long that peak-trough levels of drug are maintained within the window of pharmacologically effective blood levels. Thus, consistent absorption and bioavailability, lack of metabolism, and a relatively long plasma half-life will be required for a thrombin active site inhibitor to have pharmacokinetic and pharmacodynamic stability and to be truly competitive with heparin and warfarin.

D. Thrombin Inhibitors in Development

I. Bivalent Direct Thrombin Inhibitors

1. Hirudin

The potent anticoagulant hirudin was isolated and characterized from extracts of the leech *Hirudo medicinalis* in 1957 by MARKWARDT. The structure of hirudin bound to thrombin has been determined by X-ray crystallography (GRÜTTER et al. 1990; RYDEL et al. 1990, 1991; VITALI et al. 1992), which showed that it forms a 1:1 noncovalent complex by binding to the active site and to the fibrinogen-binding exosite. The thrombin-hirudin complex forms with slow, tight-binding kinetics; differences in the measurement of its dissociation constant (ca. 0.020 p*M*) may be due to differences in the experimental details of the measurement (STONE et al. 1990).

Hirudin was quickly shown to be effective in animal models of arterial and venous thrombosis (HERAS et al. 1990; PETERS et al. 1991; BACHER et al. 1993; HARKER 1991; KAISER et al. 1990). Based upon animal studies hirudin has been expected to be well tolerated and free of bleeding complications at therapeutic doses in humans. As a polypeptide, hirudin must be administered parenterally by intravenous, intramuscular, or subcutaneous injection. The terminal elimination half-life of hirudin in humans following intravenous or subcutaneous administration is 40 min and 2 h, respectively. Despite concerns, hirudin does not appear to be immunogenic based upon antibody generation or side effects reported in clinical trials (MÄRKI and WALLIS 1990).

The clinical potential of hirudin and the status of current clinical trials have been reviewed (JOHNSON 1994; LEFKOVITS and TOPOL 1994; WEITZ 1994). An ascending dose study of hirudin in patients undergoing total hip replacement indicated that recombinant desulfato hirudin (CGP 39393) can be safely administered before and after surgery. The results of this safety trial were highly suggestive of efficacy versus deep vein thrombosis at tolerated doses.

However, large-scale trials with hirudin have more often encountered problems with side effects (ZEYMER and NEUHAUS 1995). In the TIMI 9A (Thrombolysis In Myocardial Infarction) trial comparing heparin to hirudin as adjunctive therapy in the setting of thrombolysis by TPA or streptokinase in patients with acute myocardial infarction, unexpectedly high rates of hemorrhagic stroke led to the reconfiguration of the trial (ANTMAN 1994). The final results of TIMI 9B showed no difference in efficacy between heparin and hirudin (ANTMAN 1996).

Likewise, the GUSTO 2 (Global Use of Strategies to Open Occluded Arteries) trial, comparing heparin with hirudin in thrombolysis by TPA or streptokinase, also found unexpectedly high rates of hemorrhage in all groups treated with hirudin. This clinical trial was also restarted with significantly lower doses of hirudin. The final results of the GUSTO IIB trial demonstrated a small advantage for the use of hirudin over heparin in thrombolysis (TOPOL 1996). Finally, the HIT III (Hirudin for Improvement of Thrombolysis), using modified recombinant hirudin from Hoechst/Behringwerke, was also discontinued prematurely due to excessive bleeding complications in a setting of thrombolysis with TPA or streptokinase (NEUHAUS et al. 1994). The results and implications of these recent trials have been summarized and analyzed (ZEYMER and NEUHAUS 1995; ANTMAN 1996). The inability of the potent thrombin inhibitor hirudin to dramatically improve clinical outcomes in the above trials may call into question the athrombin hypothesis, may be due to inadequate dosing regimens, or both.

In addition to problems of efficacy, concern has also been raised regarding the high cost of both hirudin and Hirulog in comparison with heparin. In order to justify the use of these agents at ca. $1000 for a 3-day course of treatment, they will have to be clearly superior to heparin in their clinical outcome. In addition, some concern about the potential of thrombin inhibitors for rebound (vida supra) has recently been expressed in experiments in which continued thrombin generation (based on an ELISA for the thrombin-hirudin complex) was measured in spite of potent anticoagulation by hirudin (ZOLDHELYI et al. 1994). This concern is generalizable to all thrombin inhibitors and is not specific for hirudin itself; strategies for inhibiting other pathways in the coagulation cascade have also been suggested based on results of the GUSTO trial (CRENSHAW et al. 1995).

2. Hirulog

Hirulog 1 is a bivalent peptide inhibitor of thrombin that was conceived as a combined active site and exosite inhibitor analogous to, but smaller than, hirudin (FENTON 1992; MARAGANORE et al. 1990, 1991a–d; MARAGANORE 1993; WITTING et al. 1992; FENTON et al. 1990, 1993). The D-Phe-Pro-Arg N-terminus of Hirulog binds to the active site, and a hirudin-like C terminus (Asn-Gly-Asp-Phe-Glu-Glu-Ile-Pro-Glu-Glu-Tyr-Leu) binds to the positively charged anion-binding exosite. The inhibition constant for Hirulog is 1.9 n*M*. Thrombin slowly degrades Hirulog by cleaving the Arg-Pro peptide bond in the active site (WITTING et al. 1992). The data for Hirulog binding to thrombin are consistent with a mechanism in which the C-terminal region of Hirulog first binds to the anion-binding exosite, followed by occupation of the active site (PARRY et al. 1994).

The potential of Hirulog and related compounds has been presented in an overview (MARAGANORE 1993). Like hirudin, Hirulog is effective in animal models of thrombosis (HAMELINK et al. 1995; HARKER 1991) and has comparable physical and pharmacokinetic properties (MARAGANORE 1993a).

In dose-ranging studies, Hirulog was shown to be well tolerated in normal volunteers (FOX et al. 1993). It was also shown to be safe when dosed to patients with coronary artery disease, who were undergoing cardiac catheterization, and showed a dose-dependent effect on coagulation parameters without hemorrhagic or allergic complications (CANNON et al. 1993). In a pilot study of 20 patients with unstable angina, Hirulog effectively increased the APTT without hemorrhagic or other complications over the course of a 5-day continuous infusion (SHARMA et al. 1993).

A phase II study investigating the incidence of deep vein thrombosis in orthopedic patients following major hip and knee surgery demonstrated a significant decrease in DVT in the patients treated with 1.0 mg/kg Hirulog (GINSBERG et al. 1994). There were no major bleeding complications in this study. Hirulog has been shown to be safe when used in settings of coronary angioplasty (TOPOL et al. 1993), and to improve the early patency rate in adjunctive therapy with streptokinase in acute myocardial infarction (LIDON et al. 1994).

II. Reversible Inhibitors of Thrombin

Thrombin active site inhibitors that do not rely on a covalent interaction with the enzyme have the advantage of superior selectivity and lack the time-dependent effects on efficacy common to the electrophilic inhibitors above. However, the reversible inhibitors that have been described to date have short terminal elimination half-lives and poor oral bioavailability.

1. Argatroban (Novastan)

In the 1970s, researchers at Mitsubishi began publishing the detailed structure activity relationships of the arginine derivative argatroban **(1)** and related analogs (OKAMOTO et al. 1978, 1980, 1981; KIKUMOTO et al. 1980a,b; HIJIKATA-OKUNOMIYA and OKAMOTO 1992). Argatroban is a highly selective and reversible inhibitor of thrombin active site (KIKUMOTO et al. 1984) but it is limited to parenteral application. The chemistry, pharmacology, and pharmacokinetics of this compound have been reviewed BUSH (1991).

Argatroban is effective in animal models of venous (BERRY et al. 1994a) and arterial thrombosis (IKOMA et al. 1982; JANG et al. 1990). The combination of aspirin and argatroban accelerates coronary thrombolysis in dogs and prevents coronary artery reocclusion with minimal bleeding (YASUDA et al. 1990; FITZGERALD and FITZGERALD 1989; MELLOT et al. 1990). Compared with heparin, a 1-h infusion of argatroban effectively prevented occlusion of an everted arterial graft in rabbits for 24 h after the cessation of infusion (JANG et al. 1992). Argatroban is also effective at inhibiting clot-bound thrombin (BERRY et al. 1994b).

Argatroban is marketed in Japan under the names Novastan[7] and Slonnon[7] by Mitsubishi Kasei and Daiichi, respectively. Argatroban has been licensed in Europe to Synthelabo, where it is in phase II trials in acute myocardial infarction, unstable angina, and coronary angioplasty. In the United States, Novastan[7] is being evaluated in phase II/III clinical trials by Texas Biotechnology for efficacy and safety in patients with heparin-induced thrombocytopenia and heparin-induced thrombocytopenia thrombosis syndrome. In human studies, argatroban is an effective anticoagulant when administered intravenously, with a short half-life (HAUPTMANN and MARKWARDT 1992; CLARKE et al. 1991; FUJI et al. 1996). In a phase II clinical trial comparing argatroban and placebo in the setting of thrombolysis with streptokinase following acute MI, argatroban increased coronary reperfusion by 52% (1996). As previously discussed, the beneficial effects of argatroban may be compromised by a rebound procoagulant state (GOLD et al. 1993).

2. Napsagatran

For a number of years, scientists at Roche have combined the crystallographic determination of thrombin-inhibitor complexes with a medicinal chemistry effort (BANNER and HADVARY 1991). Beginning with known thrombin inhibitors and the observation that 1-amidino piperidine has thrombin inhibitory activity, the incorporation of structural information in the design of these compounds led to napsagatran **(2)** (Fig. 2), a selective and reversible inhibitor of thrombin active site with a K_i of 0.3 nM (ACKERMANN 1993, 1995a,b; BANNER et al. 1993; HILPERT et al. 1994). It is currently in phase II clinical trials to determine efficacy and safety in preventing postoperative thrombosis and for treating established venous thrombosis (TSCHOPP et al. 1995).

On a gravimetric basis, napsagatran was more potent than r-hirudin at blocking coagulation via the intrinsic or extrinsic system, and significantly increased the time to peak thrombin levels, indicating that this small molecule inhibitor may be efficacious in conditions where the velocity of thrombin formation is important (GAST et al. 1993). Unlike r-hirudin or heparin, napsagatran is more selective for clot-bound thrombin than for fluid phase thrombin (GAST et al. 1994). Taken together, these findings in vitro indicate that a reversible thrombin inhibitor such as napsagatran may be effective in settings involving rapid thrombin generation and/or clot-bound thrombin such as arterial thrombosis or thrombolysis. However, like argatroban, napsagatran apparently lacks oral bioavailability, although the human pharmacokinetics and oral bioavailability of napsagatran have not been described.

3. Inogatran

Researchers at Astra have built upon the D-Phe-Pro-Agmatine structure by alkylating the terminal nitrogen with an acetic acid moiety to give a series of compounds related to structure **(5)** (Fig. 3) (TEGER-NILSSON and BYLUND 1993). Inogatran **(5)** is a competitive and reversible inhibitor of thrombin with $K_i = 15\ \mathrm{n}M$. Inogatran is not metabolized, and has a half-life of ca. 1 h in humans (TEGER-NILSSON et al. 1995). In a phase II study comparing inogatran and heparin in unstable angina, inogatran was less effective than heparin in decreasing ischemic coronary events, particularly at the 3-day timepoint. Inogatran failed to show any dose-response relationship in this study. Discontinuation of the heparin and inogatran therapy appeared to cause rebound thrombotic complications in this clinical trial (GRIP et al. 1996).

In preclinical evaluation, inogatran was effective at inhibiting venous and arterial thrombosis in the rat (GUSTAFSSONN et al. 1995). In comparison with DuP 714 in a model of endogenous fibrinolysis induced by the snake venom batroxobin, inogatran caused no inhibition of fibrinolysis even at concentrations >40 *M*. The electrophilic boronate DuP 714 was reported to completely inhibit endogenous fibrinolysis at low concentrations, illustrating a possible difference between the two classes of compounds to induce fibrinolytic compromise. After oral dosing, inogatran had a bioavailability of 32% and 51% in male and female rats, compared with 44% and 34% in male and female dogs (ERIKSSON et al. 1995). The corresponding elimination half-life of 35 is ca. 2 h in rats, compared with ca. 1 h in dogs, consistent with the short half-life reported for humans.

III. Covalent Inhibitors of Thrombin

The electrophilic thrombin inhibitors as a class share potential problems of toxicity, lack of selectivity, and questionable efficacy. In addition to these concerns, the electrophilic functionality has the potential for chemical lability and, in the case of the aldehyde group, could provide a focus for metabolic degradation of the drug. With few exceptions, electrophilic inhibitors are *more* selective for the inhibition of trypsin compared with thrombin. It is not known whether trypsin inhibition represents a potential liability in the development of these compounds as drugs.

1. Efegatran

The most well-studied tripeptide aldehyde thrombin inhibitor is efegatran **(4)** (Fig. 3). The D-phenylalanyl series of tripeptide aldehydes was discovered by BAJUSZ at Gedeon Richter (BAJUSZ et al. 1982, 1987, 1993) and has been developed by Lilly.

The tripeptide aldehyde efegatran may be clearly distinguished from the reversible inhibitor argatroban by its relatively poor selectivity for thrombin compared with other serine proteases, as shown in Table 1 (MARYANOFF et al. 1993). A study comparing the use of efegatran with its closely related N-BOC analog and heparin in a canine model of thrombolysis in vivo found efegatran to be safer and more effective than the other agents (JACKSON et al. 1993). Efegatran **(4)** (Fig. 3) is also more selective than the *N*-BOC analog in a fibrinolysis assay in vitro (BAGDY et al. 1992). Despite the potential for "fibrinolytic compromise" shown by compounds of this class (CALLAS et al. 1994), it is not yet clear whether this concern will ultimately present a clinical problem.

Efegatran caused dose-dependent changes in the APTT, TT, and whole blood clotting time(WBCT) in several species, which rapidly returned to baseline following a bolus i.v. injection or i.v. infusions (BAJUSZ 1992). At 15 and 20 mg/kg in rabbits, efegatran increased the WBCT into the anticipated therapeutic range for 4 h. No rebound effects on the coagulation parameters were noted at any dose. In a stasis model of venous thrombosis, efegatran was found to be ca. five times less effective than r-hirudin on a weight basis (BACHER et al. 1993; SCHUMACHER et al. 1993a,b) and was demonstrated to be more effective than heparin in an electrolytic injury model of arterial thrombsis (JACKSON et al. 1992).

The oral bioavailability of efegatran in rats and dogs is 2.6% and 30%, respectively, with i.v. half-lives of 13 and 35 min (RUTERBORIES et al. 1992). After subcutaneous administration in the pig, the pharmacokinetic half-life was 55 min. The oral bioavailability and half-life of efegatran in humans has not been published. In phase II studies of patients with unstable angina, infusion of efegatran produces a rapid and dose-dependent prolongation of the APTT, which rapidly returns to baseline on cessation of the infusion (SIMOONS et al. 1994). In these studies, an increase of APTT to 2× control was obtained at 0.84 mg/kg/h. Recurrent ischemia was observed during infusion of efegatran as well as heparin and no decrease in ischemic episodes could be documented. The additional report of dose-related thrombophlebitis raises the concern that a rebound phenomenon (vida supra) may be involved (GOLD et al. 1993; SIMOONS et al. 1994).

2. Corvas: CVS 1123

Corvas has prepared a number of tripeptidyl aldehyde inhibitors related to Asp-Pro-Arg-H (PEARSON et al. 1993; VLASUK et al. 1994). From these compounds they have selected CVS 1123 **(5)** for clinical evaluation. CVS 1123 is a potent inhibitor of thrombin which contains a lipophilic C8 N-terminal amide to bind to the distal ("D") pocket of thrombin in the place of a D-phenylalanyl moiety BANNER and HADVARY (1991). Like efegatran, CVS 1123 is a slow, tight-binding inhibitor (K_i = 1.4 n*M*) and is also an inhibitor of thrombin-induced platelet aggregation in vitro at higher concentrations (EC_{50} = 112 n*M*) (ROTE et al. 1994). Studies in the pig in vivo demonstrated that CVS 1123 could increase the thrombotic occlusion time in a coronary thrombosis model from 46 to 137 min at a dose (10 mg/kg i.d.) which increased APTT between 1.5 and 2◊ over baseline. This thrombin inhibitor was also effective in an AV shunt model of venous thrombosis in the rat.

In cynomolgous monkeys, the reported oral bioavailability of CVS 1123 is 36%, with a terminal elimination half-life of 58 and 46 min following oral and intravenous dosing, respectively (RICHARD et al. 1994). These results suggest that this electrophilic aldehyde may have significant bioavailability, perhaps partially because the aldehyde and the *d*-amino group on the guanidine can form a cyclic six-membered ring hemiaminal and decrease the basicity of the guanidino moiety. The kinetics of the antithrombotic effects of structure **(5)** (Fig. 3) and related compounds in vivo have not yet been published. Schering-Plough has formed an alliance with Corvas to develop oral antithrombotic drugs such as CVS 1123 (HARRIS 1994; SCRIP 1997).

3. DuPont Merck: DuP 714

The potent boroarginine thrombin inhibitor DuP 714 **(6)** was designed to tie up the active site Ser 195 of thrombin in a covalent tetrahedral complex (KETTNER and SHENVI 1988, 1992). As one of the first potent thrombin active site inhibitors, DuP 714 has greatly influenced the direction of thrombin active site inhibitor research, and has been extensively studied in vitro and in vivo (KETTNER et al. 1990; KETTNER 1991a,b). The electrophilic boronates lack a high degree of selectivity, and have the potential for "fibrinolytic compromise" when administered in situations where plasmin activity is critical, e.g., thrombolysis. In addition, guanidine-containing boronates are slow, tight-binding inhibitors of thrombin, with all of the potential efficacy problems derived from these binding kinetics (vida supra). DuPont Merck scientists have more recently published on structural analogs of this chemotype (FEVIG et al. 1994). After evaluating the side effect profile of boroarginine thrombin inhibitors such as DuP714, electrophilic compounds with increased selectivity for thrombin over complement factor I have been developed at DuPont Merck (FEVIG et al. 1996).

E. Summary and Conclusions

Small molecule thrombin inhibitors fall into two major classes: reversible and electrophilic active site inhibitors. Reversible inhibitors related to argatroban, napsagatran, and inogatran are potent and highly selective. However, these compounds lack oral bioavailability and have short elimination half-lives, features which may be tied to the presence of arylsulfonamide moieties and/or highly basic guanidino or amidino groups. It remains to be demonstrated that surrogates of these functional groups can overcome the poor bioavailability and pharmacokinetics. The slow, tight-binding electrophilic inhibitors are generally more potent; however, they tend to lack selectivity between serine proteases. The slow on-rate for complexation with the enzyme makes these compounds lack efficacy in vivo compared to their potencies on prolonged incubation in vitro. It is not yet clear whether electrophilic compounds that possess the pharmacodynamic attributes required of a drug can be found. Possibly the most promising are compounds such as CVS 1123, which is reported to have high bioavailability, and electrophilic inhibitors lacking the basic guanidino or amidino group which may have altered kinetics of binding.

More recently, compounds that lack both the highly basic guanidino group and are also devoid of an electrophilic chemically reactive moiety have appeared in the patent literature (LUMMA et al. 1995; VON DER SAAL et al. 1996a,b; BONE et al. 1996; ILLIG et al. 1996). The above agents directly address the shortcomings of the earlier classes of thrombin inhibitors, and the further discovery, development and pharmacology of these compounds is eagerly anticipated.

None of the thrombin active site inhibitors for which the pharmacodynamics and pharmacokinetics have been reported appear to possess adequate half-lives for a once or twice daily oral dosing regimen. Neither is a compound described with the high and reproducible oral bioavailability in humans that would be highly desirable for a chronically administered anticoagulant. Taken together, it is evident that the ideal oral antithrombin/antithrombotic drug as defined by Sixma has not yet been discovered and developed. The combined issues of half-life, selectivity, pharmacodynamics, and, especially, bioavailability, will only be resolved by continuing to pursue the design, discovery, and development of small molecule direct-acting thrombin inhibitors with vigor and imagination.

References

Ackermann J (1993) Guanidine derivatives compositions and use. US patent 5,260,307
Ackermann J (1995a) Guanidine derivatives. US patent 5,393,760
Ackermann J (1995b) Sulfonamidacarboxamides. US patent 5,405,854
Agnola D, Guiomard A (1995) Anticoagulant therapy (heparins and oral vitamin K antagonists) in 1994. What's new? Sem Hop 71:413–422
Antman EM (1994) Hirudin in acute myocardial infarction. Safety report from the Thrombolysis and Thrombin Inhibition in Myocardial Infarction (TIMI) 9A trial. Circulation 90:1624–1630

Antman EM (1996) Hirudin in acute myocardial infarction. Thrombolysis and Thrombin Inhibition in Myocardial Infarction (TIMI) 9B trial Circulation 94:911–921
Bacher P, Walenga JM, Igbal O, Bajusz S, Breddin K, Fareed J (1993) The antithrombotic and anticoagulant effects of a synthetic tripeptide and recombinant hirudin in various animal models. Thromb Res 71:251–263
Bagdy D, Barabas E, Bajusz S, Szell E (1992a) In vitro inhibition of blood coagulation by tripeptide aldehydes – a retrospective screening study focused on the stable D-MePhe-Pro-Arg-H.H_2SO_4. Thromb Haemost 67:325–330
Bagdy D, Barabas E. Szabo G, Bajusz S, Szell E (1992b) In vivo anticoagulant and antiplatelet effect of D-Phe-Pro-Arg-H and D-MePhe-Pro-Arg-H. Thromb Haemost 67:357–365
Bajusz S (1992) Thromb Hemost 67:357–365
Bajusz S, Szell E, Barabas E, Bagdy D (1982) Novel peptidyl-arginine aldehyde derivatives and process for the preparation thereof. US patent 4,316,889
Bajusz S, Szell E, Bagdy D, Barabas E, Dioszegi M, Fittler Z et al (1987) Peptides-aldehydes, process for the preparation thereof and pharmaceutical compositions containing the same. US patent 4,703,036
Bajusz S, Bagdy D, Barabas E, Feher A, Szabo G, Szell G et al (1993) New anticoagulant peptide derivatives and pharmaceutical compositions containing the same as well as a process for the preparation thereof. WO patent 9318060
Bakonyi A, Kavocs L, Duong HT, Aranyi P (1994) Biochemical effect and kinetics of thrombin inhibition by GYKI-14766. Acta Physiol Hung 82:29–36
Banner DW, Hadvary P (1991) Crystallographic analysis at 3.0-C resolution of the binding to human thrombin of four active site-directed inhibitors. J Biol Chem 266:20085–20093
Banner D, Ackermann J, Gast A, Gubernator K, Hadvary P, Hilpert K et al (1993) Serine proteases: 3D structures, mechanisms of action and inhibitors. In: Testa B, Kyburz E, Fuhrer W, Giger R (eds) Perspectives in medicinal chemistry. Helvetica Chimica Acta, Basel, pp 27–43
Beaufils P (1995) What do cardiologists want from anticoagulant treatments? Sem Hop 71:409–412
Bendetowicz AV, Beguin S, Caplain H, Hemker HC (1994) Pharmacokinetics and pharmacodynamics of a low molecular-weight heparin (Enoxaparin) after subcutaneous injection, comparison with unfractionated heparin a three way cross over study in human volunteers. Thromb Haemost 71:305–313
Bergmann JF, Neuhart E (1996) A multicenter randomized double-blind study of enoxaparin compared with unfractionated heparin in the prevention of venous thromboembolic disease in elderly in-patients bedridden for an acute medical illness. The Enoxaparin in Medicine Study Group. Thromb Haemost 76:529–534
Bern MM (1992) Considerations for using lower doses of warfarin. Hematol Oncol Clin North Am 6:1105–1114
Bernard JM, Black JR, Kokseng CU, Covey DF (1992) Warfarin sodium: practitioner beware. J Am Pediatr Med Assoc 82:345–351
Berry CN, Girard D, Lochot S, Lecoffre C (1994a) Antithrombotic actions of argatroban in rat models of venous, "mixed" and arterial thrombosis, and its effects on the tail transection bleeding time. Br J Pharmacol 113:1209–1214
Berry CN, Girardot C, Lecoffre C, Lunven C (1994b) Effects of the synthetic thrombin inhibitor argatroban on fibrin- or clot-incorporated thrombin: comparison with heparin and recombinant hirudin. Thromb Haemost 72:381–386
Bertina RM, Koeleman BPC, Koster T, Rosendaal FR, Dirven RJ, DeRonde H et al (1994) Mutation in blood coagulation factor V associated with resistance to activated protein C. Nature 369:64–67
Beutler E, Lichtman MA, Coller BS, Kipps TJ (1995) Williams hematology, 5th edn. McGraw-Hill, New York
Bone RF, Soll RM, Illig CR, Lu T, Subasinghe NL (1996) Thrombin inhibitors. WO 9640118

Boneu B (1994) Low molecular weight heparin therapy: is monitoring needed? Thromb Haemost 72:330–334
Bourdon P, Jablonski JA, Chao BH (1991) Structure-function relationships of Hirulog peptide interactions with thrombin. FEBS Lett 294:163–166
Bush LR (1991) Argatroban, a selective, potent thrombin inhibitor. Cardiovasc Drug Rev 9:247–263
Callas DD, Fareed J (1995) Direct inhibition of protein Ca by site directed thrombin inhibitors: implications in anticoagulant and thrombolytic therapy. Thromb Res 78:457–460
Callas D, Bacher P, Igbal O, Hoppensteadt D, Fareed J (1994) Fibrinolytic compromise by simultaneous administration of site-directed inhibitors of thrombin. Thromb Res 74:193–205
Cannon CP, Maraganore JM, Loscalzo J, McAllister A, Eddings K, George D (1993) Anticoagulant effects of Hirulog, a novel thrombin inhibitor, in patients with coronary artery disease. Am J Cardiol 71:778–782
Claeson G (1994) Synthetic peptides and peptidomimetics as substrates and inhibitors of thrombin and other proteases in the blood coagulation system. Blood Coagul Fibrinolysis 5:411–436
Clarke RJ, Mayo G, FitzGerald GA, Fitzgerald DJ (1991) Combined administration of aspirin and a specific thrombin inhibitor in man. Circulation 83:1510–1518
Cole MS, Minifee PK, Wolma FJ (1988) Coumarin necrosis – a review of the literature. Surgery 103:271–277
Colman RW, Hirsh J, Marder VJ, Salzman EW (1987) Hemostasis and thrombosis: basic principles and clinical practice, 2nd edn. Lippincott, Philadelphia
Crenshaw BS, Becker RC, Granger CB, Tracy RP, Lambrew CT, Ross AM (1995) Thrombin activity, but not generation, is inhibited by intravenous heparin following thrombolysis: results from the GUSTO hemostasis substudy. J Am Coll Cardiol [Spec Issue]309A–310A
Cushner F, Friedman RJ (1988) Economic impact of total hip arthroplasty. South Med J 81:1379–1381
Dahlbäck B (1995) The protein C anticoagulant system: inherited defects as basis for venous thrombosis. Thromb Res 77:1–43
Das J, Kimball SD (1995) Thrombin active site inhibitors. Bioorg Med Chem 3:999–1007
Davie EW, Fujikawa K, Kisiel W (1991) The coagulation cascade: initiation, maintenance, and regulation. Biochemistry 30:10363–10370
Disch DL, Greenberg ML, Molzberger PT, Malenka DJ, Birkmeyer JD (1994) Managing chronic atrial fibrillation: a Markov decision analysis comparing warfarin, quinidine, and low-dose amiodarone. Ann Intern Med 120:449–457
Diuguid D (1997) Oral anticoagulant therapy for venous thromboembolism. N Engl J Med 336:433–434
Eichinger S, Wolzt M, Nieszpaur-Los M, Schneider B, Lechner K, Eichter HG (1994) Effects of a low molecular weight heparin (Fragmin) and of unfractionated heparin on coagulation activation at the site of plug formation in vivo. Thromb Haemost 72:831–835
Eriksson UG, Renberg L, Vedin C, Strimfors M (1995) Pharmacokinetics of inogatran, a new low molecular weight thrombin inhibitor, in rats and dogs. Thromb Haemost 73:1318
Esmon CT (1993) Molecular events that control the protein C anticoagulant pathway. Thromb Haemost 70:29–35
Falk E, Fernandez-Ortiz A (1995) Role of thrombosis in atherosclerosis and its complications. Am J Cardiol 75:5B–11B
Fauno P, Suomalalen O, Rehnberg V, Hansen TB, Kronerk, Sojmakallio S et al (1994) Prophylaxis for the prevention of venous thromboembolism after total knee arthroplasty. A comparison between unfractionated and low-molecular-weight heparin. J Bone Joint Surg A 76:1814–1818
Fenton JW II (1986) Thrombin. Ann Acad of Sci 485:5–15

Fenton JW II (1992) Coagulation disorders I. Hematology/Oncology Clinics, vol 6, pp 1121–1129
Fenton JW II, Wittig JI, Bourdon P, Maraganore JM (1990) Thrombin-specific inhibition of a novel hirudin analog. Circulation 82:III-659
Fenton JW II, Ni F, Witting JI, Brezniak DV, Andersen TT, Malik AB (1993) The rational design of thrombin-directed antithrombotics. Adv Exp Med Biol 340:1–13
Fevig JM, Kettner CA, Lee SO, Carini DJ (1994) Amido and guanidino substituted boronic acid inhibitors of trypsin-like enzymes. WO patent 9425049
Fevig JM, Buriak J, Cacciola J, Alexander RS, Kettner CA, Knabb RM, Pruitt JR, Thoolen MJ, Weber PC, Wexler RR (1996) Rational design of boropeptide thrombin inhibitors with greater selectivity over complement factor I. Book of abstracts, 212th ACS National Meeting, Orlando, FL, August 25–26, 1996. MEDI-127. American Chemical Society, Washington, DC
Fiore L, Deykin D (1995) Anticoagulant therapy. In: Williams hematology, 5th edn. McGraw-Hill, New York
Fitzgerald DJ, FitzGerald GA (1989) Role of thrombin and thromboxane A_2 in reocclusion following coronary thrombolysis with tissue-type plasminogen activator. Proc Natl Acad Sci USA 86:7585–7589
Fox I, Dawson A, Loynds P, Eisner J, Findlen K, Levin E (1993) Anticoagulant activity of Hirulog, a direct thrombin inhibitor, in humans. Thromb Haemost 69:157–163
Francis CW, Marder VJ, Evarts CM, Yaukoolbodi S (1983) Two-step warfarin therapy. Prevention of postoperative venous thrombosis without excessive bleeding. JAMA 249:374–378
Freedman MD (1992) Oral anticoagulants: pharmacodynamics, clinical indications and adverse effects. J Clin Pharmacol 32:196–209
Fuji H, Kurasako N, Takebayashi T, Tanaka T, Sakano S, Kosogabe Y et al (1996) Argatroban, a selective thrombin inhibitor, for anticoagulant therapy during and following vascular surgery (in Japanese). Masui 45:1289–1292
Garabedian HD, Gold HK, Hagstrom JN, Collen D, Bovill EG (1993) Accelerated thrombin generation accompanying specific thrombin inhibition in unstable angina patients. Circulation 88:I-264
Gast A, Hadvary P, Schmid G, Hilpert K, Ackermann J, Tschopp TB (1993) A novel, synthetic, competitive inhibitor of thrombin inhibits thrombin generation in plasma more potently than heparin and recombinant hirudin. Thromb Haemost 69:1298
Gast A, Tschopp TB, Schmid G, Hilpert K, Ackermann J (1994) Inhibition of clot-bound and free (fluid-phase thrombin) by a novel synthetic thrombin inhibitor (Ro 46–6240), recombinant hirudin and heparin in human plasma. Blood Coagul Fibrinolysis 5:879–887
Ginsberg JS, Nurmohamed MT, Gent M, MacKinnon B, Sicurella J, Brill-Edwards P (1994) Use of Hirulog in the prevention of venous thrombosis after major hip or knee surgery. Circulation 90:2385–2389
Gold HK, Torres FW, Garabedian HD, Werner W, Jang IK, Khan A et al (1993) Evidence for a rebound coagulation phenomenon after cessation of a 4-hour infusion of a specific thrombin inhibitor in patients with unstable angina pectoris. J Am Coll Cardiol 21:1039–1047
Granger CB, Miller JM, Bovill EG, Gruber A, Tracy RP, Krucoff MW (1995) Rebound increase in thrombin generation and activity after cessation of intravenous heparin in patients with acute coronary syndromes. Circulation 91:1929–1935
Grip L, Wallentin M, Dellborg P, Grande M, Halinen E, Mybre L et al (1996) Thrombin inhibition in myocardial ischaemia: low molecular weight thrombin inhibitor inogatran versus heparin for unstable angina and non-Q-wave myocardial infarction. Eur Heart J 17 [Suppl]:121
Grütter MG, Priestle J, Rahuel J, Grossenbacher H, Bode W, Hofsteenge J et al (1990) Crystal structure of the thrombin-hirudin complex: a novel mode of serine protease inhibition. EMBO J 9:2361–2365

Gustafsson D, Elg M, Lenfors S, Borjesson I, Teger-Nilsson A-C (1995) Effects of inogatran, a new low molecular weight thrombin inhibitor, on rat models of thrombosis. Thromb Haemost 73:1319
Hamelink JK, Tang DB, Barr CF, Jackson MR, Reid TJ, Gomez ER et al (1995) Inhibition of platelet deposition by combined Hirulog and aspirin in a rat carotid endarterectomy model. J Vasc Surg 21:492–498
Harpel PC (1987) Blood proteolytic enzyme inhibitors: their role in modulating blood coagulation and fibrinolytic enzyme pathways. In: Colman RW, Hirsh J, Marder VJ, Salzman EW (eds) Hemostasis and thrombosis: basic principles and clinical practice, 2nd edn. Lippincott, Philadelphia
Harker L (1990) Potent antithrombotic effects of a novel hybrid antithrombin peptide in vivo. Circulation 82:603
Harker LA (1991) Hirudin interruption of heparin-resistant arterial thrombus formation in baboons. Blood 77:1006–1012
Harker LA (1994) Strategies for inhibiting the effects of thrombin. Blood Coagul Fibrinolysis 5:S47–S58
Harker LA, Hanson SR, Runge MS (1995) Thrombin hypothesis of thrombus generation and vascular lesion formation. Am J Cardiol 75:12B–17B
Harris N (1994) Schering-Plough is creating a cardiovascular alliance. Philadelphia Inquirer 12/20/94
Hauptmann J, Markwardt F (1992) Pharmacologic aspects of the development of selective synthetic thrombin inhibitors as anticoagulants. Semin Thromb Hemost 18:200–217
Heras M, Chesebro JH, Webster MWI, Mruk JS, Grill DE, Penny WJ et al (1990) Hirudin, heparin, and placebo during deep arterial injury in the pig. The in vivo role of thrombin in platelet-mediated thrombosis. Circulation 82:1476–1484
Hijikata-Okunomiya A, Okamoto S (1992) A strategy for a rational approach to designing synthetic selective inhibitors. Semin Thromb Hemost 18:135–149
Hilpert K, Ackermann J, Banner DW, Gast A, Gubernator K, Hadvary P et al (1994) Design and synthesis of potent and highly selective thrombin inhibitors. J Med Chem 37:3889–3901
Hirsh J (1991) Oral anticoagulant drugs. N Engl J Med 324:1865–1875
Hirsh J (1993) Low molecular weight heparin. Thromb Haemost 70:204–207
Holland AE, Linker S, Bardsley WT, Kopecky S, Litin SC, Meissner I et al (1994) Warfarin versus aspirin for prevention of thromboembolism in atrial fibrillation: Stroke Prevention in Atrial Fibrillation II Study. Lancet 343:687–691
Hull R, Delmore T, Genton E, Hirsh J, Gent M, Sackett D et al (1979) Warfarin sodium versus low-dose heparin in the long-term treatment of venous thrombosis. N Engl J Med 301:855–858
Hyers TM, Hull RD, Weg JG (1992) Antithrombotic therapy for venous thromboembolic disease. Chest 102:408S–425S
Ikoma H, Ohtsu K, Tamao Y, Kikumoto R, Okamoto S (1982) Effect of a potent thrombin inhibitor, MCI-9038, on novel experimental arterial thrombosis (in Japanese). Ketsueki to Myakkan 13:72–77
Illig CR, Soll RM, Salvino JM, Tomczvk BE, Lu T, Subasinghe NL(1996) Arylsulphonylaminobenzene derivatives and the use thereof as factor Xa inhibitors. WO patent 9640100
Jackson CM (1987) Mechanisms of prothrombin activation. In: Colman RW, Hirsh J, Marder VJ, Salzman EW (eds) Hemostasis and thrombosis: basic principles and clinical practice, 2nd edn. Lippincott, Philadelphia
Jackson CM, Nemerson Y (1980) Blood coagulation. Annu Rev Biochem 49:765–811
Jackson CV, Growe VG, Frank JD, Wilson HC, Coffman WJ, Utterback BG et al (1992) Pharmacological assessment of the antithrombotic activity of the peptide thrombin inhibitor, D-methyl-phenylalanyl-prolyl-arginal (GYKI-14766), in a canine model of coronary artery thrombosis. J Pharmacol Exp Ther 261:546–552
Jackson CV, Wilson HC, Growe VG, Shuman RT, Gesellchen PD (1993) Reversible tripeptide thrombin inhibitors as adjunctive agents to coronary thrombolysis:

a comparison with heparin in a canine model of coronary artery thrombosis. J Cardiovasc Pharmacol 21:587–594
Jafri SM, Gheorghiade M, Goldstein S (1992) Oral anticoagulation for secondary prevention after myocardial infarction with special reference to the warfarin re-infarction study. Prog Cardiovasc Dis 34:317–322
Jang IK, Gold HK, Ziskind AA, Leinbach RC, Fallon JT, Collen D (1990) Prevention of platelet-rich arterial thrombosis by selective thrombin inhibition. Circulation 81:219–225
Jang IK, Gold HK, Leinbach RC, Rivera AG, Fallon JT, Bunting S et al (1992) Persistent inhibition of arterial thrombosis by a 1-hour intravenous infusion of argatroban, a selective thrombin inhibitor. Coron Artery Dis 3:407–414
Johnson PH (1994) Hirudin: clinical potential of a thrombin inhibitor. Annu Rev Med 45:165–177
Jones KC, Mann KG (1995) A model for the tissue factor pathway to thrombin. II. A mathematical simulation. J Biol Chem 269:23367–23373
Kaiser B, Hauptmann J (1992) Pharmacology of synthetic thrombin inhibitors of the tripeptide type. Cardiovasc Drug Rev 10:71–87
Kaiser B, Simon A, Markwardt F (1990) Antithrombotic effects of recombinant hirudin in experimental angioplasty and intravascular thrombolysis. Thromb Haemost 63:44–47
Kakkar VV, Boeckl O, Boneu B, Bordenave L, Brehm OA, Brucke P et al (1997) Efficacy and safety of a low-molecular-weight heparin and standard unfractionated heparin for prophylaxis of postoperative venous thromboembolism: European multicenter trial. World J Surg 21:2–8
Kalafatis M, Bertina RM, Rand MD, Mann KG (1995) Characterization of the molecular defect in factor V-R506Q. J Biol Chem 270:4053–4057
Kearon C, Hirsh J (1995) Starting prophylaxis for venous thromboembolism postoperatively. Arch Intern Med 155:366–372
Kelly A, Marzec U, Hanson S, Chao B, Maraganore J, Harker L (1990) Potent antithrombotic effects of a novel hybrid antithrombin peptide in vivo. Circulation 82:III-603
Kelly AB, Marzec UM, Krupski W, Bass A, Cadroy Y, Hanson SR et al (1991) Hirudin interruption of heparin-resistant arterial thrombus formation in baboons. Blood 77:1006–1012
Kettner CA (1991a) In vivo characterization of DuP 714, a new synthetic thrombin inhibitor. Thromb Hem 65:775
Kettner CA (1991b) Thrombin inhibition with DuP 714 accelerates reperfusion and delays reocclusion in dogs. Circulation 84:II-467
Kettner CA, Shenvi AB (1988) U.S. Patent 5,187,157
Kettner CA, Shenvi AB (1992) U.S. Patent 5,242,904
Kettner CA, Shenvi AB (1993a) Peptide boronic acid inhibitors of trypsin-like proteases. US patent 5,187,157
Kettner CA, Shenvi AB (1993b) Peptide boronic acid inhibitors of trypsin-like proteases. US patent 5,242,904
Kettner C, Mersinger L, Knabb R (1990) The selective inhibition of thrombin by peptides of boroarginine. J Biol Chem 265:18289–18297
Kikumoto R, Tamao Y, Ohkubo K, Tezuka T, Tonomura S, Okamoto S et al (1980a) Thrombin inhibitors. 2. Amide derivatives of Nα-substituted L-arginine. J Med Chem 23:830–836
Kikumoto R, Tamao Y, Ohkubo K, Tezuka T, Tonomura S, Okamoto S et al (1980b) Thrombin Inhibitors. 3. Carboxyl-containing amide derivatives of N′-substituted-l-arginine. J Med Chem 23:1293–1299
Kikumoto R, Tamao Y, Tezuka T, Tonomura S, Hara H, Ninomiya K et al (1984) Selective inhibition of thrombin by (2R,4R)-4-methyl-1-[N2-[(3-methyl-1,2,3,4-tetrahydro-8-quinolinyl) +++ sulfonyl]-L-arginyl)]-2-piperidinecarboxylic acid. Biochemistry 23:85–90

Kimball SD (1995a) Challenges in the development of orally bioavailable thrombin active site inhibitors. Blood Coagul Fibrinolysis 6:511–519
Kimball SD (1995b) Thrombin active site inhibitors. Curr Pharm Design 1:441–468
Knabb RM, Timmermans PBMWM, Kettner CA, Reilly TM (1991) In vivo characterization of DuP 714, a new synthetic thrombin inhibitor. Thromb Haemost 65:775
Koopman MM, Buller HR, Tencate JW (1995) Diagnosis of recurrent deep vein thrombosis. Haemostasis 25:49–57
Lawson JH, Kalafatis M, Stram S, Mann KG (1994) A model for the tissue factor pathway to thrombin. I. An empirical study. J Biol Chem 269:23357–23366
Lefkovits J, Topol EJ (1994) Direct thrombin inhibitors in cardiovascular medicine. Circulation 90:1522–1536
Lengfelder W (1995) Current status of antithrombotic therapy. Dtsch Med Wochenschr 120:105–110
Lidon RM, Theroux P, Lesperance J, Adelman B, Bonan R, Duval D et al (1994) A pilot, early angiographic patency study using a direct thrombin inhibitor as adjunctive therapy to streptokinase in acute myocardial infarction. Circulation 89:1567–1572
Lindmarker P, Holmstrom M (1996) Use of low molecular weight heparin (dalteparin), once daily, for the treatment of deep vein thrombosis. A feasibility and health economic study in an outpatient setting. Swedish Venous Thrombosis Dalteparin Trial Group. J Int Med 240:395–401
Litin SC, Gastineau DA (1995) Current concepts in anticoagulant therapy. Mayo Clin Proc 70:266–272
Lorand L, Radek JT (1992) Activation of human plasma factor XIII by thrombin. In: Berliner LJ (ed) Thrombin: structure and function. Plenum, New York, pp 257–271
Lowe GDO, Greer IA, Cooke TG, Dewar EP, Evans MJ, Forbes CD et al (1992) Risk of and prophylaxis for venous thromboembolism in hospital patients. Br Med J 305:567–574
Lumma WC, Freidinger RM, Brady S, Sanderson PE, Feng DM (1995) Thrombin inhibitors. WO patent 9603374
Lyle TA (1994) Small-molecule inhibitors of thrombin. Perspect Drug Discov Des 1:453–460
Maffrand JP (1992) Direct thrombin inhibitors. Nouv Rev Fr Hematologie 34:405–419
Maraganore JM (1993a) Hirudin and hirulog: advances in antithrombotic therapy. Perspect Drug Discov Des 1:461–478
Maraganore JM (1993b) Pre-clinical and clinical studies on hirulog: a potent and specific direct thrombin inhibitor. Adv Exp Med Biol 340:227–236
Maraganore JM, Bourdon P, Jablonski J, Ramachandran KL, Fenton JW II (1990) Design and characterization of hirulogs: a novel class of bivalent peptide inhibitors of thrombin. Biochemistry 29:7095–7101
Maraganore JM, Oshima T, Asai F, Sugitachi A (1991a) Comparison of anticoagulant and antithrombotic activities of Hirulog-1 and argatroban (MD-805). Thromb Haemost 65:651
Maraganore JM, Chao BH, Weitz JI, Hirsh J (1991b) Comparison of antithrombin activities of heparin and Hirulog-1: basis for improved antithrombotic properties of direct thrombin inhibitors. Thromb Haemost 65:829
Maraganore JM, Yao SK, McNatt J, Eidt J, Cui K, Buja LM et al (1991c) Hirudin-based peptides accelerate thrombolysis and delay reocclusion after treatment with recombinant tissue-type plasminogen activator. Thromb Haemost 65:1188
Maraganore JM, Fenton JW, Kline T, Bourdon P, Wittig J, Jablonski J et al (1991d) Modifications in Hirulog peptides yielding improved antithrombin activities. Thromb Haemost 65:830
Märki WE, Wallis RB (1990) The anticoagulant and antithrombotic properties of hirudins. Thromb Haemost 64:344–348
Markwardt F (1957) Die Isolierung und chemische Charakterisierung des Hirudins. Hoppe Seyler Z Physiol Chem 308:147–156

Maryanoff BE, Qui X, Padmanabhan KP, Tulinsky A, Almond HR Jr, Andrade-Gordon P et al (1993) Molecular basis for the inhibition of human a-thrombin by the macrocyclic peptide cyclotheonamide A. Proc Natl Acad Sci USA 90:8048–8052

McBride R et al (1994) Warfarin versus aspirin for prevention of thromboembolism in atrial fibrillation: Stroke Prevention in Atrial Fibrillation II Study. Lancet 343:687–691

McDonagh J (1987) Structure and function of factor XIII. In: Colman RW, Hirsh J, Marder VJ, Salzman EW (eds) Hemostasis and thrombosis: basic principles and clinical practice, 2nd edn. Lippincott, Philadelphia

Meade TW, Miller GJ (1995) Combined use of aspirin and warfarin in primary prevention of ischemic heart disease in men at high risk. Am J Cardiol 75:23B–26B

Mellot MJ, Connolly TM, York SJ, Bush LR (1990) Prevention of reocclusion by MCI-9038, a thrombin inhibitor, following t-PA-induced thrombolysis in a canine model of femoral arterial thrombosis. Thromb Haemost 64:526–534

Merli GJ (1993) Update. Deep vein thrombosis and pulmonary embolism prophylaxis in orthopedic surgery. Med Clin North Am 77:397–411

Mosesson MW, Siebenlist KR, Amrani DL, DiOrio JP (1989) Identification of covalently linked trimeric and tetrameric D domains in crosslinked fibrin. Proc Natl Acad Sci USA 86:1113–1117

Naski MC, Shafer JA (1991) A kinetic model for the α-thrombin-catalyzed conversion of plasma levels of fibrinogen to fibrin in the presence of antithrombin III. J Biol Chem 266:13003–13010

Naski MC, Shafer JA (1993) α-Thrombin within fibrin clots: inactivation of thrombin by antithrombin-III. Thromb Res 69:453–465

Neuhaus KL, von Essen R, Tebbe U, Jessel A, Heinrichs H, Maurer W et al (1994) Safety observations: a study of the Arbeitsgemeinschaft Leitender Kardiologischer Krankenhausärzte (ALKK). From the pilot phase of the randomized r-Hirudin for Improvement of Thrombolysis (HIT-III) study. Circulation 90:1638–1642

Nurmohamed MT, Rosendaal FR, Buller HR, Dekker E, Hommes DW, Vandenbroucke JP et al (1992) Low-molecular-weight heparin versus standard heparin in general and orthopaedic surgery: a meta-analysis. Lancet 340:152–156

Okamoto S, Hijikata A, Kikumoto R, Tamao Y, Ohkubo K, Tezuka T et al (1978) N_2-Arylsulfonyl-L-argininamides and the pharmaceutically acceptable salts thereof. US patent 4,117,127

Okamoto S, Kinjo K, Hijikata A, Kikumoto R, Tamao Y, Ohkubo K et al (1980) Thrombin inhibitors. 1. Ester derivatives of Nα-(arylsulfonyl)-L-arginine. J Med Chem 23:827–830

Okamoto S, Hijikata A, Kikumoto R, Tamao Y, Ohkubo K, Tezuka T et al (1981) N2-arylsulfonyl-L-argininamides and the pharmaceutically acceptable salts thereof. US patent 4,258,192

Parry MA, Maraganore JM, Stone SR (1994) Kinetic mechanism for the interaction of Hirulog with thrombin. Biochemistry 33:14807–14814

Pearson DA et al (1993) Internation Patent Application 93/15756

Peters RF, Lees CM, Mitchell KA, Tweed MF, Talbot MD, Wallis RB (1991) The characterisation of thrombus development in an improved model of arterio-venous shunt thrombosis in the rat and the effects of recombinant desulphatohirudin (CGP 39393), heparin, and iloprost. Thromb Haemost 65:268–274

Petersen P, Boysen G, Godtfredsen J, Andersen ED, Andersen B et al (1989) Placebo-controlled, randomised trial of warfarin and aspirin for prevention of thromboembolic complications in chronic atrial fibrillation. The Copenhagen AFASAK study. Lancet 1:175–179

Philippides GJ, Loscalzo J (1996) Potential advantages of direct-acting thrombin inhibitors. Coron Artery Dis 7:497–507

Pinede L, Ninet J, Boissel JP, Pasquier J (1994) Anticoagulant therapy for venous thromboembolic disease: optimal duration of oral antivitamin K therapy. Review of the literature. Presse Med 23:1817–1820

Poller L, Samama M (1994) Laboratory monitoring of oral anticoagulant therapy. Clin Lab Med 14:813–823

Poller L, McKernan A, Thomson JM, Elstein M, Hirsch PJ, Jones JB (1987) Fixed minidose warfarin: a new approach to prophylaxis against venous thrombosis after major surgery. Med J Clin Res Ed 295:1309–1312

Richard BM, Nolan TG, Tran HS, Mille MM, Nutt RG, Vlasuk GP (1994) Bioavailability of a novel direct thrombin inhibitor in non-human primates following oral administration. Circulation 90:I-180

Rote WE, Dempsey EM, Oldeschulte GL, Ardecky RJ, Tamura SY, Nutt RF et al (1994) Evaluation of a novel orally active direct inhibitor of thrombin in animal models of thrombosis. Circulation 90:I-344

Rupin A, Mennecier P, de Nanteuil G, Laubie M, Verbeuren TJ (1995) A screening procedure to evaluate the anticoagulant activity and the kinetic behaviour of direct thrombin inhibitors. Thromb Res 78:217–225

Ruterbories KJ, Hanssen BR, Lindstrom TD (1992) ISSX proceedings, 4th North American ISSX meeting, p 204

Rydel TJ, Ravichandran KO, Tulinsky A, Bode W, Huber R, Roitsche C et al (1990) The structure of a complex of recombinant hirudin and human α-thrombin. Science 249:277–280

Rydel TJ, Tulinsky A, Bode W, Huber R (1991) Refined structure of the hirudin-thrombin complex. J Mol Biol 221:583–601

Schulman S, Granquist S, Holmstrom M, Carlsson A, Lindmarker P, Nicol P et al (1997) The duration of oral anticoagulant therapy after a second episode of venous thromboembolism. The Duration of Anticoagulation Trial Study Group. N Engl J Med 336:393–398

Schumacher WA, Steinbacher TE, Heran CL, Seiler SM, Michel IM, Ogletree ML (1993a) Comparison of thrombin active site and exosite inhibitors and heparin in experimental models of arterial and venous thrombosis and bleeding. J Pharmacol Exp Ther 267:1237–1242

Schumacher WA, Steinbacher TE, Heran CL, Megill JR, Durham SK (1993b) Effects of antithrombotic drugs in a rat model of aspirin-insensitive arterial thrombosis. Thromb Haemost 69:509–514

Schumacher WA, Balasubramanian N, St. Laurent DR, Seiler SM (1994) Effect of a novel thrombin active-site inhibitor on arterial and venous thrombosis. Eur J Pharmacol 259:165–171

Schumacher WA, Heran CL, Steinbacher TE (1996) Low-molecular-weight heparin (fragmin) and thrombin active-site inhibitor (argatroban) compared in experimental arterial and venous thrombosis and bleeding time. J Cardiovasc Pharmacol 28:19–25

Scrip (1996) Novastan promising in acute MI. 2177:21

Scrip (1997) Schering-Plough further extends Corvas alliance. 2199:21

Scully MF (1992) The biochemistry of blood clotting: the digestion of a liquid to form a solid. Essays Biochem 27:17–36

Serlin MJ, Breckenridge AM (1983) Drug interactions with warfarin. Drugs 25:610–620

Sharma GV, Lapsley D, Vita JA, Sharma S, Coccio E, Adelman B et al (1993) Usefulness and tolerability of Hirulog, a direct thrombin-inhibitor, in unstable angina pectoris. Am J Cardiol 72:1357–1360

Shetty HG, Fennerty AG, Routledge PA (1989) Clinical pharmacokinetic considerations in the control of oral anticoagulant therapy. Clin Pharmacokinet 16:238–253

Siebenlist KR, Diorio JP, Budzynski AZ, Mosesson MW (1990) The polymerization and thrombin-binding properties of des-(Bβ1–42)-fibrin. J Biol Chem 265:18650–18655

Simoons M, Lenderin KT, Scheffer M, Stoel I, de Milliano P, Remme W (1994) Efegatran, a new direct thrombin inhibitor: safety and dose response in patients with unstable angina. Circulation 90:I-231
Sixma JJ, de Groot PG (1992) The ideal anti-thrombotic drug. Thromb Res 68:507–512
Smith AJ, Holt RE, Fitzpatrick JB, Palacios IF, Gold HK, Werner W et al (1996) Transient thrombotic state after abrupt discontinuation of heparin in percutaneous coronary angioplasty. Am Heart J 131:434–439
Smith P, Arnesen H, Holme I (1990) The effect of warfarin on mortality and reinfarction after myocardial infarction. N Engl J Med 323:147–152
Stone SR, Tapparelli C (1995) Thrombin inhibitors as antithrombotic agents: the importance of rapid inhibition. J Enzym Inhib 9:3–15
Stone SR, Dennis S, Wallace A, Hofsteenge J (1990) Use of protein chemistry and molecular biology to determine interaction areas between proteases and their inhibitors: the thrombin-hirudin interaction as an example. In: Festoff BW (ed) Serine proteases and their serpin inhibitors in the nervous system: regulation in development and in degenerative and malignant disease. Plenum, New York, pp 115–135
Tamaki T, Aoki N(1982) Cross-linking of "2-plasmin inhibitor to fibrin catalyzed by activated fibrin-stabilizing factor. J Biol Chem 257:14767–14772
Tapparelli C, Metternich R, Ehrhardt C, Zurini M, Claeson G, Scully MF et al (1993a) In vitro and in vivo characterization of a neutral boron-containing thrombin inhibitor. J Biol Chem 268:4734–4741
Tapparelli C, Metternich R, Ehrhardt C, Cook NS (1993b) Synthetic low-molecular weight thrombin inhibitors: molecular design and pharmacological profile. Trends Pharmacol Sci 14:366–376
Teger-Nilsson AC, Bylund R (1993) New peptide derivatives. WO patent 9311152
Teger-Nilsson A, Eriksson U, Gustafsson D, Bylund R, Fager G, Meld P et al (1995) Phase I studies on Inogatran, a new selective thrombin inhibitor. J Am Coll Cardiol 117A–118A
Theroux P, Waters D, Lam J, Juneau M, McCans J (1992) Reactivation of unstable angina after the discontinuation of heparin. N Engl J Med 327:141–145
Thomas D (1994) Venous thrombogenesis. Br Med Bull 50:803–812
Topol EJ (1995) Novel antithrombotic approaches to coronary artery disease. Am J Cardiol 75:27B–33B
Topol EJ (1996) A comparison of recombinant hirudin with heparin for the treatment of acute coronary syndromes. N Engl J Med 335:775–782
Topol EJ, Bonan R, Jewitt D, Sigwart U, Kakkar VV, Rothman M et al (1993) Use of a direct antithrombin Hirulog in place of heparin during coronary angioplasty. Circulation 87:1622–1629
Tschopp TB, Ackermann J, Gast A, Hilpert K, Kirchhofer D, Roux S et al (1995) Napsagatran. Drugs Future 20:476–479
Vitali J, Martin PD, Malkowski MG, Robertson WD, Lazar JB, Winant RC et al (1992) The structure of a complex of bovine a-thrombin and recombinant hirudin at 2.8 C resolution. J Biol Chem 267:17670–17678
Vlasuk GP (1994) Inhibitors of thrombosis. U.S. Patent 5,999,999
Vlasuk GP, Webb TA, Pearson DA, Abelman MM (1996) Inhibitors of thrombosis. US 5492895
von der Saal W, Heck R, Kucznierz R, Leinert H, Stegmeier K (1996a) Neue 4-Aminopyridazine, Verfahren zu ihrer Herstellung sowie diese Verbindungen enthaltende Arzneimittel. WO patent 9606832
von der Saal W, Leinert H, Stegmeier K (1996b) Neue Phosphanoxide, Verfahren zu ihrer Herstellung sowie diese Verbindungen enthaltende Arzneimittel. WO patent 9606849
Walenga JM, Fareed J, Pifarre R et al (1991) Current developments in the use of recombinant hirudin as an antithrombotic agent. Curr Opin Therap Pat 1:869
Wallentin L (1996) Low molecular weight heparins: a valuable tool in the treatment of acute coronary syndromes. Eur Heart J 17:1470–1476

Weinmann EE, Salzman EW (1994) Deep-vein thrombosis. N Engl J Med 331:1630–1641
Weitz J (1994) New anticoagulant strategies. Current status and future potential. Drugs 48:485–497
Weitz JI (1995) Activation of blood coagulation by plaque rupture: mechanisms and prevention. Am J Cardiol 75:18B–22B
Weitz JI, Hudoba M, Massel P, Maraganore J, Hirsh J (1990) Clot-bound thrombin is protected from inhibition by heparin-antithrombin III but is susceptible to inactivation by antithrombin III-independent inhibitors. J Clin Invest 86:385–391
Wheeldon NM (1995) Atrial fibrillation and anticoagulant therapy. Eur Heart J 16:302–312
Willerson JT, Casscells W (1993) Thrombin inhibitors in unstable angina: rebound or continuation of angina after argatroban withdrawal? J Am Coll Cardiol 21:1048–1051
Wiman B, Collen D (1979) On the mechanism of the reaction between human α_2-antiplasmin and plasmin. J Biol Chem 254:9291–9297
Wittig JI, Bourdon P, Brezniak DV, Maraganore JM, Fenton JW II (1992) Thrombin-specific inhibition by and slow cleavage of Hirulog-1. Biochem J 283:737–743
Yasuda T, Gold HK, Yaoita H, Leinbach RC, Guerrero JL, Jang IK et al (1990) Comparative effects of aspirin, a synthetic thrombin inhibitor and a monoclonal antiplatelet glycoprotein IIb/IIIa antibody on coronary artery reperfusion, reocclusion and bleeding with recombinant tissue-type plasminogen activator in a canine preparation. J Am Coll Cardiol 16:714–722
Zeymer U, Neuhaus KL (1995) Hirudin and excess bleeding. Implications for future use. Drug Saf 12:234–239
Zoldhelyi P, Bichler J, Owen WG, Grill DE, Fuster V, Mruk JS et al (1994) Persistent thrombin generation in humans during specific thrombin inhibition with hirudin. Circulation 90:2671–2678

CHAPTER 14

Inhibitors of Factor Xa

S. KUNITADA, T. NAGAHARA, and T. HARA

A. Introduction

Thrombin occupies a central position in thrombus formation and recently has been a target for the development of anticoagulant agents. As is well known, blood coagulation operates as a cascade or multiple enzymatic amplification process. In the classical intrinsic pathway, factor (f) IXa in the presence of fVIIIa, Ca^{2+} and phospholipid converts fX to fXa, while in the classical extrinsic pathway, fX is activated by fVIIa in the presence of tissue factor. In turn, fXa activates prothrombin to generate thrombin. More recently, the coagulation cascade has been divided into two stages, rather than distinct pathways (CAMERER et al. 1996). A small amount of thrombin is generated in the "initiation stage" by activation of the fVIIa/tissue factor complex. The amount of thrombin generated by the tissue factor pathway is limited by tissue factor pathway inhibitor, but enough thrombin is generated to activate factors V, VIII and XI. Recruitment of these additional factors begins the "augmentation stage" which can provide large amounts of additional thrombin that plays a dominant role in subsequent thrombus formation. In both the classical coagulation cascade and the more current tissue factor dependent stages, fXa plays a pivotal role by controlling activation of prothrombin to thrombin.

It could be speculated that the more proximal the intervention in the coagulation cascade, the more efficient the suppression of thrombin generation. It is known, for example, that one molecule of fIXa can generate 10^6 molecules of thrombin in 1 min (ELÒDI and VARADI 1979). Factor IXa inhibition (or for that matter inhibition of fXIa or fXIIa), however, would not take into account the redundancy of the various pathways proximal to fX. Inhibitors of fXa, on the other hand, could block both classically defined intrinsic and extrinsic pathways, thereby preventing the ongoing generation of thrombin, an effect not possible with thrombin inhibitors. Consequently, inhibition of fXa represents a particularly attractive target for the next generation of anticoagulants.

B. Rationale of Factor Xa Inhibitors

Heparin and warfarin are the most widely used conventional anticoagulants in the treatment of venous thrombosis. However, these anticoagulants have clinical limitations in part due to their dependence on antithrombin III or antagonism of vitamin K metabolism, respectively. In addition, heparin use may be associated with thrombocytopenia (Carreras 1980) and osteoporosis (Mätzsch et al. 1986), while warfarin treatment has a slow onset of action, requires regular clinical monitoring and is complicated by food and drug interaction (Verstrate and Wessler 1992). Considerable efforts have therefore been made over the last 2 decades to develop what might be considered a more ideal anticoagulant. Sixma and de Groot (1992) and Bagdy et al. (1992) have suggested that such an agent should have the following characteristics: (1) it should inhibit thrombosis, (2) it should not affect hemostasis, (3) it should have a long half-life, (4) it should allow for chronic oral administration, (5) it should have no important side-effects, and (6) it should have a wide therapeutic range. We would propose that an additional consideration should reflect the vastly different conditions under which thrombi form in various disease states. With this in mind, it is unlikely that a single agent can be found for all thrombotic conditions; rather the ideal antithrombotic drug should be chosen on a condition-by-condition, or even a case-by-case, basis. In terms of practical drug development, we consider that the three minimum requirements for an ideal anticoagulant would be oral bioavailability, minimum bleeding propensity, and a mechanism based on direct inhibition of an activated coagulation factor.

Direct inhibition of thrombin catalytic activity would be consistent with at least some of the criteria proposed by Sixma and Bagdy. Accordingly, potent thrombin inhibitors such as argatroban (Kumada and Abiko 1981), hirudin (Heras et al. 1989) and Hirulog (Kelly et al. 1992) have been developed as drug candidates. More recently, inogatran (Eriksson et al. 1995) and CVS-1123 (Cousins et al. 1995) have been synthesized as small-molecule inhibitors of the catalytic site of thrombin. Although oral bioavailability has been reported in animal studies for some of these compounds (Eriksson et al. 1995; Cousins et al. 1995), this has yet to be confirmed in man. The subject of oral thrombin inhibitors is elegantly covered in an earlier chapter of this handbook. One concern with direct thrombin inhibitors in general has been the tendency to prolong bleeding time at doses only slightly higher than the effective dose, which raises the concern that their therapeutic windows may be too narrow (Jackson et al. 1992). An additional consideration with direct and indirect thrombin inhibitors relates to the argument of whether it is better to neutralize thrombin activity or interfere with its production. Direct inhibition of the catalytic activity of thrombin will not prevent accumulation of thrombin during treatment, a situation which could lead to "thrombin rebound" and prothrombotic tendencies after stopping treatment. The clinical relevance of thrombin rebound has not been established with certainty but, if significant, it represents another reason for thinking that an fXa inhibitor might have an advantage over inhibitors of thrombin. In addition, given the amplification that occurs at each stage in the coagulation cascade, it may be possible to use lower doses of agents targeted to coagulation factors proximal to thrombin, such as fXa.

A variation of the phenomenon of thrombin rebound is the recurrence of thrombosis after treatment resulting from the thrombogenicity of thrombi themselves. Although initially attributed to clot-bound thrombin, recent evidence suggests that fXa may be primarily responsible for the procoagulant activity of clots in vivo (PRAGER et al. 1995). This may be an explanation for the observed efficacy of fXa inhibitors in preventing recurrent thrombosis after coronary thrombolysis (SITKO et al. 1992).

Finally, mention needs to be made of the potential antithrombotic effects of thrombin (see Chap. 16, this volume). While direct thrombin inhibitors will affect all of thrombin's activities, it is possible that the use of thrombin generation inhibitors would permit the production of enough thrombin to bind with thrombomodulin and activated protein C. Selectivity in the catalytic actions of thrombin has been described previously (BERNDT et al. 1986; HIGGINS et al. 1983).

C. Pharmacological Profile of Factor Xa Inhibitors

Despite the theoretical advantages of fXa inhibition, relatively few inhibitors have been reported, especially compared with the large number of thrombin inhibitors. It is conceivable that the three-dimensional structure of these serine proteases is too similar to enable optimization of a selective inhibitor for fXa. Although a great deal is known about the structure of thrombin, fXa has only recently been crystallized (PADMANABHAN et al. 1993) and it has proven difficult to co-crystallize inhibitors with fXa because the active site of the enzyme is blocked by other molecules of fXa. For this reason, many investigators have chosen to co-crystallize inhibitors with trypsin, which shares many of the structural features of fXa. Despite these limitations, however, a number of specific fXa inhibitors have been reported, the demonstrated efficacy of which confirms the important role of fXa in thrombosis (Table 1). The anti-Xa effects of heparin and its ATIII binding pentasaccharide fragment

Table 1. Factor Xa inhibitors (ATIII independent)

Naturally occurring recombinant	r-Tick anticoagulant Peptide (TAP)	
	r-Antistasin	
	r-NAP-5	
	Ecotin	
	TFPI	
	DEGRXa	
Peptidomimetics	SEL 2711	
Synthetic small molecule	Benzamidine	N^{α}-Naphthalenesulfonyl-3-amidino phenylalanine isopropylester
	Bisamidine	DABE
		DX-9065a(Daiichi)
		YM-60828(Yamanouchi)
	Argininal	(Corvas)
	Piperidinylpyridine	(Zeneca)

have been known for some time (CHOAY et al. 1983). Though low molecular weight heparin seems to be selective for fXa (see Chap. 10, this volume), the inhibitory activity against fXa is similar to unfractionated heparin. Therefore we did not categorize low molecular weight heparin as an fXa inhibitor. Tick anticoagulant peptide (TAP) (WAXMAN et al. 1990) and antistasin (NUTT et al. 1988), derived from the salivary glands of blood sucking animals, are ATIII-independent, peptidyl fXa inhibitors. Recently, synthetic low molecular weight fXa inhibitors have been described, some of which may have oral bioavailability. The utility of fXa inhibitors has been demon-strated in a number of experimental conditions, including models of venous thrombosis, arterial thrombosis and arterial restenosis. In the following section, we have categorized fXa inhibitors into those which are ATIII dependent, naturally occurring, and synthetic small molecule inhibitors.

I. ATIII-Dependent Inhibitors

The antithrombin III binding fragment of heparin ("pentasaccharide," Fig. 1) has been chemically synthesized and evaluated as an inhibitor of fXa (LORMEAU and HERAULT 1995). HOBBELEN et al. (1990) demonstrated the efficacy of pentasaccharide in a venous stasis model of thrombosis in rats which occurred without prolongation of bleeding time. However, they also reported that the effect was not as marked in an arterial-venous shunt model of thrombosis (HOBBELEN et al. 1990). In addition, it did not suppress platelet recruitment to the same extent as fibrinogen accumulation in the Dacron graft AV-shunt model in baboons (CADROY et al. 1993). As CHESEBRO et al. (1992) have pointed out, the amount of thrombin generated in experimental thrombosis depends on the thrombogenic stimuli. The thrombogenic severity of these models would be venous stasis, disseminated intravascular coagulation (DIC), AV shunt and arterial thrombosis. Pentasaccharide should therefore be considered a relatively weak anticoagulant because it is primarily effective in venous stasis models rather than arterial models. Reasons which might account for this include an alteration in ATIII-dependent mechanisms under differing flow conditions (PASCHE et al. 1991), and the observed protection of fXa from neutralization by the ATIII complex when fXa is associated in the prothrombinase complex (TEITEL and ROSENBERG 1983). [Note: The role

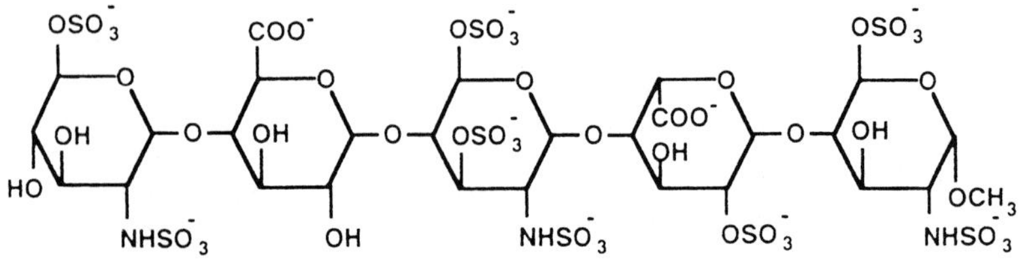

Fig. 1. Structure of pentasaccharide

of low molecular weight heparins (LMWHs) as ATIII-dependent inhibitors of fXa is thoroughly covered in Chap. 10 of this handbook.]

II. Direct Inhibitors

1. Naturally Occurring Inhibitors

The second generation of fXa inhibitors, tick anticoagulant peptide (TAP) and antistasin (ATS), have provided additional evidence showing the importance of fXa in thrombosis (NEEPER et al. 1990; NUTT et al. 1991). TAP, isolated from extracts of the soft tick *Ornithodorus mubata* (WAXMAN et al. 1990), is a 60 amino acid disulfide linked monomeric peptide having specific fXa inhibitory activity. Recombinant TAP (rTAP) (NEEPER et al. 1990) is characterized as a slow, tight-binding competitive inhibitor of human fXa based on its kinetic analysis (JORDAN et al. 1990). TAP is thought to bind to the putative exosite of fXa (JORDAN et al. 1992). Site-directed analysis of the functional domain of rTAP indicated that the N-terminal of rTAP could interact at the active site of fXa and the C-terminal region of rTAP from residues 40 to 54 could bind to a putative exosite of fXa (DUNWIDDIE et al. 1992a; MAO et al. 1993). Interactions between fVa and TAP were also postulated based on the fact that K_i values of rTAP against fXa in the presence of FVa and phospholipid (6 pM) are much lower than in the absence of these cofactors (180–300 pM) (VLASUK 1993).

The antithrombotic efficacy of TAP has been demonstrated in a wide variety of experimental models, including a rabbit model of stasis-induced venous thrombosis (BIEMOND et al. 1996), canine models of coronary and femoral artery thrombosis (LYNCH et al. 1995), and the Dacron graft AV-shunt model in baboons (SCHAFFER et al. 1991). In addition, TAP suppressed neointimal cell thickening in a coronary artery restenosis model (RAGOSTA et al. 1994). It should be noted, however, that JANG et al. did not obtain the same results in restenosis: the latter demonstrated an anti-restenotic effect with inactivated fVIIa and tissue factor pathway inhibitor (TFPI) but not with TAP in a rabbit model (JANG et al. 1995). Nonetheless, it does lend support to the suggestion that inhibitors of the early elements in coagulation may be useful targets for restenosis (KEISER and UPRICHARD 1997).

Antistasin (ATS) is a cysteine-rich polypeptide of 119 amino acid residues, which was discovered from the salivary gland of the Mexican leech (*Haementeria officinalis*) (NUTT et al. 1988; TUSZYNSKI et al. 1987). The peptide is characterized as a competitive, slow and tight-binding inhibitor of fXa with a K_i value of approximately 0.5 nM (DUNWIDDIE et al. 1989). Isoforms of ATS have also been isolated from *Haementeria ghilianii* (CONDRA et al. 1989). ATS interacts with fXa by a two-step mechanism (DUNWIDDIE et al. 1993). The Arg^{34}-Val^{35} bond of rATS (NUTT et al. 1991; HAN et al. 1989) is slowly cleaved by fXa during the inhibition process, but the resulting peptide is still active. Specific removal of the newly formed carboxy-terminal Arg^{34} residue cleaved from rATS eliminated fXa inhibitory activity (DUNWIDDIE et al. 1992c).

In addition, mutation of Arg^{34} of ATS to a nonpolar amino acid leucine abolished the activity (HOFMANN et al. 1992). These results suggest that the Arg^{34}-Val^{35} bond is the critical site for inhibition of fXa by ATS (DUNWIDDIE et al. 1992c; HOFMANN et al. 1992).

Ecotin (MCGRATH et al. 1991b), a proteinase inhibitor derived from *E. coli*, has recently been evaluated as a potent, reversible, tight-binding inhibitor of fXa (SEYMOUR et al. 1994). SHIN et al. (1993) initially carried out X-ray crystallographic analysis of ecotin which was subsequently co-crystallized with mutant trypsin (MCGRATH et al. 1991a) and the inhibition mechanism of the complex was elucidated as a macromolecular chelation (MCGRATH et al. 1994).

Recently a highly potent and specific human fXa inhibitor was isolated from hookworm, *Ancylostoma caninum*, and was named *Ancylostoma caninum* anticoagulant peptide(AcAP) (CAPPELLO et al. 1995). Although the N-terminal of this 8.7-kDa peptide did not show any homology to other serine protease inhibitors or anticoagulant peptides, AcAP suppressed the conversion of prothrombin to thrombin by inhibiting the prothrombinase complex with an IC_{50} of 336 p*M*. AcAP doubled prothrombin time (PT) and activated partial thromboplastin time (aPTT) at concentrations of 35 n*M* and 85 n*M*, respectively. Using molecular cloning, the investigators described AcAP as a family of potent small protein anticoagulants (75–84 amino acids). Two recombinant AcAP members (rAcAP5 and rAcAP6) directly inhibited the catalytic activity of fXa, while a third form (rAcAPc2) is a partial inhibitor of fXa. Recombinant AcAP5 shows much stronger fXa inhibition activity than rAcAP6 (STASSENS et al. 1996). In spite of its incomplete inhibition of the amidolytic activity of fXa, rAcAPc2 completely inhibits the conversion of thrombin from prothrombin. The observation that the prolongation of PT could not be explained on the basis of inhibition of fXa alone led STASSENS et al. (1996) to the conclusion that rAcAPc2, in the presence of fXa, predominantly inhibits the catalytic activity of the tissue factor/VIIa complex. While rAcAP5 and rAcAP6 were supposed to inhibit the active site of fXa directly, it was proposed that rAcAPc2 would bind to fXa at a distinct site away from the catalytic center. Such binding of rAcAPc2 to an fXa exosite would interfere with macromolecular interactions of this enzyme with either the substrate (prothrombin) and/or cofactor (fVa), without complete inhibition of amidolysis of the chromogenic substrate. VLASUK et al. determined the structure of rNAP5 (*N*ematode *A*nticoagulant *P*eptide, an alternative nomenclature for rAcAP5), and the inhibitory profile for free or prothrombinase assembled fXa (VLASUK et al. 1995). The authors also reported that rNAP5 significantly prolonged the time to occlusion in the porcine model of high shear, platelet-dependent arterial thrombosis after intravenous administration (VLASUK et al. 1995).

It has also been reported that the salivary glands of the female yellow fever mosquito *Aedes aegypti* contained an fXa-directed anticoagulant.

The inhibitor is a 35.5-kDa proteinaceous protease inhibitor which exhibits noncovalent inhibition kinetics (STARK and JAMES 1995).

Inactivated fXa, or Xai, is a competitive inhibitor for the assembly of fXa into the prothrombinase complex (SKOGEN et al. 1984). In this context, Xai is not an fXa inhibitor in the strictest sense, but the protein does suppress thrombin generation, albeit indirectly. BENEDICT et al. (1993) demonstrated that an intravenous infusion of bovine Xai into dogs could block electrically induced coronary thrombosis. The dose used, however, increased blood loss after a standardized abdominal incision, an apparent discrepancy between the effects of Xai and direct fXa inhibition. As BENEDICT et al. (1993) pointed out, fXa inhibitors may allow small amounts of activated enzyme to escape inhibition thereby maintaining primary hemostatic responses. It is interesting to note that, in contrast to arterial thrombosis, human Xai did not prolong bleeding time at doses that were effective in a model of venous thrombosis in rabbits (HOLLENBACH et al. 1996). It could be speculated that this finding was due to a difference in dose or to the thrombogenic stimuli, but not to the anticoagulant target as such.

2. Synthetic Small Molecule Inhibitors

Although each of the preceding peptide fXa inhibitors has added to our knowledge of the role of fXa in blood coagulation and provided evidence supporting the benefits of fXa inhibition in vivo, they clearly are not suited to the development of a long-term therapy in man. Not only is bioavailability likely to be a major hurdle, but antigenicity has proven to be more than a theoretical concern (DUNWIDDIE et al. 1992b). Also, despite low K_i values, relatively high concentrations are often required to inhibit coagulation in vivo, an observation seen with TAP, and thought to be due to slow binding kinetics (EISENBERG et al. 1993; DUNWIDDIE et al. 1992b). Accordingly, the search has been on for some time to identify and develop small molecule antagonists of fXa for long-term oral use in man.

a) Peptidomimetics

Scientists at Selectide reported the discovery and optimization of specific, active site inhibitors of fXa using synthetic combinatorial chemistry (SELIGMANN et al. 1995; AL-OBEIDI et al. 1995). The identification of SEL-1691 with a K_i value of 20 μM against fXa prompted a search for a more potent inhibitor. One of the apparent clinical candidates, SEL-2711, with K_i value of 3 nM, seems to have the structure shown in Fig. 2. The most potent agent, SEL-2684, inhibited fXa with a K_i value of 0.3 nM and demonstrated selectivity over thrombin and several other serine proteases. Although claimed to have a half-life longer than 3 h when dosed orally in rats, prolongations of clotting times were of doubtful practical significance and in vivo data with oral dosing to date have not been reported. One would have to

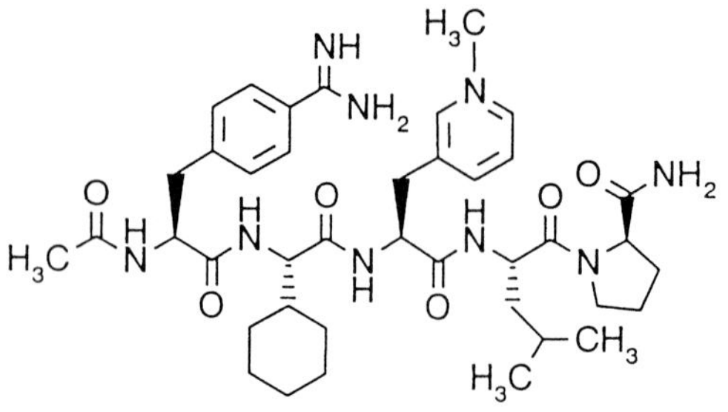

Fig. 2. Structure of SEL-2711

question if SEL 2684 is suitable for oral dosing, given the chemical structure and a molecular weight of 760.

b) Benzamidine Derivatives

Benzamidine type, arginine derivatives and aminopyridine type compounds have been categorized as synthetic low molecular weight fXa inhibitors. Although the binding mode of these inhibitors has been studied using molecular modeling with fXa and/or X-ray crystallographic analysis of inhibitor-trypsin complexes, the details of inhibition mechanisms remain unclear. However, it is still possible to glean significant information from these studies to aid development of fXa inhibitors.

A large hydrophobic group, linked appropriately to the arylamidine functionality, causes enhanced fXa activity. It is speculated that this group binds in the "aryl binding site" formed by the side chains of Trp^{215}, Tyr^{99}, and Phe^{174} in fXa. TULINSKY and coworkers, in a report on the structure of fXa and a model of its complex to DansylGluGlyArgCH_2Cl, have discussed the fit of the hydrophobic dansyl group into this "aryl binding site" (PADMANABHAN et al. 1993).

The first comparative studies on structure-activity relationships of the benzamidine derivatives against fXa and thrombin were reported by STÜRZEBECHER et al. (1976). They showed that 3-amidinophenyl ethers and ketones were more potent competitive inhibitors of fXa than thrombin. Amidinophenyl ether, amidinobenzyl esters and amidinophenyl compounds with a keto group displayed fXa inhibition in the micromolar range. On the other hand, inhibition activities of these compounds against thrombin (IC_{50}) were about 10 times weaker than against fXa. They concluded that fXa inhibition could be enhanced by hydrophobic substituents in a defined steric orientation.

Benzamidine derivatives possessing hydrophobic substituents have also been synthesized by STÜRZEBECHER'S group (1989) N^{α}-Naphthalenesulfonyl-3-amidinophenylalanine isopropylester exhibited a K_i value of 0.24 M for bovine fXa and a K_i value of 13 M for bovine thrombin. These fXa inhibitors suggested that a strong cationic group and a hydro-

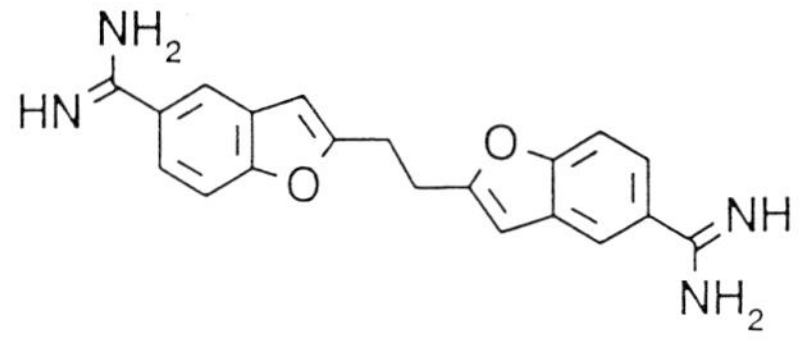

Fig. 3. Structure of 1,2-di(5-amidino-2-benzofuranyl) ethane (DABE)

Table 2. Inhibition of factor Xa[a] and thrombin[b] and prolongation of aPTT[c] by bisbenzamidine derivatives

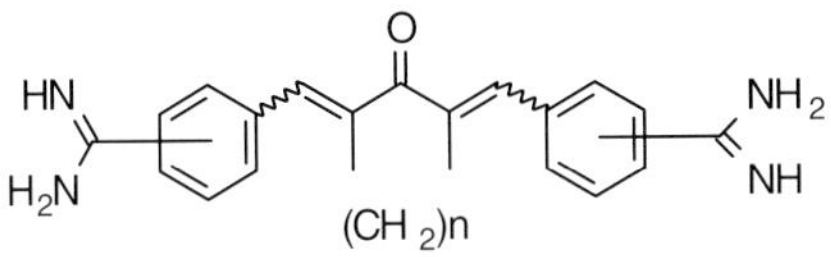

No.	*n*	Position of Amidine	K_i, (μM) Fxa[e]	K_i, (μM) Thrombin[f]	IC_E[d] (μM) aPTT
1	2	(3,3′)	0.49	5.3	25
2	2	(4,4′)	1.9	8.9	56
3	3	(3,3′)	0.32	2.4	10.5
4	3	(4,4′)	0.50	1.2	15.5
5	4	(3,3′)	0.024	8.3	8.0
6	4	(4,4′)	0.013	3.8	0.53

[a] Substrate: Bz-Ile-Glu-Gly-Arg-pNA (S-2222).
[b] Substrate: H-D-Phe-Pip-Arg-pNA (S-2238).
[c] Human plasma.
[d] The concentration needed for 60s prolongation of the clotting time (control 43.8s).
[e] Bovine.
[f] Bovine.

phobic substituent were essential elements for amidino-aryl type fXa inhibitors.

c) Bisamidine Derivatives

Bisamidine type compounds also showed modestly potent fXa inhibition activities. For example, TIDWELL et al. (1978) identified 1,2-di(5-amidino-2-benzofuranyl)ethane (DABE) as an fXa inhibitor, with a K_i value of 0.57 μM against bovine fXa (Fig. 3). They also showed that intravenous infusion of DABE prolonged aPTT in pigs (TIDWELL et al. 1980). Bisbenzamidine containing a central heptanone (BABCH) exhibited more impressive K_i values for bovine fXa (13 nM) while retaining selectivity against bovine thrombin (8.3 μM) (Table 2) (STÜRZEBECHER et al. 1989).

In an attempt to discover an orally active fXa inhibitor, we optimized bisamidine derivatives (NAGAHARA et al. 1994). An amidino-naphthyl deriva-

Table 3. Inhibitory effects of DX-9065a on human trypsin-type protease activity

DX-9065a

	Substrate	K_i (μM)
Factor Xa	S-2222	0.041
Thrombin	S-2238	>2000
Trypsin	S-2222	0.62
Chymotrypsin	S-2856	>2000
Plasmin	S-2251	23
Tissue plasminogen activator	S-2288	21
Plasma kallikrein	S-2302	2.3
Tissue kallikrein	S-2266	1000*

* 50% inhibition concentration.

tive (DX-9065a) was synthesized with fast binding kinetics for fXa, but without any effect on thrombin or other serine proteases (Table 3) (HARA et al. 1994). A docking study of DX-9065a into a modeled structure of fXa indicated that the amidinonaphthalene moiety of the inhibitor bound to the S1 site of fXa with the amidine forming a salt bridge to the carboxylic acid of Asp^{189} (KATAKURA et al. 1993). This model also explained the reduced affinity towards thrombin. The inhibitor has a carboxylic acid which was positioned adjacent to the side chain of Gln^{192} in the model complex. In the thrombin molecule, the corresponding residue is Glu^{192}, where one can expect electrostatic repulsion between the side chain carboxylic acid of Glu^{192} and the carboxylic acid present in the inhibitor, thereby explaining the lower potency against thrombin.

The structure activity relationship (SAR) of DX-9065a was studied by determining the inhibition activities against fXa, thrombin and trypsin in a battery of related compounds (Table 4) (KATAKURA et al. 1995). The SAR showed that the presence of two cationic groups in the molecule resulted in enhanced fXa inhibition, but the second cationic group did not increase trypsin or thrombin inhibition. In place of the second cationic group, a large hydrophobic group also brought about selectivity for fXa. A computationally driven docking of DX-9065a to a model of fXa showed a complex that was consistent with the SAR and provided clues to the causes for this selectivity. The pyrrolidine ring of DX-9065a, which constitutes the hydrophobic group, fits in the "aryl binding site" and the cationic group, imidopyrrolidine, hydrogen bonds to the Glu^{97} carbonyl oxygen.

However, the binding model of DX-9065a and fXa is still controversial. STUBBS et al. (1995) solved crystal structures of DX-9065a and a related

Table 4. Inhibitory activities of DX-9065a related compounds

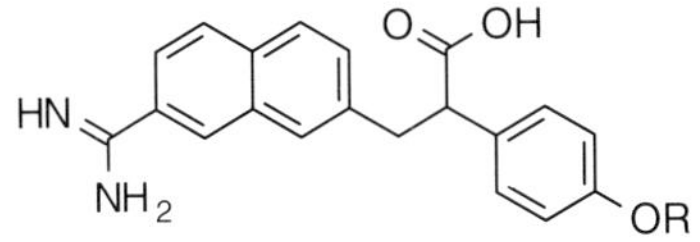

HN NH2 O OH OR

Compound	R	IC50 (μM)		
		Xa	Thrombin	Trypsin
1	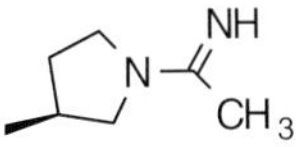 N– NH CH3	0.16	>1000	10
2	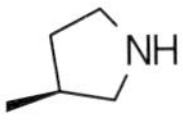 NH	0.20	>1000	5.4
3	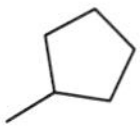	4.2	>1000	4.8
4	 NH2	7.3	>1000	10
5	—CH3	35	>1000	7.5
6	—H	26	>1000	16

compound in complex with bovine-trypsin, and then superimposed the three-dimensional structure of these inhibitors onto fXa. They postulated the existence of three interactions between the inhibitor and fXa, namely, salt bridge formation between the amidino naphthalene group and Asp[189] of fXa, a hydrophobic bond in the aryl binding site with the inhibitor's alkoxylbenzyl moiety and an electrostatic interaction of the distal basic group with an electronegative cavity on the surface of fXa at the conjunction of the carbonyl groups of Glu[97], Thr[98] and Ile[175]. The selectivity against thrombin was attributed to steric collision between the carboxylic acid of the inhibitor and the "insertion loop" of thrombin, rather than repulsion of Glu[192].

A more definitive picture of DX-9065a interactions has come about with two recent publications (Lin and Johnson 1995; Brandstetter et al. 1996). Brandstetter et al. (1996) determined the X-ray structure of the complex of DX-9065a with human des-Gla-fXa. The X-ray structure shows that the naphthamidine binds in S1 without formation of the twin hydrogen bonds normally seen for amidines. The second cationic moiety, the imidine substituted pyrrolidine, fits in a "cation hole" formed by the acid and carbonyl groups of Glu[97] (Brandstetter et al. 1996).

The K_i value of DX-9065a for fXa is 41 nM, which was one order of magnitude weaker than that of TAP. However, the concentration required to double clotting time with DX-9065a was similar to that of TAP (Hara et al. 1995c). Furthermore, since anticoagulation could be demonstrated with oral administration of DX-9065a, the inhibitor may fulfill all the criteria which SIXMA proposed. As is the case with TAP, DX-9065a has antithrombotic efficacy in a wide range of models, including stasis thrombosis (HERBERT et al. 1996), disseminated intravascular coagulation (HARA et al. 1995c; YAMAZAKI et al. 1994), arterial-venous shunt induced thrombosis (MORISHIMA et al. 1993; YOKOYAMA et al. 1995), hypercoagulable state in artificial vascular graft (KIM et al. 1996) or in the hemodialysis apparatus (HARA et al. 1995a) and $FeCl_3$-induced arterial thrombosis (FUKUDA et al. 1996) without bleeding propensity (TANABE et al. 1995; MORISHIMA et al. 1997). The anticoagulant effect of DX-9065a, however, varies across animal species depending on the differences in its inhibitory activity against fXa in each species (HARA et al. 1995b). This remains an area of active research.

The group at Yamanouchi recently synthesized a series of dibasic type fXa inhibitors (SATO et al. 1997). Compounds in Table 5 summarize the concentration required to double the fXa-activated plasma coagulation time (CT2); a thrombin-activated plasma coagulation time is shown to demonstrate selectivity for fXa inhibitors over thrombin. Although the binding mode of these compounds has not yet been reported, these inhibitors may inhibit fXa in a similar manner to the bisamidino-type compounds, such as DABE, BABCH and DX-9065a.

d) *Argininal Derivatives*

Argininal type transition state fXa inhibitors have been reported by the Corvas group. The SAR of these transition state compounds is summarized in Table 6 (RIPKA et al. 1995; BRUNCK et al. 1994). Aromatic groups are accept-

Table 5. Anticoagulant activities of Yamanouchi compounds

R H2N NH NH N N CH3 O

Compound	R	Xa Clotting Time CT2 (μM)	Fibrinogen Thrombin Time CT2 (μM)
1	$—SO_2C_2H_5$	0.05	>100
2	—H	0.09	>100
3	—COPh(2-OMe)	0.14	>100
4	$—SO_2NHCO_2C_2H_5$	0.04	>100
5	$—SO_2Ph(2\text{-}CO_2H)$	0.09	>100
6	$—SO_2NHCH_3$	0.05 (Xa IC_{50} 91 n*M*)	>100
7	$—SO_2N(CO_2C_2H_5)CH_2CO_2H$	0.04	>100
8	$—SO_2NHCH_2CO_2C_2H_5$	0.04 (Xa IC_{50} 47 n*M*)	>100

Table 6. IC50s for transition state inhibitors

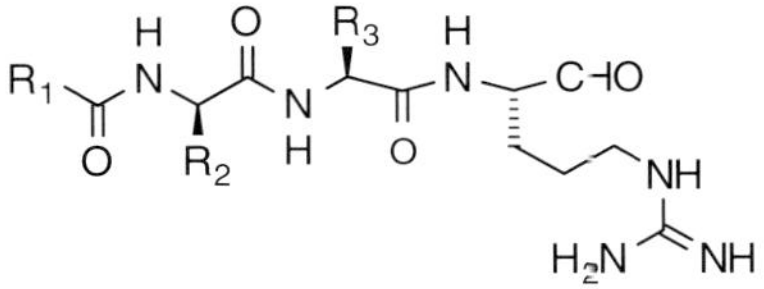

Compound	R_1	R_2	R_3	IC_{50} (μM) Xa	IXa	Thrombin	tPA
1	$—OC(CH_3)_3$	$—CH_2$-2-Np	$—CH_2Ph$	0.22	16	14	>25
2	$—OC(CH_3)_3$	—Ph	$—CH_2Ph$	0.64	13	>25	>25
3	$—OC(CH_3)_3$	$—CH_2Ph$	$—CH_2Ph$	0.21	25	>25	>25
4	$—OC(CH_3)_3$	$—CH_2Ph$	$—CH_2$-2-Np	0.89	25	25	>25
5	$—OC(CH_3)_3$	$—CH_2Ph$	$—CH_2$-1-Np	0.030	14	13	>25
6	$—CH_3$	$—CH_2Ph$	$—CH_2$-1-Np	0.025	0.32	>25	>25
7	$—(CH_2)_2COOH$	$—CH_2Ph$	$—CH_2$-1-Np	0.023	20	>25	>25
8	$R_4 = —B(OH)_2$			0.0097			
9	$R_4 = —CHO$			0.032			
10	$R_4 = —COCONHCH_2CH_2Ph$			0.055			

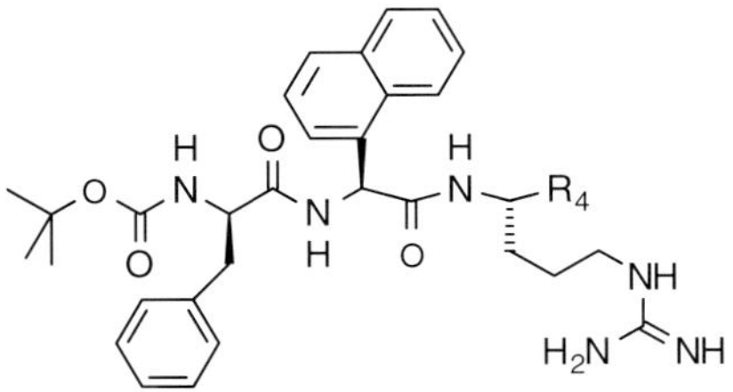

2-Np: 2-naphthyl.
1-Np: 1-naphthyl.

Table 7. IC50s of aminopyridinetype FXa inhibitors

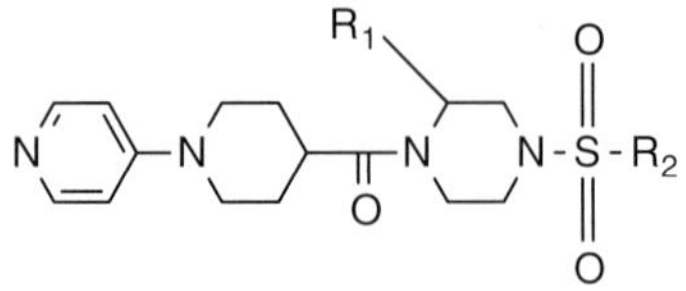

Compound	R_1	R_2	Xa IC_{50} (μM)	Thrombin IC_{50} (μM)
1	—H	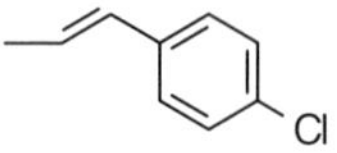	0.012	>100
2	—H	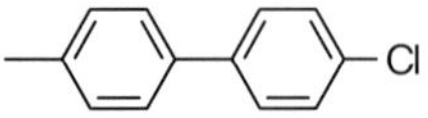	0.01	83
3	—H	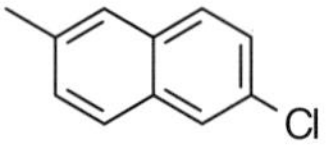	0.003	34
4	—COOEt	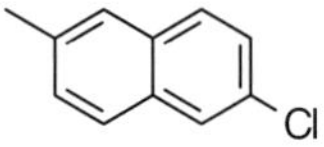	0.002	>10
5	—COOH	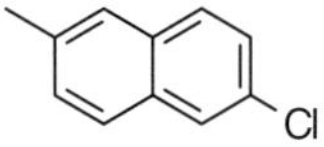	0.008	>10

able to the hydrophobic pocket (S3 site formed by Try[99], Phe[174] and Trp[215]), and a large bulky group, such as a naphthylalanine at the P2 site, showed potent inhibitory activity. Boroarginine type transition state inhibitors exhibit potent fXa inhibition activity with an IC_{50} of 9.7 n*M*, while argininal and ketoamidoarginine derivatives inhibit fXa at three- to fivefold higher concentrations.

e) Piperidinylpyridine Derivatives

Piperidinylpyridine type inhibitors with n*M* IC_{50}s against fXa have been described by researchers at Zeneca (Table 7) (FAULL et al. 1996). Since these are non-amidino, non-guanidino type FXa inhibitors not previously reported, it will be particularly interesting to see what kind of anticoagulant profile they will demonstrate in vivo.

D. ATIII-Independent Inhibition of Factor Xa on Prothrombinase

We have studied the inhibitory profile of DX-9065a against FXa within the prothrombinase complex (ISHIHARA et al. 1996). As is well known, fXa is assembled into prothrombinase as a complex with fVa, phospholipids and Ca^{2+}. Under physiological conditions, this complex can generate thrombin

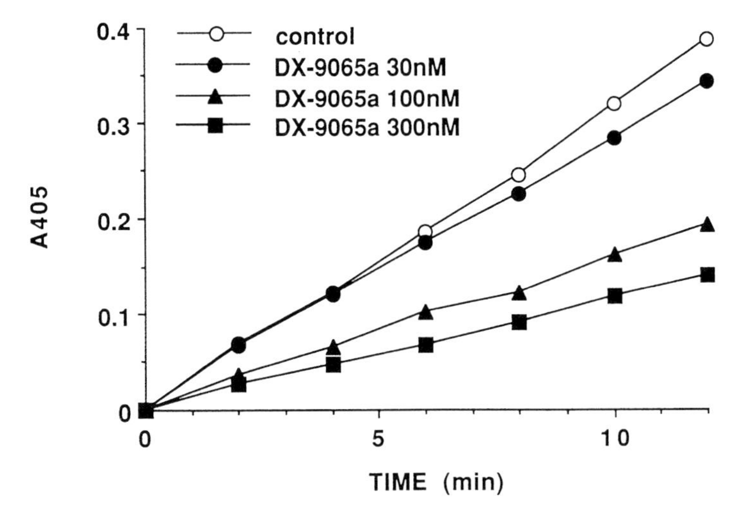

Fig. 4. Inhibition of FXa on prothrombinase complex with DX-9065a. DX-9065a was added (0 nM; 30 nM; 100 nM; 300 nM). The aliquots were collected to measure thrombin generation with amidolytic method

approximately 300 000 times more than fXa alone. Although DX-9065a inhibits fXa selectively in an amidolytic assay system, the determination as to whether DX-9065a would inhibit the fXa on prothrombinase complex was considered essential to characterize fully its antithrombotic potential. It has been suggested that fXa assembled in prothrombinase could be resistant to inhibition by ATIII (TEITEL and ROSENBERG 1983). We speculated that the inhibition profile of a direct fXa inhibitor against prothrombinase would be different and that such an inhibitor could suppress thrombin generation under any conditions.

The prothrombinase complex was reconstituted with phosphatidylcholine/phosphatidylserine containing liposomes, bovine fVa, human fXa and Ca^{2+}. Various concentrations of DX-9065a were preincubated in the system and the first reaction was initiated with the addition of human prothrombin. Every 2 min the aliquots were diluted in EGTA containing TRIS buffer. The thrombin so generated was measured by the hydrolysis of a chromogenic substrate. DX-9065a inhibited thrombin generation in a concentration dependent manner (Fig. 4). The prothrombinase activity was inhibited 50% with 98 n*M* DX-9065a, similar to the IC_{50} value against free fXa in the amidolytic assay system. Thus the inhibitory effect of DX-9065a on fXa activity was retained when fXa was assembled in the prothrombinase complex.

We examined the difference in inhibitory activity between the simultaneous addition of inhibitor during the assembly of prothrombinase and the addition of inhibitor after preincubation with the assembled prothrombinase complex. DX-9065a could inhibit prothrombinase activity equally before or after the assembly of the complex (Fig. 5). This is in contrast to the findings

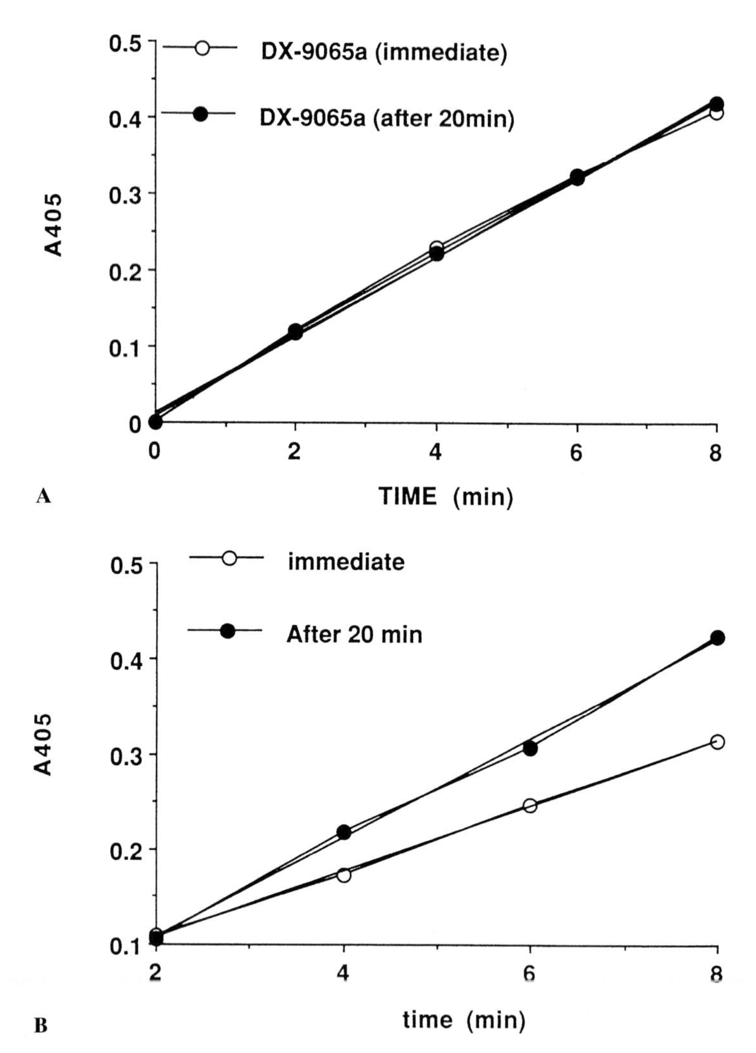

Fig. 5A,B. Resistance of FXa against inhibitor after the assembly of prothrombinase complex. **A** DX-9065a (100 nM) was added simultaneously (○) or 20 min after (●) the assembly of prothrombinase. **B** Heparin and ATIII were added simultaneously (○) or 20 min after (●) the assembly of prothrombinase

with heparin-ATIII, where activity is reduced only when the drug is pre-incubated with the prothrombinase complex.

In an attempt to interrupt prothrombin activation during the reaction, the inhibitors were added either prior to or 4 min after prothrombin activation. fXa activity was measured with the prothrombin fragment, F_{1+2}. Prothrombin activation was completely inhibited in the presence of DX-9065a. As was the case with prior addition, prothrombin activation was also inhibited with the addition of DX-9065a 4 min after the initiation of the reaction, and the veloc-

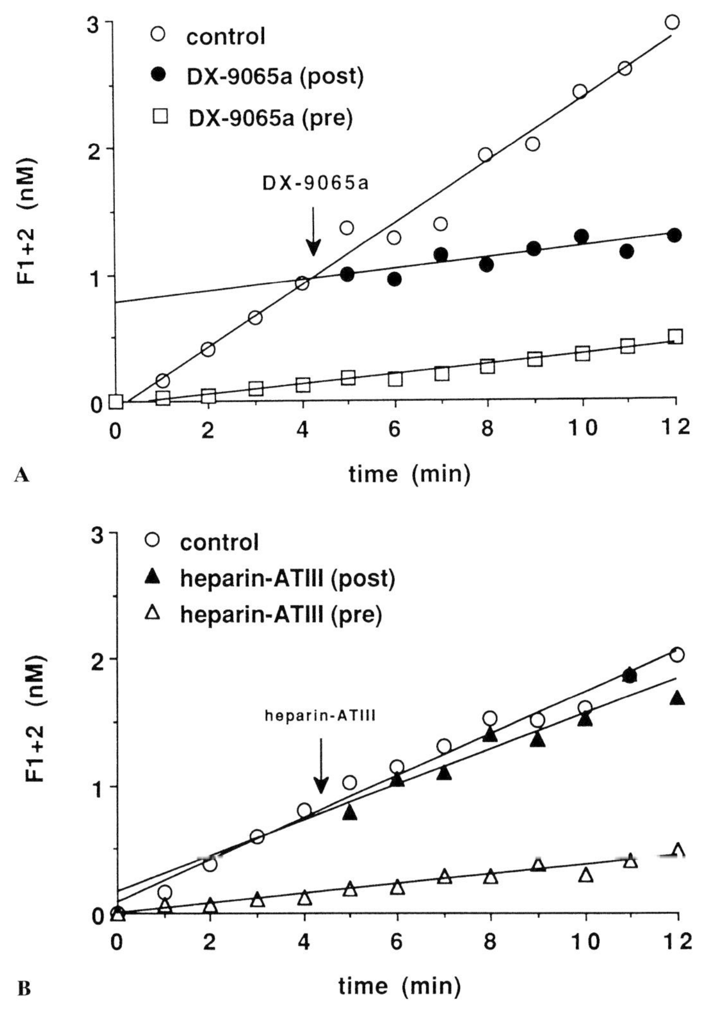

Fig. 6A,B. Interruption of prothrombin activation during the ongoing reaction. **A** DX-9065a (300 n*M*) was added before (□) or 4 min after (●) the addition of prothrombin to the prothrombinase complex control (○). **B** Heparin (3 mU/ml) and ATIII (20 n*M*) were added before or (△) 4 min after (▲) the addition of prothrombin to the assembled prothrombinase complex control (○).

ity of the activation declined to the same extent. When heparin with ATIII was added prior to the prothrombinase activation, thrombin generation was suppressed to the same degree as with DX-9065a. However, heparin and ATIII could not interrupt thrombin generation once the activation was initiated (Fig. 6). Incomplete inactivation of prothrombinase activity was also observed with the addition of low molecular weight heparin or pentasaccharide in the presence of ATIII.

Table 8. Comparison of effective doses (ED_{50}) in thrombosis model of DX-9065a with that of heparin and LMWH

	Stasis[a]	TF-CDIC[b]	AV shunt[c]	Hemodialysis[d]	$FeCl_3$[e]
DX-9065a	1.2 mg/kg	0.09 mg/kg/hr	0.27 mg/kg/hr	0.06 mg/kg + 0.03 mg/kg/hr	3 mg/kg + 6 mg/kg/hr
Heparin	2.9 U/kg	18.1 U/kg/hr	46.4 U/kg/hr	40 U/kg + 20 U/kg/hr	1000 U/kg + 2000 U/kg/hr
LMWH	N.D.	33.3 U/kg/hr	84.6 U/kg/hr	N.D.	>1000 U/kg + 2000 U/kg/hr

[a] Stasis-induced thrombosis after injection of FXa in rats.
[b] Thromboplastin (0.27 mg/hr) was i.v. infused into rats.
[c] Copper wire inserted plastic catheter was connected between the carotide artery and the jugular vein in rats.
[d] Hemodialyser was connected between the carotide artery and the jugular vein in monkeys.
[e] Filter paper soaked with $FeCl_2$ solution was attached on the femoral artery of rats for 15 minutes.

Taken together, these results demonstrate that a direct fXa inhibitor can inhibit fXa in fluid phase, fXa on prothrombinase before or after assembly, and fXa on prothrombinase during the initiation of prothrombin activation. In contrast, ATIII-dependent inhibitors could not inhibit fXa in the prothrombinase complex after the assembly or during activation. It is conceivable that this difference may explain some of the limitations of ATIII dependent anticoagulants.

E. Comparative Antithrombotic Efficacy of Direct Factor Xa Inhibitors

We have summarized the effective doses of DX-9065a and heparin in Table 8. As CHESEBRO et al. (1992) demonstrated using hirudin, the more severe the thrombotic stimulus, the higher the doses of DX-9065a or heparin required to prevent coagulation. For example, to prevent thrombosis in hemodialysis it required 50 times the efficacious dose in the AV shunt model for heparin. In the model of $FeCl_3$-induced arterial thrombosis, heparin could not prevent coagulation. Since this was the case also for LMWH, it suggests that ATIII-dependent anticoagulants may have limited utility in arterial thrombosis models. DX-9065a, however, was efficacious in all the models tested, with only slightly higher doses required for the arterial model. More recently we have evaluated DX-9065a in a model of AV shunt thrombosis in squirrel monkeys (MORISHIMA et al. 1998). At effective doses of 0.01 mg/kg/h i.v. or 1.0 mg/kg orally, the drug did not prolong bleeding time, an encouraging finding in anticipation of clinical development. This, in conjunction with the fast on and off kinetics and inhibition of the FXa on prothrombinase complex at all stages of assembly, places DX-9065a as a promising candidate in the search for the next generation of antithrombotic drugs.

F. Summary and Conclusions

In conclusion, we have reviewed and summarized current information on inhibition of fXa. Particular emphasis was placed on an ATIII-independent, nonpeptide, fast binding fXa inhibitor which we feel offers promise as a new class of antithrombotic agent. Advantages of such an agent could arguably include less bleeding risk, given the apparent dissociation between doses required for efficacy and those associated with prolongation of bleeding time. Apart from the usual antithrombotic/anticoagulant indications for such an agent, the demonstrated inhibition of clot-bound fXa suggests the drug could be useful in preventing the recurrence of thrombus formation after thrombolytic therapy. In addition, the demonstration of activity against fXa even after assembly of the prothrombinase complex may provide an additional benefit by inhibiting pathophysiological thrombin generation in high shear blood flow conditions.

References

Al-Obeidi F, Lebl M, Ostrem J, Safar P, Stierandova A, Strop P et al (1995) Factor Xa inhibitors. WO929189

Bagdy D, Barabás É, Szabó G, Bajusz S, Széll E (1992) In vivo anticoagulant and antiplatelet effect of D-Phe-Pro-Arg-H and D-Me-Phe-Pro-Arg-H. Thromb Haemost 67:357–365

Berndt MC, Gregory C, Dowden G, Castaldi PA (1986) Thrombin interactions with platelet membrane proteins. Ann NY Acad Sci 485:374–386

Benedict CR, Ryan J, Todd J, Kuwabara K, Tijburg P, Cartwright J Jr et al (1993) Active site-blocked factor Xa prevents thrombus formation in the coronary vasculature in parallel with inhibition of extravascular coagulation in a canine thrombosis model. Blood 81:2059–2066

Biemond BJ, Friederich PW, Levi M, Vlasuk GP, Büller HR, ten Cate JW (1996) Comparison of sustained antithrombotic effects of inhibitors of thrombin and factor Xa in experimental thrombosis. Circulation 93:153–160

Brandstetter H, Kuhne A, Bode W, Huber R, von der Saal W, Wirthensohn K et al (1996) X-ray structure of active site-inhibited clotting factor Xa. Implications for drug design and substrate recognition. J Biol Chem 47: 29988–29992

Brunck TK, Webb TR, Ripka WC (1994) Novel inhibitors of factor Xa. WO94113693

Cadroy Y, Hanson SR, Harker LA (1993) Antithrombotic effects of synthetic pentassacharide with high affinity for plasma antithrombin III in non-human primates. Thromb Haemost 70:631–635

Camerer E, Kolstø A-B, Prydz H (1996) Cell biology of tissue factor, the principal initiator of blood coagulation. Thromb Res 81:1–41

Cappello M, Vlasuk GP, Bergum PW, Huang S, Hotez PJ (1995) Ancylostoma caninum anticoagulant peptide: a hookworm-derived inhibitor of human coagulation factor Xa. Proc Natl Acad Sci USA 92:6152–6156

Carreras LO (1980) Thrombosis and thrombocytopenia induced by heparin. Scand J Haematol [Suppl] 25:64–80

Chesebro JH, Webster MW, Zoldhelyi P, Roche PC, Badimon L, Badimon JJ (1992) Antithrombotic therapy and progression of coronary artery disease. Antiplatelet versus antithrombins. Circulation 86:III-100–III-110

Choay J, Petitou M, Lormeau JC, Sinay P, Casu B, Gatti G (1983) Structure-activity relationship in heparin: a synthetic pentasaccharide with high affinity for

antithrombin III and eliciting high anti-factor Xa activity. Biochem Biophys Res Commun 116:492–499
Condra C, Nutt E, Petroski CJ, Simpson E, Freidman PA (1989) Isolation and structural characterization of a potent inhibitor of coagulation factor Xa from the leech Haementeria ghilianii. Thromb Haemost 61:437–441
Cousins GR, Friedrichs GS, Sudo Y, Rote WE, Vlasuk GP, Nolan T et al (1995) Orally effective CVS-1123 prevents coronary artery thrombosis in the conscious canine. Circulation 92:I-303
Dunwiddie C, Thornberry NA, Bull HG, Sardana M, Friedman PA, Jacobs JW et al (1989) Antistasin, a leech-derived inhibitor of factor Xa. Kinetic analysis of enzyme inhibition and identification of the reactive site. J Biol Chem 264: 16694–16699
Dunwiddie CT, Neeper MP, Nutt EM, Waxman L, Smith DE, Hofmann KJ (1992a) Site-directed analysis of the functional domains in the factor Xa inhibitor tick anticoagulant peptide: identification of two distinct regions that constitute the enzyme recognition sites. Biochemistry 31:12126–12131
Dunwiddie CT, Nutt EM, Vlasuk GP, Siegl PK, Schaffer LW (1992b) Anticoagulant efficacy and immunogenicity of the selective factor Xa inhibitor antistasin following subcutaneous administration in the rhesus monkey. Thromb Haemost 67: 371–376
Dunwiddie CT, Vlasuk GP, Nutt EM (1992c) The hydrolysis and resynthesis of a single reactive site peptide bond in recombinant antistasin by coagulation factor Xa. Arch Biochem Biophys 294:647–653
Dunwiddie CT, Waxman L, Vlasuk GP, Friedman PA (1993) Purification and characterization of inhibitors of blood coagulation factor Xa from hematophagous organisms. Methods Enzymol 223:291–312
Eisenberg PR, Siegel JE, Abendschein DR, Miletich JP (1993) Importance of factor Xa in determining the procoagulant activity of whole-blood clots. J Clin Invest 91:1877–1883
Elódi S, Varadi K (1979) Optimization of conditions for the catalytic effect of the factor IXa-factor VIII complex: probable role of the complex in the amplification of blood coagulation. Thromb Res 15:617–629
Eriksson UG, Renberg L, Vedin C, Strimfors M (1995) Pharmacokinetics of inogatran, a new low molecular weight thrombin inhibitor, in rats and dogs. Thromb Haemost 73:1318
Faull AW, Mayo CM, Preston J, Stocker A (1996) Aminoheterocyclic derivatives as antithrombotic or anticoagulant agents. WO96110022
Fukuda T, Morishima Y, Hara T, Kunitada S (1996) Beneficial effect of DX-9065a, a selective factor Xa inhibitor, in a ferric chloride-induced arterial thrombosis model in rats. Jpn J Pharmacol 71:327P
Han JH, Law SW, Keller PM, Kniskern PJ, Silberklang M, Tung JS et al (1989) Cloning and expression of cDNA encoding antistasin, a leech-derived protein having anticoagulant and anti-metastatic properties. Gene 75:47–57
Hara T, Yokoyama A, Ishihara H, Yokoyama Y, Nagahara T, Iwamoto M (1994) DX-9065a, a new synthetic, potent anticoagulant and selective inhibitor for factor Xa. Thromb Haemost 71:314–319
Hara T, Morishima Y, Kunitada S (1995a) Selective factor Xa inhibitor, DX-9065a, suppressed hypercoagulable state during haemodialysis in cynomolgus monkeys. Thromb Haemost 73:1311
Hara T, Yokoyama A, Morishima Y, Kunitada S (1995b) Species differences in anticoagulant and anti-Xa activity of DX-9065a, a highly selective factor Xa inhibitor. Thromb Res 80:99–104
Hara T, Yokoyama A, Tanabe K, Ishihara H, Iwamoto M (1995c) DX-9065a, an orally active, specific inhibitor of factor Xa, prevents thrombosis without affecting bleeding time in rats. Thromb Haemost 74:635–639
Heras M, Chesebro JH, Penny WJ, Bailey KR, Badimon L, Fuster V (1989) Effects of thrombin inhibition on the development of acute platelet-thrombus deposition

during angioplasty in pigs. Heparin versus recombinant hirudin, a specific thrombin inhibitor. Circulation 79:657–665

Herbert JM, Bernat A, Dol BF, Hérault JP, Crépon B, Lormeau JC (1996) DX-9065a, a novel, synthetic, selective and orally active inhibitor of factor Xa: in vitro and in vivo studies. J Pharmacol Exp Ther 276:1030–1038

Higgins DL, Lewis SD, Shafer JA (1983) Steady state kinetic parameters for the thrombin-catalyzed conversion of human fibrinogen to fibrin. J Biol Chem 258:9276–9282

Hobbelen PM, van Dinther TG, Vogel GM, van Boeckel CA, Moelker HC, Meuleman DG (1990) Pharmacological profile of the chemically synthesized antithrombin III binding fragment of heparin (pentasaccharide) in rats. Thromb Haemost 63: 265–270

Hofmann KJ, Nutt EM, Dunwiddie CT (1992) Site-directed mutagenesis of the leech-derived factor Xa inhibitor antistasin. Probing of the reactive site. Biochem J 287:943–949

Hollenbach SJ, Wong AG, Ku P, Needham KM, Lin P-H, Sinha U (1996) Efficacy of "fXa inhibitors" in a rabbit model of venous thrombosis. Circulation 92:I-486–I-487

Ishihara H, Hara T, Kunitada S (1996) Antithrombin III-independent mechanism of factor Xa inhibition on prothrombinase with DX-9065a. Haemostasis 26 [Suppl 3]:580

Jackson CV, Crowe VG, Frank JD, Wilson HC, Coffman WJ, Utterback BG et al (1992) Pharmacological assessment of antithrombotic activity of the peptide thrombin inhibitor, D-methyl-phenylalanyl-prolyl-arginal (GYKI-14766), in a canine model of coronary artery thrombosis. J Pharmacol Exp Ther 261:546–552

Jang Y, Guzman LA, Lincoff M, Gottsauner-Wolf M, Forudi F, Hart CE et al (1995) Influence of blockade at specific levels of the coagulation cascade on restenosis in a rabbit atherosclerotic femoral artery injury model. Circulation 92:3041–3050

Jordan SP, Waxman L, Smith DE, Vlasuk GP (1990) Tick anticoagulant peptide: kinetic analysis of the recombinant inhibitor with blood coagulation factor Xa. Biochemistry 29:11095–11100

Jordan SP, Mao SS, Lewis D, Shafer JA (1992) Reaction pathway for inhibition of blood coagulation factor Xa by tick anticoagulant peptide. Biochemistry 31: 5374–5380

Katakura S, Nagahara T, Hara T, Iwamoto M (1993) A novel factor Xa inhibitor: structure-activity relationships and selectivity between factor Xa and thrombin. Biochem Biophys Res Commun 197:965–972

Katakura S, Nagahara T, Hara T, Kunitada S, Iwamoto M (1995) Molecular model of an interaction between factor Xa and DX-9065a, a novel factor Xa inhibitor: contribution of the acetimidoylpyrrolidine moiety of the inhibitor to potency and selectivity for serine proteases. Eur J Med Chem 30:387–394

Keiser JA, Uprichard ACG (1997) Restenosis: is there a pharmacologic fix in the pipeline? Adv Pharmacol 39:313–351

Kelly AB, Maraganore JM, Bourdon P, Hanson SR, Harker LA (1992) Antithrombotic effects of synthetic peptides targeting various functional domains of thrombin. Proc Natl Acad Sci USA 89:6040–6044

Kim DI, Kambayashi J, Shibuya T, Sakon M, Kawasaki T (1996) In vivo evaluation of DX-9065a, a synthetic factor Xa inhibitor, in experimental vein graft. J Atheroscler Thromb 2:110–116

Kumada T, Abiko Y (1981) Comparative study of heparin and a synthetic thrombin inhibitor no. 805 (MD-805) in experimental antithrombin III-deficient animals. Thromb Res 24:285–298

Lin Z, Johnson ME (1995) Proposed cation-mediated binding by factor Xa: a novel enzymatic mechanism for molecular recognition. FEBS Lett 370:1–5

Lormeau JC, Herault JP (1995) The effect of the synthetic pentasaccharide SR90107/ORG31540 on thrombin generation ex vivo is uniquely due to ATIII-mediated neutralization of factor Xa. Thromb Haemost 74:1474–1477

Lynch JJ Jr, Sitko GR, Lehman ED, Vlasuk GP (1995) Primary prevention of coronary arterial thrombosis with the factor Xa inhibitor rTAP in a canine electrolytic injury model. Thromb Haemost 74:640–645
Mao S-S, Huang J, Neeper MP, Shafer JA (1993) Construction of rTAP variants with improved inhibitory potency toward human factor Xa. Thromb Haemost 69:1046
Mätzsch T, Bergqvist D, Hedner U, Nilsson B, Østergaard P(1986) Heparin-induced osteoporosis in rats. Thromb Haemost 56:293–294
McGrath ME, Erpel T, Browner MF, Fletterick RJ (1991a) Expression of the protease inhibitor ecotin and its co-crystallization with trypsin. J Mol Biol 222: 139–142
McGrath ME, Hines WM, Sakanari JA, Fletterick RJ, Craik CS (1991b) The sequence and reactive site of ecotin. A general inhibitor of pancreatic serine proteases from Escherichia coli. J Biol Chem 266:6620–6625
McGrath ME, Erpel T, Bystroff C, Fletterick RJ (1994) Macromolecular chelation as an improved mechanism of protease inhibition: structure of the ecotin–trypsin complex. EMBO J 13:1502–1507
Morishima Y, Tanabe K, Hara T, Kunitada S (1993) Antithrombotic efficacy of a novel selective factor Xa inhibitor, DX-9065a: lack of bleeding-inducing capacity. Circulation 88:I-265
Morishima Y, Tanabe K, Terada Y, Hara T, Kunitada S (1997) Antithrombotic and hemorrhagic effects of DX-9075a, a direct and selective factor Xa inhibitor: comparison with a direct thrombin inhibitor and antithrombin II-dependent anticoagulants. Thromb Haemost 78:1366–1377
Morishima Y, Hara T, Kunitada S (1998) Selective inhibition of activated factor X by orally active and direct factor Xa inhibitor, DX-9065a, in nonhuman primates (manuscript in preparation)
Nagahara T, Yokoyama Y, Inamura K, Katakura S, Komoriya S, Yamaguchi H et al (1994) Dibasic (amidinoaryl)propanoic acid derivatives as novel blood coagulation factor Xa inhibitors. J Med Chem 37:1200–1207
Neeper MP, Waxman L, Smith DE, Schulman CA, Sardana M, Ellis RW et al (1990) Characterization of recombinant tick anticoagulant peptide. A highly selective inhibitor of blood coagulation factor Xa. J Biol Chem 265: 17746–17752
Nutt EM, Gasic T, Rodkey J, Gasic GJ, Jacobs JW, Friedman PA et al (1988) The amino acid sequence of antistasin. A potent inhibitor of factor Xa reveals a repeated internal structure. J Biol Chem 263:10162–10167
Nutt EM, Jain D, Lenny AB, Schaffer L, Siegl PK, Dunwiddie CT (1991) Purification and characterization of recombinant antistasin: a leech-derived inhibitor of coagulation factor Xa. Arch Biochem Biophys 285:37–44
Padmanabhan K, Padmanabhan KP, Tulinsky A, Park CH, Bode W, Huber R et al (1993) Structure of human des(1–45) factor Xa at 2.2 Å resolution. J Mol Biol 232:947–966
Pasche B, Elgue G, Olsson P, Riesenfeld J, Rasmuson A (1991) Binding of antithrombin to immobilized heparin under varying flow conditions. Artif Organs 15:481–491
Prager NA, Abendschein DR, McKenzie CR, Eisenberg PR (1995) Role of thrombin compared with factor Xa in the procoagulant activity of whole blood clots. Circulation 92:962–967
Ragosta M, Gimple LW, Gertz SD, Dunwiddie CT, Vlasuk GP, Haber HL et al (1994) Specific factor Xa inhibition reduces restenosis after balloon angioplasty of atherosclerotic femoral arteries in rabbits. Circulation 89:1262–1271
Ripka W, Brunck T, Stassens P, LaRoche Y, Lauwereys M, Lambeir AM et al (1995) Strategies in the design of inhibitors of serine proteases of the coagulation cascade–Factor Xa. Eur J Med Chem 30:87s–100s
Sato K, Kawasaki I, Taniuchi Y, Hirayama F, Koshio H, Matsumoto Y (1997) YM-60828, a novel factor Xa inhibitor: separation of its antithrombotic effects from its prolongation of bleeding time. Eur J Pharmacol 339:141–146

Schaffer LW, Davidson JT, Vlasuk GP, Siegl PK (1991) Antithrombotic efficacy of recombinant tick anticoagulant peptide. A potent inhibitor of coagulation factor Xa in a primate model of arterial thrombosis. Circulation 84:1741–1748
Seligmann B, Abdul-Latif F, Al-Obeidi F, Flegelova Z, Issakova O, Kocis P et al (1995) The construction and use of peptide and non-peptidic combinatorial libraries to discover enzyme inhibitors. Eur J Med Chem 30:319s–335s
Seymour JL, Lindquist RN, Dennis MS, Moffat B, Yansura D, Reilly D et al (1994) Ecotin is a potent anticoagulant and reversible tight-binding inhibitor of factor Xa. Biochemistry 33:3949–3958
Shin DH, Hwang KY, Kim KK, Lee HR, Lee CS, Chung CH et al (1993) Crystallization and preliminary X-ray crystallographic analysis of the protease inhibitor ecotin. J Mol Biol 229:1157–1158
Sitko GR, Ramjit DR, Stabilito II, Lehman D, Lynch JJ, Vlasuk GP (1992) Conjunctive enhancement of enzymatic thrombolysis and prevention of thrombotic reocclusion with the selective factor Xa inhibitor, tick anticoagulant peptide. Comparison to hirudin and heparin in a canine model of acute coronary artery thrombosis. Circulation 85:805–815
Sixma JJ, de Groot PTG (1992) The ideal anti-thrombotic drug. Thromb Res 67: 305–311
Skogen WF, Esmon CT, Cox AC (1984) Comparison of coagulation factor Xa and des-(1–44) factor Xa in the assembly of prothrombinase. J Biol Chem 259:2306–2301
Stassens P, Bergum PW, Gansemans Y, Jespers L, Laroche Y, Huang S et al (1996) Anticoagulant repertoire of the hookworm Ancylostoma caninum. Proc Natl Acad Sci USA 93:2149–2154
Stark KR, James AA (1995) A factor Xa-directed anticoagulant from the salivary glands of the yellow fever mosquito Aedes aegypti. Exp Parasitol 81:321–331
Stubbs MT, Huber R, Bode W (1995) Crystal structures of factor Xa specific inhibitors in complex with trypsin: structural grounds for inhibition of factor Xa and selectivity against thrombin. FEBS Lett 375:103–107
Stürzebecher J, Markwardt F, Walsmann P (1976) Synthetic inhibitors of serine proteinases XIV. Inhibition of factor Xa by derivatives of benzamidine. Thromb Res 9:637–646
Stürzebecher J, Stürzebecher U, Vieweg H, Wagner G, Hauptmann J, Markwardt F (1989) Synthetic inhibitors of bovine factor Xa and thrombin: comparison of their anticoagulant efficiency. Thromb Res 54:245–252
Tanabe K, Honda Y, Kunitada S (1995) An orally active, specific inhibitor of factor Xa does not facilitate haemorrhage at the effective doses in rat thrombosis model. Thromb Haemost 73:1312
Teitel JM, Rosenberg RD (1983) Protection of factor Xa from neutralization by the heparin-antithrombin complex. J Clin Invest 71:1383–1391
Tidwell RR, Geratz JD, Dann O, Volz G, Zeh D, Loewe H (1978) Diarylamidine derivatives with one or both of the aryl moieties consisting of an indole or indole-like ring. Inhibitors of arginine-specific esteroproteases. J Med Chem 21: 613–623
Tidwell RR, Webster WP, Shaver SR, Geratz JD (1980) Strategies for anticoagulation with synthetic protease inhibitors. Xa inhibitors versus thrombin inhibitors. Thromb Res 19:339–349
Tuszynski GP, Gasic TB, Gasic GJ (1987) Isolation and characterization of antistasin. An inhibitor of metastasis and coagulation. J Biol Chem 262:9718–9723
Verstraete M, Wessler S (1992) Drug interference with heparin and oral anticoagulants. In: Fuster V, Verstraete M (eds) Thrombosis in cardiovascular disorders. Saunders, Philadelphia, p 141
Vlasuk GP (1993) Structure and functional characterization of tick anticoagulant peptide (TAP): a potent and selective inhibitor of blood coagulation factor Xa. Thromb Haemost 70:212–216
Vlasuk GP, Dempsey EM, Oldeschulte GL, Bernardino VT, Richard BM, Rote WE (1995) Evaluation of a novel small protein inhibitor of blood coagulation factor Xa (rNAP-5) in animal models of thrombosis. Circulation 92:I-685

Waxman L, Smith DE, Arcuri KE, Vlasuk GP (1990) Tick anticoagulant peptide (TAP) is a novel inhibitor of blood coagulation factor Xa. Science 248:593–596

Yamazaki M, Asakura H, Aoshima K, Saito M, Jyokaji H, Uotani C et al (1994) Protective effects of DX-9065a, an orally active, newly synthesized and specific inhibitor of factor Xa, against experimental disseminated intravascular coagulation in rats. Thromb Haemost 72:392–396

Yokoyama T, Kelly AB, Marzec UM, Hanson SR, Kunitada S, Harker LA (1995) Antithrombotic effects of orally active synthetic antagonist of activated factor X in nonhuman primates. Circulation 92:485–491

CHAPTER 15

Inhibitors of Tissue Factor/Factor VIIa

K.P. GALLAGHER, T.E. MERTZ, L. CHI, J.R. RUBIN, and A.C.G. UPRICHARD

A. Introduction

Tissue factor (TF) is the prime mover in the coagulation cascade. Unlike other factors in the cascade which circulate in the plasma, TF is anchored in the membrane of cells. It has a large extracellular domain to which FVII binds, a single transmembrane domain and a small intracellular domain. In contrast to other elements of the coagulation cascade, no cleavage of TF occurs to activate enzymatic activity. Rather TF provides a scaffold for FVII that also activates and/or amplifies the actions of this factor initiating the extrinsic pathway of coagulation. The TF/VIIa complex activates FX to FXa which, as part of the prothrombinase complex, activates prothrombin to thrombin. The amounts of factor Xa and thrombin generated by the extrinsic pathway are relatively small but the thrombin so produced recruits the coagulation factors in the intrinsic pathway which dramatically amplify formation of additional thrombin and, thereby, fibrin. Although the TF/VIIa complex is not directly responsible for the production of all of the thrombin needed to form a complete thrombus, it plays the crucial role of initiating the process.

The importance of TF as the "scaffold" for FVII/FVIIa is well appreciated. Much less is known about the nonhemostatic functions of TF in physiology and pathophysiology. Binding of FVII to TF on cells transfected to overexpress TF can evoke calcium transients, an effect which may involve the intracellular domain, suggesting that TF is a true receptor. Thus far the signaling pathway and the cellular responses to TF receptor activation have not been defined, so they represent particularly intriguing aspects of TF biology. Clues may be available, however, from members of the cytokine superfamily to which TF belongs. For example, TF acts like a potent chemotactic agent for vascular smooth muscle cells (SATO et al. 1996) which may contribute directly to development of atherosclerosis or intimal proliferation. TF can also initiate the local production of fibrin in developing atherosclerotic plaques (SMITH et al. 1990; BINI et al. 1989; DE BUYZERE et al. 1993; LASSILA et al. 1993) thereby indirectly promoting atherosclerosis. This is an effect that could be compounded by recurrent episodes of clinically silent plaque rupture followed by nonocclusive thrombus formation in coronary and other arterial vessels. Admittedly, this is a conjectural possibility, but it does emphasize the potential importance of TF in conditions above and beyond those usually ascribed to simple thrombus formation.

In this review, the evidence supporting the rationale for targeting TF/VIIa therapeutically will be summarized. Most of the evidence supporting the idea that inhibition of TF/VIIa is beneficial has been derived using antibodies and polypeptide inhibitors. This has provided valuable information on potential efficacy largely in acute circumstances, but to explore the effects of inhibitors on progression of atherosclerosis or long term treatment of intravascular thrombotic problems would be greatly facilitated by development of potent and specific compounds that can be administered chronically. There are no low molecular weight (much less orally bioavailable) inhibitors of TF or the TF/VIIa complex commercially available as yet.

B. Role in Hemostasis

Coagulation begins when FVII in the blood binds to TF on the surface of subendothelial cells (see Chap. 1). Although classically identified as the starting point in the "extrinsic" pathway of coagulation, it is now widely accepted that formation of the TF/VIIa complex is the principal means of starting the cascade that leads to fibrin formation under most conditions. The "intrinsic" pathway, driven largely by thrombin, represents an amplification loop that accelerates and augments additional thrombin formation.

Normally, TF is isolated from the blood by endothelial cells. Tissue factor, however, is localized to sites close to endothelial cells which puts it in an ideal position to initiate coagulation when and if it is needed. In effect, the vasculature is "surrounded" by TF but, under normal circumstances, the TF remains separated from the blood by the endothelium. Tissue factor has been identified on the vascular adventitia and media, the connective tissue surrounding most organs, the mucosa, the epithelium of the skin (DRAKE et al. 1989b; FLECK et al. 1990) and the astrocytes in the brain (FLOSSEL et al. 1994; DEL ZOPPO et al. 1992). It is not normally expressed by endothelial cells (WILCOX et al. 1989; DRAKE et al. 1989a) but when the integrity of the endothelium is disrupted by trauma, the underlying subendothelial structures are exposed to blood (DRAKE et al. 1989b) and FVII circulating in the plasma. Factor VII binds to TF and is rapidly activated to a two chain serine protease, FVIIa (BUTENAS and MANN 1996), which amplifies the activity of the enzyme approximately 100-fold. Factor VIIa then catalyzes the next steps in the coagulation cascade, activation of FXa and FIXa. The activation of FXa, it is widely accepted, is the most important in terms of generating thrombin which, by activating factors in the "intrinsic" pathway, explosively amplifies the formation of additional thrombin and ultimately the fibrin clot.

Factor VIIa remains bound to TF, keeping it close to the phospholipid bilayer when activation of FXa and FIXa occurs. This is important because optimal alignment of the substrates, FX and FIX, is facilitated by proximity to a phospholipid surface (NEMERSON 1988). Likewise, evidence indicates that maintaining a high level of FVIIa activity depends on proximity to anionic phospholipids although the precise nature of these protein-phospholipid interactions remains to be determined (Rapaport and Rao 1995; Morrissey et al. 1997).

How FVII is "activated" when it binds to TF is not entirely clear. With the development of precision assays, it has been reported that approximately 1% of FVII in the plasma is in the form of FVIIa leading to the contention that it is this FVIIa which is primarily responsible for initiating coagulation (Morrissey et al. 1993). This is an appealing idea, consistent with the notion that the hemostatic system is always activated at a low level (Bauer 1997), but it does not explain where the FVIIa originates or how it is activated. The potential importance of elevated FVIIa levels in predisposing individuals to acute coronary syndromes has also been proposed but the clinical evidence, so far, is somewhat contradictory (Bauer 1997). Consequently the significance, pathophysiologically as well as physiologically, of circulating FVIIa remains controversial.

C. Structural Biology of Tissue Factor and Factor VIIa

Tissue factor is anchored in the plasma membrane of fibroblasts, smooth muscle cells, and macrophages in blood vessel walls. In contrast to most other components of the coagulation cascade, it is not circulating in the plasma as a zymogen and it does not require proteolytic cleavage before it can be activated. Human TF is composed of a single 263 amino acid polypeptide chain, the first 219 of which comprise the soluble extracellular region (sTF) to which FVII binds. TF also has a single membrane-spanning region (residues 220–242) and a small cytoplasmic domain (residues 243–263) (Morrissey et al. 1987; Scarpati et al. 1987; Spicer et al. 1987). Various aspects of the crystal structure of sTF have been solved (Harlos et al. 1994; Muller et al. 1994; Ashton et al. 1995; Muller et al. 1996; Banner et al. 1996).

For example, in the report by Banner et al. (1996), crystals were obtained representing a complex of human recombinant FVIIa inhibited with the irreversible peptide chloromethyl ketone inhibitor, D-Phe-L-Phe-L-Arg-chloromethylketone (Banner et al. 1996). The inhibited FVIIa was complexed with human recombinant sTF which had been subjected to limited proteolysis by subtilisin, a protease treatment that removes a pentapeptide fragment (residues 85–89) from the sTF structure. The ternary complex was crystallized by the hanging drop vapor diffusion method.

The crystals so obtained were orthorhombic (space group P212121) with unit cell dimensions a = 70.65 Å, b = 82.55 Å and c = 126.5 Å. X-ray diffraction data to 1.95 Å resolution was collected using an image plate X-ray detector at the Daresbury Synchrotron X-ray Source. The structure was solved by molecular replacement methods using a thrombin-based model of the catalytic domain of FVIIa and the published structure of sTF (Kirchofer et al. 1995). The remainder of the structure, including the Gla domain and EGF-1 and

EGF-2 domains were determined from difference electron density maps. In addition, nine calcium ions and 198 water molecules were subsequently located. The entire structure was refined to a crystallographic R-factor of 23.2% using X-ray data to 2.0 Å resolution.

As shown in Fig. 1, the sTF/inhibited FVIIa ternary complex forms an elongated structure, approximately 115 Å in length and 45 Å in thickness. The FVIIa molecule is held in a rigid linear structure by stabilizing interactions with the rigid sTF molecule. Soluble TF makes extensive contact (1800 $Å^2$ buried surface area) with all four of the FVIIa domains, thereby maintaining the otherwise flexible FVIIa molecule in a rigid extended conformation (Fig. 2). The FVIIa structure itself is elongated and consists of four clearly defined structural domains (Gla, EGF-1, EGF-2, and the catalytic domain) connected by flexible linking peptide segments. The membrane anchoring Gla domain is at one end of the structure and the catalytic domain is at the opposite end. There is very little contact between the individual domains of FVIIa with the exception of extensive surface contacts and a disulfide linkage between EGF-2 and the catalytic domain. As a general rule, the EGF-like domains are involved in protein-protein contacts. In the TF/FVIIa structure, EGF-2 binds to the catalytic domain, whereas EGF-1 binds to TF.

The gamma carboxyglutamic acid domain (Gla) domain of FVIIa is so named because it contains 17 modified glutamic acid residues which are involved in chelating 7 tightly bound Ca^{2+} ions. The Ca^{2+} ions are responsible for maintaining the conformation of the Gla domain as well as for the attachment of FVIIa to the cell membrane. The Gla domain is linked to the EGF-1 domain through an amphipathic alpha-helical segment which forms part of a modified "E-F Hand" calcium ion binding structural motif.

The first EGF-like domain (EGF-1) of FVIIa contains the second half of the modified "E-F Hand" calcium ion binding site. This site is comprised of a calcium ion binding loop which links the C-terminal alpha helix of the Gla domain to the remainder of the EGF-1 domain. The "E-F Hand"-like structure linking the Gla and EGF-1 domains of FVIIa represents what could be called a helix-loop-EGF domain in which the binding of a single calcium ion is responsible for inducing an active molecular conformation. The remainder of the EGF-1 structure contains the three disulfide bonds which are characteristic of all EGF-like domains. One surface of the EGF-1 domain is very hydrophobic and is involved in extensive contacts with TF (Fig. 1). These contacts are primarily hydrophobic and are responsible for the majority of the binding energy and selectivity in the TF/FVIIa interaction. The EGF-2 domain of FVIIa is involved in extensive interaction and very tight contact with the catalytic domain of FVIIa (Fig. 2).

The catalytic domain of FVIIa (Fig. 3) has a fold which is very similar to that of trypsin. Within the fold is the catalytic triad characteristic of all serine proteases (Asp-102, His-57 and Ser-195). In addition, the recognition site residue, Asp-189, is present in the P1 substrate recognition pocket of the enzyme. This is responsible for the specificity of FVIIa for cleavage adjacent to

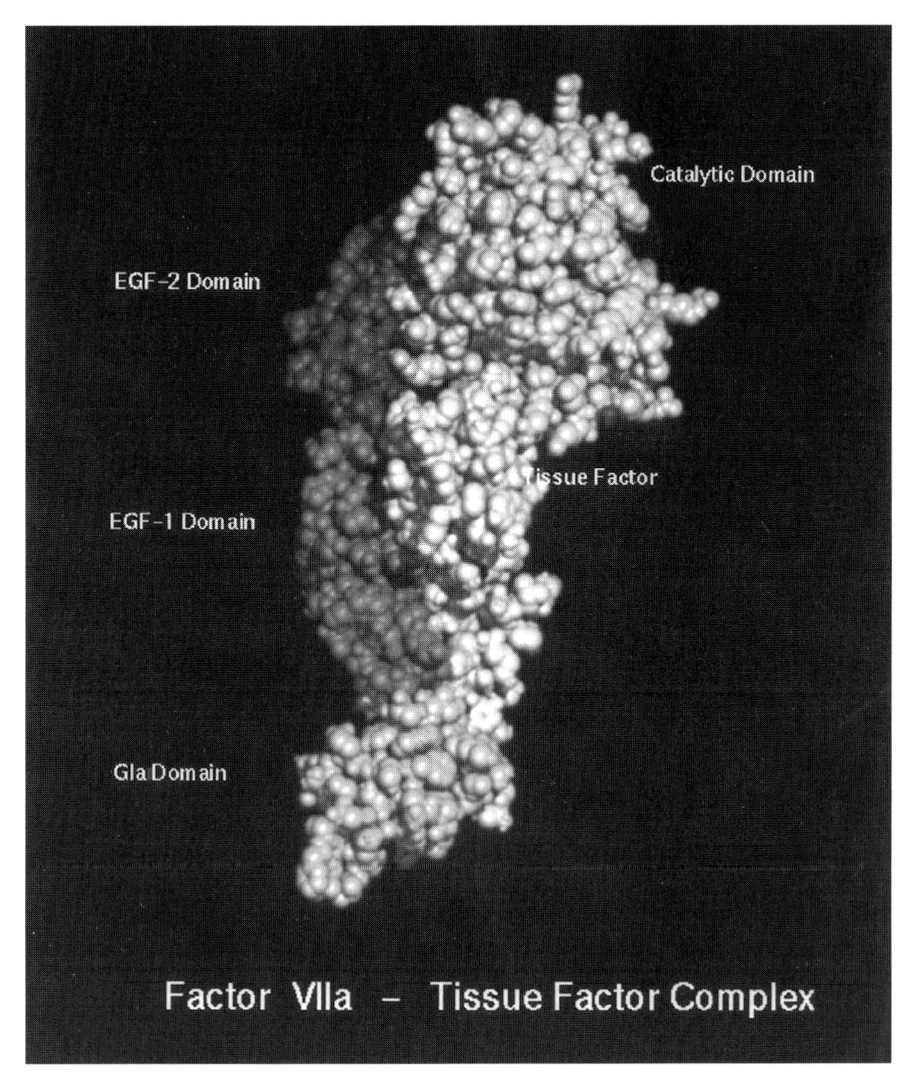

Fig. 1. A space filling model of the FVIIa/sTF complex structure based on the findings of BANNER et al. (1996), showing the principal domains of the complex

arginine residues in peptide substrates. The active site in the structure by BANNER et al. (1996) contains the bound peptide inhibitor, D-Phe-L-Phe-L-Arg-chloromethylketone which is covalently attached to the catalytic residues, His-57 and Ser-195. The arginine side chain of the inhibitor is bound in the P1 recognition pocket of the enzyme and the D-Phe side chain is bound in the hydrophobic P3 binding pocket of the catalytic domain (Fig. 3). The FVIIa catalytic domain also contains the same calcium binding loop that is found in trypsin. This explains the absolute requirement of calcium ions for catalytic

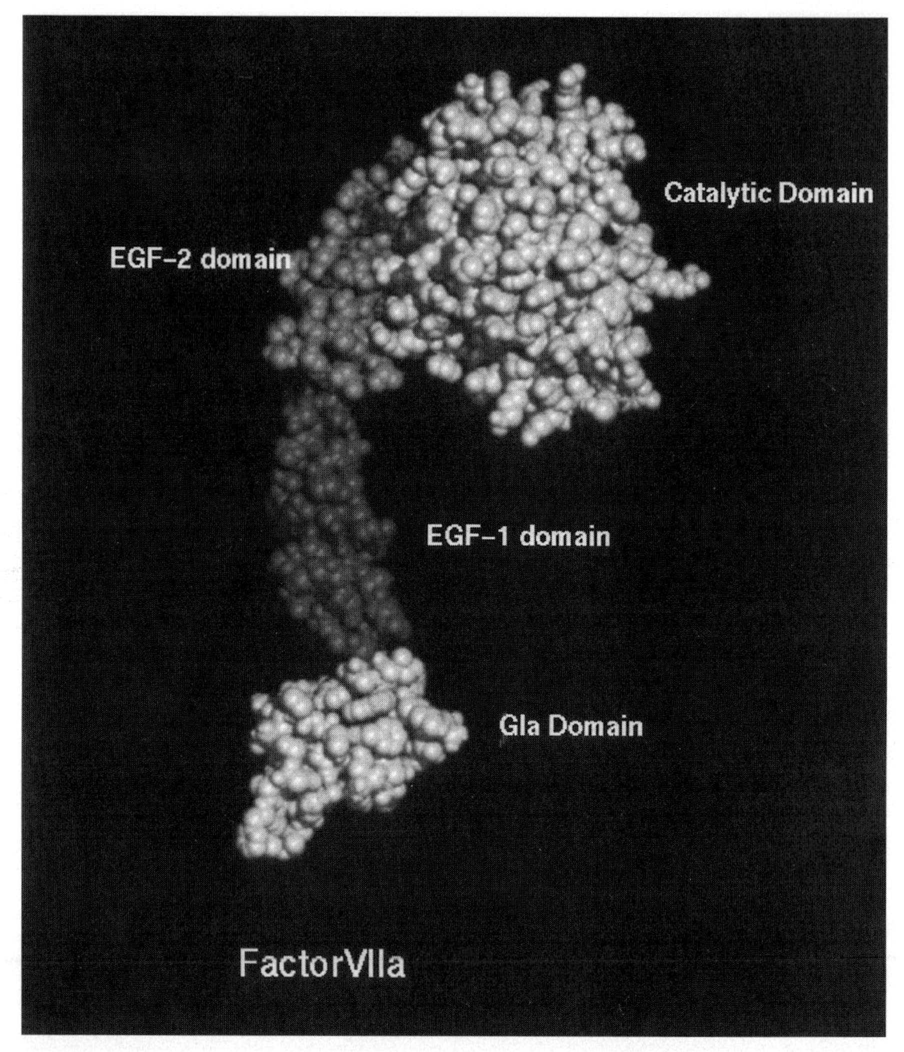

Fig. 2. A space filling model of the FVIIa, a structure with the sTF component deleted

activity of FVIIa. The major differences between the FVIIa catalytic domain and the catalytic domains of other serine proteases occurs in the EGF-2 binding loop which forms specific contacts, as well as a disulfide linkage, with the EGF-2 domain of the FVIIa light chain.

There are three critical regions on FVIIa which contain calcium binding sites and divalent calcium ion binding to all three regions is required for FVIIa activity. The highest affinity calcium binding site is located in the catalytic domain (Fig. 3). The second site is located in the modified "E-F hand" calcium binding motif at the junction of the Gla and EGF-1 domains. Binding of

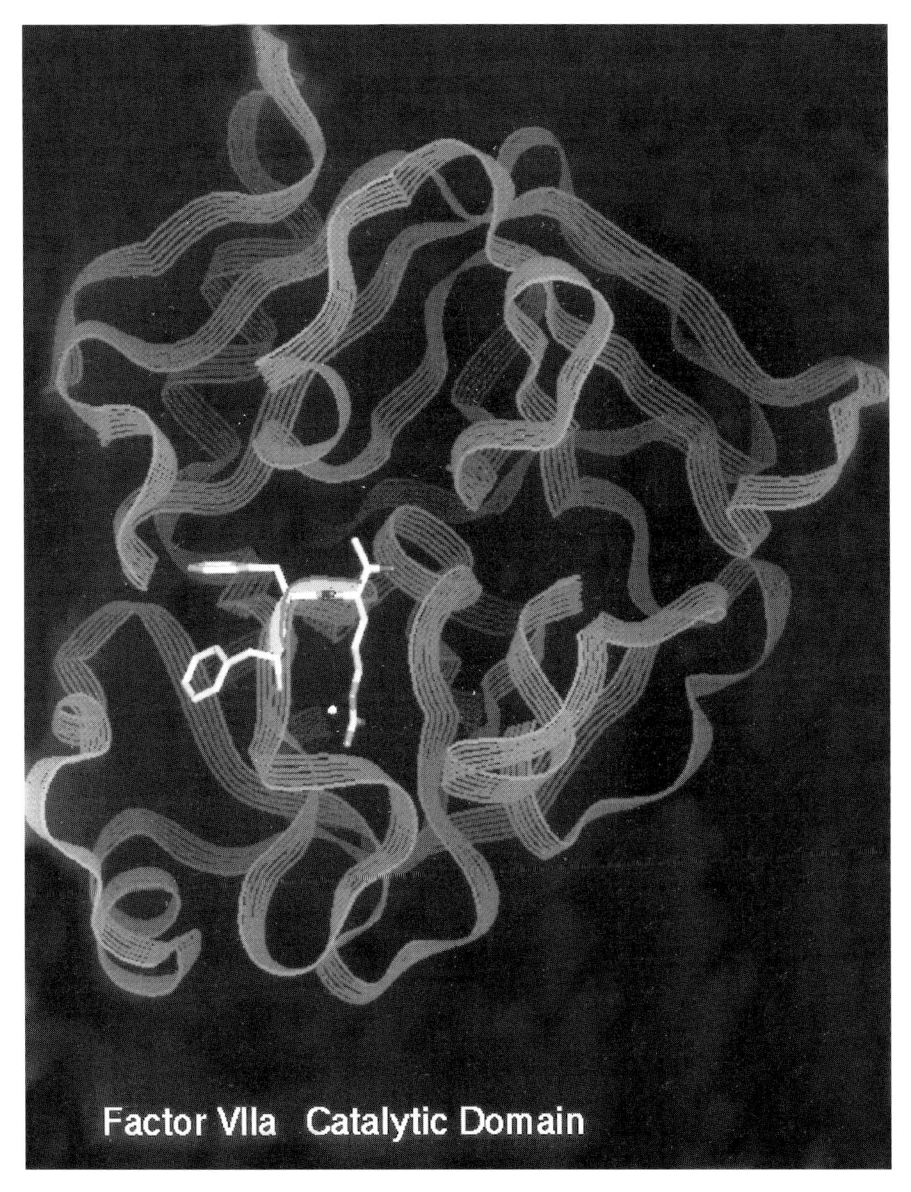

Fig. 3. A ribbon diagram of the polypeptide backbone of the catalytic domain of FVIIa showing the D-Phe-L-Phe-I-Arg tripeptide inhibitor bound in the active site cleft

calcium ions at this site makes rigid the inter-domain linkage and places the FVIIa in an active conformation which is required for FVIIa binding to TF. The third site for calcium is located in the Gla domain where seven calcium ions are bound and they are required for correct folding of the Gla domain as well as binding of FVIIa to TF and the phospholipids of the cell membrane.

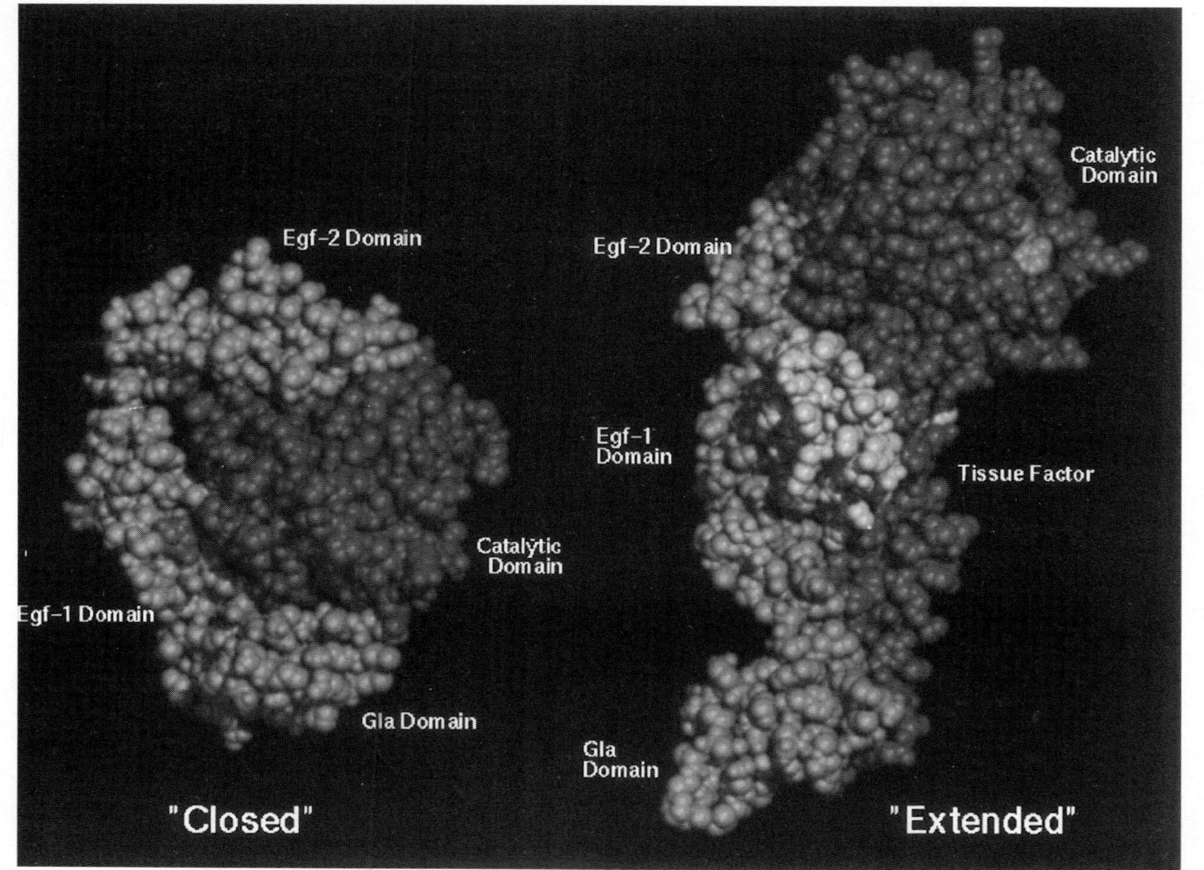

Fig. 4. *Left side* A computer generated model of the "closed" structure of FVIIa that we speculate may exist in the absence of TF. It may explain, in part, the relatively low enzymatic activity of FVII because the active site in the catalytic domain is masked by the gla domain, making it inaccessible to substrates. *Right side* The "extended" crystal structure of FVIIa bound to sTF (as shown in Fig. 1) in which the active site is readily accessible to its substrates, FX and FIX

One of the great mysteries of the TF/FVIIa complex is how binding to TF activates (or, at least, augments the catalytic activity of) FVIIa. Summarized in Fig. 4 is a completely speculative suggestion as to how this may occur. In the absence of TF, the flexible link between EGF domains 1 and 2 could allow the EGF-1 and Gla domains to bind to the catalytic domain and effectively inhibit the enzyme by blocking access to the active site of the enzyme (Fig. 4, left). The binding of FVIIa to TF, we propose, opens and extends the molecule thereby exposing the active site of the catalytic domain (Fig. 4, right) allowing protein substrates to access the active site. Verifying this possibility will require independent crystallization of FVII/FVIIa which has not been reported as yet.

In addition to the soluble portion of TF that binds FVIIa, TF includes a transmembrane spanning domain and an intracellular domain. The existence of the intracellular domain implies that TF may be a true membrane receptor and not merely a docking site for FVIIa. Supporting this view are data from cells in culture that overexpressed TF. When TF was activated by specific high affinity ligands (FVII/FVIIa), an intracellular signal (cytosolic calcium transients) was generated (ROTTINGEN et al. 1995; CAMERER et al. 1996). What the biological importance of this signal is, or for that matter, what the signaling pathway is, however, remain to be determined. Likewise, the fact that a calcium signal was generated in cultured cells overexpressing TF does not guarantee that normal cells would do the same thing. It has been suggested that TF is a member of the cytokine superfamily based on evidence that TF appears to contribute to immunological responses (COLE et al. 1985; BLAKEY et al. 1994) and that it is a potent chemotactic factor for cultured aortic smooth muscle cells (SATO et al. 1996). Other potential associations with TF activation have also been identified, such as promotion of tumor angiogenesis (CONTRINO et al. 1996; ZHANG et al. 1994), tumor-induced hypercoagulation (RAO 1992), and metastasis (FISCHER et al. 1995; MUELLER et al. 1992; BROMBERG et al. 1995). These findings support the contention that the receptor related functions of TF may be important but also emphasize that additional investigation on the TF signaling pathway is needed to find out why. The effect of TF/VIIa inhibitors on TF signaling (if there is signaling) has not been investigated.

D. Endogenous Regulators of Tissue Factor

I. Tissue Factor Pathway Inhibitor

Physiologically, TF and TF/VIIa are regulated by tissue factor pathway inhibitor (TFPI) and antithrombin. TFPI is a plasma protein which is released primarily from endothelial cells (WERLING et al. 1993). It is a Kunitz-type serine protease with a molecular weight of approximately 40 kDa and is composed of 276 amino acids with three tandemly arranged Kunitz domains (WUN et al. 1988). TFPI complexes with FXa (inhibiting FXa activity) which, in turn, binds to the TF/VIIa complex to form a quaternary complex that inhibits activation of additional FXa by TF/VIIa (BROZE et al. 1988; GIRARD and BROZE 1993). Most TFPI circulating in plasma is bound to lipoproteins such as LDL and HDL (LESNICK et al. 1993). Studies with mutated forms and fragments of TFPI demonstrate that Kunitz domain 1 inhibits TF/VIIa, domain 2 inhibits factor Xa, and domain 3 contains one of two heparin-binding sites (GIRARD et al. 1989; PETERSEN et al. 1995; JESKE et al. 1996). Secondary-site interactions also contribute to FXa inhibition, since full length TFPI has 1000-fold higher affinity for FXa than does the separated domain 2 alone (PETERSEN et al. 1995).

Administration of unfractionated heparin or low molecular weight heparin leads to an increase in the plasma concentration of TFPI (ABILDGAARD 1993) suggesting that part of heparin's antithrombotic effect may be due to the release of TFPI resulting in FXa inhibition. Therapeutic concentrations (approximately 100 nM) of TFPI can reduce prothrombinase activity but lower, more physiologic, concentrations do not exert this effect (MAST and BROZE 1996). The possibility that patients with deep venous thrombosis have abnormal levels of TFPI has been investigated, but no abnormality in circulating TFPI concentrations was identified (HOLST et al. 1993).

II. Antithrombin

Another endogenous regulator of TF is antithrombin (AT) which is the major plasma inhibitor of coagulation proteases. AT has little effect on FVIIa alone, but in the presence of heparin, it inhibits the catalytic activity of the TF/VIIa (RAO et al. 1993) complex, it is speculated, by inducing an change in FVIIa that impedes binding to TF (RAO et al. 1995). Thus AT and TFPI share the common feature of inhibiting FVIIa activity when it is complexed with TF. Tissue factor pathway inhibitor, however, requires the presence of FXa in order to inhibit TF/VIIa (BROZE et al. 1990; RAPAPORT 1991) but the complex of AT/heparin does not (SHIGEMATSU et al. 1992; RAO et al. 1993).

E. Pathophysiology of TF/VIIa

Normally, there are no TF/VIIa complexes because TF and FVII are separated from one another by endothelial cells. Disruption of the endothelium due to trauma brings the two elements together to start the coagulation cascade. In pathophysiologic conditions, however, TF may be expressed in the "wrong place at the wrong time", potentially promoting thromboembolic consequences.

For example, the fibrous cap of lipid rich plaques are prone to erosion or rupture, exposing the contents to blood flowing in the vessel (DAVIES and THOMAS 1985; FARB et al. 1996). Angiographic studies show that thrombotic occlusion occurs in 80%–95% of patients with acute myocardial infarction.

Pathological results from human patients have been interpreted to suggest that recurrent but clinically silent plaque rupture leading to nonocclusive thrombus formation is a fairly common event which accelerates the development of clinically significant lesions. If plaque rupture exposes enough TF to blood it may trigger formation of an intimal thrombus large enough to occlude the vessel and result in acute myocardial infarction (DAVIES and THOMAS 1985; FARB et al. 1996). The content of TF in atherectomy samples from human patients and human pathological specimens is high (MARMUR et al. 1996; TOSCHI et al. 1997), especially in patients with unstable angina (MORENO et al. 1996), supporting the potential importance of TF in acute coronary syndromes.

The TF/VIIa complex is reported to play a role in a number of other conditions that involve promotion of both thrombotic and nonthrombotic effects. The primary therapy for myocardial infarction is rapid removal of the clot either by thrombolytic therapy or balloon angioplasty. Reocclusion is a frequent complication following either of these treatments, as well as with angioplasty performed for other indications. Reocclusion occurs in 5%–10% of patients during the acute phase (within hours to several days) after reperfusion due to formation of another occlusive thrombus (DE FEYTER et al. 1992; DELIGONUL et al. 1988). Over a longer timeframe (6 months), restenosis occurs in approximately 30%–50% of patients due to intimal proliferation and arterial remodeling (HOLMES et al. 1991).

Recent studies suggest that TF may play an important role in both reocclusion and restenosis. In addition to exposing TF in the underlying media and adventitia as a result of endothelial damage caused by the procedure, angioplasty rapidly induces expression of new TF mRNA and increased TF activity in arterial smooth muscle (MARMUR et al. 1993; TAUBMAN 1993). The precise mediator(s) responsible for TF induction after angioplasty have not been established, but in cultured vascular smooth muscle cells (VSMC) mRNA for TF is rapidly induced by a number of agents known to be involved in the response to vessel wall injury, including growth factors, vasoactive agonists, and alpha-thrombin (TAUBMAN et al. 1993). Thrombin also plays a potentially important role in restenosis, acting as a mitogen for smooth muscle cells (CHEN and BUCHANAN 1975; BAR-SHAVIT et al. 1990) and stimulating them to release growth factors (WILSON et al. 1993; BENEZRA et al. 1993).

JANG et al. (1995) showed that treatment with recombinant TFPI (rTFPI) for three days following balloon angioplasty in rabbits with femoral artery atherosclerosis (induced by air-dessication injury and a hypercholesterolemic diet) reduced the degree of angiographic restenosis and decreased neointimal hyperplasia 21 days later. Similar salutary effects were achieved with inactivated FVIIa (FVIIai). These investigators saw no effects on restenosis with recombinant tick anticoagulant peptide (rTAP), which blocks FXa, or with hirudin, which blocks thrombin, but others have shown protection in the same model using higher doses of these agents (SAREMBOCK et al. 1991; RAGOSTA et al. 1994). Similar beneficial effects with TF blockade have also been reported in baboons. Administration of FVIIai immediately prior to femoral artery balloon angioplasty and carotid endarterectomy prevented thrombus formation and decreased vascular lesion formation by approximately 50% at 30 days (HARKER et al. 1996), and notably, there were no hemorrhagic complications associated with FVIIai treatment. Collectively, these studies suggest that blockade at different levels of the coagulation cascade for a relatively short time after angioplasty or endarterectomy may be useful for preventing or minimizing restenosis. There is substantial evidence that thrombosis contributes to the pathophysiology of atherosclerosis itself. Early atherosclerotic lesions are characterized by increased endothelial permeability and an influx of plasma macromolecules into the intima, including low density lipoprotein (LDL). Some of the LDL is modified, probably by oxidation (YLA-HERTTUALA et al. 1991), and endocytosed by monocyte-derived macrophages, thereby giving rise to the lipid-filled foam cells that are characteristic of atherosclerosis (FALK and FERNANDEZ-ORTIZ 1995). Oxidized LDL can also induce the expression of TF in endothelial cells and monocytes, which may account for the higher concentration of TF in atherosclerotic lesions than in normal vessels (WILCOX et al. 1989; MARMUR et al. 1996; MORENO et al. 1996; TOSCHI et al. 1997).

Tissue factor has both direct and indirect effects that may promote the development of atherosclerosis. It is a strong chemotactic factor for cultured aortic smooth muscle cells (SATO et al. 1996), and thus may act directly on smooth muscle cells, attracting them into developing atherosclerotic lesions. Tissue factor may also act indirectly by promoting the production of fibrin. Exposure of blood to TF expressed by endothelial cells at the plaque surface or by insudation of blood components into the TF-rich plaque may account for the increased production and degradation of fibrin which occurs in atherosclerosis. Fibrinogen, fibrin, and their degradation products have chemotactic and mitogenic effects which have been proposed to contribute to the progression of atherosclerosis (WILCOX 1993). Increased deposition of fibrin and its breakdown products within the arterial wall of atherosclerotic vessels has been documented in samples taken at postmortem examination or during reconstructive surgery (SMITH et al. 1990; BINI et al. 1989) supporting this idea. Similarly, in patients with peripheral artery occlusive disease there is a significant correlation between the plasma concentrations of both fibrinogen and the degradation products of cross-linked fibrin with the severity of atherosclerosis (DE BUYZERE et al. 1993; LASSILA et al. 1993). SMITH (1986) reported that fibrinogen and fibrin concentrations are ten times higher in advanced plaques than in normal vessels.

F. Experimental Inhibitors of TF/VIIa

The are a number of strategies that can be pursued to inhibit TF and TF/VIIa. One would be to block the binding of FVIIa to sTF. Since the binding area between FVIIa and TF is quite large, however, it may not be easy to design inhibitors of this extensive protein-protein interaction. To do so with molecules small enough to be orally bioavailable (e.g., for chronic administration) may be very difficult. An alternative approach would be to concentrate on active site inhibition of FVIIa. Given the apparent change in conformation of FVIIa and the clear-cut change in its activity when bound to TF, active site inhibitors of FVIIa should be tested using TF/VIIa complexes.

Most of the work conducted to evaluate the effects of blocking TF and TF/VIIa has relied on antibodies or other, relatively large polypeptides. In studies directed at TF blockade, monoclonal antibodies to TF or FVII/FVIIa and inactivated FVIIa (which bind to TF) have been used. In studies focused on inhibition of TF/VIIa activity, recombinant forms of TFPI (rTFPI) or an anticoagulant protein derived from nematodes have been utilized. The inactivated form of FVIIa, designated as FVIIai, is made by irreversibly interacting FVIIa with Glu-Gly-Arg chloromethylketone (HARKER et al. 1996) or Phe-Phe-Arg chloromyethylketone (GOLINO et al. 1998) which blocks the catalytic site. The inactivated form of FVIIa binds to TF readily, so it can act like a competitive inhibitor of TF-dependent FXa or FIXa formation.

I. Recombinant TFPI (rTFPI) and Truncated rTFPI

Recombinant human TFPI has been produced in quantity by expression in *Escherichia coli* as a non-glycosylated protein with an additional alanine attached to the aminoterminus of the wild type molecule (DIAZ-COLLIER et al. 1994). $TFPI_{1-161}$, a truncated recombinant TFPI lacking the C-terminal region of native TFPI, is produced by expression in yeast cells (PETERSEN et al. 1993). Like endogenous TFPI, recombinant or truncated forms of TFPI inhibit the TF/VIIa complex rather than TF per se. They operate by binding to FXa which results in inhibition of both FXa and the TF/VIIa complex.

The two forms of rTFPI (full length and truncated) have different pharmacokinetics and activity profiles. Optimal inhibition of TF/VIIa activity appears to be achieved better with full length TFPI compared to the C-terminal truncated TFPI variants, since full-length rTFPI binds to FXa at a much faster rate, resulting in faster inhibition of the TF/VIIa complex. In an in vitro study, TFPI inhibited activated protein C (aPC) amidolytic activity in the presence of pharmacological levels of heparin, whereas a truncated form of TFPI lacking the third Kunitz domain and C-terminal region ($TFPI_{1-161}$), failed to inhibit aPC amidolytic activity with or without heparin (HAMAMOTO and KISIEL 1995). Due to its binding to glycosaminoglycans on the cell surface, however, full-length rTFPI is cleared rapidly from circulation so it may be difficult to target it to a focal area of vessel wall injury (ELSAYED et al. 1996; BAJAJ and BAJAJ 1997).

Development of a practical local delivery system for rTFPI may be worth pursuing based on experimental results obtained in pigs (WANG et al. 1996). In this study, normal pigs underwent carotid arterial injury by balloon angioplasty, and local intramural infusion of rTFPI (500 μg/kg) inhibited mural platelet thrombosis for up to 12 h. If TF-mediated coagulation is primarily responsible for prolonged procoagulant activity of balloon-injured arteries, as proposed by SPEIDEL et al. (1995), local delivery of rTFPI may be useful in PTCA patients to reduce the likelihood of reocclusion. It would be interesting to see a comparison of different antithrombotic and antiplatelet regimens in experimental models of reocclusion to help determine which would be the most effective.

Truncated rTFPI has potent anti-FXa effects but tends to exert less inhibition on the TF/VIIa complex since it does not prolong PT or diluted-PT assays (a standard PT assay modified to allow TF to be the rate limiting reagent) (JESKE et al. 1996; HOLST et al. 1996a). In a rat gastric mucosa model, both a two-domain non-glycosylated TFPI (117QTFPI$_{1-161}$) and low molecular weight heparin (LMWH, tinzaparin) significantly increased the bleeding time and the amount of blood loss compared to controls. Tinzaparin also produced greater aPTT prolongation and changes in the diluted-PT assay (HOLST et al. 1996a). The glycosylation of TFPI did not appear to contribute significantly to the antithrombotic effect of TFPI, because the non-glycosylated 117QTFPI$_{1-161}$ had an antithrombotic effect similar to the glycosylated TFPI$_{1-161}$ (HOLST et al. 1996b).

Tissue factor pathway inhibitor has been demonstrated to be effective in various models of venous and arterial thrombosis. In rabbit models of thrombosis induced by stasis or vessel wall damage with partial stasis in the jugular vein, both full-length rTFPI (JESKE et al. 1996; KAISER and FAREED 1996) and TFPI$_{1-161}$ (HOLST et al. 1994) dose-dependently prevented the primary formation of venous thrombi. Full-length rTFPI at higher doses also inhibited thrombotic reocclusion after lysis with tPA (KAISER and FAREED 1996). Only minimal effects were produced in terms of aPTT, PT, and blood loss. The antithrombotic efficacy of TFPI$_{1-161}$ was equivalent to LMWH (Logiparin) in another study but smaller increases in aPTT and PT were produced (HOLST et al. 1994), supporting the idea that the antithrombotic benefit-hemorrhagic risk ratio may be better with inhibition of TF/VIIa.

In other experiments, microvascular vessels were protected from occlusive thrombosis by topical application of rTFPI in a rabbit ear model of vascular trauma (KOURI et al. 1993). Thrombosis produced by vascular damage and stasis in the vena cava of rabbits was blocked by rTFPI administered intravenously as a bolus 20 min before stasis was induced (SPOKAS and WUN 1992). In a similar model using the rabbit jugular vein, TFPI$_{1-161}$ was as effective as LMWH in preventing thrombosis, and no hemorrhagic side effects were noted with either agent (HOLST et al. 1994). Both TFPI$_{1-161}$ and FVIIai reduced thrombus formation by 80%–90% on endarterectomized segments of aorta incorporated into arteriovenous shunts in baboons (LINDAHL et al. 1993a,b). TFPI$_{1-161}$ also produced moderate inhibition of thrombus formation in the absence of TF on dacron grafts (LINDAHL et al. 1993a), reflecting its moderate, direct anti-FXa effect. FVIIai had no effect on thrombus formation on dacron grafts, however, suggesting that blockade was specific to TF and that the ability of the blood to coagulate in response to other stimuli remained intact. This was substantiated by the observation that bleeding time was unaffected by either agent. The lack of effect on bleeding time, again, is particularly notable because it adds support to the contention that inhibiting TF/VIIa activity may pose a smaller bleeding risk than some other antithrombotic strategies.

Reocclusion of coronary arteries after thrombolytic therapy or angioplasty also represents an important clinical problem which may be related to exposure of TF to blood flow after dissolution of the original blood clot. In an experimental study on reocclusion (ABENDSCHEIN et al. 1995), coronary artery thrombosis was first induced in dogs by anodal current injury of the intima of the left anterior descending coronary artery, then recanalization was produced by intravenous infusion of recombinant tissue-type plasminogen activator (tPA) over 1 h. Reocclusion due to thrombus formation occurred within 90 minutes after discontinuing the tPA in all control dogs. Intravenous infusion of rTFPI concomitant with and for 1 h after tPA prolonged the time to reocclusion in a dose-related manner. There was no effect on template bleeding time and only modest increases in PT (1.4-fold) and aPTT (2.1-fold). Prevention of reocclusion has also been shown with rTFPI in a similar study in the femoral arteries of dogs (HASKEL et al. 1991).

Recombinant TFPI is also effective in inhibiting the consumption of fibrinogen, platelets and FVIIa in experimental models of disseminated intravascular coagulation (DIC) (ELSAYED et al. 1996). Full length rTFPI is currently being used in clinical trials for patients with DIC secondary to sepsis and in patients following microvascular surgery. In a phase I trial, SC-59735, a recombinant analog of TFPI, was administered to healthy volunteers. It was well tolerated and caused no severe adverse events. In another clinical study, SC-59735 was a safe and effective additive to irrigation solutions for the prevention of thrombosis during free flap surgery. A multicenter clinical trial is ongoing to evaluate the efficacy of topical SC-59735 in 600 free flap transfer patients (KOURI et al. 1997). Other potential therapeutic indications for TFPI, such as coronary artery disease, restenosis, stroke, and deep venous thrombosis, are still being evaluated in preclinical studies.

II. Inactivated Factor VIIa: FVIIai

Inactivated FVIIa or FVIIai is a competitive antagonist of FVIIa for binding to TF. In experimental studies, the effect of FVIIai on thrombosis appears to be model-dependent. In dogs with electrically-induced coronary artery thrombosis undergoing thrombolysis with tPA, tick anticoagulant peptide (TAP), a FXa inhibitor, reduced the incidence of reocclusion, but FVIIai and TFPI did not, although they each accelerated the time to reperfusion (LEFKOVITS et al. 1996). The lack of effectiveness of FVIIai and TFPI in this model may simply indicate that reocclusion due to renewed thrombus formation after thrombolysis is less dependent on TF. Contrasting results were obtained in a model focused on the problem of restenosis. In this study, an atherosclerotic rabbit arterial injury model was used in which FVIIai (1 mg/kg followed by infusion at 50 μg/kg/h for 3 days) and rTFPI (1 mg/kg followed by infusion at 15 μg/kg/h for 3 days) reduced angiographic restenosis and decreased neointimal hyperplasia 21 days after balloon angioplasty, whereas TAP and hirudin (an inhibitor of thrombin) did not show significant effects (JANG et al. 1995). It should be pointed out, however, that TAP was effective in a pig model of restenosis emphasizing that there may be important species differences that influence the effects of these agents.

Novo Nordisk is developing a recombinant active site inhibited FVIIa (FFR-rFVIIa) for the potential treatment of restenosis. Experimentally, it has proved to be an effective antithrombotic agent with minimal elevation of bleeding parameters. Local application of FFR-rFVIIa significantly reduced thrombus weight and improved patency in an experimental venous thrombosis model in the rabbit, without affecting the aPTT and diluted-PT (HOLST et al. 1997). In a rabbit nail cuticle bleeding model, FFR-rFVIIa, at four different doses, combined with heparin or aspirin, or both, did not significantly prolong the total bleeding time (KRISTENSEN et al. 1997). In a rabbit model of recurrent arterial occlusion, FFR-rFVIIa (0.1 mg/kg/min, intracarotid administration) abolished cyclical flow reductions and reduced production of FPA for up to 3 h after a 10 min infusion (GOLINO et al. 1998).

III. Recombinant Nematode Anticoagulant Peptide (rNAPc2)

Nematode anticoagulant peptide (NAP) is a TF/VIIa inhibitor being developed by Corvas. This developmental candidate is NAPc2, a small protein (MW = 9500) originally isolated from the canine hookworm, *Ancylostoma caninum* (STASSENS et al. 1996). The recombinant form of this protein, rNAPc2, inhibits the catalytic complex of TF/VIIa in the presence of FXa, in a manner similar to TFPI. The sequence of NAPc2 as well as the binding site at which it interacts with FXa, however, are different from that of TFPI. In addition, it is not cleared from circulation as quickly, with elimination half-lives in conscious rats, dogs, and cynomolgus monkeys of 15.6, 18.0 and 17.1 h, respectively. The absolute bioavailability after subcutaneous administration is >90% in dogs and monkeys (VLASUK et al. 1997). Given the long duration of action and subcutaneous bioavailability, rNAPc2 appears to be a promising antithrombotic candidate, mitigating the large size of the molecule and lack of oral bioavailability.

Another strong feature of rNAPc2 is impressive in vivo potency which has been evaluated in recent studies. In a pig model of acute coronary thrombosis, ROTE et al. (1996) reported that rNAPc2 dose-dependently increased time to occlusion, prolonged PT, and reduced the incidence of coronary occlusion at an intravenous dose of 100 μg/kg. The intravenous ED_{50} was 30 μg/kg in a rat model of $FeCl_3$-induced carotid artery thrombosis. When administered subcutaneously to rats in a chronic model of deep vein thrombosis, rNAPc2 reduced thrombus weights at 24 h by >50% at doses below 3 μg/kg, attesting to the impressive potency of this compound. In a TF-induced DIC model in the rat, pretreatment with rNAPc2, rNAP5 (a FXa inhibitor), and LMWH completely prevented the reduction of fibrinogen concentration and platelet counts at doses of 0.003, 0.03, and 3.0 mg/kg, iv., respectively (Berkeley et al. 1996). A clinical trial with rNAPc2 was started in the UK in April 1997. This single center clinical trial is a dose escalation study designed to assess the safety, pharmacokinetic properties and systemic effects of rNAPc2 in healthy male volunteers. The feasibility of subcutaneous administration of rNAPc2 will also be evaluated. Assuming there are no problems with immunogenicity, and subcutaneous pharmacokinetics in the human volunteers are favorable, rNAPc2 could be an exciting new drug.

IV. TF Antibodies

Another inhibitor of the TF/VIIa complex is AP-1, a monoclonal anti-rabbit TF antibody. Unlike TFPI and rNAPc2, AP-1 acts by preventing the binding of FVII and FVIIa to TF. This ultimately results in inhibition of new thrombin generation and the positive feedback loop that amplifies thrombin generation, by stopping the first step in the cascade.

That this novel approach works effectively has been demonstrated in experimental studies. In a rabbit model of carotid artery thrombosis, AP-1 abolished CFRs (Himber et al. 1997), shortened thrombolysis time by tPA and reduced the incidence of reocclusion (Ragni et al. 1996). In contrast to the comparators heparin or the direct thrombin inhibitor napsagatran, AP-1, at effective antithrombotic doses, did not cause changes in PT, aPTT, and bleeding time (Himber et al. 1997) or ex vivo platelet aggregation (Ragni et al. 1996). Compared with control rabbits, AP-1 (0.15 mg/kg, intravenous bolus) completely inhibited thrombin formation, as reflected in unaltered FPA levels during thrombolysis. In terms of drug development, relatively little has been done thus far with AP-1, so it will be interesting to learn more about applications of this approach to antithrombotic therapy.

Disseminated intravascular coagulation (DIC) is another thrombotic disorder that may be amenable to inhibition of TF and TF/VIIa, a possibility which has been tested in an *Escherichia coli*-induced septic shock model in baboons. In this model, an intravenous infusion of *E. coli* causes a DIC-like condition characterized by widespread deposition of fibrin in the blood vessels and extensive organ damage, a condition thought to be initiated at least in part by exposure of the blood to TF (Muller-Berghaus 1989). Administration of a monoclonal antibody against TF (TF9–5B7) 30 min before the start of a lethal infusion of *E. coli* resulted in all animals surviving for the entire 7-day observation period. In contrast, the mean survival time for the controls was 12.1 h (Taylor et al. 1991a). It should be noted that other agents are also effective in this model, including rTFPI (Creasey et al. 1993), activated protein C (Taylor et al. 1987), and an antithrombin (Taylor et al. 1988).

While the results with TF9–5B7 suggest that DIC initiated by TF plays an important role in the septic shock model, studies with other agents suggest that other mechanisms are also involved. Specifically, treatment with antibodies to tissue necrosis factor after the *E. coli* challenge did not prevent the coagulopathic response but did prevent death (HINSHAW et al. 1990). Conversely, treatment with active site blocked factor Xa (DEGR-Xa) prevented fibrinogen and factor V consumption (markers of DIC) similar to the other anticoagulant agents, but did not prevent shock, organ damage or death (TAYLOR et al. 1991b). Although the mechanisms responsible for the toxicity in this model of septic shock are not fully understood, activation of the coagulation system by TF appears to play an important role and inhibition of TF/VIIa is clearly beneficial. Moreover, protective effects have also been documented with TF blockade in other species in similar types of studies. Endotoxin-induced DIC in chimpanzees has been blocked by an anti-TF antibody (TF8–5G9) (LEVI et al. 1994) and by a different anti-factor VII/VIIa antibody (Corsevin M, also known as 12D10) (BIEMOND et al. 1995). Endotoxin-induced DIC in rabbits has also been blocked by an anti-TF antibody (anti-TF IgG) (WARR et al. 1990). These results suggest that blockade or antagonism of TF may provide a new, potentially powerful strategy for the treatment of septic shock and DIC.

G. Summary and Conclusions

Tissue factor is the primary initiator of in vivo coagulation making it an important target for preventing or minimizing thromboembolic disorders. Because TF acts at the beginning of the coagulation cascade, it has been suggested that the bleeding risk associated with its inhibition may be less than that observed with other antithrombotics such as heparin, antithrombins, or GP IIb/IIIa antagonists. Experimental evidence supports this contention (LINDAHL et al. 1993b; HARKER et al. 1996) making an even stronger case for developing inhibitors of TF and the TF/VIIa complex because they may have a better safety margin than alternative antithrombotic strategies. Clinical trials with protein inhibitors of TF/VIIa are underway, so we should know soon if the apparently wide safety margin will hold up in humans.

The cellular effects of TF activation are poorly understood but the beneficial effects of inhibiting TF in experimental models of restenosis and septic shock with disseminated intravascular coagulation suggests that there may a variety of indications in which such agents could be useful. Experimental evidence also supports the contention that TF may play a role in the development of atherosclerosis. It seems unlikely that atherosclerosis will be developed as an important new indication for TF inhibitors but they may prove to be useful for preventing the contributions of thrombosis (e.g., in recurrent plaque rupture) in the progression of the atherosclerotic process.

The inhibitors tested to date have largely been high molecular weight, polypeptide compounds (antibodies, inactivated FVIIa, recombinant forms of the endogenous regulator TFPI, and rNAPc2). These agents have been used to demonstrate the potential therapeutic utility of TF and TF/VIIa inhibition experimentally. In general, high molecular weight inhibitors will be limited to intravenous administration in clinical settings making it unlikely they will be candidates for chronic clinical indications. Low molecular weight agents with oral bioavailability would be ideal to exploit the potential benefits of long term TF and TF/VIIa inhibition. Since no such agents exist as yet, perhaps the most promising TF/FVIIa inhibitor currently available is rNAPc2. Although relatively large (MW 9500), its high bioavailability when administered subcutaneously may make rNAPc2 the first TF/FVIIa inhibitor to see widespread use.

References

Abendschein DR, Meng YY, Torr-Brown S, Sobel BE (1995) Maintenance of coronary patency after fibrinolysis with tissue factor pathway inhibitor. Circulation 92: 944–949

Abildgaard U (1993) Heparin/low molecular weight heparin and tissue factor pathway inhibitor. Haemostasis 23 Suppl 1:103–106

Ashton AW, Kemball-Cook G, Johnson DJ, Martin DM, O'Brien DP, Tuddenham EG et al (1995) Factor VIIa and the extracellular domains of human tissue factor form a compact complex: a study by X-ray and neutron solution scattering. FEBS Lett 374:141–146

Bajaj MS, Bajaj SP (1997) Tissue factor pathway inhibitor: potential therapeutic applications. Thromb Haemost 78:471–477

Banner DW, D'Arcy A, Chene C, Winkler FK, Guha A, Konigsberg WH et al (1996) The crystal structure of the complex of blood coagulation factor VIIa with soluble tissue factor. Nature 380:41–46

Bar-Shavit R, Benezera M, Eldor A, Hy-Am E, Fenton JW 2d, Wilner GD et al (1990) Thrombin immobilized to extracellular matrix is a potent mitogen for vascular smooth muscle cells: nonenzymatic mode of action. Cell Regul 1:453–463

Bauer KA (1997) Activation of the factor VII-tissue factor pathway. Thromb Haemost 78:108–111

Benezra M, Vlodavsky I, Ishai-Michaeli R, Neufeld G, Bar-Shavit R (1993) Thrombin-induced release of active basic fibroblast growth factor-heparan sulfate complexes from subendothelial extracellular matrix. Blood 81:3324–3331

Berkeley CJ, Vlasuk GP, Rote WE (1996) Comparison of nematode anticoagulant proteins (NAPs) with low molecular weight heparin in a model of disseminated intravascular coagulation (DIC). Circulation 94:1–696

Biemond BJ, Levi M, ten Cate H, Soule HR, Morris LD, Foster DL et al (1995) Complete inhibition of endotoxin-induced coagulation activation in chimpanzees with a monoclonal Fab fragment against factor VII/VIIa. Thromb Haemost 73:223–230

Bini A, Fenoglio JJ Jr, Mesa-Tejada R, Kudryk B, Kaplan KL (1989) Identification and distribution of fibrinogen, fibrin, and fibrin(ogen) degradation products in atherosclerosis. Use of monoclonal antibodies. Arteriosclerosis 9:109–121

Blakely ML, Van der Werf WJ, Berndt MC, Dalmasso AP, Bach FH, Hancock WW (1994) Activation of intragraft endothelial and mononuclear cells during discordant xenograft rejection. Transplantation 58:1059–1066

Bromberg ME, Garen A, Koningsberg WH (1995) Increasing expression of tissue factor in human melanoma cells induces metastasis in a murine model. Thromb Haemost 73:1177

Broze GJ Jr, Girard TJ, Novotny WF (1990) Regulation of coagulation by a multivalent Kunitz-type inhibitor. Biochemistry 29:7539–7546

Broze GJ Jr, Warren LA, Novotny WF, Higuchi DA, Girard JJ, Miletich JP (1988) The lipoprotein-associated coagulation inhibitor that inhibits the factor VII-tissue factor complex also inhibits factor Xa: insight into its possible mechanism of action. Blood 71:335–343

Butenas S, Mann KG (1996) Kinetics of human factor VII activation. Biochemistry 35:1904–1910

Camerer E, Rottingen JA, Iversen JG, Prydz H (1996) Coagulation factors VII and X induce Ca2+ oscillations in Madin-Darby canine kidney cells only when proteolytically active. J Biol Chem 271:29034–29042

Chen LB, Buchanan JM (1975) Mitogenic activity of blood components. I. Thrombin and prothrombin. Proc Natl Acad Sci USA 72:131–135

Cole EH, Schulman J, Urowitz M, Keystone E, Williams C, Levy GA (1985) Monocyte procoagulant activity in glomerulonephritis associated with systemic lupus erythematosus. J Clin Invest 75:861–868

Contrino J, Hair G, Kreutzer DL, Rickles FR (1996) In situ detection of tissue factor in vascular endothelial cells: correlation with the malignant phenotype of human breast disease. Nat Med 2:209–215

Creasey AA, Chang AC, Feigen L, Wun TC, Taylor FB Jr, Hinshaw LB (1993) Tissue factor pathway inhibitor reduces mortality from Escherichia coli septic shock. J Clin Invest 91:2850–2856

Davies MJ, Thomas AC (1985) Plaque fissuring–the cause of acute myocardial infarction, sudden ischaemic death, and crescendo angina. Br Heart J 53: 363–373

De Buyzere M, Philippe J, Duprez D, Baele G, Clement DL (1993) Coagulation system activation and increase of D-dimer levels in peripheral arterial occlusive disease. Am J Hematol 43:91–94

De Feyter PJ, de Jaegere PP, Murphy ES, Serruys PW (1992) Abrupt coronary artery occlusion during percutaneous transluminal coronary angioplasty. Am Heart J 123:1633–1642

Del Zoppo GJ, Yu JQ, Copeland BR, Thomas WS, Schneiderman J, Morrissey JH (1992) Tissue factor localization in non-human primate cerebral tissue. Thromb Haemost 68:642–647

Deligonul U, Gabliani GI, Caralis DG, Kern MJ, Vandormael MG (1988) Percutaneous transluminal coronary angioplasty in patients with intracoronary thrombus. Am J Cardiol 62:474–476

Diaz-Collier JA, Palmier MO, Kretzmer KK, Bishop BF, Combs RG, Obukowicz MG et al (1994) Refold and characterization of recombinant tissue factor pathway inhibitor expressed in Escherichia coli. Thromb Haemost 71: 339–346

Drake TA, Ruf W, Morrissey JH, Edgington TS (1989a) Functional tissue factor is entirely cell surface expressed on lipopolysaccharide-stimulated human blood monocytes and a constitutively tissue factor-producing neoplastic cell line. J Cell Biol 109:389–395

Drake TA, Morrissey JH, Edgington TS (1989b) Selective cellular expression of tissue factor in human tissues. Implications for disorders of hemostasis and thrombosis. Am J Pathol 134:1087–1097

Elsayed YA, Nakagawa K, Kamikubo YI, Enjyoji KI, Kato H, Sueishi K (1996) Effects of recombinant human tissue factor pathway inhibitor on thrombus formation and its in vivo distribution in a rat DIC model. Am J Clin Pathol 106: 574–583

Falk E, Fernandez-Ortiz A (1995) Role of thrombosis in atherosclerosis and its complications. Am J Cardiol 75:3B–11B

Farb A, Burke AP, Tang AL, Liang TY, Mannan P, Smialek J et al (1996) Coronary plaque erosion without rupture into a lipid core. A frequent cause of coronary thrombosis in sudden coronary death. Circulation 93:1354–1363

Fischer EG, Ruf W, Mueller BM (1995) Tissue factor-initiated thrombin generation activates the signaling thrombin receptor on malignant melanoma cells. Cancer Res 55:1629–1632

Fleck RA, Rao LV, Rapaport SI, Varki N (1990) Localization of human tissue factor antigen by immunostaining with monospecific, polyclonal anti-human tissue factor antibody. Thromb Res 59:421–437

Flossel C, Luther T, Muller M, Albrecht S, Kasper M (1994) Immunohistochemical detection of tissue factor (TF) on paraffin sections of routinely fixed human tissue. Histochemistry 101:449–453

Girard TJ, Broze GJ Jr (1993) Tissue factor pathway inhibitor. Methods Enzymol 222:195–209

Girard TJ, Warren LA, Novotny WF, Likert KM, Brown SG, Miletich JP et al (1989) Functional significance of the Kunitz-type inhibitory domains of the lipoprotein-associated coagulation inhibitor. Nature 338:518–520

Golino P, Ragni M, Cirillo P, D'Andrea D, Scognamiglio A, Ravera A et al (1998) Antithrombotic effects of recombinant human, active site-blocked factor VIIa in a rabbit model of recurrent arterial thrombosis. Circ Res; 82: 39–46

Hamamoto T, Kisiel W (1995) Full-length human tissue factor pathway inhibitor inhibits human activated protein C in the presence of heparin. Thromb Res 80:291–297

Harker LA, Hanson SR, Wilcox JN, Kelly AB (1996) Antithrombotic and antilesion benefits without hemorrhagic risks by inhibiting tissue factor pathway. Haemostasis 26:76–82 Suppl 1

Harlos K, Martin DM, O'Brien DP, Jones EY, Stuart DI, Polikarpov I et al (1994). Crystal structure of the extracellular region of human tissue factor. Nature 370:662–666

Haskel EJ, Torr SR, Day KC, Palmier MO, Wun TC, Sobel BE et al (1991) Prevention of arterial reocclusion after thrombolysis with recombinant lipoprotein-associated coagulation inhibitor. Circulation 84:821–827

Himber J, Kirchhofer D, Riederer M, Tschopp TB, Steiner B, Roux SP (1997) Dissociation of antithrombotic effect and bleeding time prolongation in rabbits by inhibiting tissue factor function. Thromb Haemost 78:1142–1149

Hinshaw LB, Tekamp-Olson P, Chang AC, Lee PA, Taylor FB Jr., Murray CK, Peer GT et al (1990) Survival of primates in LD100 septic shock following therapy with antibody to tumor necrosis factor (TNFα). Circ Shock 30: 279–292

Holmes DR Jr, Schwartz RS, Webster MW (1991) Coronary restenosis: what have we learned from angiography? J Am Coll Cardiol 17:14B–22B

Holst J, Kristensen AT, Kristensen HI, Bak H, Ezban M, Hedner U (1997) Local application of active site inhibited factor VIIa significantly reduces thrombus weight and improves patency in a venous experimental thrombosis model. Thromb Haemost Suppl: 687

Holst J, Lindblad B, Bergqvist D, Nordfang O, Ostergaard PB, Petersen JG et al (1994) Antithrombotic effect of recombinant truncated tissue factor pathway inhibitor (TFPI1–161) in experimental venous thrombosis-a comparison with low molecular weight heparin. Thromb Haemost 71:214–219

Holst J, Lindblad B, Matthiasson SE, Stjernquist U, Ezban M, Ostergaard PB et al (1996a) Experimental haemorrhagic effect of two-domain non-glycosylated tissue factor pathway inhibitor compared to low molecular weight heparin. Thromb Haemost 75:585–589

Holst J, Lindblad B, Nordfang O, Ostergaard PB, Hedner U (1996b) Does glycosylation influence the experimental antithrombotic effect of a two-domain tissue factor pathway inhibitor? Haemostasis 26:23–30

Holst J, Lindblad B, Wedeberg E, Bergqvist D, Nordfang O, Ostergaard PB et al (1993) Tissue factor pathway inhibitor (TFPI) and its response to heparin in patients with spontaneous deep vein thrombosis. Thromb Res 72:467–470

Jang Y, Guzman LA, Lincoff AM, Gottsauner-Wolf M, Forudi F, Hart CE et al (1995) Influence of blockade at specific levels of the coagulation cascade on restenosis in a rabbit atherosclerotic femoral artery injury model. Circulation 92:3041–3050

Jeske W, Hoppensteadt D, Callas D, Koza MJ, Fareed J (1996) Pharmacological profiling of recombinant tissue factor pathway inhibitor. Semin Thromb Hemost 22:213–219

Kaiser B, Fareed J (1996) Recombinant full-length tissue factor pathway inhibitor (TFPI) prevents thrombus formation and re-thrombosis after lysis in a rabbit model of jugular vein thrombosis. Thromb Haemost 76:615–620

Kirchofer D, Guha A, Nemerson Y, Konigsberg WH, Vilbois F, Chene C et al (1995) Activation of blood coagulation factor VIIa with cleaved tissue factor estracellular domain and crystallization of the active complex. Proteins 22:419–425

Kouri RK, Koudsi B, Kaiding F, Ornberg RL, Wun TC (1993) Prevention of thrombosis by topical application of tissue factor pathway inhibitor in a rabbit model of vascular trauma. Ann Plast Surg 30:398–402

Kouri RK, Benes CO, Ingram D, Natarajan N, Sherman JW, Yeramian P (1997) Topical human recombinant tissue factor pathway inhibitor to prevent thrombosis: Experimental studies and Phase I trial results. Thromb Haemost Abstract Suppl: 688

Kristensen AT, Ezban M, Villumsen A, Groes L, Elm T, Medner U (1997) The effect of simultaneous administration of aspirin, heparin, and the FFR-ck-active-site inhibited rFVIIa on nail cuticle bleeding in rabbits. Thromb Haemost: 688

Lassila R, Peltonen S, Lepantalo M, Saarinen O, Kauhanen P, Manninen V (1993) Severity of peripheral atherosclerosis is associated with fibrinogen and degradation of cross-linked fibrin. Arterioscler Thromb 13:1738–1742

Lefkovits J, Malycky JL, Rao JS, Hart CE, Plow EF, Topol EJ et al (1996) Selective inhibition of Factor Xa is more efficient than factor VIIa-Tissue factor complex blockade at facilitating coronary thrombolysis in the canine model. J Am Coll Cardiol 28:1858–1865

Lesnick P, Vonica A, Guerin M, Moreau M, Chapman MJ (1993) Anticoagulant activity of tissue factor pathway inhibitor in human plasma is preferentially associated with dense subspecies of LDL and HDL and with Lp(a). Arterioscler Thromb 13:1066–1075

Levi M, ten Cate H, Bauer KA, van der Poll T, Edgington TS, Buller HR et al (1994) Inhibition of endotoxin-induced activation of coagulation and fibrinolysis by pentoxifylline or by a monoclonal anti tissue factor antibody In chimpanzees. J Clin Invest 93:114–120

Lindahl AK, Nordfang O, Wildgoose P, Kelly AB, Harker LA, Hanson SR (1993a) Antithrombotic effects of a truncated tissue factor pathway inhibitor (TFPI) in baboons. Thromb Haemost 69:742

Lindahl AK, Wildgoose P, Lumsden AB, Allen R, Kelly AB, Harker LA et al (1993b) Active site-inhibited factor VIIa blocks tissue factor activity and prevents arterial thrombus formation in baboons. Circulation 88:1–417

Marmur JD, Rossikhina M, Guha A, Fyfe B, Friedrich V, Mendlowitz M et al (1993) Tissue factor is rapidly induced in arterial smooth muscle after balloon injury. J Clin Invest 91:2253–2259

Marmur JD, Thiruvikraman SV, Fyfe BS, Guha A, Sharma SK, Ambrose JA et al (1996) Identification of active tissue factor in human coronary atheroma. Circulation 94:1226–1232

Mast AE, Broze GJ Jr (1996) Physiological concentrations of tissue factor pathway inhibitor do not inhibit prothrombinase. Blood 87:1845–1850

Moreno PR, Bernardi VH, Lopez-Cuellar J, Murcia AM, Palacios IF, Gold HK et al (1996) Macrophages, smooth muscle cells, and tissue factor in unstable angina. Implications for cell-mediated thrombogenicity in acute coronary syndromes. Circulation 94:3090–3097

Morrissey JH, Fakhrai H, Edgington TS (1987) Molecular cloning of the cDNA for tissue factor, the cellular receptor for the initiation of the coagulation protease cascade. Cell 50:129–135

Morrissey JH, Macik BG, Neuenschwander PF, Comp PC (1993) Quantitation of activated factor VII levels in plasma using a tissue factor mutant selectivity deficient in promoting factor VII activation. Blood 81:734–744

Morrissey JH, Neuenschwander PF, Huang Q, McCallum CD, Su B, Johnson AE (1997) Factor VIIa-tissue factor: functional importance of protein-membrane interactions. Thromb Haemost 78:112–116

Mueller BM, Reisfeld RA, Edgington TS, Ruf W (1992) Expression of tissue factor by melanoma cells promotes efficient hematogenous metastasis. Proc Natl Acad Sci USA 89:11832–11836

Muller YA, Ultsch MH, de Vos AM (1996) The crystal structure of the extracellul ard-omain of human tissue factor refined to 1.7Å resolution. J Mol Biol 256:144–159

Muller YA, Ultsch MH, Kelley RF, de Vos AM (1994) Structure of the extracellular domain of human tissue factor: location of the factor VIIa binding site. Biochemistry 33:10864–10870

Muller-Berghaus G (1989) Pathophysiologic and biochemical events in disseminated intravascular coagulation: dysregulation of procoagulant and anticoagulant pathways. Semin Thromb Hemost 15:58–87

Nemerson Y (1988) Tissue factor and hemostasis. Blood 71:1–8

Petersen JG, Meyn G, Rasmussen JS, Petersen J, Bjorn SE, Jonassen I et al (1993) Characterization of human tissue factor pathway inhibitor variants expressed in Saccharomyces cerevisiae. J Biol Chem 268:13344–13351

Petersen LC, Valentin S, Hedner U (1995) Regulation of the extrinsic pathway system in health and disease: the role of factor VIIa and tissue factor pathway inhibitor. Thromb Res 79:1–47

Ragni M, Cirillo P, Pascucci I, Scognamiglio A, D'Andrea D, Eramo N et al (1996) A monoclonal antibody against tissue factor shortens tissue plasminogen activator lysis time and prevents reocclusion in a rabbit model of carotid artery thrombosis. Circulation 93:1913–1918

Ragosta M, Gimple LW, Gertz SD, Dunwiddie CT, Vlasuk GP, Haber HL et al (1994) Specific factor Xa inhibition reduces restenosis after balloon angioplasty of athero-sclerotic femoral arteries in rabbits. Circulation 89:1262–1271

Rao LV (1992) Tissue factor as a tumor procoagulant. Cancer Metastasis Rev 11: 249–266

Rao LV, Nordfang O, Hoang AD, Pendurthi UR (1995) Mechanism of antithrombin III inhibition of factor VIIa/tissue factor activity on cell surfaces. Comparison with tissue factor pathway inhibitor/factor Xa-induced inhibition of factor VIIa/tissue factor activity. Blood 85:121–129

Rao LV, Rapaport SI, Hoang AD (1993) Binding of factor VIIa to tissue factor permits rapid antithrombin III/heparin inhibition of factor VIIa. Blood 81:2 600–2607

Rapaport SI (1991) The extrinsic pathway inhibitor: a regulator of tissue factor-dependent blood coagulation. Thromb Haemost 66:6–15

Rapaport SI, Rao LV (1995) The tissue factor pathway: how it has become a "prima ballerina." Thromb Haemost 74:7–17

Rivers RP, Hathaway WE, Weston WL (1975) The endotoxin-induced coagulant activity of human monocytes. Brit J Haematol 30:311–316

Rote WE, Oldeschulte GL, Dempsey EM, Vlasuk GP (1996) Evaluation of a novel small protein inhibitor of the blood coagulation factor VIIa/tissue factor complex in animal models of arterial and venous thrombosis. Circulation 94: 1–695

Rottingen JA, Enden T, Camerer E, Iversen JG, Prydz H (1995) Binding of human factor VIIa to tissue factor induces cytosolic Ca^{2+} signals in J82 cells, transfected COS-1 cells, Madin-Darby canine kidney cells and in human endothelial cells induced to synthesize tissue factor. J Biol Chem 270:4650–4660

Sarembock IJ, Gertz SD, Gimple LW, Owen RM, Powers ER, Roberts WC (1991) Effectiveness of recombinant desulphatohirudin in reducing restenosis after balloon angioplasty of atherosclerotic femoral arteries in rabbits. Circulation 84: 232–243

Sato Y, Asada Y, Marutsuka K, Hatakeyama K, Sumiyoshi A (1996) Tissue factor induces migration of cultured aortic smooth muscle cells. Thromb Haemost 75:389–392

Scarpati EM, Wen D, Broze GJ Jr, Miletich JP, Flandermeyer RR, Siegel NR et al (1987) Human tissue factor: cDNA sequence and chromosome localization of the gene. Biochemistry 26:5234–5238

Shigematsu Y, Miyata T, Higashi S, Miki T, Sadler JE, Iwanaga S (1992) Expression of human soluble tissue factor in yeast and enzymatic properties of its complex with factor VIIa. J Biol Chem 267:21329–21337

Smith EB (1986) Fibrinogen, fibrin and fibrin degradation products in relation to atherosclerosis. Clin Haematol 15:355–370

Smith EB, Keen GA, Grant A, Stirk C (1990) Fate of fibrinogen in human arterial intima. Arteriosclerosis10:263–275

Speidel CM, Eisenberg PR, Ruf W, Edgington TS, Abendschein DR (1995) Tissue factor mediates prolonged procoagulant activity on the luminal surface of balloon-injured aortas of rabbits. Circulation 92:3323–3330

Spicer EK, Horton R, Bloem L, Bach R, Williams KR, Guha A et al (1987) Isolation of cDNA clones coding for human tissue factor: primary structure of the protein and cDNA. Proc Natl Acad Sci USA 84:5148–5152

Spokas EG, Wun TC (1992) Venous thrombosis produced in the vena cava of rabbits by vascular damage and stasis. J Pharmacol Toxicol Methods 27:225–232

Stassens P, Bergum PW, Ganseman Y, Jespers L, Laroche Y, Huang S et al (1996) Anticoagulant repertoire of the hookworm Ancyclostoma caninum. Proc Natl Acad Sci USA 93:2149–2154

Taubman MB (1993) Tissue factor regulation in vascular smooth muscle: a summary of studies performed using in vivo and in vitro models. Am J Cardiol 72:55C–60 C

Taubman MB, Marmur JD, Rosenfield CL, Guha A, Nichtberger S, Nemerson Y (1993) Agonist-mediated tissue factor expression in cultured vascular smooth muscle cells. Role of Ca^{2+} mobilization and protein kinase C activation. J Clin Invest 91:547–552

Taylor FB Jr, Chang A, Esmon CT, D'Angelo A, Vigano-D'Angelo S, Blick KE (1987) Protein C prevents the coagulopathic and lethal effects of Escherichia coli infusion in the baboon. J Clin Invest 79:918–925

Taylor FB Jr, Chang A, Ruf W, Morrissey JH, Hinshaw L, Catlett R (1991a) Lethal E. coli septic shock is prevented by blocking tissue factor with monoclonal antibody. Circ Shock 33:127–134

Taylor FB Jr, Chang AC, Peer GT, Mather T, Blick K, Catlett R et al (1991b) DEGR-Factor Xa blocks disseminated intravascular coagulation initiated by Escherichia coli without preventing shock or organ damage. Blood 78:364–368

Taylor FB Jr, Emerson TE Jr, Jordan R, Chang AK, Blick KE (1988) Antithrombin-III prevents the lethal effects of Escherichia coli infusion in baboons. Circ Shock 26:227–235

Toschi V, Gallo R, Lettino M, Fallon JT, Gertz SD, Fernandez-Ortiz A et al (1997) Tissue factor modulates the thrombogenicity of human atherosclerotic plaques. Circulation 95:594–599

Vlasuk GP, Bergum PW, Rote WE, Ruf W (1997) Mechanistic and pharmacological evaluation of NAPc2, a novel small protein inhibitor of the factor VIIa/tissue factor complex. Thromb Haemost Abstract Suppl: 688

Wang Z, Hebert D, Kaplan AV, Creasy A, Galluppi GR (1996) Persistent inhibition to 12 hours of mural platelet thrombosis after a single local infusion of tissue factor pathway inhibitor at the site of angioplasty. Circulation 94:1–268

Warr TA, Rao LV, Rapaport SI (1990) Disseminated intravascular coagulation in rabbits induced by administration of endotoxin or tissue factor: effect of anti-tissue factor antibodies and measurement of plasma extrinsic pathway inhibitor activity. Blood 75:1481–1489

Werling RW, Zacharski LR, Kisiel W, Bajaj SP, Memoli VA, Rousseau SM (1993) Distribution of tissue factor pathway inhibitor in normal and malignant human tissues. Thromb Haemost 69:366–369

Wilcox JN (1993) Molecular biology: insight into the causes and prevention of restenosis after arterial intervention. Am J Cardiol 72:88E–95E

Wilcox JN, Smith KM, Schwartz SM, Gordon D (1989) Localization of tissue factor in the normal vessel wall and the atherosclerotic plaque. Proc Natl Acad Sci USA 86:2839–2843

Wilson E, Mai Q, Sudhir K, Weiss RH, Ives HE (1993) Mechanical strain induces growth of vascular smooth muscle cells via autocrine action of PDGF. J Cell Biol 123:741–747

Wun TC, Kretzmer KK, Girard TJ, Miletich JP, Broze GJ Jr (1988) Cloning and characterization of a cDNA coding for the lipoprotein-associated coagulation inhibitor shows that it consists of three tandem Kunitz-type inhibitory domains. J Biol Chem 263:6001–6004

Yla-Herttuala S, Rosenfeld ME, Pathasarathy S, Sigal E, Sarkioja T, Witztum JL (1991) Gene expression in macrophage-rich human atherosclerotic human lesions. 15-Lipoxygenase and acetyl low density lipoprotein receptor messenger RNA colocalize with oxidation specific lipid-protein adducts. J Clin Invest 87: 1146–1152

Zhang Y, Deng Y, Luther T, Muller M, Ziegler R, Waldherr R et al (1994) Tissue factor controls the balance of angiogenic and antiangiogenic properties of tumor cells in mice. J Clin Invest 94:1320–1327

CHAPTER 16

Natural Anticoagulants and Their Pathways

C.T. ESMON

A. Introduction

Blood coagulation is a complex process involving blood cell surfaces, plasma proteins, and inhibitory mechanisms. Many of the inhibitory mechanisms are associated with the blood vessel wall. The protein C pathway is one of the major mechanisms involved in controlling blood coagulation. The interaction of the protein C pathway with a highly simplified version of the major coagulation cascade is illustrated in Fig. 1. The blood clotting process occurs when tissue factor is exposed to blood. This triggers either factor X or factor IX activation. Factor IXa complexes with factor VIIIa to provide an alternative activation mechanism for factor X. Factor Xa, in complex with factor Va, generates thrombin which causes fibrin clots and platelet activation. All of these reactions proceed on the surface of activated or damaged cells when the membrane phospholipid asymmetry is perturbed, resulting in exposure of negatively charged phospholipids. These clotting reactions are offset by natural anticoagulant mechanisms, the most important of which are thought to include the tissue factor pathway inhibitor (Chap. 15), antithrombin-heparin (Chap. 9) and the protein C anticoagulant pathway (depicted in Fig. 1).

The protein C anticoagulant pathway is initiated when thrombin binds to thrombomodulin (TM) on the surface of the endothelial cell. This complex generates activated protein C (APC) from its plasma precursor, protein C. This activation can be facilitated by an endothelial cell protein C receptor, EPCR, that binds protein C or APC. APC can inactivate factor Va or factor VIIIa on membrane surfaces, a process that is facilitated by protein S. When bound to EPCR, APC does not inactivate factor Va and presumably targets an alternative, yet to be identified, substrate. The net effect of this process is an "on demand" anticoagulant response that senses changes in thrombin concentration.

The natural anticoagulants function by distinctly different and complementary mechanisms. Tissue factor pathway inhibitor inhibits factor Xa and the factor Xa-inhibitor complex then feeds back to inactivate tissue factor-factor VIIa complexes, thus shutting down the initiation of coagulation (BROZE et al. 1990; RAPAPORT and RAO 1992; see also Chap. 15). The antithrombin-heparin mechanism contributes to inhibition of thrombin, factor Xa, factor IXa and tissue factor VIIa, with its major targets probably including

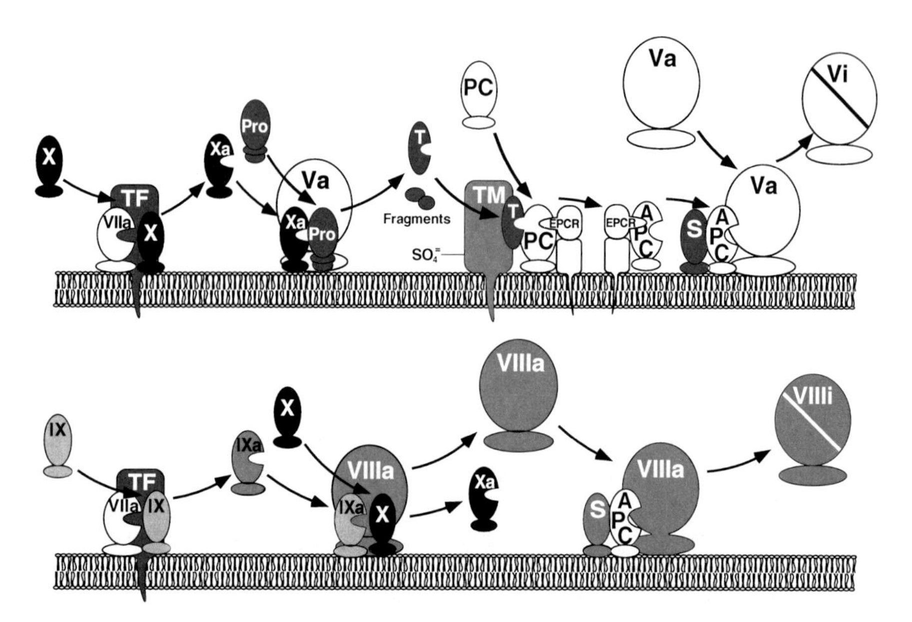

Fig. 1. The function of membranes and cofactors in blood coagulation. The enzymes associate with cofactors on membrane surfaces. Factor VIIa associates with tissue factor to activate either factor X or factor IX. Factor IXa associates with factor VIIIa to activate factor X. Factor Va associates with factor Xa to activate prothrombin. Thrombin associates with TM to activate protein C, a process which is enhanced on some blood vessels by the presence of the endothelial cell protein C receptor, EPCR. Once activated protein C is formed it dissociates relatively slowly from EPCR to interact with protein S to inactivate factors Va and VIIIa, thereby blocking the coagulation cascade. See the text for discussions of mechanism. (Modified from Esmon, CT: Thrombomodulin as a model of molecular mechanisms that modulate protease specificity and function at the vessel surface. FASEB J. 9:946, 1995. Copyright 8 1995 The FASEB Journal)

thrombin and factor Xa (MARCUM and ROSENBERG 1988; BOURIN and LINDAHL 1993). Thus, the antithrombin-heparin mechanism is primarily responsible for controlling amplification of the coagulation reactions and the terminal clotting enzyme, thrombin. These mechanisms are focused on the protease components of coagulation. The protein C pathway plays two roles, regulation of thrombin activity and inactivation of the two key cofactors, factors Va and VIIIa. Not surprisingly, these inhibitors work much better in combination than separately when added to purified coagulation reactions in vitro. Indeed, thrombotic risk seems to increase when deficiencies exist in more than one pathway, or in more than one component of a pathway (LANE et al. 1996a,b).

In principle, hereditary thrombosis might be associated with increased coagulation reactions or with impairment of the inhibitory mechanisms. Current information points to impairment of the anticoagulant reactions as the major cause of hereditary thrombosis (thrombophilia) (LANE et al. 1996a,b).

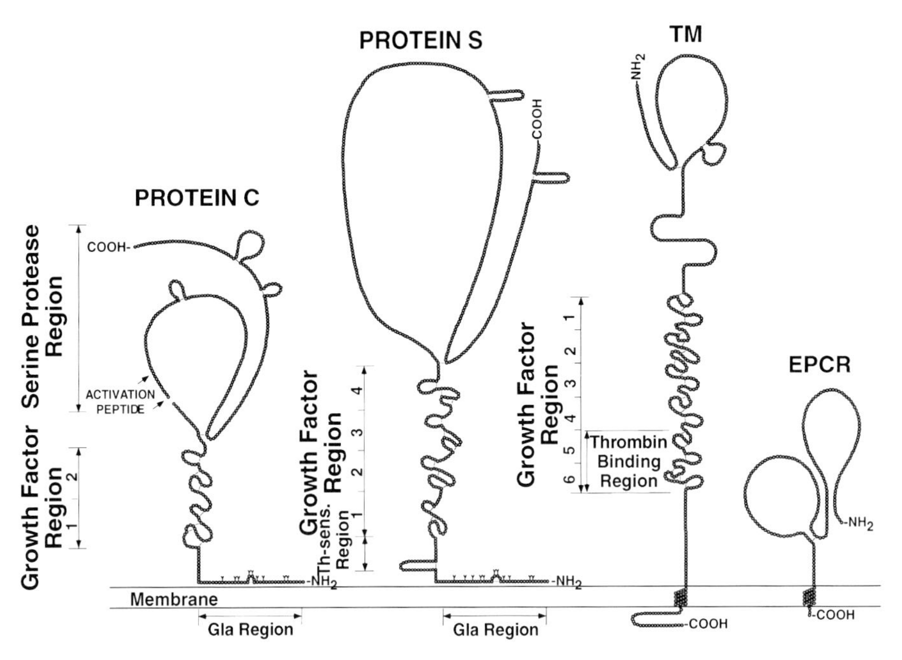

Fig. 2. Schematic representation of protein C, protein S, the endothelial cell protein C receptor (EPCR) and TM. The vitamin K dependent Gla residues of protein C and protein S are indicated by small Y-shaped symbols. Formation of these vitamin K-dependent residues is essential to full activity of protein C and protein S. Gla, (-carboxyglutamic acid; Th.-sens., thrombin sensitive. The EPCR domain structure is based on homology to the CD1 family. (Modified from Esmon CT: The roles of protein C and thrombomodulin in the regulation of blood coagulation. J Biol Chem 1989; 264:4743. Copyright 8 1989 the American Society for Biochemistry and Molecular Biology, Inc.)

Of the natural anticoagulant pathways, abnormalities in the protein C pathway are most commonly associated with thrombophilia (Dahlbäck 1994; Lane et al. 1996a,b; Florell and Rodgers 1997).

Each of the components of the protein C pathway functions by a unique mechanism, and therefore their application to the treatment of thrombotic disease will vary. To understand this pathway, it is useful to consider our current knowledge of the mechanisms of action of each of these components and their regulation. In considering the mechanism of action, a basic understanding of the protein structure and its interaction with the membrane surface is essential. A simplified depiction of the components of the pathway as they interact with membranes is presented in Fig. 2. For those desiring more information about the clinical or basic aspects of the pathway, there have been a number of reviews written by this author (Esmon and Schwarz 1995; Esmon 1995; Esmon and Fukudome 1995) and others (Fulcher et al. 1984; Koedam et al. 1988; Eaton et al. 1986; Griffin et al. 1993; Halbmayer et al. 1994;

Bertina et al. 1994; Dahlbäck 1994; Dahlbäck 1991; Davie et al. 1991; Walker and Fay 1992; Pabinger et al. 1992; Bourin and Lindahl 1993; Alving and Comp 1992; Reitsma et al. 1993; Castellino 1995; Reitsma et al. 1995; Florell and Rodgers 1997).

I. The Protein C Activation Complex

The simplest view of the protein C activation complex is shown in Fig. 3. The model illustrates the key feature of the activation complex that will be dis-

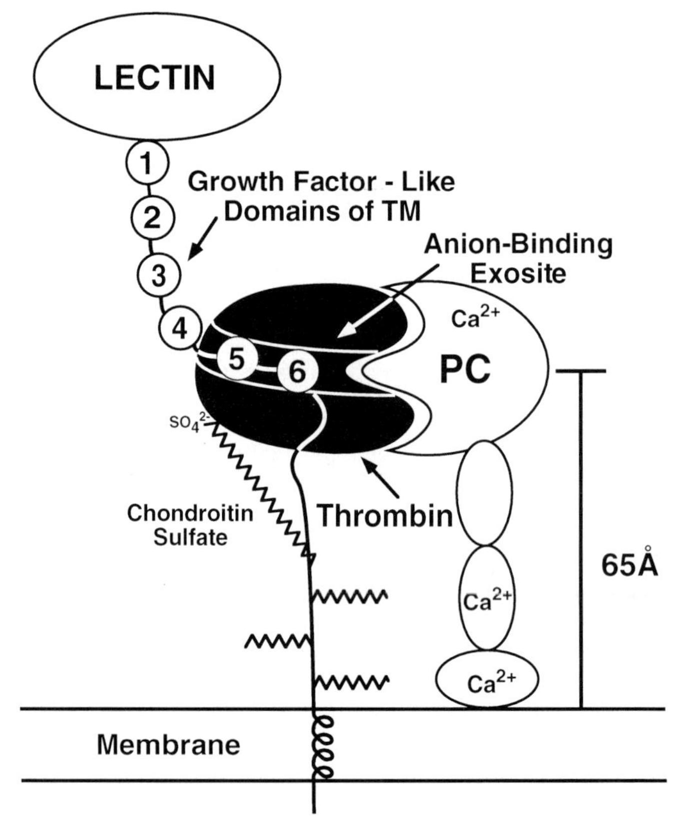

Fig. 3. A model of the protein C activation complex. The small balls labeled 1–6 correspond to the EGF like repeats in thrombomodulin, the extended structure rising from the membrane contains O linked glycosylation sites (shown as zig zag lines) and is the site of attachment of the chondroitin sulfate, shown as the zig-zag line with the terminal sulfate. This complex attacks protein C to release a 12 residue peptide from the amino terminus of the heavy chain of protein C (not shown) that leads to activation. Bothy protein C and activated protein C interact with the activation complex [39,40]. Lectin refers to the lectin like amino terminal domain of thrombomodulin. (Esmon CT: Molecular events that control the protein C anticoagulant pathway. Thromb Haemostas 1993;70(1):29. Copyright 8 1993 F.K. Schattauer Verlagsgesellschaft mbH, Stuttgart)

cussed in more detail below. TM interaction with anion binding exosite 1, a site shared with other thrombin substrates and receptors, is mediated by the protein portion of TM. A second interaction with thrombin is mediated by the chondroitin sulfate moiety. Forms of TM exist with and without the chondroitin sulfate, and the presence of this moiety has major influences on the functions of the activation complex. A critical role for TM is suggested by gene deletion in the mouse which results in early embryonic lethality (HEALY et al. 1995). Furthermore, a TM mutation has been identified in a patient with thrombosis (ÖHLIN and MARLAR 1995), and mutations in the TM gene appear to co-segregate with thrombotic disease in families (ÖHLIN et al. 1997).

The endothelium is the major site of protein C activation (ESMON 1989; ESMON and OWEN 1981). Thrombin can activate protein C by itself, but the rate of activation is too slow to be physiologically relevant. On the endothelial cell surface, thrombin interacts with TM, and this complex activates protein C rapidly. Protein C activation is augmented by EPCR on endothelial cell surfaces (STEARNS-KUROSAWA et al. 1996). The EPCR enhancement on cultured endothelium is about four- to fivefold and is due primarily to an apparent increase in protein C affinity (decrease in *K*m) (STEARNS-KUROSAWA et al. 1996). Soluble recombinant EPCR inhibits this protein C activation on cell surfaces, but has no effect in solution (REGAN et al. 1996). In addition to EPCR, human factor Va can interact with endothelium and promote protein C activation about threefold (MARUYAMA et al. 1984). It can also facilitate protein C activation in solution (SALEM et al. 1983; SALEM et al. 1984). Thus, factor Va can be a substrate for APC and participate in APC formation.

TM is found at high levels on all vascular endothelium except brain capillaries (ISHII et al. 1986) and liver sinusoidal endothelium (LASZIK et al. 1997), at low levels in platelets (SUZUKI et al. 1988), and as soluble proteolytic degradation products which circulate in normal plasma at about 10 ng/ml (TAKANO et al. 1990; ISHII and MAJERUS 1985). Some of these degradation products are capable of complexing with thrombin to activate protein C (TAKANO et al. 1990).

EPCR, the other integral membrane component of the activation complex, exhibits a vascular distribution different from TM. In general, EPCR is present at high levels on arteries, veins, arterioles and some postcapillary venules, but is absent or at low levels on capillary endothelium, except in the vasa recta of the kidney medulla, the liver sinusoids, and the subcapsular tubules of the adrenal cortex (LASZIK et al. 1997). This difference in distribution suggests that the functional properties of the protein C activation complexes differ between large vessels and capillaries. The differences in the capillary and large vessel activation complexes are illustrated schematically in Fig. 4. The complex to the left depicts the capillary activation complex which has a low affinity for protein C due to the paucity of EPCR, while the complex to the right represents the complex on large vessels which has a high affinity for protein C. In the capillaries, the effective endothelial cell surface area to blood volume ratio is in excess of 1000 cm^2/ml which would result in TM

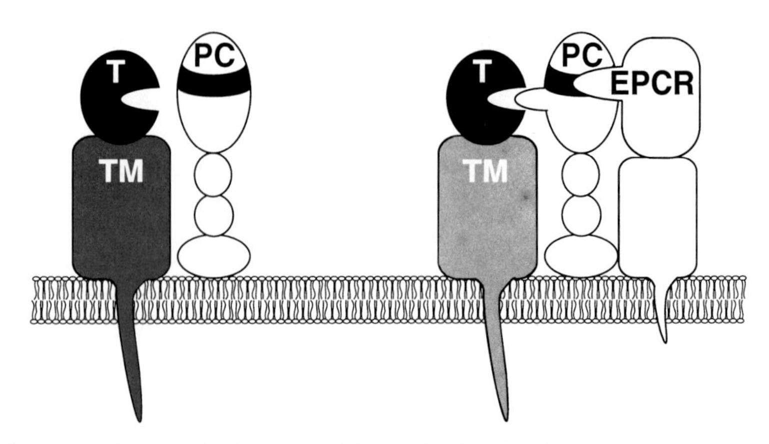

Fig. 4. Comparison of the hypothetical protein C activation complexes on large and small vessels. The complex on the left lacking EPCR would be representative of the capillary bed where TM concentrations are high but EPCR is low. This complex would be anticipated to have low affinity for protein C. The complex on the right is representative of large vessels where the effective TM concentration is low but EPCR is abundant. This complex would be higher affinity for protein C and might be less influenced by changes in circulating protein C concentrations than that in the capillary. See the text for discussion

concentrations in excess of 100 nM (Esmon 1989), assuming 50000 TM molecules/endothelial cell (Maruyama and Majerus 1985). Because of the much greater effective concentration of TM in the capillaries, one might expect that this would be the primary site of protein C activation. Clinically, protein C deficiency is often associated with microvascular thrombosis, consistent with a major role of protein C in controlling microvascular thrombotic processes. Since EPCR is absent or at low levels in the capillaries, and EPCR improves the affinity of protein C for the activation complex, changes in protein C levels probably have a greater effect on protein C activation rates in capillary endothelial cell activation complexes than in those of the larger vessels.

TM has other anticoagulant effects in addition to the activation of protein C. Thrombin-binding to TM blocks the ability of thrombin to clot fibrinogen (Esmon et al. 1982), activate platelets (Esmon et al. 1983) or activate factors V (Esmon et al. 1982) or XIII (Polgar et al. 1986). The ability of TM to block these reactions can be explained mechanistically by the observation that these substrates and receptors all interact with anion binding exosite 1 on thrombin (Mathews et al. 1994a,b; Tsiang et al. 1990; Ye et al. 1992; Hofsteenge et al. 1986; reviewed in Esmon 1995; and see Fig. 3). Hence, TM, which binds tightly to thrombin ($Kd \leq 10$ nM), blocks alternative substrate binding and functions as an effective competitive inhibitor.

The affinity for thrombin is modulated by the presence of a chondroitin sulfate moiety on TM. Recombinant TM exhibits about a 10- to 20-fold increase in affinity when the chondroitin sulfate is present. Virtually all of the

TM molecules isolated from rabbit lung contain the chondroitin sulfate (Bourin and Lindahl 1993), but the majority of TM isolated from human placenta appears to lack the chondroitin sulfate moiety (Lin et al. 1994). This probably accounts for the observation that TM isolated from placenta is a much weaker inhibitor of thrombin clotting than rabbit TM (Maruyama et al. 1985) and some recombinant soluble human TM preparations. Whether the chondroitin sulfate is absent from most TM on human vessels or is lost during TM isolation remains an open question.

TM containing the chondroitin sulfate can stimulate thrombin inactivation by antithrombin (Bourin and Lindahl 1993). The thrombin-antithrombin complex then rapidly dissociates from TM. In addition, TM enhances thrombin inactivation by the protein C inhibitor. This acceleration does not require the chondroitin sulfate moiety, but the chondroitin sulfate stimulates the reaction about twofold. Based on rate calculations, it is likely that these inhibitors contribute approximately equally to the inactivation of thrombin bound to TM (Rezaie et al. 1995).

From the above discussion of TM function, it is clear that TM mediates at least three functionally distinct anticoagulant mechanisms: the acceleration of thrombin-dependent protein C activation, direct inhibition of thrombin clotting activities, and enhanced inhibition of TM-bound thrombin by plasma proteinase inhibitors. Physiologically, the relative importance of these activities on the net antithrombotic effects of TM is uncertain.

In addition to the antithrombotic activities described above, TM has activities which seem to be coagulant in nature. For instance, TM accelerates the activation of factor XI by thrombin about 20-fold (Gailani and Broze 1991). However, since TM accelerates thrombin inhibition more than 20-fold, the net effect on coagulation may be neutral. TM has also been shown to accelerate the inhibition of pro-urokinase by thrombin (de Munk et al. 1991; Molinari et al. 1992). Soluble TM accelerates the activation of TAFI, a thrombin-dependent fibrinolytic inhibitor, which is the zymogen of a plasma procarboxypeptidase B (Bajzar et al. 1996). The acceleration of TAFI and protein C activation by TM are on the same order of magnitude (Bajzar et al. 1996). Thus, it would appear that TM might function to inhibit fibrinolysis. Alternatively, TAFI has major in vivo functions yet to be identified which dominate its inhibitory activity toward fibrinolysis. Even if one assumes that TM does function physiologically as an inhibitor of fibrinolysis, as predicted by compelling in vitro data, it would appear that its dominant activity is the anticoagulant response, since TM infusion protects against thrombosis/DIC (disseminated intravascular coagulation) in a variety of animal models of thrombotic injury (see Sect. C).

II. The APC Anticoagulant Complexes

Once APC is formed, it can complex with protein S to inactivate either factor Va or factor VIIIa. Protein S functions include enhancing the affinity of APC for the membrane surface (Walker 1981), increasing the rate of factor Va and VIIIa inactivation to a small (3-fold) extent (Dahlbäck 1986; Koedam et al. 1988), and preventing coagulation enzymes from complexing with factor Va (Solymoss et al. 1988) or VIIIa (Regan et al. 1994) to prevent APC inactivation of these cofactors. Factor V has been reported to augment APC inactivation of factor VIII and to work synergistically with protein S (Varadi et al. 1996; Shen and Dahlbäck 1994). Factor Va does not appear to possess this activity. Thus, under conditions where most of the factor V has been activated, one would presume that the APC anticoagulant activity might diminish. In plasma, the APC anticoagulant response is augmented approximately tenfold by protein S (D'Angelo et al. 1988). There is little doubt that protein S is important to the inhibition of coagulation since deficient patients exhibit thrombotic complications (Pabinger et al. 1992, 1994) which can include skin lesions (purpura fulminans) (Mahasandana et al. 1990) in the homozygous state.

Protein S circulates in two forms, free and in complex with C4bBP, a regulatory protein of the complement cascade. Only the free form of protein S can function as a cofactor for APC (Comp et al. 1984; Dahlbäck 1986). In vivo, administration of excess C4bBP to baboons results in a dramatically increased coagulation response to infusion of low doses of *Escherichia coli*, and this response can be blocked by co-infusion of excess protein S (Taylor et al. 1991a). Low levels of free protein S are commonly observed in clinical conditions associated with thrombotic disease (Sacco et al. 1989; Vigano-D'Angelo et al. 1987; Boerger et al. 1987; D'Angelo et al. 1988), and are apparently in part due to increased complex formation with C4bBP.

It is unclear what positive role protein S-C4bBP complex formation plays in regulating biological processes. It has been observed that C4bBP can interact with inflammatory cells in a protein S-dependent fashion (Furmaniak-Kazmierczak et al. 1993), possibly playing a role in protecting cells from complement-mediated damage.

In addition to augmenting APC anticoagulant activity, protein S has been shown to inhibit procoagulant reactions in purified systems (Heeb et al. 1993, 1994; Hackeng et al. 1994; Koppelman et al. 1995). The physiological importance of these findings is uncertain since immunoadsorption of protein S from plasma has little effect on coagulation reactions, at least as judged by clotting times.

III. Inhibition of the Anticoagulant Complex

APC differs from most other serine proteases by being quite resistant to inactivation by the plasma proteinase inhibitors. In vivo studies have suggested that the half-life of APC is approximately 10–15 min (Comp et al. 1982; Gruber et al. 1989, 1991, Espana et al. 1991; Okajima et al. 1990) as compared to seconds for thrombin (Lollar and Owen 1980). The major inhibitors appear to be protein C inhibitor, α_1-antitrypsin inhibitor, and $\alpha 2$-macroglobulin (SCULLY et al. 1993; HEEB et al. 1989a, 1991).

B. Modulation of the Protein C Pathway in Disease

Several lines of evidence suggest that the protein C anticoagulant pathway function may be down-regulated in disease states. Some of the changes that are believed to occur in the protein C system between normal (Fig. 5a) and inflammatory situations (Fig. 5b) are illustrated here and outlined briefly below, to provide a framework for the more detailed discussion to follow. Central features of the differences are: loss of TM from the vasculature, reduced free protein S concentrations, and recruitment of inflammatory cells to the vessel wall, which may inhibit the system due to proteolysis of TM or release of inflammatory cytokines. With regard to coagulation, membrane perturbation can occur under inflammatory circumstances that can promote amplification of coagulation (WIEDMER et al. 1986) and lead to expression of monocyte tissue factor (EDGINGTON et al. 1991; CARSON and BROZNA 1993; MORRISSEY and DRAKE 1993; RAPAPORT and RAO 1992). The vascular damage, the increase in procoagulant substances and the decrease in anticoagulant pathway function probably provide insights into the basis for the association between inflammation and thrombosis. This situation may be aggravated by the fact that people at risk of, or with, thrombotic disease are often treated with vitamin K antagonists. Unlike the other natural anticoagulants, protein C and protein S are both vitamin K-dependent proteins, and hence their ability to bind to membrane surfaces and function in this pathway are impaired by the vitamin K antagonists used in oral anticoagulant therapy (VIGANO-D'ANGELO et al. 1986; D'ANGELO et al. 1988).

Many of the processes associated with thrombotic disease have inflammatory components. Based on in vitro data, endotoxin (MOORE et al. 1987) and the inflammatory cytokines, tumor necrosis factor (TNFα) or interleukin-1, can down-regulate TM (NAWROTH and STERN 1986; NAWROTH et al. 1986) and EPCR (FUKUDOME and ESMON 1994) expression on endothelial cells in culture. TM levels have been observed to decrease in vivo, under some conditions such as allograft rejection (TSUCHIDA et al. 1992) and villitis (LABARRERE et al. 1990). These may be more complex situations than acute inflammatory injury caused by endotoxin shock, since in rat kidney, TM antigen and activity was not altered in kidneys infarcted with thrombi (LASZIK et al. 1994), and baboon endothelial cells did not appear to have reduced TM levels based on qualitative immunohistochemical analysis (DRAKE et al. 1993). TM down-regulation can be prevented by many factors including interleukin-4 (KAPIOTIS et al. 1991) and retinoic acid (DITTMAN et al. 1994; KOYAMA et al. 1994), making our understanding of the pathophysiologic conditions under which this potentially important phenomenon may occur very incomplete.

Systemic assays for the presence of TM degradation products would favor the concept that TM is down-regulated by inflammatory processes. Specifically, many inflammatory human diseases, including septic shock, result in large increases in circulating plasma TM levels (TAKANO et al. 1990; TANAKA et al. 1991; TAKAHASHI et al. 1991, 1992; ASAKURA et al. 1991; WADA et al. 1992; OHDAMA et al. 1994), probably resulting from neutrophil elastase-mediated proteolytic cleavage of endothelial cell-associated TM (BOEHME et al. 1996). TM activity can be inhibited by eosinophil release products, in

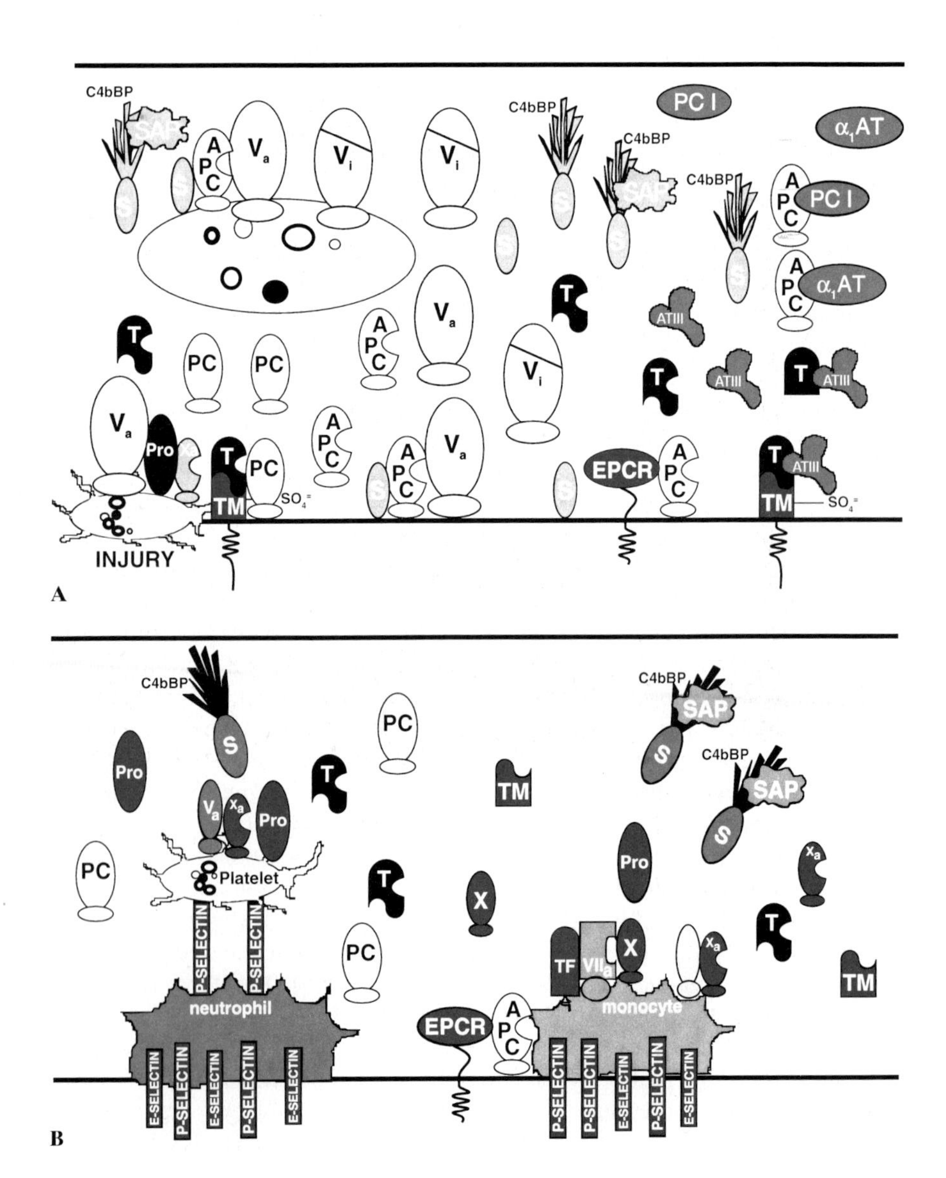

particular, major basic protein (SLUNGAARD et al. 1993), thereby providing an additional mechanism by which inflammation can inhibit the anticoagulant function of the pathway.

Protein S levels decrease in DIC and autoimmune disease. This decrease may be due to changes in synthesis rate, increased binding to C4bBP, or proteolytic degradation of protein S (SACCO et al. 1989; VIGANO-D'ANGELO et al. 1987; BOERGER et al. 1987; D'ANGELO et al. 1988, 1993; HEEB et al. 1989b; GIROLAMI et al. 1989; SHETH and CARVALHO 1991; SCOTT et al. 1991; TAKAHASHI et al. 1989; FOURRIER et al. 1992; TSUCHIDA et al. 1992; CHAFA et al. 1992; PRINCE et al. 1995). In some forms of septic shock, particularly meningococcemia, protein C consumption correlates well with the onset of necrotic skin lesions (purpura fulminans) and a negative clinical outcome (FIJNVANDRAAT et al. 1995; POWARS et al. 1993).

Taken together, the prevalence of deficiencies in thrombophilic patients and the impairment of the protein C system by inflammatory events and common drugs provide a framework for proposing that augmentation of this

Fig. 5A,B. Changes between control of coagulation in normal vs inflammed vasculature. **A** The protein C anticoagulant pathway under normal conditions. Vascular injury initiates prothrombin (Pro) activation that results in thrombin (T) formation. Prothrombin activation involves complex formation between factor Va (Va) and factor Xa (Xa). Thrombin then binds to thrombomodulin (TM) on the lumen of the endothelium, illustrated by the heavy line, and the thrombin- TM complex converts protein C (PC) to activated protein C (APC). Thrombin bound to TM can be inactivated very rapidly by antithrombin III (ATIII), at which time the thrombin-antithrombin III complex rapidly dissociates from TM. Activated protein C (APC) then binds to protein S (S) on cellular surfaces. The activated protein C-protein S complex then converts factor Va to an inactive complex (Vi), illustrated by the slash through the larger part of the two-subunit factor Va molecule. Protein C and activated protein C (PAC) interact with an endothelial cell protein C receptor (EPCR). This may concentrate the zymogen and enzyme near the cell surface and facilitate the function of the pathway, but this is yet to be shown directly. Protein S circulates in complex with C4bBP, which may in turn bind serum amyloid P (SAP). APC is inhibited by forming complexes with either the protein C inhibitor (PCI), “$_1$-antitrypsin (“$_1$AT) or “$_2$-macroglobulin (not shown). See text for a more complete discussion. (Modified from Esmon CT: Cell mediated events that control blood coagulation and vascular injury: Annu Rev Cell Biol 1993; 9:1. Copyright 8 1993 Annual Review Inc.)
B The protein C pathway after inflammation. In this model, inflammatory mediators lead to the disappearance of thrombomodulin from the endothelial cell surface. Endothelial cell leukocyte adhesion molecules (P-selectin) or E-selectin are synthesized or expressed on endothelial or platelet surfaces. Tissue factor (TF) is expressed on monocytes and binds factor VIIa (VIIa), and this complex converts factor X(X) to factor Xa(Xa), which forms complexes with factor Va(Va) to generate thrombin (T) from prothrombin (Pro). Because little activated protein C (APC) is formed and the little that forms does not function well because of low protein S (S), factor Va is not inactivated and prothrombin complexes are stabilized. Elevation in circulating C4bBP concentration results in little free protein S. See text for discussion., SAP, serum amyloid P. (Modified from Esmon CT: Cell mediated events that control blood coagulation and vascular injury: Annu Rev Cell Biol 1993; 9:1. Copyright 8 1993 Annual Review Inc.)

natural anticoagulant pathway might be a safe and effective approach to treatment of thrombotic disease.

I. APC Resistance and Factor V Leiden

A potential complication in application of this system to the treatment of thrombotic disease is that there is a common factor V variant (Factor V Leiden) that is resistant to proteolytic inactivation by APC (FULCHER et al. 1984; KOEDAM et al. 1988; EATON et al. 1986; GRIFFIN et al. 1993; HALBMAYER et al. 1994; BERTINA et al. 1994; DAHLBÄCK 1994). This variant results in a complication commonly referred to as APC resistance and is caused by a substitution of Gln at residue 506 in factor V. This corresponds to the first cleavage site in factor Va (KALAFATIS et al. 1995; BILLY et al. 1995; ROSING et al. 1995). APC resistance is less severe, even in the homozygous form, than protein C deficiencies. The probable basis for the reduced severity is that failure to cleave factor Va at Arg506 only slows the inactivation partially, and this difference is partially eliminated by protein S (ROSING et al. 1995). Nonetheless, future applications of the protein C system to the treatment of thrombotic disease will have to consider this variant, which is very prevalent in Caucasians (0.5%) (LANE et al. 1996a,b), but rare in Black and Oriental populations.

C. Thrombomodulin as an Antithrombotic Agent

The multiple anticoagulant functions of TM, including the ability to block fibrinogen clotting, activate protein C, and promote thrombin inhibition, suggest that TM could be useful as an antithrombotic. The most likely candidates are TM forms truncated above the membrane spanning region (see Fig. 2). Thesc retain their ability to perform all of the above functions if they contain the chondroitin sulfate moiety. As predicted, based on its properties, immobilizing TM on thrombogenic surfaces renders the surfaces non-thrombogenic in vitro (KISHIDA et al. 1994). Whether this approach to preparing artificial surfaces for implantation in vivo will be effective remains to be tested.

In addition to its multiple anticoagulant effects, the potential utility of TM as a soluble antithrombotic is suggested by the observation that TM activity is likely reduced in many inflammatory diseases associated with thrombotic risk (see references in Sect. B.I), and hence, replacement therapy would target the acquired deficiency state, potentially offering optimal response with minimal risk. The author is unaware of published data with TM in humans at this time.

Most of the animal experimentation with TM has focused on DIC rather than occlusive thrombotic models. For instance, in rats, soluble TM has been shown to block tissue factor (NAWA et al. 1992) or endotoxin-induced DIC (GONDA et al. 1993). In addition, TM was able to block the pulmonary vascular injury that results from endotoxin exposure (UCHIBA et al. 1995). The potent antithrombotic agent, active site-blocked factor Xa (TAYLOR et al. 1991b) (see Sect. G), blocked the coagulation response, but failed to protect from lung injury, suggesting a protein C pathway-specific response in addition to the anticoagulant effect (UCHIBA et al. 1995). APC also blocked lung injury in this model (MURAKAMI et al. 1996), suggesting that the TM effect was mediated by increased APC formation. Since fibrin deposition is a component of injury in all of the above models, and TM blocked fibrin deposition, it follows that the net effect of TM supplementation, at least in these experimental settings, is antithrombotic. Although TM has been shown to have antifibrinolytic activities, these in vivo results suggest that the majority of the physiological activity is anticoagulant, possibly anti-inflammatory, and that these outweigh the antifibrinolytic contributions. The putative antifibrinolytic activities observed in vitro should be considered, however, in contemplating the use of TM derivatives for the treatment of diseases in which thrombi are present. No studies of this type have been reported to date.

TM with or without the chondroitin sulfate was effective in tissue factor-mediated DIC models (NAWA et al. 1992). The TM containing chondroitin sulfate was more effective on a mass basis, but was also cleared from the circulation with a $T_{1/2}$ of 20 min vs 1 h. TM appeared to have less effect on bleeding time than heparin. When expressed as the concentration required to double bleeding time vs the concentration required to block the decrease in platelet count by 50%, TM containing chondroitin sulfate was about threefold, and without chondroitin, twofold better than standard heparin. TM was also shown to be effective in a rat arteriovenous shunt thrombosis model. In this model, approximately 0.1 mg/kg TM gave inhibition of thrombus mass equivalent to 10 units/kg standard heparin (AOKI et al. 1994). Thrombus formation in the injured, ligated inferior vena cava of the rat was also prevented with soluble TM, in this case lacking chondroitin sulfate (SOLIS et al. 1991, 1994). These investigators found that TM blocked thrombosis at concentrations that doubled the bleeding time. Effective antithrombotic doses of hirudin or heparin resulted in bleeding times more than six times the control.

D. Protein C as an Antithrombotic Agent

The initial information about the use of protein C in the treatment of thrombotic disease in humans came from the treatment of congenital deficiencies. Homozygous protein C deficiency usually results in life threatening thrombotic complications in infancy (SELIGSOHN et al. 1984; GLADSON et al. 1987; BRANSON et al. 1983; DREYFUS et al. 1991). The initial symptoms usually involve microvascular thrombosis of the skin capillaries (purpura fulminans). Replacement therapy with protein C has been shown to prevent further progression of these lesions, which subsequently heal rapidly (DREYFUS et al. 1995; MULLER et al. 1996). These lesions are not prevented by heparinization (SILLS et al. 1984).

The symptoms of the homozygous protein C deficiency are reminiscent of other disease processes. For instance, a similar lesion occurs in Warfarin-induced skin necrosis. Protein C concentrate administration has halted lesion progression rapidly in the patients treated to date (MUNTEAN et al. 1991; ESMON and SCHWARZ 1995; SCHRAMM et al. 1993). A biochemical basis for the association between protein C and Warfarin-induced skin necrosis appears to be the vitamin K dependence of protein C and the rapid decline in protein C activity following Warfarin treatment. Specifically, protein C activity declines more rapidly than all other vitamin K dependent factors except factor VII, which has a similar half-life, i.e., 6–8 h (VIGANO-D'ANGELO et al. 1986). This temporal relationship presumably favors coagulation. In support of this concept, coagulation factor activation peptide levels, a marker of ongoing systemic coagulation, have been observed to increase transiently after the administration of Warfarin (CONWAY et al. 1987). In addition to indicating that protein C might be a logical treatment for patients with Warfarin-induced skin necrosis, the data suggest that supplementation with protein C might be an effective means of safely increasing the antithrombotic effectiveness of Warfarin with less bleeding risk than when increasing the oral anticoagulant dose. This proposal is theoretical and remains to be tested. Support for the concept comes from work with experimental animals in which APC administration has resulted in minimal bleeding complications even at antithrombotic doses (GRUBER et al. 1989; EMERICK et al. 1987). This observation has held even in animals subjected to open chest surgery (SNOW et al. 1991). The basis for the apparently favorable antithrombotic to bleeding profile is uncertain, but it may be related to the observation that APC is most effective as an anticoagulant at low doses of tissue factor. Low levels of tissue factor may be more common at damage sites within the vessel, while high levels are likely to be present in the extravascular space surrounding the vessel (DRAKE et al. 1989).

DIC and septic shock are additional situations in which microvascular thrombosis can occur. This is particularly common in meningococcemia (POWARS et al. 1987, 1993). Protein C consumption in humans with meningococcemia correlates with the formation of the purpura-like lesions and death more than other markers examined (POWARS et al. 1993; et al. 1995). Although suggestive, this correlation between protein C levels and disease progression does not prove a cause and effect relationship. The concept that protein C or APC might be of use in this system comes from an early observation that thrombin infusion into dogs, which were subsequently challenged with lethal numbers of *E. coli*, resulted in survival, and prevention of DIC (TAYLOR et al. 1984). Since thrombin infusion leads to protein C activation in vivo (COMP et al. 1982), we tested the ability of APC to prevent the lethal response to *E. coli* in baboons. APC protected the animals from death, DIC and organ dysfunction (TAYLOR et al. 1986). Blocking the protein C pathway made animals hyper-responsive to sublethal numbers of *E. coli* (TAYLOR et al. 1991a; TAYLOR et al. 1986). Blocking the pathway not only increased the DIC response to the sublethal *E. coli*, but it also increased the circulating TNFα

levels compared to control animals given the same numbers of *E. coli*. Restoration of the protein C system prevented the DIC, organ damage, and elaboration of elevated cytokine levels. Taken together, these results suggest that protein C is a major regulator of microvascular thrombosis, and that the system modulates the inflammatory response by as yet unknown mechanisms. The links presented above between protein C deficiency and microvascular thrombosis provided a rationale for the use of protein C in the prevention of some complications of septic shock. To date, treatment of 11 severely ill children with relatively advanced septic shock, usually due to meningococcemia, with protein C has been reported (RIVARD et al. 1995; ESMON and SCHWARZ 1995). In general, protein C infusion has been associated with normalization of circulating protein C levels and reversal of organ dysfunction, including a rapid regain of consciousness and kidney function (RIVARD et al. 1995; and reviewed in ESMON and SCHWARZ 1995). In addition to meningococcemia, one patient with a group A β-hemolytic streptococcal infection and varicella, developed septic shock, DIC and purpura, and had undetectable protein C levels, probably due to consumption (GERSON et al. 1993). His condition improved rapidly following protein C supplementation. Although the rationale and preliminary anecdotal clinical results appear promising, a larger clinical trial is needed to verify the validity of this approach.

Protein C might inhibit the septic shock process by several mechanisms including the regulation of thrombin formation at the vessel surface, or modulation of the inflammatory response. Several observations suggest that this system can modulate inflammation. APC has been reported to inhibit tumor necrosis factor elaboration by monocytes in vitro (GREY et al. 1994), to bind the monocyte cell surface, probably by interacting with a specific receptor, and to prevent interferon gamma mediated Ca^{2+} transients and cellular proliferation (HANCOCK et al. 1995). Protein C has also been observed to inhibit leukocyte adhesion to selectins (GRINNELL et al. 1994). EPCR on the endothelium is structurally related to the major histocompatibility (MHC) class 1 molecules. This class of molecules is involved in inflammation and EPCR is regulated by inflammatory cytokines, suggesting the possible, but not yet proven, role of this receptor in the control of inflammatory processes (FUKUDOME and ESMON 1994). It remains unclear whether any of these in vitro findings contribute significantly to the protective effects observed in septic shock.

I. Protein C and Arterial Thrombosis

APC has been shown to be effective in preventing platelet and fibrin accretion on vascular grafts at arterial flow rates in baboons (GRUBER et al. 1989). Concentrations of APC that inhibited platelet accretion had little effect on bleeding time (GRUBER et al. 1989). Previous studies from this group have shown that this model is highly thrombin dependent and direct thrombin inhibitors can block platelet accretion. Interestingly, when thrombin is infused systemically, platelet accretion decreases in this model despite the fact that thrombin is critical to the platelet accretion (Hanson et al. 1993). This apparent paradox is due to thrombin-dependent protein C activation since the protective effect of thrombin in this study was eliminated by an antibody that blocks protein C activation. These studies suggest the possibility that APC might be effective in preventing arterial thrombosis. Clinically, however, protein C deficiency is usually associated with venous or microvascular, rather than arterial thrombosis. Therefore, the clinical link between the deficiency and a particular form of thrombosis associated with septic shock, venous thrombosis and Warfarin skin necrosis is absent for arterial thrombosis. It is possible that very low levels of APC are sufficient to minimize arterial thrombotic complications in the absence of advanced atherosclerosis or a thrombogenic surface, and that these levels are present even in individuals with heterozygous deficiencies.

II. Reperfusion Injury

Thrombin generation in the microvasculature leads to protein C activation, and thrombin formation probably occurs in ischemic injury (Snow et al. 1991). Results from animal experiments have shown that when the left anterior descending coronary artery was occluded in pigs (Snow et al. 1991) or dogs, protein C activation occurred within 2 min and activation was restricted to the region at risk. Blocking protein C activation with a monoclonal antibody in this model resulted in impaired recovery of the animal's left ventricular function. Ventricular fibrillation was often observed in the animals in which protein C activation was blocked, but not in the controls. APC infusion appeared to improve recovery of left ventricular function, but these responses failed to reach statistical significance. These data suggest that APC or protein C supplementation might diminish reperfusion injury. This view has been challenged by the observation that infusion of human APC at sufficient levels to produce an anticoagulant response in the dog failed to decrease the ischemic injury or the amount of leukocyte infiltration (Hahn et al. 1996). The basis for the discrepancy in these results is uncertain, but they could be related to species specificity since human APC was used in the latter studies and porcine or canine in the former. In ischemia, protein C activation from endogenous sources is robust, and hence the maximal protective effect of APC on the injury might have been satisfied by the endogenous activation. Clearly, additional experimentation is needed to address the potential utility of APC in reperfusion injury.

Theoretically, APC might be very attractive for the treatment of patients with chronic coronary disease. These patients have been reported to have elevated TNF levels (Maury and Teppo 1989) possibly allowing for local TM down-regulation. TM down-regulation in the coronary bed has been observed in transplant rejection (Hancock et al. 1991). These theoretical considerations assume that protein C activation in ischemia is due to the thrombin-TM complex. This has not been documented, and other activators like plasmin (Bajaj et al. 1983; Varadi et al. 1994) which can activate and subsequently inactivate protein C, may be involved.

E. Protein S as an Antithrombotic Agent

Protein S as a therapeutic has not been studied extensively. Protein S deficiency has been described in patients with Warfarin-induced skin necrosis (Goldberg et al. 1991), and patients with homozygous protein S deficiency may develop purpura fulminans (Marlar and Neumann 1990). Free and total protein S levels are often low in septic shock or following thrombosis (Heeb et al. 1989b; Fourrier et al. 1992; Nguyen et al. 1994). Therefore, it is reasonable to infer that protein S supplementation might be antithrombotic, especially when the levels are low due to the disease process.

In a baboon model of *E. coli*-induced septic shock, modulation of protein S activity with C4bBP has been shown to exacerbate the response to sublethal levels of *E. coli*. Specifically, when protein S levels are normal, 10% of a lethal number of *E. coli* produces only an acute phase response, but when free protein S is reduced by infusion of C4bBP or by an antibody to protein S, the same dose of bacteria leads to either DIC or microvascular thrombosis, organ failure, and death (Taylor et al. 1991a). Protein S supplementation protected the animals from DIC and death. While protein S levels have been observed to decrease with septic shock, the preliminary clinical experience with protein C infusion in septic shock patients suggests that it is effective without simultaneous protein S supplementation. Whether protein S supplementation would improve the efficacy of protein C, or whether it may be required in some patients for protein C to be effective may be clarified during clinical trials.

F. Mutations to Modulate Natural Anticoagulant Responses

Our current understanding of structure-function relationships of serine proteases and their zymogens makes it possible to rationally design mutants of thrombin or protein C that have altered properties. In the case of thrombin, desirable features might be high protein C activation even in the absence of TM, and low coagulant activity. For protein C, attractive mutations might be those that allow thrombin to activate protein C rapidly in the absence of TM. Both approaches have now been employed to modulate this system.

I. Mutations in Protein C

A protein C mutant has been designed that activates rapidly with thrombin in the absence of TM (Richardson et al. 1992). This was accomplished by mutation of two acidic residues near the scissile bond and an additional modification was incorporated to alter the Ca^{2+} inhibition of activation by free thrombin. This mutant is activated sufficiently rapidly that it anticoagulates plasma, something wild type protein C does not do in the absence of TM. Recently, this mutant has been shown to be 10 times more effective than wild type protein C and about 10% as effective as APC in a guinea pig model of extracorporeal thrombosis (Kurz et al. 1997). This in vivo demonstration would seem to validate this approach to antithrombotic therapy.

II. Mutations in Thrombin

A second approach is to modify thrombin. In this case several functions are potential aids in eliciting a selective response. In principle, this could be accomplished by selectively diminishing fibrinogen clotting activity or increasing protein C activation activity. Enhanced protein C activation could involve either TM dependent or independent mechanisms, and several thrombin mutants with one or more of these properties have been identified (Wu et al. 1991; Le Bonniec and Esmon 1991; Gibbs et al. 1995; Dang et al. 1997). A particularly promising thrombin mutant has been identified and studied in vivo (Gibbs et al. 1995). This mutant has at least a 40-fold decrease in activity toward fibrinogen and platelets, but it retains about 50% of its protein C activating capacity in the presence of TM (Gibbs et al. 1995). Another potentially useful feature is that the rate of inhibition by antithrombin is slowed about sevenfold. In vivo, this thrombin mutant generates the anticoagulant response observed previously with wild type thrombin (Comp et al. 1982; Hanson et al. 1993) but without as much fibrinogen consumption.

A variation on the modulation of thrombin specificity through mutagenesis was suggested by the recent finding that small molecules could bind to thrombin and accelerate protein C activation, partially mimicking TM function but also inhibiting fibrinogen clotting activity (Berg et al. 1996). This approach would have the advantage over the recombinant thrombins of avoiding the infusion of mutant thrombin molecules which could elicit an immune response, and in the worst case scenario, could lead to production of autoantibodies against thrombin.

G. Inactive Coagulation Factors as Antithrombotics

Examination of Fig. 1 reveals that each step in the coagulation cascade involves obligate complex formation between the protease component and a cofactor to amplify the coagulation response. The formation of these complexes does not require the active site of the enzyme, as would be predicted since the active site has to be more or less available to process the substrates. It follows that coagulation enzymes such as factors IXa, VIIa or Xa that have been modified in the active site would still bind to their respective cofactors, factor VIIIa, tissue factor or factor Va. Under conditions where the cofactor is limiting, these modified enzymes serve as competitive inhibitors for binding of the corresponding active enzymes and thus block the coagulation response. Available evidence suggests that factors Va and VIIIa are quite specific for activated factors X and IX and hence, there is little cross competition with the zymogens. With tissue factor, the enzyme and the zymogen bind with similar affinity (MORRISSEY and DRAKE 1993; CARSON and BROZNA 1993; EDGINGTON et al. 1991).

Inactive forms of these enzymes can be prepared by insertion of small, covalent inhibitors or by mutation of active site residues such as the active site serine (SINHA et al. 1991; TAYLOR 1996; LOLLAR and FASS 1984). Blocking the active site prevents reactivity with proteinase inhibitors in the plasma, and can increase the circulating half-life of the modified enzyme from minutes to hours (TAYLOR et al. 1991b).

Inactive or inactivated forms of factor VIIa (TAYLOR 1996; HARKER et al. 1996), factor Xa (TAYLOR et al. 1991; HOLLENBACH et al. 1994), and factor IXa (BENEDICT et al. 1991; TIJBURG et al. 1991) have all been examined as antithrombotics and been shown to have good in vivo antithrombotic activity and a favorable antithrombotic to bleeding profile relative to heparin. For ease of description, inactive or inactivated factors are referred to with a small i, such that factor Xai indicates inactive/inactivated factor Xa.

Factor VIIai has been examined in a baboon model of arterial thrombosis, and in an endarterectomy model of restenosis (HARKER et al. 1996) as well as in a model of septic shock (TAYLOR 1996). The factor VIIai was effective in preventing arterial thrombosis in the model and also minimized the intimal thickening. Factor VIIai blocked DIC in the *E. coli* model of septic shock in the baboon and improved survival when given prior to *E. coli* infusion (TAYLOR 1996). In contrast, although factor Xai blocked DIC in this shock model, it did not improve survival. The basis for the differences in the shock model is unclear. No reports on the influence of factor Xai or factor IXai on restenosis models are available. Factor IXai has been shown to be effective in blocking fibrin deposition in Meth A sarcoma tumors in mice treated with TNFα suggesting that this approach might be useful in blocking tumor-associated coagulation abnormalities.

In addition to their relative positions in the cascade, one potential difference between the in vivo effects of these proteins is the recent observation that there are cellular receptors for each that may involve signaling. Cellular signaling in the case of factor VIIa appears to involve tissue factor. It has recently been shown that factor VIIa interaction with tissue factor can mobilize intracellular Ca^{2+} and presumably activate cells (CAMERER et al. 1996). This signaling requires proteolytic activity and hence the factor VIIai may modulate the capacity of factor VIIa to signal cells. In the case of factor Xa, a new receptor, effector protease receptor 1 (EPR-1), was recently identified (ALTIERI 1994). Factor Xa interaction with this receptor apparently stimulates inflammatory responses (ALTIERI and STAMNES 1994). Binding to this receptor is active site independent, but cellular signaling is active site dependent (Nicholson et al. 1996). Thus, factor Xai would be anticipated to interfere with EPR-1-mediated cellular signaling. Finally, factor IX and IXa have endothelial cell receptors, one of which has recently been identified as collagen type IV (Cheung et al. 1996). It is not known whether factor IXa interaction with this receptor causes any unique signaling mechanisms. The identification and functional understanding of these and other cellular receptors for coagulation factors may allow a greater appreciation of the mechanisms by which these inactivated coagulation factors can function.

H. Summary

Our developing understanding of the coagulation system offers promise for the design of safer anticoagulant therapy. It is likely that each of the anticoagulant approaches described in this chapter will prove effective in some particular clinical setting. A future challenge will be to understand which pathway is most responsible for particular thrombotic complications. This will provide the basis for tailoring therapy to the patient. The protein C system offers promise as a novel antithrombotic agent and appears to be especially important in the control of microvascular thrombosis. The limited clinical data suggest that protein C supplementation is safe and effective for the treatment of congenital deficiencies, Warfarin-induced skin necrosis and gram negative sepsis. Whether protein C or APC will prove beneficial in larger, randomized clinical trials or to treat forms of septic shock in the absence of purpura lesions remains to be determined. If the protein C system does elicit anti-inflammatory activity as inferred from the basic and physiological studies, modulation of this pathway may constitute a unique approach to diminish morbidity and mortality associated with septic shock. New receptors are being identified, such as EPCR, that can modulate APC activity. As we gain a better understanding of the system, our ability to apply the components of the system to the treatment of thrombotic disease should improve significantly.

The active site-blocked coagulation factors may also prove to have unique activities not previously recognized. The apparent ability of the coagulation factor-receptor complexes to activate cells and/or contribute to the inflammatory process could in turn propagate the thrombotic response. The ability to block cellular activation with active site-blocked coagulation factors may provide novel mechanisms for intervening in thrombotic processes. Future understanding of the role of coagulation factors in modulating cellular responses will likely contribute to our selection of the most appropriate antithrombotic agent for particular disease processes.

References

Altieri DC (1994) Molecular cloning of effector cell protease receptor-1, a novel cell surface receptor for the protease factor Xa. J Biol Chem 269:3139–3142

Altieri DC, Stamnes SJ (1994) Protease-dependent T cell activation: ligation of effector cell protease receptor-1 (EPR-1) stimulates lymphocyte proliferation. Cell Immunol 155:372–383

Alving BM, Comp PC (1992) Recent advances in understanding clotting and evaluating patients with recurrent thrombosis. Am J Obstet Gynecol 167:1184–1191

Aoki Y, Takei R, Mohri M, Gonda Y, Gomi K, Sugihara T et al (1994) Antithrombotic effects of recombinant human soluble thrombomodulin (rhs-TM) on arteriovenous shunt thrombosis in rats. Am J Hematol 47:162–166

Asakura H, Jokaji H, Saito M, Uotani C, Kumabashiri I, Morishita E et al (1991) Plasma levels of soluble thrombomodulin increase in cases of disseminated intravascular coagulation with organ failure. Am J Hematol 38:281–287

Bajaj SP, Rapaport SI, Maki SL, Brown SF (1983) A procedure for isolation of human protein C and protein S as by-products of the purification of factors VII, IX, X and prothrombin. Prep Biochem 13:191–214

Bajzar L, Morser J, Nesheim M (1996) TAFI, or plasma procarboxypeptidase B, couples the coagulation and fibrinolytic cascades through the thrombin-thrombomodulin complex. J Biol Chem 271:16603–16608

Benedict CR, Ryan J, Wolitzky B, Ramos R, Gerlach M, Tijburg P et al (1991) Active site-blocked factor IXa prevents intravascular thrombus formation in the coronary vasculature without inhibiting extravascular coagulation in a canine thrombosis model. J Clin Invest 88:1760–1765

Berg DT, Wiley MR, Grinnell BW (1996) Enhanced protein C activation and inhibition of fibrinogen cleavage by a thrombin modulator. Science 273:1389–1391

Bertina RM, Koeleman BP, Koster T, Rosendaal FR, Dirven RJ, de Ronde H et al (1994) Mutation in blood coagulation factor V associated with resistance to activated protein C. Nature 369:64–67

Billy D, Willems GM, Hemker HC, Lindhout T (1995) Prothrombin contributes to the assembly of the factor Va–factor Xa complex at phosphatidylserine-containing phospholipid membranes. J Biol Chem 270:26883–26889

Boehme MW, Deng Y, Raeth U, Bierhaus A, Ziegler R, Stremmel W et al (1996) Release of thrombomodulin from endothelial cells by concerted action of TNF-alpha and neutrophils: in vivo and in vitro studies. Immunology 87:134–140

Boerger LM, Morris PC, Thurnau GR, Esmon CT, Comp PC (1987) Oral contraceptives and gender affect protein S status. Blood 69:692–694

Bourin MC, Lindahl U (1993) Glycosaminoglycans and the regulation of blood coagulation. Biochem J 289:313–330

Branson HE, Katz J, Marble R, Griffin JH (1983) Inherited protein C deficiency and a coumarin-responsive chronic relapsing purpura fulminans syndrome in a newborn infant. Lancet 2:1165–1168

Broze GJ Jr, Girard TJ, Novotny WF (1990) Regulation of coagulation by a multivalent Kunitz-type inhibitor. Biochemistry 29:7539–7546

Camerer E, Rottingen JA, Iversen JG, Prydz H (1996) Coagulation factors VII and X induce Ca^{2+} oscillations in Madin-Darby canine kidney cells only when proteolytically active. J Biol Chem 271:29034–29042

Carson SD, Brozna JP (1993) The role of tissue factor in the production of thrombin. Blood Coagul Fibrinolysis 4:281–292

Castellino FJ (1995) Human protein C and activated protein C: components of the human anticoagulation system. Trends Cardiovasc Med 5:55–62

Chafa O, Fischer AM, Meriane F, Chellali T, Sternberg C, Otmani F et al (1992) BehHet syndrome associated with protein S deficiency. Thromb Haemost 67:1–3

Cheung WF, van den Born J, Kuhn K, Kjellen L, Hudson BG, Stafford DW (1996) Identification of the endothelial cell binding site for factor IX. Proc Natl Acad Sci USA 93:11068–11073

Comp PC, Jacocks RM, Ferrell GL, Esmon CT (1982) Activation of protein C in vivo. J Clin Invest 70:127–134

Comp PC, Nixon RR, Cooper MR, Esmon CT (1984) Familial protein S deficiency is associated with recurrent thrombosis. J Clin Invest 74:2082–2088

Conway EM, Bauer KA, Barzegar S, Rosenberg RD (1987) Suppression of hemostatic system activation by oral anticoagulants in the blood of patients with thrombotic diatheses. J Clin Invest 80:1535–1544
D'Angelo A, Vigano-D'Angelo S, Esmon CT, Comp PC (1988) Acquired deficiencies of protein S. Protein S activity during oral anticoagulation, in liver disease and in disseminated intravascular coagulation. J Clin Invest 81:1445–1454
D'Angelo A, Della-Valle P, Crippa L, Pattarini E, Grimaldi LM, Vigano-D'Angelo SV (1993) Brief report: Autoimmune protein S deficiency in a boy with severe thromboembolic disease. N Engl J Med 328:1753–1757
Dahlbäck B (1986) Inhibition of protein Ca cofactor function of human and bovine protein S by C4b-binding protein. J Biol Chem 261:12022–12027
Dahlbäck B (1991) Protein S and C4b-binding protein: components involved in the regulation of the protein C anticoagulant system. Thromb Haemost 66:49–61
Dahlbäck B (1994) Physiological anticoagulation. Resistance to activated protein C and venous thromboembolism. J Clin Invest 94:923–927
Dang QD, Guinto ER, di Cera E (1997) Rational engineering of activity and specificity in a serine protease. Nat Biotechnol 15:146–149
Davie EW, Fujikawa K, Kisiel W (1991) The coagulation cascade: initiation, maintenance and regulation. Biochemistry 30:10363–10370
de Munk GA, Groeneveld E, Rijken DC (1991) Acceleration of the thrombin inactivation of single chain urokinase-type plasminogen activator (pro-urokinase) by thrombomodulin. J Clin Invest 88:1680–1684
Dittman WA, Nelson SC, Greer PK, Horton ET, Palomba ML, McCachren SS (1994) Characterization of thrombomodulin expression in response to retinoic acid and identification of a retinoic acid response element in the human thrombomodulin gene. J Biol Chem 269:16925–16932
Drake TA, Morrissey JH, Edgington TS (1989) Selective cellular expression of tissue factor in human tissues: Implications for disorders of hemostasis and thrombosis. Am J Pathol 134:1087–1097
Drake TA, Cheng J, Chang A, Taylor FB Jr (1993) Expression of tissue factor, thrombomodulin, and E-selectin in baboons with lethal Escherichia coli sepsis. Am J Pathol 142:1458–1470
Dreyfus M, Magny JF, Bridey F, Schwarz HP, Planch JC, Dehan M et al (1991) Treatment of homozygous protein C deficiency and neonatal purpura fulminans with a purified protein C concentrate. N Engl J Med 325:1565–1568
Dreyfus M, Masterson M, David M, Rivard GE, Muller FM, Kreuz W et al (1995) Replacement therapy with a monoclonal antibody purified protein C concentrate in newborns with severe congenital protein C deficiency. Semin Thromb Hemost 21:371–381
Eaton D, Rodriguez H, Vehar GA (1986) Proteolytic processing of human factor VIII. Correlation of specific cleavages by thrombin, factor Xa, and activated protein C with activation and inactivation of factor VIII coagulant activity. Biochemistry 25:505–512
Edgington TS, Mackman N, Brand K, Ruf W (1991) The structural biology of expression and function of tissue factor. Thromb Haemost 66:67–79
Emerick SC, Murayama H, Yan SB, Long GL, Harms CS, Marks CA et al (1987) Preclinical pharmacology of activated protein C. In: Holcenberg JS, Winkelhake JL (eds) The Pharmacology and Toxicology of Proteins, UCLA Symposia on Molecular and Cellular Biology. Liss, New York, pp 351–367
Esmon CT (1989) The roles of protein C and thrombomodulin in the regulation of blood coagulation. J Biol Chem 264:4743–4746
Esmon CT (1993a) Molecular events that control the protein C anticoagulant pathway. Thromb Haemost 70:29–35
Esmon CT (1993b) Cell mediated events that control blood coagulation and vascular injury. Annu Rev Cell Biol 9:1–26
Esmon CT (1995) Thrombomodulin as a model of molecular mechanisms that modulate protease specificity and function at the vessel surface. FASEB J 9:946–955

Esmon CT, Fukudome K (1995) Cellular regulation of the protein C pathway. Semin Cell Biol 6:259–268

Esmon CT, Owen WG (1981) Identification of an endothelial cell cofactor for thrombin-catalyzed activation of protein C. Proc Natl Acad Sci USA 78:2249–2252

Esmon CT, Schwarz HP (1995) An update on clinical and basic aspects of the protein C anticoagulant pathway. Trends Cardiovasc Med 5:141–148

Esmon CT, Esmon NL, Harris KW (1982) Complex formation between thrombin and thrombomodulin inhibits both thrombin-catalyzed fibrin formation and factor V activation. J Biol Chem 257:7944–7947

Esmon NL, Carroll RC, Esmon CT (1983) Thrombomodulin blocks the ability of thrombin to activate platelets. J Biol Chem 258:12238–12242

Espana F, Gruber A, Heeb MJ, Hanson SR, Harker LA, Griffin JH (1991) In vivo and in vitro complexes of activated protein C with two inhibitors in baboons. Blood 77:1754–1760

Fijnvandraat K, Derkx B, Peters M, Bijlmer R, Sturk A, Prins MH et al (1995) Coagulation activation and tissue necrosis in meningococcal septic shock: severely reduced protein C levels predict a high mortality. Thromb Haemost 73:15–20

Florell SR, Rodgers GM (1997) Inherited thrombotic disorders: an update. Am J Hematol 54:53–60

Fourrier F, Chopin C, Goudemand J, Hendrycx S, Caron C, Rime A, Marey A, Lestavel P (1992) Septic shock, multiple organ failure, and disseminated intravascular coagulation. Compared patterns of antithrombin III, protein C, and protein S deficiencies. Chest 101:816–823

Fukudome K, Esmon CT (1994) Identification, cloning and regulation of a novel endothelial cell protein C/activated protein C receptor. J Biol Chem 269:26486–26491

Fulcher CA, Gardiner JE, Griffin JH, Zimmerman TS (1984) Proteolytic inactivation of human factor VIII procoagulant protein by activated human protein C and its analogy with factor V. Blood 63:486–489

Furmaniak-Kazmierczak E, Hu CY, Esmon CT (1993) Protein S enhances C4b binding protein interaction with neutrophils. Blood 81:405–411

Gailani D, Broze GJ Jr (1991) Factor XI activation in a revised model of blood coagulation. Science 253:909–912

Gerson WT, Dickerman JD, Bovill EG, Golden E (1993) Severe acquired protein C deficiency in purpura fulminans associated with disseminated intravascular coagulation: treatment with protein C concentrate. Pediatrics 91:418–422

Gibbs CS, Coutre SE, Tsiang M, Li WX, Jain AK, Dunn KE et al (1995) Conversion of thrombin into an anticoagulant by protein engineering. Nature 378:413–416

Girolami A, Simioni P, Lazzaro AR, Cordiano I (1989) Severe arterial cerebral thrombosis in a patient with protein S deficiency (moderately reduced total and markedly reduced free protein S): a family study. Thromb Haemost 61:144–147

Gladson CL, Groncy P, Griffin JH (1987) Coumarin necrosis, neonatal purpura fulminans, and protein C deficiency. Arch Dermatol 123:1701a–1706a

Goldberg SL, Orthner CL, Yalisove BL, Elgart ML, Kessler CM (1991) Skin necrosis following prolonged administration of coumarin in a patient with inherited protein S deficiency. Am J Hematol 38:64–66

Gonda Y, Hirata S, Saitoh K, Aoki Y, Mohri M, Gomi K et al (1993) Antithrombotic effect of recombinant human soluble thrombomodulin on endotoxin-induced disseminated intravascular coagulation in rats. Thromb Res 71:325–335

Grey ST, Tsuchida A, Hau H, Orthner CL, Salem HH, Hancock WW (1994) Selective inhibitory effects of the anticoagulant activated protein C on the responses of human mononuclear phagocytes to LPS, IFN-gamma, or phorbol ester. J Immunol 153:3664–3672

Griffin JH, Evatt B, Wideman C, Fernandez JA (1993) Anticoagulant protein C pathway defective in majority of thrombophilic patients. Blood 82:1989–1993

Grinnell BW, Hermann RB, Yan SB (1994) Human protein C inhibits selectin-mediated cell adhesion: Role of unique fucosylated oligosaccharide. Glycobiology 4:221–225

Gruber A, Griffin JH, Harker LA, Hanson SR (1989) Inhibition of platelet-dependent thrombus formation by human activated protein C in a primate model. Blood 73:639–642

Gruber A, Harker LA, Hanson SR, Kelly AB, Griffin JH (1991) Antithrombotic effects of combining activated protein C and urokinase in nonhuman primates. Circulation 84:2454–2462

Hackeng TM, van't Veer C, Meijers JC, Bouma BN (1994) Human protein S inhibits prothrombinase complex activity on endothelial cells and platelets via direct interactions with factors Va and Xa. J Biol Chem 269:21051–21058

Hahn RA, MacDonald BR, Chastain M, Grinnell BW, Simpson PJ (1996) Evaluation of activated protein C on canine infarct size in a nonthrombotic model of myocardial reperfusion injury. J Pharmacol Exp Ther 276:1104–1110

Halbmayer WM, Haushofer A, Schon R, Fischer M (1994) The prevalence of poor anticoagulant response to activated protein C (APC resistance) among patients suffering from stroke or venous thrombosis and among healthy subjects. Blood Coagul Fibrinolysis 5:51–57

Hancock WW, Tanaka K, Salem HH, Tilney NL, Atkins RC, Kupiec-Weglinski JW (1991) TNF as a mediator of cardiac transplant rejection, including effects on the intragraft protein C/protein S/thrombomodulin pathway. Transplant Proc 23:235–237

Hancock WW, Grey ST, Hau L, Akalin E, Orthner C, Sayegh MH, et al (1995) Binding of activated protein C to a specific receptor on human mononuclear phagocytes inhibits intracellular calcium signaling and monocyte-dependent proliferative responses. Transplantation 60:1525–1532

Hanson SR, Griffin JH, Harker LA, Kelly AB, Esmon CT, Gruber A (1993) Antithrombotic effects of thrombin-induced activation of endogenous protein C in primates. J Clin Invest 92:2003–2012

Harker LA, Hanson SR, Wilcox JN, Kelly AB (1996) Antithrombotic and antilesion benefits without hemorrhagic risks by inhibiting tissue factor pathway. Haemostasis 26 Suppl 1:76–82

Healy AM, Rayburn HB, Rosenberg RD, Weiler H (1995) Absence of the blood-clotting regulator thrombomodulin causes embryonic lethality in mice before development of a functional cardiovascular system. Proc Natl Acad Sci USA 92:850–854

Heeb MJ, Espana F, Griffin JH (1989a) Inhibition and complexation of activated protein C by two major inhibitors in plasma. Blood 73:446–454

Heeb MJ, Mosher D, Griffin JH (1989b) Activation and complexation of protein C and cleavage and decrease of protein S in plasma of patients with intravascular coagulation. Blood 73:455–461

Heeb MJ, Gruber A, Griffin JH (1991) Identification of divalent metal ion-dependent inhibition of activated protein C by alpha$_2$-macroglobulin and alpha$_2$-antiplasmin in blood and comparisons to inhibition of factor Xa, thrombin, and plasmin. J Biol Chem 266:17606–17612

Heeb MJ, Mesters RM, Tans G, Rosing J, Griffin JH (1993) Binding of protein S to Factor Va associated with inhibition of prothrombinase that is independent of activated protein C. J Biol Chem 268:2872–2877

Heeb MJ, Rosing J, Bakker HM, Fernandez JA, Tans G, Griffin JH (1994) Protein S binds to and inhibits factor Xa. Proc Natl Acad Sci USA 91:2728–2732

Hofsteenge J, Taguchi H, Stone SR (1986) Effect of thrombomodulin on the kinetics of the interaction of thrombin with substrates and inhibitors. Biochem J 237:243–251

Hogg PJ, Öhlin AK, Stenflo J (1992) Identification of structural domains in protein C involved in its interaction with thrombin-thrombomodulin on the surface of endothelial cells. J Biol Chem 267:703–706

Hollenbach S, Sinha U, Lin PH, Needham K, Frey L, Hancock T et al (1994) A comparative study of prothrombinase and thrombin inhibitors in a novel rabbit model of non-occlusive deep vein thrombosis. Thromb Haemost 71:357–362
Ishii H, Majerus PW (1985) Thrombomodulin is present in human plasma and urine. J Clin Invest 76:2178–2181
Ishii H, Salem HH, Bell CE, Laposata EA, Majerus PW (1986) Thrombomodulin, an endothelial anticoagulant protein, is absent from the human brain. Blood 67:362–365
Kalafatis M, Bertina RM, Rand MD, Mann KG (1995) Characterization of the molecular defect in factor VR506Q. J Biol Chem 270:4053–4057
Kapiotis S, Besemer J, Bevec D, Valent P, Bettelheim P, Lechner K et al (1991) Interleukin-4 counteracts pyrogen-induced downregulation of thrombomodulin in cultured human vascular endothelial cells. Blood 78:410–415
Kishida A, Ueno Y, Maruyama I, Akashi M (1994) Immobilization of human thrombomodulin onto biomaterials. Comparison of immobilization methods and evaluation of antithrombogenicity. ASAIO J 40:M840–M845
Koedam JA, Meijers JC, Sixma JJ, Bouma BN (1988) Inactivation of human factor VIII by activated protein C. Cofactor activity of protein S and protective effect of von Willebrand factor. J Clin Invest 82:1236–1243
Koppelman SJ, van't Veer C, Sixma JJ, Bouma BN (1995) Synergistic inhibition of the intrinsic factor X activation by protein S and C4b-binding protein. Blood 86:2653–2660
Koyama T, Hirosawa S, Kawamata N, Tohda S, Aoki N (1994) All-trans retinoic acid upregulates thrombomodulin and downregulates tissue-factor expression in acute promyelocytic leukemia cells: distinct expression of thrombomodulin and tissue factor in human leukemic cells. Blood 84:3001–3009
Kurz KD, Smith T, Wilson A, Gerlitz B, Richardson MA, Grinnell BW (1997) Antithrombotic efficacy in the guinea pig of a derivative of human protein C with enhanced activation by thrombin. Blood 89:534–540
Labarrere CA, Esmon CT, Carson SD, Faulk WP (1990) Concordant expression of tissue factor and Class II MHC antigens in human placental endothelium. Placenta 11:309–318
Lane DA, Mannucci PM, Bauer KA, Bertina RM, Bochkov NP, Boulyjenkov et al (1996a) Inherited thrombophilia: Part 1. Thromb Haemost 76:651–662
Lane DA, Mannucci PM, Bauer KA, Bertina RM, Bochkov NP, Boulyjenkov et al (1996b) Inherited thrombophilia. Part 2. Thromb Haemost 76:824–834
Laszik Z, Carson CW, Nadasdy T, Johnson LD, Lerner MR, Brackett DJ et al (1994) Lack of suppressed renal thrombomodulin expression in a septic rat model with glomerular thrombotic microangiopathy. Lab Invest 70:862–867
Laszik Z, Mitro A, Taylor FB, Jr., Ferrell G, Esmon CT (1997) The human protein C receptor is present primarily on endothelium of large blood vessels: Implications for the control of the protein C pathway. J Clin Invest (Abstract)
Le Bonniec BF, Esmon CT (1991) Glu-192–Gln substitution in thrombin mimics the catalytic switch induced by thrombomodulin. Proc Natl Acad Sci USA 88:7371–7375
Lin JH, McLean K, Morser J, Young TA, Wydro RM, Andrews WH et al (1994) Modulation of glycosaminoglycan addition in naturally expressed and recombinant human thrombomodulin. J Biol Chem 269:25021–25030
Lollar P, Fass DN (1984) Inhibition of activated porcine factor IX by dansyl-glutamyl-glycyl-arginyl-chloromethylketone. Arch Biochem Biophys 233:438–446
Lollar P, Owen WG (1980) Clearance of thrombin from the circulation in rabbits by high-affinity binding sites on endothelium. J Clin Invest 66:1222–1230
Mahasandana C, Suvatte V, Chuansumrit A, Marlar RA, Manco-Johnson MJ, Jacobson LJ et al (1990) Homozygous protein S deficiency in an infant with purpura fulminans. J Pediatr 117:750–753
Marcum JA, Rosenberg RD (1988) The biochemistry and physiology of anticoagulantly active heparin-like molecules. In: Simionescu N, Simionescu M

(eds) Endothelial cell biology in health and disease. Plenum , New York, pp 207–228
Marlar RA, Neumann A (1990) Neonatal purpura fulminans due to homozygous protein C or protein S deficiencies. Semin Thromb Hemost 16:299–309
Maruyama I, Majerus PW (1985) The turnover of thrombin-thrombomodulin complex in cultured human umbilical vein endothelial cells and A549 lung cancer cells. Endocytosis and degradation of thrombin. J Biol Chem 260:15432–15438
Maruyama I, Salem HH, Majerus PW (1984) Coagulation factor Va binds to human umbilical vein endothelial cells and accelerates protein C activation. J Clin Invest 74:224–230
Maruyama I, Salem HH, Ishii H, Majerus PW (1985) Human thrombomodulin is not an efficient inhibitor of procoagulant activity of thrombin. J Clin Invest 75:987–991
Mathews II, Padmanabhan KP, Ganesh V, Tulinsky A, Ishii M, Chen J et al (1994a) Crystallographic structures of thrombin complexed with thrombin receptor peptides: existence of expected and novel binding modes. Biochemistry 33:3266–3279
Mathews II, Padmanabhan KP, Tulinsky A, Sadler JE (1994b) Structure of a nonadecapeptide of the fifth EGF domain of thrombomodulin complexed with thrombin. Biochemistry 33:13547–13552
Maury CP, Teppo AM (1989) Circulating tumour necrosis factor-alpha (cachectin) in myocardial infarction. J Intern Med 225:333–336
Molinari A, Giogetti C, Lansen J, Vaghi F, Orsini G, Faioni EM et al (1992) Thrombomodulin is a cofactor for thrombin degradation of recombinant single-chain urokinase plasminogen activator "in vitro" and in a perfused rabbit heart model. Thromb Haemost 67:226–232
Moore KL, Andreoli SP, Esmon NL, Esmon CT, Bang NU (1987) Endotoxin enhances tissue factor and suppresses thrombomodulin expression of human vascular endothelium in vitro. J Clin Invest 79:124–130
Morrissey JH, Drake TA (1993) Procoagulant response of the endothelium and monocytes. In: Schlag G, Redl H (eds) Pathophysiology of shock, sepsis, and organ failure. Springer, Berlin Heidelberg New York, pp 564–574
Muller FM, Ehrenthal W, Hafner G, Schranz D (1996) Purpura fulminans in severe congenital protein C deficiency: monitoring of treatment with protein C concentrate. Eur J Pediatr 155:20–25
Muntean W, Finding K, Gamillscheg A, Schwarz HP (1991) Multiple thromboses and coumarin-induced skin necrosis in a child with anticardiolipin antibodies: effects of protein C concentrate administration. Thromb Haemost 65:1254
Murakami K, Okajima K, Uchiba M, Johno M, Nakagaki T, Okabe H et al (1996) Activated protein C attenuates endotoxin-induced pulmonary vascular injury by inhibiting activated leukocytes in rats. Blood 87:642–647
Nawa K, Itani T, Ono M, Sakano K, Marumoto Y, Iwamoto M (1992) The glycosaminoglycan of recombinant human soluble thrombomodulin affects antithrombotic activity in a rat model of tissue factor-induced disseminated intravascular coagulation. Thromb Haemost 67:366–370
Nawroth PP, Stern DM (1986) Modulation of endothelial cell hemostatic properties by tumor necrosis factor. J Exp Med 163:740–745
Nawroth PP, Handley DA, Esmon CT, Stern DM (1986) Interleukin1 induces endothelial cell procoagulant while suppressing cell-surface anticoagulant activity. Proc Natl Acad Sci USA 83:3460–3464
Nguyen P, Reynaud J, Pouzol P, Munzer M, Richard O, Francois P (1994) Varicella and thrombotic complications associated with transient protein C and protein S deficiencies in children. Eur J Pediatr 153:646–649
Nicholson AC, Nachman RL, Altieri DC, Summers BD, Ruf W, Edgington TS et al (1996) Effector cell protease receptor-1 is a vascular receptor for coagulation factor Xa. J Biol Chem 271:28407–28413

Ohdama S, Matsubara O, Aoki N (1994) Plasma thrombomodulin in Wegener's granulomatosis as an indicator of vascular injuries. Chest 106:666–671
Öhlin AK, Marlar RA (1995) The first mutation identified in the thrombomodulin gene in a 45-year-old man presenting with thromboembolic disease. Blood 85:330–336
Öhlin AK, Norlund L, Marlar RA (1997) Thrombomodulin gene variations and thromboembolic disease. Thromb Haemost 78:396–400
Okajima K, Koga S, Kaji M, Inoue M, Nakagaki T, Funatsu A et al (1990) Effect of protein C and activated protein C on coagulation and fibrinolysis in normal human subjects. Thromb Haemost 63:48–53
Olsen PH, Esmon NL, Esmon CT, Laue TM (1992) The Ca^{2+}-dependence of the interactions between protein C, thrombin and the elastase fragment of thrombomodulin. Analysis by ultracentrifugation. Biochemistry 31:746–754
Pabinger I, Brucker S, Kyrle PA, Schneider B, Korninger HC, Niessner H et al (1992) Hereditary deficiency of antithrombin III, protein C and protein S: prevalence in patients with a history of venous thrombosis and criteria for rational patient screening. Blood Coagul Fibrinolysis 3:547–553
Pabinger I, Kyrle PA, Heistinger M, Eichinger S, Wittmann E, Lechner K (1994) The risk of thromboembolism in asymptomatic patients with protein C and protein S deficiency: a prospective cohort study. Thromb Haemost 71:441–445
Polgar J, Lerant I, Muszbek L, Machovich R (1986) Thrombomodulin inhibits the activation of factor XIII by thrombin. Thromb Res 43:685–690
Powars DR, Rogers ZR, Patch MJ, McGehee WG, Francis RB Jr (1987) Purpura fulminans in meningococcemia: association with acquired deficiencies of proteins C and S. N Engl J Med 317:571–572
Powars D, Larsen R, Johnson J, Hulbert T, Sun T, Patch MJ et al (1993) Epidemic meningococcemia and purpura fulminans with induced protein C deficiency. Clin Infect Dis 17:254–261
Prince HM, Thurlow PJ, Buchanan RC, Ibrahim KM, Neeson PJ (1995) Acquired protein S deficiency in a patient with systemic lupus erythematosus causing central retinal vein thrombosis. J Clin Pathol 48:387–389
Rapaport SI, Rao LV (1992) Initiation and regulation of tissue factor-dependent blood coagulation. Arterioscler Thromb 12:1111–1121
Regan LM, Lamphear BJ, Huggins CF, Walker FJ, Fay PJ (1994) Factor IXa protects factor VIIIa from activated protein C. Factor IXa inhibits activated protein C-catalyzed cleavage of factor VIIIa at Arg562. J Biol Chem 269:9445–9452
Regan LM, Stearns-Kurosawa DJ, Kurosawa S, Mollica J, Fukudome K, Esmon CT (1996) The endothelial cell protein C receptor. Inhibition of activated protein C anticoagulant function without modulation of reaction with proteinase inhibitors. J Biol Chem 271:17499–17503
Reitsma PH, Poort SR, Bernardi F, Gandrille S, Long GL, Sala N, Cooper DN (1993) Protein C deficiency: a database of mutations. The Protein C and S Subcommittee of the Scientific and Standardization Committee of the International Society on Thrombosis and Haemostasis. Thromb Haemost 69:77–84
Reitsma PH, Bernardi F, Doig RG, Gandrille S, Greengard JS, Ireland H et al (1995) Protein C deficiency: a database of mutations, 1995 update. The Subcommittee on Plasma Coagulation Inhibitors of the Scientific and Standardization Committee of the ISTH. Thromb Haemost 73:876–879
Rezaie AR, Cooper ST, Church FC, Esmon CT (1995) Protein C inhibitor is a potent inhibitor of the thrombin-thrombomodulin complex. J Biol Chem 270:25336–25339
Richardson MA, Gerlitz B, Grinnell BW (1992) Enhancing protein C interaction with thrombin results in a clot-activated anticoagulant. Nature 360:261–264
Rivard GE, David M, Farrell C, Schwarz HP (1995) Treatment of purpura fulminans in meningococcemia with protein C concentrate. J Pediatr 126:646–652

Rosing J, Hoekema L, Nicolaes GA, Thomassen MC, Hemker HC, Varadi K et al (1995) Effects of protein S and factor Xa on peptide bond cleavages during inactivation of factor Va and factor Va^{R506Q} by activated protein C. J Biol Chem 270:27852–27858

Sacco RL, Owen J, Mohr JP, Tatemichi TK, Grossman BA (1989) Free protein S deficiency: a possible association with cerebrovascular occlusion. Stroke 20:1657–1661

Salem HH, Broze GJ, Miletich JP, Majerus PW (1983) Human coagulation factor Va is a cofactor for the activation of protein C. Proc Natl Acad Sci USA 80:1584–1588

Salem HH, Esmon NL, Esmon CT, Majerus PW (1984) Effects of thrombomodulin and coagulation factor Va-light chain on protein C activation in vitro. J Clin Invest 73:968–972

Schramm W, Spannagl M, Bauer KA, Rosenberg RD, Birkner B, Linnau Y et al (1993) Treatment of coumarin-induced skin necrosis with a monoclonal antibody purified protein C concentrate. Arch Dermatol 129:753–756

Scott BD, Esmon CT, Comp PC (1991) The natural anticoagulant protein S is decreased in male smokers. Am Heart J 122:76–80

Scully MF, Toh CH, Hoogendoorn H, Manuel RP, Nesheim ME, Solymoss S et al (1993) Activation of protein C and its distribution between its inhibitors, protein C inhibitor, α1-antitrypsin and α2-macroglobulin, in patients with disseminated intravascular coagulation. Thromb Haemost 69:448–453

Seligsohn U, Berger A, Abend M, Rubin L, Attias D, Zivelin A et al (1984) Homozygous protein C deficiency manifested by massive venous thrombosis in the newborn. N Engl J Med 310:559–562

Shen L, Dahlbäck B (1994) Factor V and protein S as synergistic cofactors to activated protein C in degradation of factor VIIIa. J Biol Chem 269:18735–18738

Sheth SB, Carvalho AC (1991) Protein S and C alterations in acutely ill patients. Am J Hematol 36:14–19

Sills RH, Marlar RA, Montgomery RR, Desphande GN, Humbert JR (1984) Severe homozygous protein C deficiency. J Pediatr 105:409–413

Sinha U, Hancock TE, Esmon CT, Lin PH, Wolf DL (1991) Inactive recombinant human factor $Xa(N^{282}A^{379})$; a novel coagulation inhibitor. Thromb Haemost 65: 941

Slungaard A, Vercellotti GM, Tran T, Gleich GJ, Key NS (1993) Eosinophil cationic granule proteins impair thrombomodulin function. A potential mechanism for thromboembolism in hypereosinophilic heart disease. J Clin Invest 91:1721–1730

Snow TR, Deal MT, Dickey DT, Esmon CT (1991) Protein C activation following coronary artery occlusion in the in situ porcine heart. Circulation 84:293–299

Solis MM, Cook C, Cook J, Glaser C, Light D, Morser J et al (1991) Intravenous recombinant soluble human thrombomodulin prevents venous thrombosis in a rat model. J Vasc Surg 14:599–604

Solis MM, Vitti M, Cook J, Young D, Glaser C, Light D et al (1994) Recombinant soluble human thrombomodulin: A randomized, blinded assessment of prevention of venous thrombosis and effects on hemostatic parameters in a rat model. Thromb Res 73:385–394

Solymoss S, Tucker MM, Tracy PB (1988) Kinetics of inactivation of membrane-bound factor Va by activated protein C. J Biol Chem 263:14884–14890

Stearns-Kurosawa DJ, Kurosawa S, Mollica JS, Ferrell GL, Esmon CT (1996) The endothelial cell protein C receptor augments protein C activation by the thrombin-thrombomodulin complex. Proc Natl Acad Sci USA 93:10212–10216

Suzuki K, Nishioka J, Hayashi T, Kosaka Y (1988) Functionally active thrombomodulin is present in human platelets. J Biochem (Tokyo) 104:628–632

Takahashi H, Tatewaki W, Wada K, Shibata A (1989) Plasma protein S in disseminated intravascular coagulation, liver disease, collagen disease, diabetes mellitus, and under oral anticoagulant therapy. Clin Chim Acta 182:195–208

Takahashi H, Hanano M, Wada K, Tatewaki W, Niwano H, Tsubouchi J et at (1991) Circulating thrombomodulin in thrombotic thrombocytopenic purpura. Am J Hematol 38:174–177
Takahashi H, Ito S, Hanano M, Wada K, Niwano H, Seki Y et al (1992) Circulating thrombomodulin as a novel endothelial cell marker: comparison of its behavior with von Willebrand factor and tissue-type plasminogen activator. Am J Hematol 41:32–39
Takano S, Kimura S, Ohdama S, Aoki N (1990) Plasma thrombomodulin in health and diseases. Blood 76:2024–2029
Tanaka A, Ishii H, Hiraishi S, Kazama M, Maezawa H (1991) Increased thrombomodulin values in plasma of diabetic men with microangiopathy. Clin Chem 37:269–272
Taylor FB Jr (1996) Role of tissue factor and factor VIIa in the coagulant and inflammatory response to LD_{100} Escherichia coli in the baboon. Haemostasis 26 Suppl 1:83–91
Taylor FB Jr, Chang A, Hinshaw LB, Esmon CT, Archer LT, Beller BK (1984) A model for thrombin protection against endotoxin. Thromb Res 36:177–185
Taylor FB Jr, Stern DM, Nawroth PP, Esmon CT, Hinshaw LB, Blick KE (1986) Activated protein C prevents E. coli induced coagulopathy and shock in the primate. Circulation 74:64
Taylor F, Chang A, Ferrell G, Mather T, Catlett R, Blick K et al (1991a) C4b-binding protein exacerbates the host response to Escherichia coli. Blood 78:357–363
Taylor FB Jr, Chang AC, Peer GT, Mather T, Blick K, Catlett R et al (1991b) DEGR-factor Xa blocks disseminated intravascular coagulation initiated by Escherichia coli without preventing shock or organ damage. Blood 78:364–368
Tijburg PNM, Ryan J, Stern DM, Wollitzky B, Rimon S, Rimon A et al (1991) Activation of the coagulation mechanism on tumor necrosis factor-stimulated cultured endothelial cells and their extracellular matrix. The role of flow and factor IX/IXa. J Biol Chem 266:12067–12074
Tsiang M, Lentz SR, Dittman WA, Wen D, Scarpati EM, Sadler JE (1990) Equilibrium binding of thrombin to recombinant human thrombomodulin: effect of hirudin, fibrinogen, factor Va, and peptide analogues. Biochemistry 29:10602–10612
Tsuchida A, Salem H, Thomson N, Hancock WW (1992) Tumor necrosis factor production during human renal allograft rejection is associated with depression of plasma protein C and free protein S levels and decreased intragraft thrombomodulin expression. J Exp Med 175:81–90
Uchiba M, Okajima K, Murakami K, Nawa K, Okabe H, Takatsuki K (1995) Recombinant human soluble thrombomodulin reduces endotoxin-induced pulmonary vascular injury via protein C activation in rats. Thromb Haemost 74:1265–1270
Varadi K, Philapitsch A, Santa T, Schwarz HP (1994) Activation and inactivation of human protein C by plasmin. Thromb Haemost 71:615–621
Varadi K, Rosing J, Tans G, Pabinger I, Keil B, Schwarz HP (1996) Factor V enhances the cofactor function of protein S in the APC-mediated inactivation of factor VIII: influence of the factor V^{R506Q} mutation. Thromb Haemost 76:208–214
Vigano-D'Angelo S, Comp PC, Esmon CT, D'Angelo A (1986) Relationship between protein C antigen and anticoagulant activity during oral anticoagulation and in selected disease states. J Clin Invest 77:416–425
Vigano-D'Angelo S, D'Angelo A, Kaufman CE Jr, Sholer C, Esmon CT, Comp PC (1987) Protein S deficiency occurs in the nephrotic syndrome. Ann Intern Med 107:42–47
Wada H, Ohiwa M, Kaneko T, Tamaki S, Tanigawa M, Shirakawa S et al (1992) Plasma thrombomodulin as a marker of vascular disorders in thrombotic thrombocytopenic purpura and disseminated intravascular coagulation. Am J Hematol 39:20–24
Walker FJ (1981) Regulation of activated protein C by protein S. The role of phospholipid in factor Va inactivation. J Biol Chem 256:11128–11131

Walker FJ, Fay PJ (1992) Regulation of blood coagulation by the protein C system. FASEB J 6:2561–2567
Wiedmer T, Esmon CT, Sims PJ (1986) On the mechanism by which complement proteins C5b-9 increase platelet prothrombinase activity. J Biol Chem 261:14587–14592
Wu QY, Sheehan JP, Tsiang M, Lentz SR, Birktoft JJ, Sadler JE (1991) Single amino acid substitutions dissociate fibrinogen-clotting and thrombomodulin-binding activities of human thrombin. Proc Natl Acad Sci USA 88:6775–6779
Ye J, Liu L, Esmon CT, Johnson AE (1992) The fifth and sixth growth factor-like domains of thrombomodulin bind to the anion-binding exosite of thrombin and alter its specificity. J Biol Chem 267:11023–11028

Subject Index

Q

R

Printing: Saladruck, Berlin
Binding: Buchbinderei Lüderitz & Bauer, Berlin